NEUROMETHODS

Series Editor
Wolfgang Walz
University of Saskatchewan
Saskatoon, SK, Canada

For further volumes:
http://www.springer.com/series/7657

Isolated Central Nervous System Circuits

Edited by

Klaus Ballanyi

Faculty of Medicine and Dentistry, Department of Physiology and Centre for Neuroscience, University of Alberta, Edmonton, AB, Canada

Humana Press

Editor
Klaus Ballanyi
Faculty of Medicine and Dentistry
Department of Physiology
and Centre for Neuroscience
University of Alberta
Edmonton, AB, Canada

ISBN 978-1-4939-6183-2 ISBN 978-1-62703-020-5 (eBook)
DOI 10.1007/978-1-62703-020-5
Springer New York Heidelberg Dordrecht London

Printed on acid-free paper

Humana Press is a brand of Springer
Springer is part of Springer Science+Business Media (www.springer.com)

Preface to the Series

Under the guidance of its founders Alan Boulton and Glen Baker, the Neuromethods series by Humana Press has been very successful since the first volume appeared in 1985. In about 27 years, 37 volumes have been published. In 2006, Springer Science+Business Media made a renewed commitment to this series. The new program will focus on methods that are either unique to the nervous system and excitable cells or which need special consideration to be applied to the neurosciences. The program will strike a balance between recent and exciting developments like those concerning new animal models of disease, imaging, in vivo methods, and more established techniques. These include immunocytochemistry and electrophysiological technologies. New trainees in neurosciences still need a sound footing in these older methods in order to apply a critical approach to their results. The careful application of methods is probably the most important step in the process of scientific inquiry. In the past, new methodologies led the way in developing new disciplines in the biological and medical sciences. For example, physiology emerged out of anatomy in the nineteenth century by harnessing new methods based on the newly discovered phenomenon of electricity. Nowadays, the relationships between disciplines and methods are more complex. Methods are now widely shared between disciplines and research areas. New developments in electronic publishing also make it possible for scientists to download chapters or protocols selectively within a very short time of encountering them. This new approach has been taken into account in the design of individual volumes and chapters in this series.

Neuherberg, Germany *Wolfgang Walz*

Preface

Advances in methodologies and experimental models are pivotal to furthering our understanding of central nervous system (CNS) functions in mammals. A most important technology in that regard is "patch-clamp," which was originally developed for monitoring currents through single ion channels in "cell-attached" or "excised patch" configurations. However, current neuroscience studies use patch-clamp primarily for the analysis of membrane potential changes or underlying ion currents in the "whole-cell" mode, with concomitant (defined) dialysis of the cytoplasm, while the "perforated" patch configuration can be applied to retain an intact cellular milieu. Patch-clamp techniques were adapted about two decades ago for studying CNS cells in their natural in situ environment and have since mostly replaced technically more challenging sharp microelectrode recording. Similarly, exciting advances were established for optical approaches using fluorescent dyes, e.g., for monitoring membrane potential, mitochondrial potential, or cytosolic pH. Yet, the most common optical approach is to visualize dynamic changes of the pivotal cytosolic "second-messenger" Ca^{2+}, either in single CNS cells or neural networks that are comprised of neurons and neighboring (micro)glia. Compared to initially used (charge-coupled device) video cameras, optical spatial resolution is notably improved by confocal microscopy while multiphoton imaging allows visualization of cells in deeper CNS layers. Besides, infrared differential interference contrast optics are a convenient low-cost tool for visually targeted patch recording in tissue depths of up to 150 μm. Recently developed genetic tools enable "knock-out" of particular cellular features, such as glutamate receptor subtypes, or allow expression and subsequent imaging of intrinsically fluorescent Ca^{2+}-sensitive or structural proteins in identified CNS cells. "Optogenetics" makes use of genetically inserted, light-activated ion channels that change the activity of specific cells to reveal their functions. Finally, local expression of cell-type-specific proteins is studied using immunohistochemical approaches or molecular tools such as "western blots" or "polymerase chain reaction" for analyzing cytoplasm of individual CNS cells, obtained by extraction after whole-cell recording, or of homogenized nervous tissue.

Until about 30 years ago, most studies on mammalian CNS functions were carried out using in vivo models of diverse mammalian species, mostly cats and rats, despite the fact that various in vitro CNS models were developed already in the 1950s. It was not until the mid-1970s that in vitro conditions were sufficiently well developed to keep these isolated CNS models, including cultures, en bloc ("slab") preparations, and brain slices viable for several hours in solutions that mimicked the composition of cerebrospinal fluid (CSF) (or rather the fluid in the interstitial space within CNS tissues). At the same time, electrical stimulation and single- or multiunit extracellular recording approaches plus sharp micro-electrode intracellular recording techniques, such as "single electrode voltage-clamp," were adapted to these models. Until the end of the last millennium, work on acutely isolated brain slices dominated CNS research with emphasis on studying mechanisms of synaptic plasticity associated with long-term potentiation or depression evoked by afferent axon tract stimulation. In these "classical" brain slices, electrical or pharmacological stimulation was typically needed for evoking neuronal responses, contrary to often pronounced spontaneous activity of the same cells in vivo. It was believed that this limitation is primarily related to the fact that the thickness of brain slices is mostly less than 500 μm for providing sufficient diffusional supply of cells with the energy substrates oxygen and glucose contained in superfused artificial CSF (ACSF). Consequently, neuronal dendrites and axons are partly sectioned, which presumably attenuates network connectivity and thus depresses spontaneous activity. To circumvent this, various laboratories used mainly newborn rodents already since about 30 years ago to develop en bloc models with active neural networks. During the same time period, others succeeded to keep large CNS aspects (up to entire rodent brains) viable by arterial perfusion. During the last decade, procedures for generating active isolated CNS tissues have improved further, e.g., by using an ionic ACSF composition that reflects more closely that of the fluid in the extracellular space of neural networks instead of that in CSF of the subarachnoid space and brain ventricles.

This contribution to the *Neuromethods* series exemplifies the application of a majority of the above-mentioned and other technologies to mostly active in vitro preparations from basically different CNS regions with a diversity of functions. Specifically, Chapter 1 by *Trapp and Ballanyi* deals with neurons in rodent brainstem slices that control vagal outflow. It outlines how "tonic" activity of these cells is modulated by metabolic processes and how underlying mechanisms are studied with single channel and whole-cell patch-clamp techniques, gas- and ion-sensitive microelectrodes, optical photomultiplier-based techniques, and diverse molecular approaches. Further emphasis is on how methods for slice generation and storage plus superfusate composition affect properties of these vagal neurons and neurons in general. Chapter 2 by *Ruangkittisakul et al.* delves deeper into the latter topic by pointing out the particular importance of superfusate K^+, Ca^{2+}, and glucose, and also of physical dimensions of newborn rodent en bloc and slice preparations for spontaneous activity of respiratory neurons in the lower brainstem. Further, they exemplify that whole-cell and suction electrode recording (for neuronal population activity) combined with multiphoton/confocal Ca^{2+} imaging is used for investigating contributions of neurons *versus* glia to respiratory rhythm. Chapter 3 by *Moore et al.* summarizes techniques for studying (spontaneous) activity in neurons of human fetal cortex slices, focusing on how slices of this almost gel-like tissue can be generated and how developing electrical properties such as immature Na^+ action potentials can be discriminated from imperfect whole-cell recording conditions in these delicate cells. Chapter 4 by *Fish et al.* deals with histological characterization of physiologically determined fast spiking interneurons in slices of the dorsolateral prefrontal cortex of monkeys. It outlines particularly how high-resolution confocal imaging is combined with sophisticated optical analyses for elucidating structure-function relationships for these cells. Chapter 5 by *Nakamura et al.* describes neural networks in the suprachiasmatic nucleus of the hypothalamus that continue to generate circadian rhythm in acute slices or slice cultures. It shows how these circuits depend on experimental conditions, such as time of day for their generation, and how they are analyzed with patch-clamp plus multiunit activity recording, molecular approaches, and Ca^{2+} or genetic bioluminescence imaging. Chapter 6 by *Stachniak et al.* deals with other hypothalamic networks that regulate plasma osmolarity in the body and whose responses to osmotic stimuli are studied with patch-clamp and immunohistochemical approaches in conventional and thick hypothalamic slices. In Chapter 7 by *McKay et al.*, patch-clamp recording and histological analysis are used to show that repetitive synaptic input establishes in vivo–like activity in rat cerebellar slice neurons and that biophysical neuron properties change during postnatal development. The latter findings are important for comparing in vivo data from adult animals with in vitro findings that are often obtained in preparations from juvenile animals. Chapter 8 by *Sanchez-Vives* shows, using primarily whole-cell and multiunit activity recording, that patterns of spontaneous rhythmic activities in slices of adult cerebral cortex depend on animal species, superfusate composition, and temperature. Chapter 9 by *Broicher and Speckmann* reports how spontaneous and evoked neuronal activities in acute cortical slices from patients who needed surgical removal of brain tissue are analyzed by combining electrophysiological approaches with voltage imaging. Chapter 10 by *Luhmann and Kilb* outlines how cellular properties and network activity are analyzed in intact in vitro preparations of neonatal rodent cerebral cortex. Chapter 11 by *Kantor et al.* deals with the use of suction electrode recording and Ca^{2+} plus morphological multiphoton/confocal imaging for studying spontaneous network oscillations in hippocampal, neocortical, and *locus ceruleus* slices from newborn rats and piglets. Chapter 12 by *De Curtis et al.* describes methods for arterial perfusion of isolated guinea pig brains that retain functional and interacting neural networks. Examples are given for spontaneous and electrically evoked activities that are analyzed under normal conditions and upon evoked seizures or ischemia with extra- plus intracellular electrophysiological approaches, ion-sensitive microelectrodes, and voltage plus Ca^{2+} imaging. Chapter 13 by *Day and Wilson* describes a juvenile rat model for independent dual perfusion of carotid bodies and lower brainstem for analysis of contribution to respiratory rhythm of peripheral and central chemoreceptors, respectively. Chapter 14 by *Biggs et al.* outlines how to generate organotypic spinal cord slices for investigating with electrophysiologic and Ca^{2+} imaging approaches mechanisms of pain-related central sensitization. Chapter 15 by Mandadi et al. reviews slice and en bloc cord preparations for studying locomotor networks with electrophysiologic and Ca^{2+} imaging approaches.

Due to space limitation, other established or recently developed isolated CNS preparations and their applications could not be dealt with, such as tonically active *substantia nigra* networks, isolated optic nerves or *corpus callosum* slices for studying axon-glia interactions, or (organotypic) brain slices with intact connectivity of distinct regions, e.g., between the thalamus and the cortex. However, the in vitro approaches and methodologies described here are most likely applicable to further improve the latter models and to develop corresponding models of not yet intensively studied structures such as *nucleus ruber*, *superior colliculus*, or basal ganglia.

Klaus Ballanyi

Contents

Contributors

DUSTIN M. ANDERSON • *Hotchkiss Brain Institute, HRIC 1AA14, University of Calgary, Calgary, AB, Canada*
SRDJAN D. ANTIC • *Department of Neuroscience, University of Connecticut Health Center, Farmington, CT, USA*
KLAUS BALLANYI • *Faculty of Medicine & Dentistry, Department of Physiology and Centre for Neuroscience, University of Alberta, Edmonton, AB, Canada*
JAMES E. BIGGS • *Faculty of Medicine & Dentistry, Department of Pharmacology and Centre for Neuroscience, University of Alberta, Edmonton, AB, Canada*
GENE D. BLOCK • *Department of Psychiatry and Biobehavioral Sciences, University of California, Los Angeles, CA, USA*
NICOLETA BOBOCEA • *Faculty of Medicine & Dentistry, Department of Physiology and Centre for Neuroscience, University of Alberta, Edmonton, AB, Canada*
CHARLES W. BOURQUE • *Montreal General Hospital Campus, Centre for Research in Neuroscience, Research Institute of the McGill University Health Center, Montreal, QC, Canada*
TILMAN BROICHER • *Department of Bioengineering, University of Utah, Salt Lake City, UT, USA*
PO-YIN CHEUNG • *Faculty of Medicine & Dentistry, Departments of Pediatrics, Pharmacology & Surgery, University of Alberta, Edmonton, AB, Canada*
KATRINA Y. CHOE • *Centre for Research in Neuroscience, Research Institute of the McGill University Health Center, Montreal, QC, Canada*
SORANA CIURA • *Centre for Research in Neuroscience, Research Institute of the McGill University Health Center, Montreal, QC, Canada*
WILLIAM F. COLMERS • *Faculty of Medicine & Dentistry, Department of Pharmacology and Centre for Neuroscience, University of Alberta, Edmonton, AB, Canada*
CHRISTOPHER S. COLWELL • *Department of Psychiatry and Biobehavioral Sciences, University of California, Los Angeles, CA, USA*
TREVOR A. DAY • *Department of Biology, Mount Royal University, Calgary, AB, Canada*
MARCO DE CURTIS • *Unit of Experimental Neurophysiology and Epileptology, Fondazione Istituto Neurologico Carlo Besta, Milano, Italy*
JORDAN D.T. ENGBERS • *Hotchkiss Brain Institute, HRIC 1AA14, University of Calgary, Calgary, AB, Canada*
FERNANDO R. FERNANDEZ • *Hotchkiss Brain Institute, HRIC 1AA14, University of Calgary, Calgary, AB, Canada*
KENNETH N. FISH • *Department of Psychiatry, Western Psychiatric Institute and Clinic, University of Pittsburgh, Pittsburgh, PA, USA*
VADYM GNATKOVSKY • *Unit of Experimental Neurophysiology and Epileptology, Fondazione Istituto Neurologico Carlo Besta, Milano, Italy*

GUILLERMO GONZALEZ-BURGOS • *Department of Psychiatry, Western Psychiatric Institute and Clinic, University of Pittsburgh, Pittsburgh, PA, USA*
PENGCHENG HAN • *Department of Comparative Biology and Experimental Medicine, Hotchkiss Brain Institute, University of Calgary, Calgary, AB, Canada*
JENNIFER M. HUMPHREYS • *Department of Comparative Biology and Experimental Medicine, Hotchkiss Brain Institute, University of Calgary, Calgary, AB, Canada*
MIRCEA IFTINCA • *Hotchkiss Brain Institute, HRIC 1AA14, University of Calgary, Calgary, AB, Canada*
MAKITO IIZUKA • *Center for Medical Sciences, Ibaraki Prefectural University of Health Sciences, Ami, Ibaraki, Japan*
IGOR JAKOVCEVSKI • *Department of Neuroscience, University of Connecticut Health Center, Farmington, CT, USA*
JACK H. JHAMANDAS • *Faculty of Medicine & Dentistry, Department of Medicine (Neurology) and Centre for Neuroscience, University of Alberta, Edmonton, AB, Canada*
IGOR JOVANOVIC • *Faculty of Medicine & Dentistry, Department of Physiology and Centre for Neuroscience, University of Alberta, Edmonton, AB, Canada*
CHASE KANTOR • *Faculty of Medicine & Dentistry, Department of Physiology and Centre for Neuroscience, University of Alberta, Edmonton, AB, Canada*
WERNER KILB • *Institute of Physiology and Pathophysiology, University Medical Center, Johannes Gutenberg University, Mainz, Germany*
HELENA J. KIM • *Faculty of Medicine & Dentistry, Department of Pharmacology and Centre for Neuroscience, University of Alberta, Edmonton, AB, Canada*
JUNYA KURIBAYASHI • *Faculty of Medicine & Dentistry, Department of Physiology and Centre for Neuroscience, University of Alberta, Edmonton, AB, Canada*
AARON LAI • *Faculty of Medicine & Dentistry, Department of Pharmacology and Centre for Neuroscience, University of Alberta, Edmonton, AB, Canada*
TZE-FUN LEE • *Faculty of Medicine & Dentistry, Departments of Pediatrics, Pharmacology & Surgery, University of Alberta, Edmonton, AB, Canada*
VICTORIA LEUNG • *Faculty of Medicine & Dentistry, Department of Physiology and Centre for Neuroscience, University of Alberta, Edmonton, AB, Canada*
DAVID A. LEWIS • *Departments of Psychiatry and Neuroscience, University of Pittsburgh, Pittsburgh, PA, USA*
LAURA LIBRIZZI • *Unit of Experimental Neurophysiology and Epileptology, Fondazione Istituto Neurologico Carlo Besta, Milano, Italy*
VAN B. LU • *Faculty of Medicine & Dentistry, Department of Pharmacology and Centre for Neuroscience, University of Alberta, Edmonton, AB, Canada*
HEIKO J. LUHMANN • *Institute of Physiology and Pathophysiology, University Medical Center, Johannes Gutenberg University, Mainz, Germany*
DAVE MACTAVISH • *Faculty of Medicine & Dentistry, Department of Medicine (Neurology) and Centre for Neuroscience, University of Alberta, Edmonton, AB, Canada*
SRAVAN MANDADI • *Department of Comparative Biology and Experimental Medicine, Hotchkiss Brain Institute, University of Calgary, Calgary, AB, Canada*
BRUCE E. MCKAY • *Department of Psychology, Wilfrid Laurier University, Waterloo, ON, Canada*
STEPHAN MICHEL • *Laboratory for Neurophysiology, Department of Molecular Cell Biology, Leiden University Medical Center, Leiden, The Netherlands*
ANNA R. MOORE • *Department of Neuroscience, University of Connecticut Health Center, Farmington, CT, USA*

TAKAHIRO J. NAKAMURA • *Faculty of Pharmaceutical Sciences, Teikyo Heisei University, Ichihara, Japan*
STAN T. NAKANISHI • *Department of Comparative Biology and Experimental Medicine, Hotchkiss Brain Institute, University of Calgary, Calgary, AB, Canada*
BOGDAN PANAITESCU • *Faculty of Medicine & Dentistry, Department of Physiology and Centre for Neuroscience, University of Alberta, Edmonton, AB, Canada*
ARAYA RUANGKITTISAKUL • *Faculty of Medicine & Dentistry, Department of Physiology and Centre for Neuroscience, University of Alberta, Edmonton, AB, Canada*
MARIA V. SANCHEZ-VIVES • *ICREAÐIDIBAPS, Barcelona, Spain*
LUCIA SECCHIA • *Faculty of Medicine & Dentistry, Department of Physiology and Centre for Neuroscience, University of Alberta, Edmonton, AB, Canada*
PETER A. SMITH • *Faculty of Medicine & Dentistry, Department of Pharmacology and Centre for Neuroscience, University of Alberta, Edmonton, AB, Canada*
ERWIN-JOSEF SPECKMANN • *Westfälische Wilhelms-Universität Münster, Institut für Physiologie I, Münster, Germany*
TEVYE J. STACHNIAK • *Centre for Research in Neuroscience, Research Institute of the McGill University Health Center, Montreal, QC, Canada*
JESSICA R. SUDBURY • *Centre for Research in Neuroscience, Research Institute of the McGill University Health Center, Montreal, QC, Canada*
REZA TADAYONNEJAD • *Hotchkiss Brain Institute, HRIC 1AA14, University of Calgary, Calgary, AB, Canada*
KATHRYN G. TODD • *Faculty of Medicine & Dentistry, Department of Psychiatry and Centre for Neuroscience, University of Alberta, Edmonton, AB, Canada*
STEFAN TRAPP • *Department of Surgery and Cancer, Imperial College London, London, UK*
ERIC TRUDEL • *Centre for Research in Neuroscience, Research Institute of the McGill University Health Center, Montreal, QC, Canada*
RAY W. TURNER • *Hotchkiss Brain Institute, HRIC 1AA14, University of Calgary, Calgary, AB, Canada*
LAURA UVA • *Unit of Experimental Neurophysiology and Epileptology, Fondazione Istituto Neurologico Carlo Besta, Milano, Italy*
PATRICK J. WHELAN • *Department of Comparative Biology and Experimental Medicine, Hotchkiss Brain Institute, University of Calgary, Calgary, AB, Canada*
RICHARD J.A. WILSON • *Departments of Physiology and Pharmacology, Hotchkiss Brain Institute, University of Calgary, Calgary, AB, Canada*
ALEKSEY V. ZAITSEV • *I.M. Sechenov Institute of Evolutionary Physiology and Biochemistry RAS, St. Petersburg, Russia*
NADA ZECEVIC • *Department of Neuroscience, University of Connecticut Health Center, Farmington, CT, USA*
WEN-LIANG ZHOU • *Department of Neuroscience, University of Connecticut Health Center, Farmington, CT, USA*

Chapter 1

Autonomic Nervous System In Vitro: Studying Tonically Active Neurons Controlling Vagal Outflow in Rodent Brainstem Slices

Stefan Trapp and Klaus Ballanyi

Abstract

Central mammalian neurons often show spontaneous discharge of single action potentials at regular rate. Several types of such 'tonic' neurons are involved in sensing or control of the intake and processing of food. Brain slices containing tonically active hypothalamic neurons are one established in vitro model for studying interneuronal and cellular signalling pathways underlying metabolism-related autonomous nervous functions. Here, we focus on brainstem neurons of the dorsal vagal (pre)motor nucleus (DMNX) and the nucleus tractus solitarius (NTS), which both represent pivotal relay areas for control of nutrition and digestion. We compare tonic (pre)motor dorsal vagal neurons (DVN) with similarly active glucagon-like peptide-1 (GLP-1) neurons that comprise a subgroup of NTS cells. This chapter has four aims: (1) to show that central neurons with different functions are capable of generating tonic discharge upon isolation in a brain slice; (2) to exemplify electrophysiological, optical and molecular tools applicable for analysing responses of such neurons to anoxia and aglycaemia; (3) to exemplify biophysical membrane processes involved in the latter metabolism-related or neuromodulator-induced variation of their activity; and (4) to point out methodological strategies for preserving their (sensory) functions and also pitfalls in that regard. We expect that some of our findings using improved in vitro conditions for studying metabolism-sensitive tonic brainstem neurons are applicable, if not necessary, for in vivo-like spontaneous activities and cellular properties of neural networks in slices from other brain regions.

Key words: Action potential, Appetite, ATP-sensitive potassium channels (K_{ATP} channels), Dorsal vagal motoneurons, Glucose, Leptin, Nutrition, Nucleus of the solitary tract, Preproglucagon

1. Introduction

The autonomic nervous system regulates many vital functions of the human body, be it breathing, heart rate, food intake and digestion, or reproduction. For a long time, autonomous regulatory neural circuits were investigated on a systems level only. Cellular functional

Klaus Ballanyi (ed.), *Isolated Central Nervous System Circuits*, Neuromethods, vol. 73,
DOI 10.1007/978-1-62703-020-5_1, © Springer Science+Business Media New York 2012

studies in isolated neural networks focused on higher brain structures, mainly cortex and hippocampus. This might have partly been due to 'less clear' organisation of the autonomic neural circuits, with neuron somata being organised in nuclei rather than laminae in 'lower', that is, more caudal, brain regions. In view of homeostatic control of body function, the vagal system is of pivotal interest. Its best known function entails the modulation of heart rate that led to discovery of the 'Vagus-Stoff' acetylcholine by Otto Loewi (1). Additional functions include gastrointestinal homeostatic processes such as the cephalic phase of stimulation of insulin release upon food intake (2), but might also include a role in neuronal disease. For instance, vagus nerve stimulation produces therapeutic benefits in the treatment of epilepsy (3). The efferent vagal nerve originates from two nuclei located in the lower brainstem ('medulla oblongata'). These are the nucleus ambiguus and the dorsal vagal (pre)motor nucleus (DMNX) which contain the cell bodies of the preganglionic vagal neurons (Figs. 1 and 2).

Intrinsic (i.e. non-synaptic) electrical properties of dorsal vagal (pre)motor neurons (DVN) in guinea pig brainstem slices were first reported in 1985 (4, 5). This prompted various groups, including ourselves, to study DVN in slices, mostly from juvenile or mature rat (e.g. (6–11)). For example, Sah and co-workers analysed membrane currents involved in the multiphase action potential ('spike') after-hyperpolarisation which greatly determines the rate of their typical regular ('tonic') discharge (10–12). This and related work established that tonic spiking of DVN in slices is similar to their activity in vivo (13–16). Moreover, extensive work by Travagli's group has established that such spontaneous action potential firing of DVN is influenced by multiple neuromodulators including serotonin (5-HT), noradrenaline and appetite-regulating peptides such as the incretin glucagon-like peptide-1 (GLP-1) or neuropeptide Y (17–21). Our previous studies on DVN in juvenile rat brainstem slices focused on analysing (sub)cellular mechanisms involved in their tolerance to oxygen depletion ('anoxia'). This research revealed, for example, that anoxic activation of ATP-sensitive K^+ (K_{ATP}) channels causes sustained hyperpolarisation that abolishes DVN firing (8, 22). Based on our finding that these neurons are also 'functionally inactivated' by K_{ATP} channels in response to glucose depletion ('aglycaemia') (22), we hypothesised that these plasmalemmal proteins constitute a central glucose sensor as in other neurons, for example, in hypothalamus (23, 24). This is discussed here in view of our more recent studies that indicated various subtypes of glucose-sensitive neurons in the DMNX and the adjacent nucleus tractus solitarius (NTS) (25, 26).

The DMNX contains primarily cholinergic DVN whereas the NTS hosts diverse neuron populations of different function and neurochemical phenotypes which cannot easily be distinguished in slices. However, increased availability of transgenic animals which

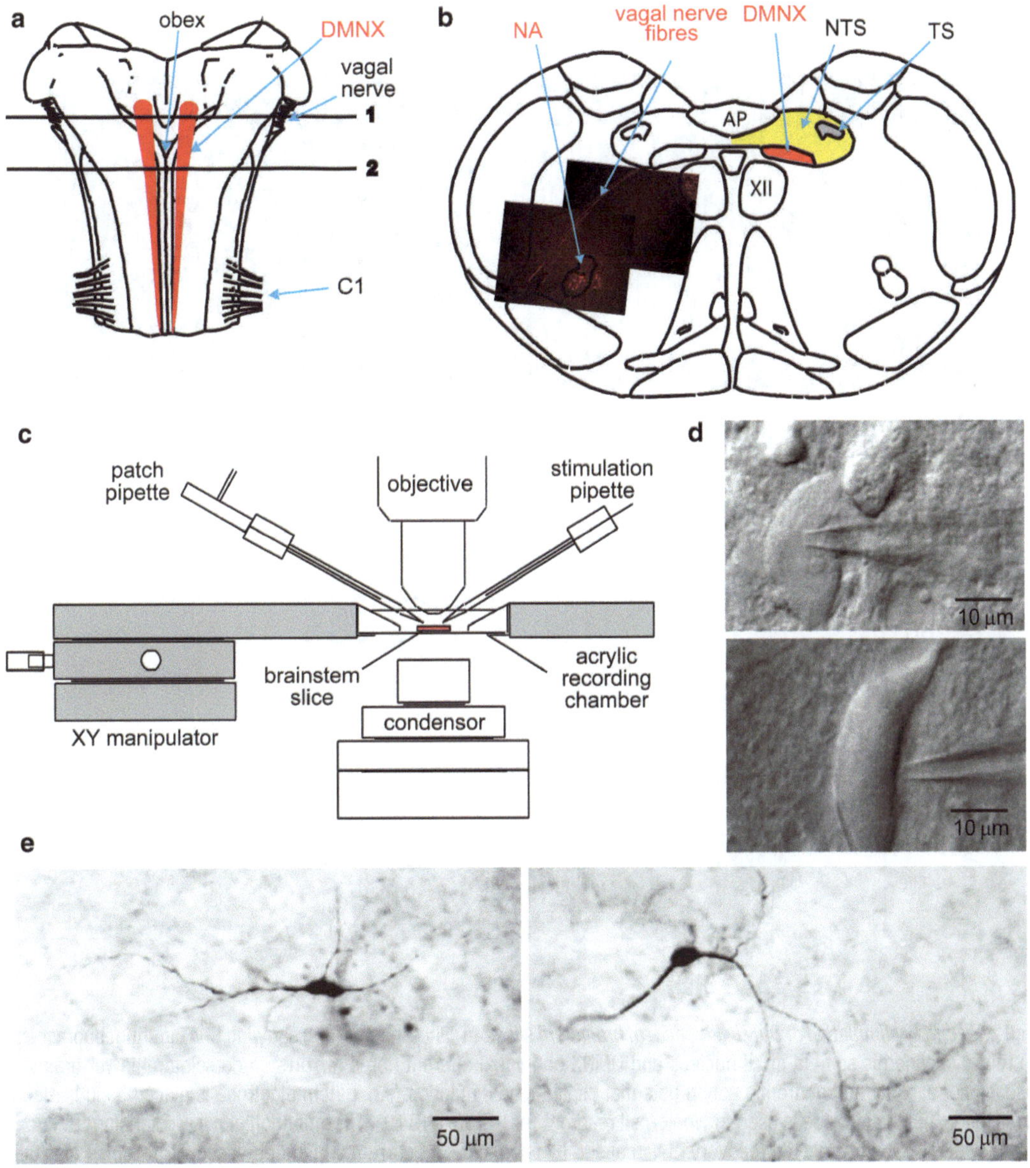

Fig. 1. *Metabolism-sensing and appetite-controlling neurons in rat brainstem.* (**a**) Schematic dorsal view of lower brainstem with cerebellum removed for better visibility. 1 and 2 delineate region from which transversal ('coronal') slices (**b**) are taken. Rostrocaudal extent of dorsal vagal (motor) nucleus (DMNX) is indicated in *red.* (**b**) Schematic coronal section at level bregma −13.68 mm (rat) or bregma −7.64 mm (mouse). Embedded photomicrographs reveal vagal nerve fibres originating from DMNX and cell bodies in nucleus ambiguus (NA) with immunofluorescence for choline acetyltransferase (ChAT). Abbreviations: AP, area postrema; XII, hypoglossal motor nucleus; NTS, nucleus tractus solitarius; TS, solitary tract. (**c**) Set-up for recording from transversal slices in (**b**) under visual control via an upright microscope attached to an xy-moveable microscope stage 'fixed' for movement in z-axis. The xy stage contains an acrylic recording chamber in which the brainstem slice is fixed with metal harp (compare Fig. 4). Patch pipettes or other types of (metal) electrodes can be positioned on slice surface for electrical stimulation. (**d**) Differential interference contrast (DIC) images of preganglionic dorsal vagal neurons (DVN) in juvenile rats that can be approached in set-up in **c** either from top (*upper panel*) or from side (*lower panel*) with patch pipette for recording of membrane potential (V_m) or current (I_m). These are actual screen shots at 80× magnification using a 40× water immersion lens plus 2× magnifying lens in front of a charge-coupled device (CCD) camera. Also note that nucleus and nucleolus start to show in upper cell as a sign of stress/deterioration. (**e**) Photomicrographs of juvenile mouse DVN labelled during whole-cell recording with the dye biocytin, which was processed for visualisation after experiments in chemically fixed brainstem slices ((**c**) With kind permission from (35); (**e**) from K. Ballanyi and J. Brockhaus, unpublished).

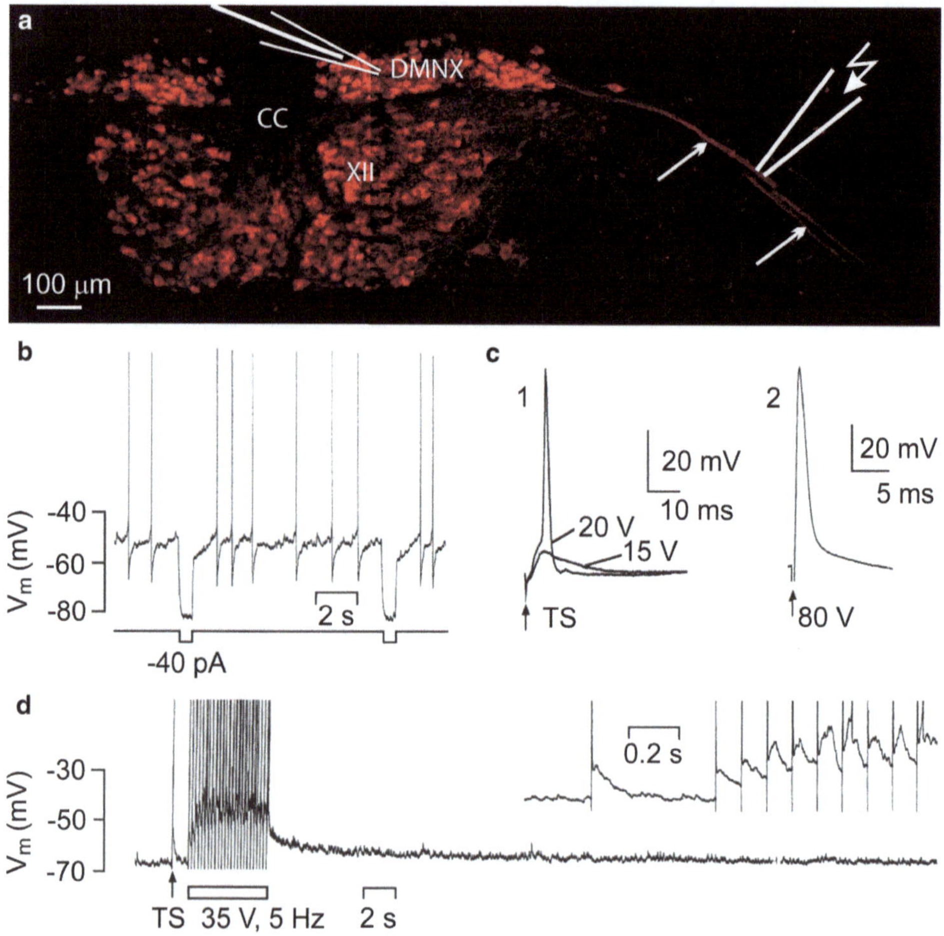

Fig. 2. *Electrical stimulation of DVN in juvenile rat brainstem slices*. (**a**) Photomicrograph showing ChAT immunofluorescence to reveal cholinergic somata in XII nucleus and DMNX and vagal efferent fibres (*arrows*) in dorsolateral lower brainstem aspect. For evoking an antidromic action potential ('spike'), a stimulation (e.g. patch) electrode is placed as indicated by flash, and a patch electrode is used to whole-cell record V_m or I_m from a DVN. (**b**) Termination of hyperpolarising DC current pulse was followed by delayed recovery of V_m caused by an 'A-type' K^+ current which depressed 'tonic' spike discharge typical for DVN. (**c**) Electrical stimulation of tractus solitarius (TS, in 1, compare Fig. 1) with 0.1 ms short 15 V pulse evoked an excitatory postsynaptic potential, which led to orthodromic spike at slightly higher stimulus strength (20 V), whereas stimulation of vagal efferent fibres (see **a**) with 80 V pulse evoked antidromic spike (in 2). (**d**) Subsequent to recovery from a single electrical pulse, repetitive ('tetanic' or 'train') TS stimulation led to summating postsynaptic potentials and concomitant repetitive spike discharge of a DVN, in which tonic discharge was depressed by injection of hyperpolarising DC current ((**a**) From S. Trapp, unpublished; (**b–d**) with kind permission from (22)).

express fluorescent markers in specific cell populations now allows targeted analyses, for example, of catecholaminergic (27, 28) or pro-opiomelanocortin neurons (29, 30). Here we exemplify this for mice that express enhanced yellow fluorescent protein (eYFP) in NTS neurons producing GLP-1 (31, 32).

In this chapter, we summarise our studies on DVN and, to a lesser extent, GLP-1 NTS neurons in rodent brainstem slices that made use of a wide range of techniques to identify signalling pathways underlying nutrition- or metabolism-related cellular responses. Studies on DVN have been so successful because firstly, these neurons remain vital for several hours in brainstem slices and tolerate long-term recording of synaptic or intrinsic biophysical membrane properties with either fine-tipped or patch-clamp microelectrodes. Secondly, cellular signalling factors such as mitochondrial membrane potential ($\Delta\Psi$) or the free concentration, that is, activity, of cytosolic Ca^{2+} (Ca_i) and H^+, the latter determining intracellular pH (pH_i), can be monitored with optical techniques while their extracellular microenvironment can be assessed with oxygen- or ion-sensitive microelectrodes. Moreover, the molecular identity of their functional membrane proteins such as K_{ATP} or other types of K^+ channels and enzymes like glucose-sensing glucokinase can be determined with molecular tools including in situ hybridisation or single-cell polymerase chain reaction (PCR). Finally, electrical stimulation can be used to identify DVN neurons via antidromic action potentials or evoke excitatory and inhibitory synaptic responses.

The strong emphasis on critical appraisal of these methodologies in this chapter should help to facilitate studies of similar systems, for example, specific (metabolism-sensing) cell populations in the lower brainstem or the hypothalamus, in order to widen our knowledge of the autonomic nervous system at a cellular and sub-cellular level. Finally, while this chapter focuses on brainstem in vitro preparations, we envisage the information given here to be a useful guide for setting up or optimising in vitro preparations from CNS regions ranging from olfactory bulb to spinal cord.

2. Materials

2.1. Animal Models

Brainstem and other neurons are mostly studied in slices from rodents, traditionally Wistar or Sprague-Dawley rats and sometimes guinea pigs. Increasingly, mice are being used due to the rapidly progressing development of transgenic strains. Here, we refer to a mouse strain (mGLU-eYFP) that expresses yellow fluorescent protein (YFP) under the control of the promoter for the mouse glucagon gene (mGLU) as an example (32). The vast majority of patch-clamp studies under visual control (Figs. 1 and 3) are performed on rat brain slices obtained from juvenile animals of <6 weeks of age because of perceived difficulty in visualising neurons from older animals (33–35). In our recent experience, this limitation is not present in the DVN or NTS of adult (>8 weeks) mice (31). Additional anecdotal evidence from our lab suggests that recordings are easily obtained from visualised neurons in brainstem slices from mice as old as >1 year.

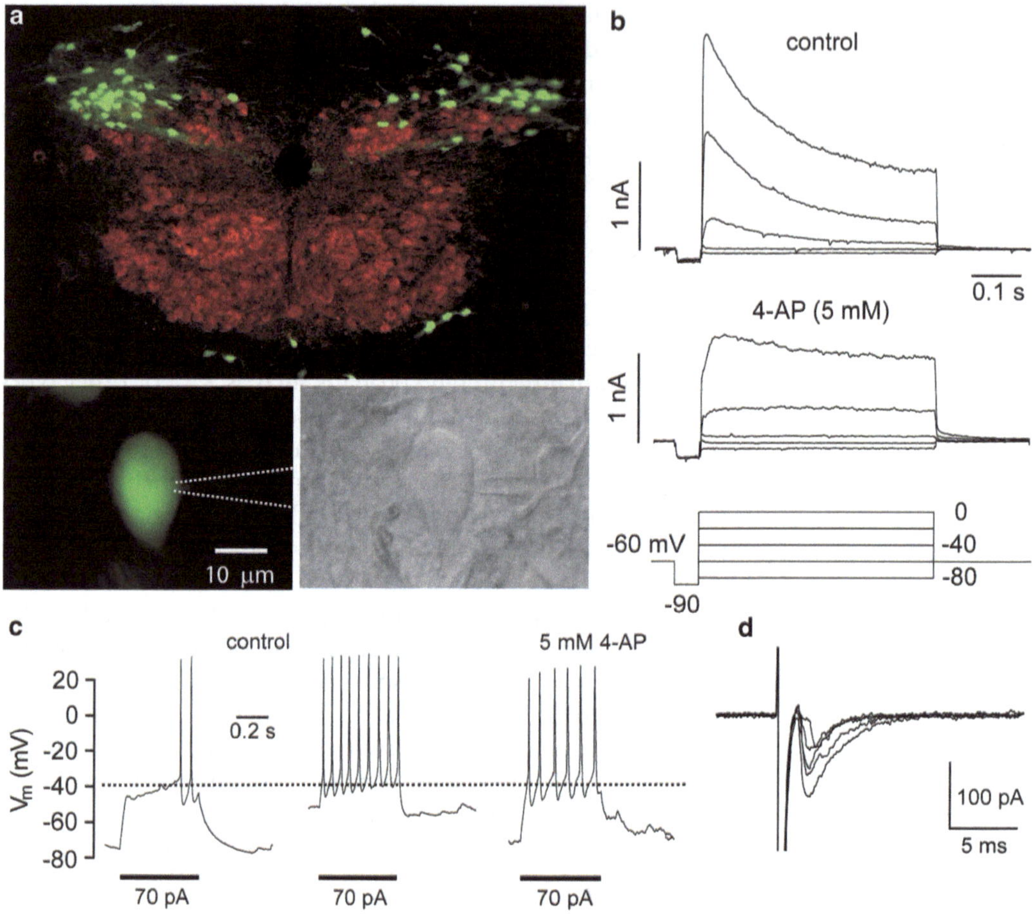

Fig. 3. *Properties of glucagon-like peptide-1 (GLP-1) producing preproglucagon (PPG) neurons.* (**a**) Photomicrograph in top showing location of preproglucagon (PPG) neurons that express enhanced *yellow* fluorescence protein (eYFP; *green* fluorescence), in relation to cholinergic ChAT-marked (*red* fluorescence) DVN and XII neurons in 30-μm-thick coronal brainstem slice at position bregma −7.75 mm. *Bottom*; eYFP fluorescence (*left*) and DIC image (*right*) of PPG neuron during whole-cell recording (*left*, patch pipette position indicated by *dotted lines*). (**b**, **c**) A-current characterization. (**b**) *Top trace* shows activation and time-dependent inactivation of A-current at various voltages, whereas *bottom trace* shows remaining current in presence of the specific blocker 4-aminopyridine (4-AP). *Bottom panel* shows the stimulation protocol used. (**c**) Demonstrates A-current effect on spiking in current clamp. When cell is hyperpolarised by constant DC current injection, a short depolarizing DC current pulse (*black bar*) first activates the A-current and spike generation is delayed (*left trace*). When cell is not hyperpolarised, the same depolarising current pulse does not activate A-current and elicits spikes straight away (*middle trace*). When cell is hyperpolarised and A-current is inhibited by 4-AP, the depolarising pulse leads to spike firing without delay (*right trace*). (**d**) TS stimulation elicits excitatory postsynaptic currents in a voltage-clamped PPG neuron; shown is an overlay of five consecutive stimulations; initial vertical traces indicate stimulus artefact ((**a**, **d**) Modified with kind permission from (31); (**b**, **c**) from S. Trapp, unpublished).

2.2. Generation of Brainstem Slices

Brain slices are typically generated with a vibrating microtome ('vibratome') such as the VT1000S (Leica Biosystems, Nussloch, Germany; www.leica-microsystems.com) or the 5,000 mz (Campden Instruments Ltd, Loughborough, UK; www.campden-instruments.com). They can also be obtained using a tissue chopper such as the McIlwain (Warner Instruments, Hamden, CT,

USA; www.warneronline.com). Tissue choppers are less expensive than vibratomes and slices can be generated faster than upon sectioning with an optimally slowly progressing vibrating blade. However, due to reduced precision in cutting, there is likely trauma to (superficial layers of) acutely generated slices which could be detrimental to their quality. Consequently, tissue choppers are currently almost exclusively used for generation of organotypic slice cultures, where initial damage to the slice surface does not matter, but where ease of keeping it aseptic or to obtain a large number of slices per brain area is desirable.

After sectioning, slices are stored for up to 10 h on a net of fabric or nylon covering a plastic Petri dish with a broken out wedge (Fig. 4). The net is positioned close to the meniscus of standard superfusate (Sect. 3.1) by attaching it to a smaller Pyrex® glass beaker (25 ml) inserted bottom up into a 100 ml Pyrex® beaker. The solution is gassed with carbogen (95% O_2, 5% CO_2) via an injection cannula positioned under the net and kept at constant temperature, commonly between 30°C and 35°C, by inserting the beakers into a water bath, for example, the Isotemp (Fisher Scientific, Canada; www.fishersci.com) (34) (Fig. 4). An alternative system that we have developed to keep the holding chamber smaller for mouse brainstem slices and to eliminate the water bath as a potential source of infection uses a standard block heater (e.g. Fisher) kept at 30–34°C with a heating block that holds 50-ml glass beakers (31). The beaker is filled with 25 ml of standard solution that is gassed with carbogen ('carbogenated') via an inverted plastic Pasteur pipette with an array of 100–200 μm holes punched into its handle to allow passage of small carbogen bubbles into the solution (Fig. 4). On top of the Pasteur pipette, the slice-holding chamber is placed to keep slices submerged while being gassed with carbogen from beneath. The slice-holding chamber is made of a 50-ml skirted Falcon tube that is cut to a height of 5 cm and has its conical bottom removed. The bottom of the chamber is formed by a piece of nylon mesh (200 μm pore size) that is held in place on the skirt of the Falcon tube by a ring made from the lid of the electrode glass container (Harvard Apparatus, Saint-Laurent, QC, Canada; www.harvardapparatus.ca) that had its centre removed. This assembly is easy to take apart for cleaning and put together again (Fig. 4).

2.3. Recording Chamber and Superfusion System

Brain(stem) slices are mostly studied using a 'submerged'-type acrylic recording chamber with glass bottom (34, 36) (Fig. 4). A small chamber volume (<1 ml) is desirable for fast solution exchange, or if substances need to be tested that are either extremely expensive or only available in small quantities, such as synthesised peptides. One example for a sophisticated small chamber system is given by Wilson et al. (37). A larger chamber (>1.5 ml) is advantageous if multiple microelectrodes are to be positioned, for example,

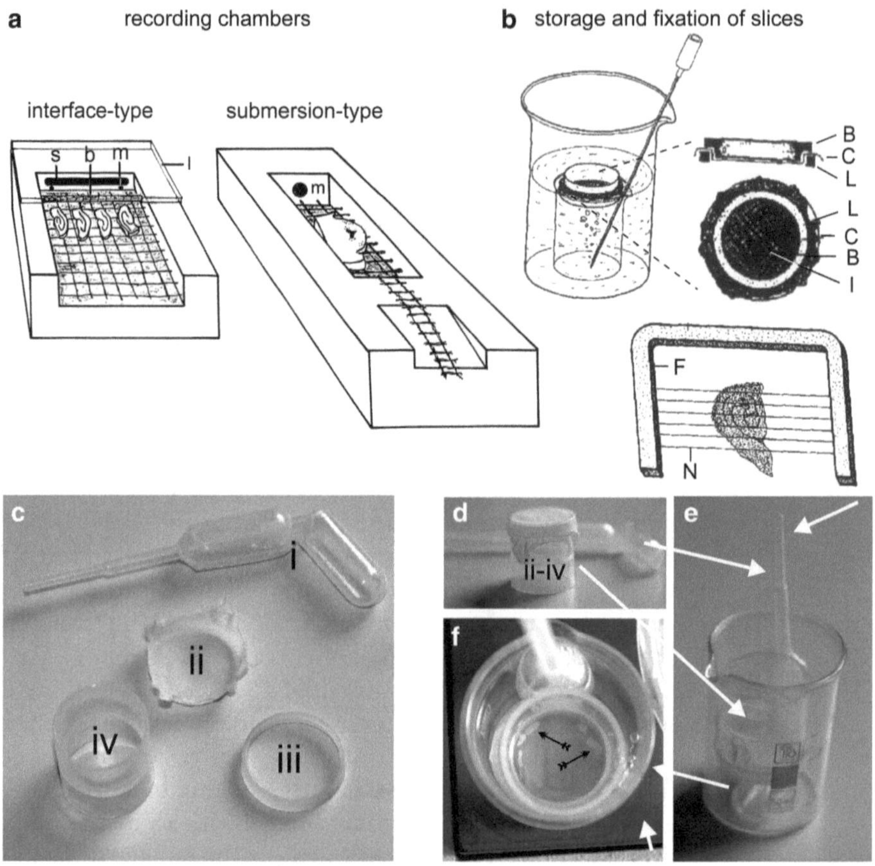

Fig. 4. *Chambers for storing and subsequent recording from brainstem slices*. (**a**) In 'interface'-type recording chamber (*left*), slices rest on net with fluid level high enough to keep their surface moist, and low enough to avoid mechanical disturbance. Warmed and moistened carbogen (95% O_2, 5% CO_2) is blown over slices from a slot (s) and directed to their surface by lid (l; wet paper weighted by a glass slide). Carbogenated standard solution (see Table 1 for composition) also enters in back (m) and crosses a barrier (b). Another barrier may be used at front end; fluid level is adjusted by amount of draining material (e.g. fabric or filter paper) at front end, from which fluid drips. In 'submersion'-type chambers (*right*), slices are either stabilised with platinum or tantal harp ('grid', see **b**) or metal weights of the same materials. They can also be fixed with insect needles pinned into a Silgard® bottom layer of chamber. (**b**) Slice-holding chamber is placed on top of 50-ml beaker inserted into 250-ml beaker both filled with standard solution to level of top of holding chamber. A hypodermic needle inserted through spout of inner beaker to carbogenate slices. Assembly is placed in water bath (at 25–37°C, depending on slice type) and covered with lid (e.g. plastic Petri dish) to prevent evaporation. A suitable holding chamber is made by breaking top (l) and bottom (b) out of small (10 × 35 mm) plastic Petri dish, forming two rings. When inverted, ring formed by l fits tightly onto lip of b. Piece of fine cotton mesh (c) is stretched over b and can be clamped in place by l. Grids for fixation of slices consist of nylon threads (n), glued to a platinum or tantalum frame (f). (**c–f**) Alternative assembly of slice-holding chamber used with heating block. (**c**) Individual components of inner chamber (ii–iv) and gassing (with carbogen) device (i) made of disposable Pasteur pipette that had its head perforated with 25-gauge injection needle to allow passing of small gas bubbles. Head of second Pasteur pipette was separated from pipette and another hole was cut into top end to allow sliding over stem of first pipette. This produces more stability once entire chamber is assembled. Inner (slice-holding) chamber consists of nylon mesh (ii), cap of electrode glass container (Harvard apparatus), that had centre cut out (iii) and bottom quarter (cut off, and conical bottom removed) of 50-ml skirted Falcon tube (iv). Electrode glass container lid fits precisely onto skirted bottom of falcon tube to keep mesh in place in between. (**d**) Assembled inner chamber, which is then inserted into 50-ml beaker in parallel with gassing device. (**e**) Gassing device is placed so that gas bubbles out below inner chamber. Low pressure carbogen supply is connected to top of bubbler (*white arrow*). (**f**) Once filled with standard solution to cover slices, without reaching rim of inner chamber, whole assembly is placed into standard heating block (*short arrow*, +/− 1°C accuracy; designed to fit 50-ml beakers) and is kept at 32°C ((**a**) With kind permission from (36); (**b**) with kind permission from (33); (**c–f**) from S. Trapp as used in (31)).

for simultaneous recording of temperature, partial pressure of oxygen (PO_2) and pH in the superfusate and/or next to a neuron, even in combination with electrical stimulation and intracellular (plus extracellular) electrophysiological recording or imaging. Alternatively, 'interface' chambers are being used in which slices, positioned on a net, have their upper margin exposed to the interface between superfusate (or rather 'subfusate') and a humidified carbogen atmosphere. This requires that a major portion of the recording chamber is covered with an (acrylic) lid (34, 36) (Fig. 4). As one major advantage of interface-type chambers, field potentials have larger amplitudes presumably due to a greater extracellular resistance (34). On the other hand, these chambers do not allow patch-clamp recordings under visual control. Either type of recording chamber is available from companies such as Warner Instruments, ALA Scientific Instruments (Farmingdale, NJ, USA; www.alascience.com) or Luigs & Neumann (Ratingen, Germany; www.luigs-neumann.com).

Superfusate is conveniently administered via a peristaltic roller pump, for example, the Watson-Marlow Alitea (Sin-Can, Calgary, AB, Canada; www.watson-marlow.com), and typically kept at constant temperature of 30–35°C in the recording chamber using a control device such as the TC-324B (Harvard Apparatus). Alternatively, a closed bottle system driven by the carbogen gas pressure plus gravity produces continuous flow without the typical, though normally small, 'spurts' produced by peristaltic pumps (38). However, flow rate needs to be adjusted related to changes in the fluid level in the closed bottles for control and experimental solutions. The simplest way to administer solutions is via a solely gravity-driven system, although this also requires continuous adjustment for constant flow rate. Removal of solution from the recording chamber is usually achieved by means of a syringe needle, connected to either (a further channel of) a peristaltic roller pump or a vacuum system, for example, an aquarium pump. Other systems use a strip of fabric positioned across the rim of the recording chamber (Fig. 4). Tubings for the superfusion system should be impermeable to gases, in particular O_2 and carbon dioxide (CO_2), to ensure that superfusate PO_2 and pH, respectively, do not change during transport to the recording chamber. Stainless steel or Tygon® tubing (e.g. from Fisher) is highly recommended for this purpose. For optimal viability and (spontaneous) activity of mammalian brain slices, flow rates should be at least 3–5 ml/min for recording chambers with a volume of ~1 ml. Larger chambers require a higher flow rate for a similar exchange rate of superfusate within the chamber. Lower flow rates may cause interstitial accumulation of neuromodulators such as adenosine in slices (34).

2.4. Microelectrodes for Recording and Stimulation

Patch electrodes with outer tip diameters of 1.5–3 μm are used for 'whole-cell' recording (39–41) of membrane potential (V_m) or membrane current (I_m) of DVN and GLP-1 neurons for up to several hours. For DVN and most types of NTS neurons plus neighbouring glial cells, V_m or I_m recording is also feasible with fine-tipped ('sharp') microelectrodes (outer tip diameter <0.2 μm). It can be challenging to obtain stable microelectrode recording from small neurons because impalement causes a leak associated with notable (ion) fluxes into and out of the cell. Both types of intracellular electrodes, and also 'theta-type' double-barrelled ion-sensitive microelectrodes, are typically produced from (thin-walled) borosilicate glass capillaries provided, for example, by Warner Instruments or World Precision Instruments ('WPI'; Sarasota, FL, USA; www.wpiinc.com). WPI also distributes O_2-sensitive (micro) electrodes and compatible amperometers. A recent study reported generation and use of custom-made microelectrodes for amperometric PO_2 measurements in brain tissue (42).

Electrical stimulation of fibre tracts in brain(stem) slices, is reliably achieved using concentric platinum/iridium PTM3CC02INS electrodes (50–200 μm outer diameter) or related models from WPI. Often, single axons or local circuits can also be stimulated effectively using a large patch electrode (inner tip diameter 5–10 μm) filled with standard superfusate. If this size cannot be achieved with the electrode puller, a smaller diameter electrode can be broken to the appropriate size under microscopic control (33, 35). Patch electrodes, sharp microelectrodes and ion-sensitive electrodes are fabricated using microelectrode pullers such as the P-1000 (Sutter Instrument Company, Novato, CA, USA; www.sutter.com) or the DMZ/Zeitz Universal Puller (Dagan Corporation, Minneapolis, MI, USA; www.dagan.com). The PC-10 (Narishige International Inc, East Meadow, NY, USA; www.narishige-group.com) is a notably less expensive alternative, but is not designed for producing sharp microelectrodes.

2.5. Micromanipulators and Vibration Isolation Tables

Commonly used (electrical) manipulators to position microelectrodes for whole-cell patch-clamp recording are the Sutter MP-285, the Scientifica PatchStar and the Narishige MHW-3 (all distributed by ALA Scientific; www.alascience.com) as well as the Unit Mini (Luigs & Neumann) or the MX7600 (Siskiyou Corporation, Grants Pass, OR, USA; www.siskiyou.com). All these devices use high-precision step motors. For impalement of DVN and NTS neurons with sharp microelectrodes, fast-step piezo-driven micromanipulators are recommended, such as the Märzhäuser PM-10 (ALA Scientific). Frequently used coarser, less expensive, mechanical manipulators are the Märzhäuser MM33, the Narishige MMN-1 or M-152 and the Scientifica LBM-7 (all distributed by ALA Scientific). The micrometre precision required for intracellular recordings usually necessitates that the equipment is mounted on a vibration isolation table to avoid that floor vibrations disrupt the measurements. These tables, produced by a number of manufacturers, for example, Newport (Irvine, CA, USA; www.newport.com)

or Kinetic Systems (Boston, MA, USA; www.kineticsystems.com), work by having a heavy table top 'float' on an air cushion provided by pneumatic cylinders located in the table legs. The pneumatic cylinders are connected to a pressurised air or gas supply (e.g. a nitrogen bottle). Floor vibrations are dampened by the air cushion in combination with the inertia of the heavy table top. Moreover, it is advisable, and for single-channel recordings often essential, to use a Faraday cage supplied, for example, by Newport or Kinetic Systems that is most commonly mounted on the vibration isolation table. This means that the microscope, experimental chamber and manipulators are located within the cage while 'peripheral' equipment, such as amplifiers, or computers, is placed outside. A simple, often sufficient, solution for an anti-vibration system is a heavy metal or stone plate resting on rubber or silicone dampers while metal (tool) cabinets can serve as Faraday cage.

2.6. Amplifiers, Stimulators and Computer-Based Data Acquisition

V_m or I_m are recorded with a patch-clamp amplifier, such as the 'EPC-family' (HEKA, Lambrecht, Germany, www.heka.com), or a 'sample-and-hold' single-electrode voltage-clamp amplifier like the 'SEC family' (npi electronic, Tamm, Germany; www.npielectronic.com). Similar patch-clamp and sample-and-hold amplifiers are the Axopatch 200B and Axoclamp 900A, respectively (Molecular Devices, Sunnyvale, CA, USA; www.moleculardevices.com). A high impedance amplifier (npi electronic) is needed for ion-sensitive microelectrode recording because of the high resistance of ion-sensitive resins (Sect. 3.6.2). Electrical signals are sampled into a computer via an interface and acquisition software. Convenient hardware/software packages are available from HEKA, npi electronic or Axon Instruments (Molecular Devices). Commonly used multichannel systems for data acquisition are PowerLab (ADInstruments, Colorado Springs, CO, USA; www.adinstruments.com) or Spike2 (Cambridge Electronic Design Ltd, Cambridge, UK; www.ced.co.uk). In most of these systems, software options are implemented for single-pulse or repetitive ('tetanic' or 'train') electrical stimulation. For such stimulation, the output of the corresponding interface needs to be connected to an electrical isolation unit that generates current or voltage pulses applied to the stimulation electrode. Such isolation units are produced, for example, by WPI or A.M.P.I. (Jerusalem, Israel; www.ampi.co.il) and A-M SYSTEMS (Sequim, WA, USA; www.a-msystems.com). Also hardware-based programmable stimulation units are available from the latter companies and others like Warner Instruments and ALA Scientific.

2.7. Microscopes and Imaging Devices

For 'blind' patch or microelectrode recording (43) of V_m or I_m of DVN in deeper brainstem (slice) layers, stereomicroscopes can be used such as the Stemi-2000 (Zeiss, Oberkochen, Germany; www.zeiss.com), SZX (Olympus, Center Valley, PA,

USA; www.olympusamerica.com) or the ELZ (Scienscope, Chino, CA, USA; www.scienscope.com). Using such stereomicroscopes, visualisation of slice areas and microelectrode positioning are facilitated by simple video cameras (e.g. supplied by WPI) that are attached either to one ocular tube or to a separate optical port. The video signal is displayed on a separate monitor or, using a frame grabber, on the monitor of the computer used for controlling the (intracellular) recording (see also Chap. 3-Moore et al. (44)).

For recording from neurons in slices under visual control (33, 35) (Fig. 1), an upright microscope is needed such as the Zeiss Axioskop 2 FS (or successor) and the Olympus BX61WI. This type of microscope requires a water immersion lens (usually 40× or 60× magnification) with a working distance >1 mm to facilitate pipette positioning and should also be equipped with differential interference contrast (DIC) optics (45, 46). These optics produce a pseudo relief image of slice layers, where individual cells appear to look like 'hills' protruding from surrounding tissue (Figs. 1, 3, and 5), thus aiding targeting of an identified cell. Precision can be further improved, and cells can be visualised within deeper slice layers (up to 80 μm) by attaching a charge-coupled device (CCD) camera with sensitivity into the near-infrared (IR) spectrum to the microscope and displaying the resulting IR-DIC image on a monitor (45, 46). Examples for such CCD cameras are the ImagEM or ORCA-R2 from Hamamatsu (Bridgewater, NJ, USA; www.hamamatsu.com) or the CoolSNAP from Roper Scientific (Ottobrunn, Germany; http://www.roperscientific.com). In our systems used for single-cell reverse transcription (RT)-PCR, another 2× magnification lens is placed in front of the CCD camera for observing flow of cytoplasm into the pipette when harvesting mRNA (Fig. 5). For 'patch-clamping' cells tagged with a fluorescent label such as (e)YFP or green fluorescent protein (GFP) (Fig. 3), an epifluorescence system is required which consists of a light source, for example, a halogen lamp, and excitation plus emission filters specific to the fluorescent label.

For live cell recording, these components have to be chosen carefully because intense (short wavelength) excitation light can cause damage (47, 48). Mercury lamps emit high-intensity bands of short wavelengths, which need to be blocked within the optical path to avoid cell damage, whereas more expensive xenon lamps lack these peaks and might be safer in this regard. In our system, this problem did not appear with a mercury lamp where the light was fed into the microscope via a light fibre (X-Cite® 120; EXFO Photonic Solutions Inc.; supplied by Zeiss). Similarly, for imaging of fluorescent dyes, such microscopes need to be equipped with epifluorescence optics, a light source such as a monochromator, for example, Polychrome V (TILL Photonics, Victor, NY, USA;

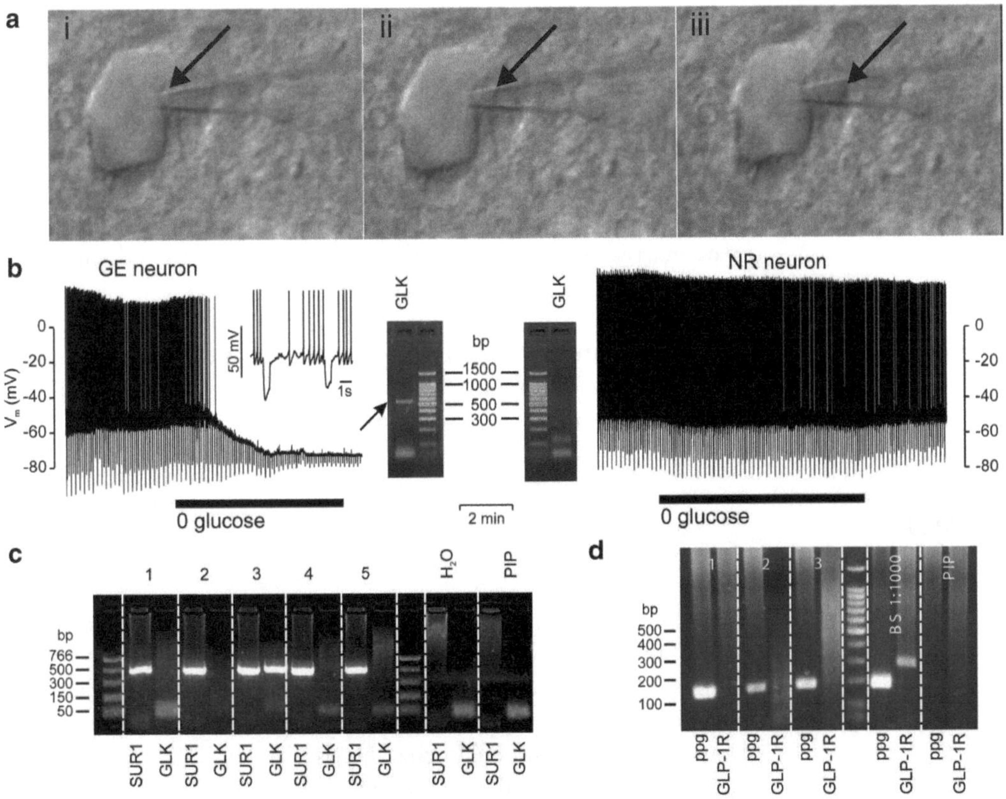

Fig. 5. *Single-cell reverse transcriptase polymerase chain reaction (RT-PCR) for expression of key molecular components of glucose-sensing brainstem neurons*. (**a**) Photomicrographs showing how cytoplasm (*arrows*) is aspirated into patch pipette used for whole-cell recording. (**b**) Current-clamp recordings from glucose-excited (GE) cell (*left*) and glucose-non-responsive (NR) cell (*right*). GE cell hyperpolarised and ceased spike firing within few minutes of changing superfusate glucose concentration from 10 to 0 mM contrary to lack of change in firing in the NR cell. Phenotypes of both neurons were confirmed by single-cell RT-PCR detection of glucokinase. GE cell expressed glucokinase (525 bp PCR product; *arrow*), whereas NR cell did not. (**c**) Single-cell RT-PCR analysis for glucokinase (GLK) and ATP-sensitive K^+ (K_{ATP}) channel subunit sulfonylurea receptor type-1 (SUR1) of 5 neurons. 1,2,4,5: NR neurons; 3: GE neuron. Controls: PIP, pipette solution without cytoplasm extracted from cell. H_2O, no cDNA. (**d**) Single-cell RT-PCR analysis for expression of PPG (ppg) and GLP-1 receptor (GLP-1R) in yellow fluorescent protein (YFP) positive neurons of mGLU-YFP mice. Controls: RT-PCR performed on brainstem cDNA (BS; 1:1,000 dilution) and pipette solution without cytoplasm extracted from neurons (PIP) ((**a**) From S. Trapp, unpublished; (**b**, **c**) with kind permission from (26); (**d**) with kind permission from (31)).

www.till-photonics.com) or the Lambda DG-4 (Sutter Instrument Company). The simplest and least expensive option for optical recording uses photomultipliers such as those distributed, by Luigs & Neumann. The photomultiplier approach is very sensitive to modest changes of optical signals. However, it represents a 'point source' measurement (49) and thus does not enable analyses of spatio-temporal patterns of optical signals, contrary to CCD camera- or multiphoton/confocal microscope-based imaging (47, 48) (see also Chap. 2-Ruangkittisakul et al. (50)).

2.8. Materials for Molecular Biology

Genotyping of transgenic animals is most conveniently done with PCR from ear biopsies using an ear puncher obtained from a veterinary supplier. A thermal cycler for PCR can be obtained from various companies. We use a PCR System 9700 (Applied Biosystems, Carlsbad, CA, USA; www.appliedbiosystems.com) for both genotyping and for processing samples for single-cell RT-PCR. Additionally, an electrophoresis system, for example, Sub-Cell system with PowerPac300 (Bio-Rad Laboratories, Hercules, CA, USA; www3.bio-rad.com), is required to separate PCR products on an agarose gel, and a fluorescent light table or a gel documentation station is needed to visualise them on the gel.

2.9. Other Materials

Sources for materials such as drugs or dyes that cannot be obtained from established suppliers like Sigma, VWR, Fisher or Tocris are referred to specifically in the text.

3. Methods

In this section, techniques are described for electrophysiological, optical and gas- or ion-sensitive microelectrode recording of biophysical membrane properties and the cytosolic or extracellular milieu of DVN. Also, techniques for electrical identification and stimulation plus molecular characterisation of DVN and GLP-1 neurons are referred to, and most recent methods are noted for retaining their metabolism-sensing mechanisms in slices. Finally, molecular methods such as single-cell RT-PCR and genotyping by PCR are briefly covered.

3.1. Procedures to Obtain and Maintain Brainstem Slices

Generation of viable brainstem slices from rodent brains is not easily achieved for several reasons. For example, some brain tissue is very vulnerable to insults due to arrest of blood flow ('complete ischemia'), and culling an animal in order to extract the tissue is obviously causing this. Various strategies have been developed over the years to minimise the impact of the surgical trauma on the quality of brainstem slices. The critical time period for damage is from the cessation of microvascular blood flow to the brain until brainstem slices are finally maintained in carbogenated storage solution. This time period typically ranges from one to several minutes depending on the type of brain(stem) region to be studied (34).

In our initial studies, juvenile (10–22 days old) Wistar or Sprague-Dawley rats or NMRI mice were anaesthetised until disappearance of the paw withdrawal reflex with ether, which was replaced by halothane or isoflurane in more recent studies. The animals were then decapitated and the brain removed within <2 min and transferred into solution with ion and glucose contents similar to those of the interstitial space of brain tissue and arterial blood, respectively (51) (Sect. 3.2), for 5 min before sectioning on a vibratome. In order to reduce the electrical activity of the tissue

Table 1
Extracellular solutions for generating and studying rodent brainstem slices

	'Low Ca'	'Low Na'	Recovery	'Standard'
Temperature	0–4°C	0–4°C	32–34°C	28–34°C
NaCl	118 mM	0 mM	118 mM	118 mM
Sucrose	0 mM	200 mM	0 mM	0 mM
KCl	3 mM	2.5 mM	3 mM	3 mM
$MgCl_2$	1 mM	7 mM	7 mM	1 mM
$CaCl_2$	0.5 mM	0.5 mM	0.5 mM	1.5 mM
$NaHCO_3$	25 mM	28 mM	25 mM	25 mM
NaH_2PO_4	1.2 mM	1.25 mM	1.2 mM	1.2 mM
D-glucose	10 mM	7 mM	2.5 mM	10 mM
Pyruvate	0 mM	3 mM	0 mM	0 mM

and thus ongoing energy demand during preparation, the solution is ice-cold (0–4°C) and contains a reduced (0.5 mM) $CaCl_2$ concentration (8, 9, 22, 34, 52–60) (for details of superfusate composition, see Table 1).

In our later studies, we performed a transcardial perfusion with ice-cold solution under terminal anaesthesia in order to cool the rat brain more rapidly and effectively (25, 26) (Table 1). For this, rats are anaesthetised with an injectable anaesthetic such as sodium pentobarbital (60 mg/kg; intraperitoneal). Sufficient depth of anaesthesia is assured upon the loss of the toe pinch reflex. The sternum is then exposed by lifting the skin over the chest with forceps and cutting with scissors. It is then lifted and the diaphragm and rib cage cut open to expose the heart. The sternum is lifted up further and clamped back (and not cut to avoid bleeding!) with a needle holder to allow unobstructed access to the heart. A blunt catheter (18 gauge) is inserted into the left ventricle with its tip placed in the aorta and clamped in place with a needle holder. The right atrium is lesioned and the animal perfused with 15–40 ml of ice-cold low Na^+ solution (Table 1), where NaCl is replaced with sucrose to further limit bioelectrical activity during surgery and tissue slicing. These measures allowed us to shift the age of rats to 18–30 days without compromising the viability of superficial neurons within the DMNX. We observed a larger number of healthy-looking DVN and NTS neurons under visual control and found clearer electrical responses to hypoglycaemia (Sect. 4.6) (25, 26).

Our latest studies have been performed on mice. Presumably due to their smaller size, brain cooling by transcardial perfusion proved unnecessary. Submerging the cranium in ice-cold solution while removing the brain and using the sucrose-based solution for preparation were sufficient to allow us to perform the entire study on adult (>8 weeks old; routinely up to 5 months old) animals without compromising slice quality and success rates in patch-clamp recordings from 'healthy-looking' neurons (31).

Because the DMNX extends rostrocaudally by several millimetres around the opening of the fourth ventricle ('obex') (13, 61) (Fig. 1), 6–8 transversal (also termed 'coronal') brainstem slices of 150–200 μm thickness, or correspondingly fewer 400–500 μm-thick slices, are obtained by sectioning in ice-cold solution using a vibratome (Sect. 2.2). To ensure that the solution remains ice-cold during sectioning, a stainless steel block that has been cooled down to −80°C and acts as a heat sink is added to the slicing chamber. Compared to the use of frozen 'slushy' solution, this has the advantage that no pieces of ice float in the slicing chamber, which could mechanically damage the brain tissue. Straight after cutting, each individual slice is transferred with a cut and heat-polished Pasteur pipette (opening diameter 3–5 mm) into 'standard' solution for storage up to 10 h at 30–34°C (Sect. 2.2). An additional incubation period of 30–60 min in 'recovery solution' prior to long-term storage in standard solution was introduced in our later studies (25, 26, 31). The latter solution returns normal Na^+ levels but maintains Ca^{2+} levels low and limits the amount of glucose available (Table 1). This should minimise potential 'reperfusion injury'. The pH of all solutions is adjusted to a value of 7.4 by gassing with carbogen.

3.2. Relevance of In Vitro Conditions for Neuronal Properties

Particularly, when planning to evaluate responses of in vitro neural networks to metabolic challenges, such as hypoxia/anoxia or hypoglycaemia/aglycaemia, the question arises as to what levels of O_2 and glucose supply are appropriate to mimic in vivo conditions as closely as possible. When slice preparations were initially developed, it was seen as most important to avoid a lack of either O_2 or glucose (34). Since then, most studies on brain slices apply solutions that contain 10–30 mM glucose and are saturated with carbogen resulting in a partial PO_2 of typically >600 mmHg in the recording chamber. However, recordings with O_2-sensitive microelectrodes indicated an average interstitial brain PO_2 of <100 mmHg (62, 63). According to the latter studies, the exact level of PO_2 in the vicinity of a neuron in the mammalian brain may depend on both the distance of its location from arterioles and the activity of this neuron or adjacent structures. For similar reasons, the level of glucose next to a neuron may also depend on various factors. This may contribute to the fact that measurements of interstitial glucose concentration in vivo provide a rather large range, specifically from 0.1 to 5 mM (34, 64–69) (see also Chap. 2-Ruangkittisakul et al. (50)).

Because microcirculation in brain is extremely efficient in supplying fuel to the individual neuron, low concentrations of O_2 and glucose in vivo probably mirror meticulously matched supply to demand. This is reflected in the ability to use functional magnetic resonance imaging, which visualises the hemodynamic response to demonstrate which parts of the brain are active at any given point in time, and also in the fact that disruption of blood supply to the brain in vivo leads to loss of consciousness within few seconds. The lack of the highly efficient in vivo microcirculation is particularly problematic for thicker brain slices which require a higher superfusate PO_2 and glucose concentration because of notable distances for diffusional supply. The question what concentrations are required to ensure normal functioning of cells within slices is further complicated by the fact that any hyperoxia or hyperglycaemia might also interfere with their putative O_2 or glucose sensor properties (34).

Most early studies analysing neuronal responses to glucose in vitro in freshly dissociated cells or brain slices from hypothalamus (70–72) used a superfusate glucose concentration around 10 mM and employed aglycaemia (via application of glucose-free superfusate) as the challenge to elicit a sensor response. Recently, some laboratories investigating glucose sensing in hypothalamic neurons started using standard solutions with lower glucose levels, such as 2.5 mM (73, 74). Moreover, Burdakov and colleagues (75) kept mouse hypothalamic slices in potentially hypoglycaemic conditions at 0.2 mM glucose and raised glucose levels to 4.5 mM to elicit an electrical response in orexin-producing hypothalamic neurons. However, when storing juvenile rat brainstem slices in ≤5 mM glucose, we observed deterioration of the slice surface occurring over the storage period of 5 h (26). Contrary, another laboratory studying glucose responses in the NTS kept rat brainstem slices in 5 mM and noticed no deterioration (76, 77). Recently, we revisited this issue using brainstem slices from transgenic mice with eYFP-tagged GLP-1 neurons (31, 32). When storing the slices in 1 mM glucose over a period of 5 h, we noticed deterioration of the slice surface but did not observe any noticeable change in the morphology or viability of GLP-1 neurons (Hisadome and Trapp, unpublished).

Our recent findings on newborn rat brainstem slices showed that rhythmic respiratory activity in newborn rat brainstem slices depends greatly on superfusate glucose content (Chap. 2-Ruangkittisakul et al. (50)). Specifically, in solution with physiological (3 mM) K^+ rhythmic bursting of the isolated pre-Bötzinger complex inspiratory centre is rather weak and short-lived when the glucose content of solution used for generating 'breathing slices' is 30 mM compared to 5 mM, whereas raising glucose from 5 to 30 mM in the recording chamber does not affect long-term stable inspiratory rhythm in slices generated in 5 mM glucose. Furthermore, inspiratory-related rhythm in 600 µm-thick pre-Bötzinger complex slices is stable for >6 h in standard solution with 1 mM glucose and 7 mM K^+. Contrary, 20 mM glucose is needed for stable inspiratory rhythm in the newborn rat 'en bloc' brainstem-spinal cord model which is of substantially larger size than the rhythmic slice (Chap. 2-Ruangkittisakul et al. (50)).

The latter observations indicate that even within a small area such as the vagal complex in brainstem slices, the response and/or vulnerability of cells to hypoglycaemia can vary strongly, and the in vitro conditions chosen should be matched to the cell population or neuronal circuit studied. In turn, this also highlights the advantage of the use of genetic markers (e.g. GFP-tags) to single out a specific cell population to be studied functionally in vitro (Sect. 4.8).

3.3. Electrophysiological Intracellular Recordings

For experiments, a single brainstem slice is transferred to the recording chamber (Fig. 4) (Sect. 2.3). The slice is held in place by a nylon net fixed to a platinum frame (33). The frame needs to be heavy and inert. In our hands, silver was not heavy enough to substitute for platinum, but tantalum turned out to be a viable, cheaper alternative to platinum, both available at Goodfellow (Huntington, United Kingdom; www.goodfellow.com). Superfusate is administered at a flow rate of 5 ml/min via a roller pump and kept at 28–32°C in the recording chamber (Sect. 2.3).

Either sharp microelectrodes filled with a high concentration of KCl, K-acetate or Mg-acetate (concentration typically between 0.15 and 3 M) or whole-cell patch electrodes (39–41) can be used for recording of V_m or I_m of DVN or NTS neurons for time periods of up to several hours. Patch-clamp recordings from visually identified neurons in superficial layers of brain slices have become the mainstay of analysis of electrical activity at cellular level over the past 20 years (33–35) (Figs. 1, 3, and 5). Patch electrodes are often filled with solution that aims to resemble closely the ionic composition of the cytoplasm. This is because the low impedance electrical access to the cell in whole-cell recordings comes at a price of high diffusional exchange between cytoplasm and pipette solution. This exchange can be of advantage, for example, to fill the cell with a fluorescent label or ion indicator dye. But this can also be of disadvantage because it can lead to 'washout' into the pipette of second-messenger systems or of ATP for metabolic fuel. If the disadvantages prevail, 'perforated' patch-clamp can be chosen as an alternative recording configuration as described below. While most positive electrical charges in the cytoplasm are provided by K^+, negative charges originate primarily from polyvalent proteins, with a typically minor contribution of other anions such as Cl^-, HCO_3^- or SO_4^-. For the majority of central neurons in mature mammals, the concentration of free intracellular Cl^- is ~5 mM (78). Thus, for most experiments dealt with in the Results section of this chapter (Sect. 4), standard 'low Cl^-' patch pipette solution was used, where KCl was substituted with K-gluconate.

Modifying the patch pipette solution can gain additional information regarding contribution of specific ion species to a membrane current under study. For example, replacing K-gluconate in the patch electrode with KCl shifts the equilibrium potential for Cl^- from about −80 mV to 0 mV. Thus, at a holding potential (V_h) in voltage clamp of −50 mV, opening of Cl^- channels causes an outward current with 'low Cl^-' electrodes, but an inward current when recorded with KCl 'high Cl^-' electrodes (60, 79). Low Cl^- patch electrodes typically have a DC resistance of 3–8 MΩ, and those filled with high Cl^- solutions have a DC resistance of 2–4 MΩ. Sharp microelectrodes, filled with high (e.g. 3-M KCl) or low (e.g. 1-M K-acetate or Mg-acetate) Cl^- solution have a DC resistance of 20–60 or 50–120 MΩ, respectively.

The composition of pipette solution can also be modified to 'unmask' specific currents in whole-cell recordings. The prevalent conductance ('g') of most neurons at rest is gK^+ which thus primarily determines resting V_m (39). Also, upon depolarisation, large K^+ currents are activated to repolarise V_m after neuronal firing. It can be difficult to investigate Ca^{2+} currents in the presence of these strong K^+ currents. To alleviate this problem, K^+ in the pipette solution is often substituted with 130–150 mM Cs^+ or with a mixture of 100 mM Cs^+ plus 30 mM tetraethylammonium (TEA) for depressing K^+ currents. Using this solution, sustained 'L'-, 'N'- or 'P/Q'-type voltage-activated Ca^{2+} currents can be analysed (39–41), which was though not a specific topic of our studies on DVN or NTS neurons yet. However, we used such a solution in one of our previous reports on DVN (60) for generation of plateau potentials with duration of up to several seconds to augment activity-related decreases of pH_i (Fig. 6).

For whole-cell recordings, it is imperative to buffer Ca_i to nanomolar levels because notably higher levels are unphysiological and can be toxic (56, 80–83). Such Ca^{2+} buffering is achieved by the addition to the patch electrode of either ethylene glycol tetraacetic acid (EGTA) or 1,2-bis(o-aminophenoxy)ethane-N,N,N′,N′-tetraacetic acid (BAPTA), usually 1–10 mM, together with 1 mM $CaCl_2$ and 1 mM $MgCl_2$ as the second physiologically important divalent cation. Further, 10 mM 4-(2-hydroxyethyl)-1-piperazineethanesulfonic acid (HEPES) is normally used to buffer pH_i. Finally, varying concentrations of Na_2ATP or K_2ATP (usually 1–2 mM) and Na_2-GTP (0.2–1 mM) are added to compensate for potential washout of these important metabolic fuels and 'second-messengers'. Intracellular solutions for whole-cell recording typically have an osmolarity of 270–290 mosmol/L, similar to the intracellular and interstitial fluids, and are adjusted to a pH of 7.4 with 1 N KOH. Importantly, as revealed by our fluorometric pH_i

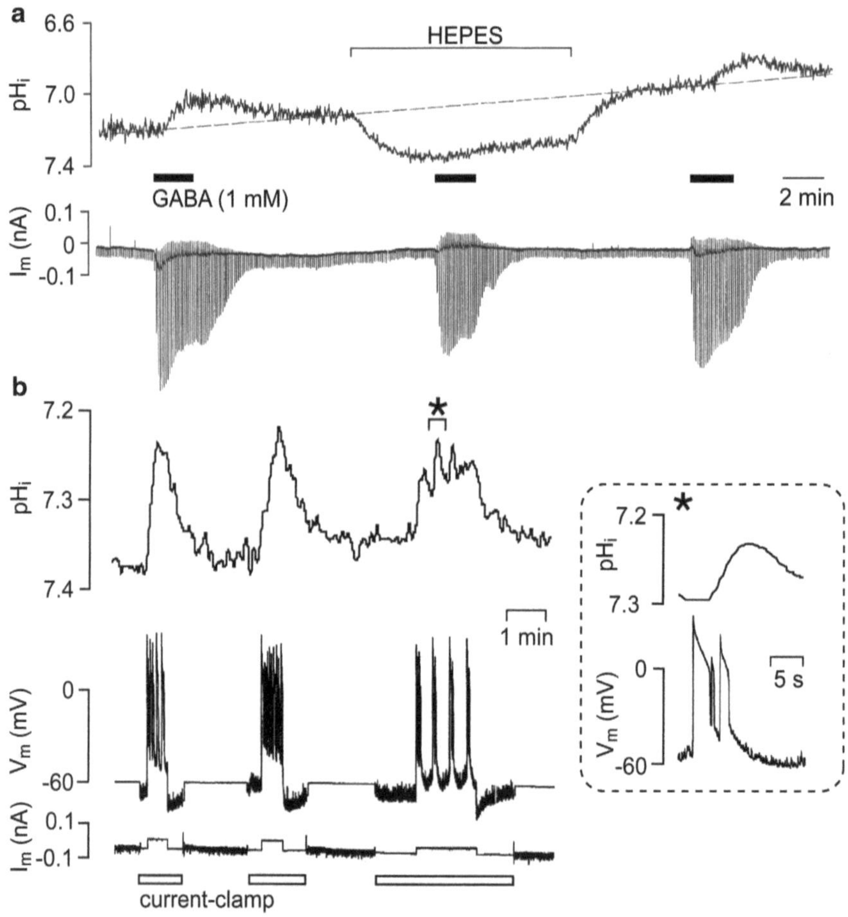

Fig. 6. *Activity-related fall of cytosolic pH (pH_i) monitored with photomultiplier-based optical system in soma of DVN from juvenile rat brainstem slice.* (**a**) A DVN was dialysed during whole-cell recording at a holding potential of –50 mV with the pH-sensitive fluorescent dye 2′,7′-bis-(2-carboxyethyl)-5-(and-6)-carboxyfluorescein (BCECF, 50 μM) added to patch electrode solution containing, among other (ionic) constituents, 5 mM Cl^-. Bath application of the inhibitory neurotransmitter γ-aminobutyric acid (GABA) notably increased membrane conductance (measured upon repetitive hyperpolarisation pulses) and decreased pH_i by ~0.2 pH units. This intracellular acidosis was abolished after changing to CO_2/HCO_3^--free, piperazineethanesulfonic acid (HEPES) pH-buffered superfusate, whereas the GABA-induced conductance increase was only slightly reduced. *Dashed line* indicates slow drift of optical pH signal, indicating apparent ongoing intracellular acidification. (**b**) For whole-cell recording from this DVN patch electrode solution contained K^+ channel blockers, specifically Cs^+ (100 mM) and tetraethylammonium$^-$ (30 mM). Current-induced bursts of long duration (>2 s) spikes evoked pronounced and rapid pH_i fall. Note appearance of rhythmic spike bursting upon application of 90 s depolarising DC current pulse. Inset on right shows that even small numbers of prolonged spikes evoke major cellular acidosis (With kind permission from (60)).

measurements (Sect. 3.5.2) (60), the major relevant physiological intracellular pH buffer is comprised by (dissociated) intracellular CO_2 and HCO_3^- (24–28 mM) in DVN in carbogenated standard solution. Accordingly, as we showed in the latter study, a change from the CO_2/HCO_3^- pH-buffered standard solution to HEPES pH-buffered saline caused a rise of baseline pH_i and abolished the pH_i fall due to activation of A-type receptors for γ-aminobutyric acid ($GABA_A$) (Fig. 6). The fluorescent dye Lucifer yellow (0.5–2%) or the tracers NEUROBIOTIN (0.2–0.5%) and biocytin (0.2–0.5%) (Fig. 1) can also be added to the pipette solution for offline histological reconstruction of cell morphology.

To avoid major modification of the intracellular milieu, perforated patch-clamp techniques were developed (39–41). Instead of rupturing the cell membrane in front of the pipette by a strong suction pulse once a 'gigaseal' (>1 GΩ) is formed, the pipette solution for a 'perforated patch' contains ionophores such as nigericin, gramicidin or amphotericin. These antibiotics can integrate into the plasma membrane to form pores that are primarily permeable for K^+ (and H^+) but do not have a major permeability for Na^+, Ca^{2+} and most anions or larger molecules, thus keeping the integrity of the cytoplasm largely intact. The input impedance of perforated patch recordings is considerably higher than that of whole-cell recordings. This can sometimes be problematic for space clamp in (fast) voltage-clamp recordings. Perforated patch recordings are also more time consuming. We usually wait for 10–30 min from sealing onto the cell until incorporation of the ionophore into the membrane has reached a sufficient and stable level to allow V_m or I_m recording (31). The most critical step is to figure out the optimal concentration of the ionophore in the pipette solution, which can vary on a daily basis and needs to be empirically established. We use solubilised amphotericin B (Sigma) and add it to the pipette solution at a final concentration of ~137 μg/ml freshly every day.

Patch-clamp techniques were originally developed to monitor current flow through single ion channels (39–41). Such single-channel recordings are also feasible in DVN, and some findings are exemplified below (Sect. 4.5). For single-channel recordings, patch pipettes were pulled from thick-walled borosilicate glass (GC150F-10, Warner Instruments) to obtain lower noise levels. Filled with 140 mM KCl, 2 mM $CaCl_2$ and 10 mM HEPES (pH 7.4), their resistance was 7–10 MΩ. Single-channel recordings in the cell-attached and inside-out configuration were performed at 25°C to improve stability of membrane patches. Because of the characteristic low activity of K_{ATP} channels, the true number of channels 'N' in the patch could not be determined.

$$Np_o = \sum_{i=1}^{n} \frac{it_i}{t} \tag{1}$$

According to Eq. 1, we therefore determined Np_o rather than calculating the open probability p_o of a single channel. Channel activity was quantified from 30-s segments by determining the time t_i spent in the respective level (all closed, one channel open, two channels open, etc.). Open/closed durations were determined from 60-s time segments according to the threshold crossing method, with a 50% threshold set in between closed and open state.

Changes in Np_o during ATP administration or drug treatment were referred to control conditions. Inward currents are shown as downward deflections, and given V_h are defined as $V_h = V_{bath} - V_{pipette}$, that is, inverted pipette potentials. Further detailed information on methods for intracellular electrophysiological recording and determination of biophysical membrane properties is given in (Chap. 3-Moore et al. (44)).

3.4. Electrical Stimulation

DVN in 400–500 μm-thick transversal slices can routinely be identified by an electrically evoked antidromic action potential. For this purpose, a stimulation electrode (Sect. 2.4) is positioned in the dorso-/mediolateral slice aspect which contains efferent axon tracts that originate from the DMNX (4, 5, 22) (Figs. 1 and 2). However, this can be problematic in thin 150–200 μm-thick slices because the likelihood that the axon is severed increases with a decrease in slice thickness and with the location of the visually identified cell near the slice surface. Similar electrical stimulation of the solitary tract or the mediolateral medulla evokes an orthodromic postsynaptic potential which can induce an action potential when its amplitude reaches or exceeds its threshold (Fig. 2). A further convenient way for identifying DVN is recording their 'A-type' K^+ current (39–41). In ~90% of DVN, the 4-aminopyridine (4-AP)-sensitive A-current delays recovery to resting V_m, which is due to a transient outward current revealed in voltage clamp following termination of a hyperpolarising voltage pulse (Figs. 2 and 3). However, both an A-current and synaptic input from the solitary tract are also revealed in a notable number of catecholaminergic neurons (27) and GLP-1 neurons (31) that are both located in an NTS area that is adjacent to the DNMX and may even partially overlap (Fig. 3).

3.5. Optical Recording

Previous optical recording from DVN by our groups was mostly based on single-detector photomultiplier techniques (Sect. 2.7). More advanced (and more expensive) imaging approaches, which enable analysis of spatio-temporal patterns of optical signals, are based on CCD camera technology or confocal/multiphoton microscopy (47, 48) (Chap. 2-Ruangkittisakul et al. (50)). We have used the latter optical approaches mostly for assessment of cytosolic Ca_i, pH_i and $\Delta\Psi$ in various types of isolated mammalian nervous tissues. Specifically, we have applied these approaches to neurons in the dorsal or ventral horn of (organotypic) newborn rat spinal cord slices (84–86), pyramidal neurons in rat hippocampal or neocortical slices (87), Purkinje neurons plus Bergmann glial cells in juvenile mouse cerebellar slices (80, 88) and inspiratory neurons in newborn rat brainstem slices (47, 89) (see also Chaps. 2-Ruangkittisakul et al. (50), 3-Moore et al. (44), 5-Nakamura et al. (90), 11-Kantor et al. (91), 12-De Curtis et al. (92) and 14-Biggs et al. (93)). Based on our experience, DVN or GLP-1 neurons in brainstem slices should be easily accessible to optical recording with CCD camera and confocal/multiphoton techniques.

3.5.1. Optical Ca_i Recording

For recording of dynamic changes of Ca_i in the soma region, DVN were loaded in our previous studies with 100 μM of the Ca^{2+} sensitive fluorescent dye Fura-2 (Molecular Probes, Eugene, OR, USA; www.invitrogen.com) via diffusion from the whole-cell patch electrode. Ca_i was measured with an Axio skop 1 connected to a photomultiplier by Luigs & Neumann and a Polychrome V by TILL Photonics (Sect. 2.7) for alternating fluorescent excitation at 360 and 390 nm. Emission was measured at 510 nm through a pinhole diaphragm, which limits the region from which light is collected to a circular spot of 20–40 μm diameter.

Fluorescence ratios were converted into Ca_i using

$$Ca_i = K(R - R_{min}) / (R_{max} - R) \quad (2)$$

in which R is the fluorescence ratio (360 nm/390 nm) and K is the effective dissociation constant of Fura-2 (94). In vivo calibration for determination of R_{min}, R_{max} and K was done according to the method described by Neher (49). For this purpose, three pipette solutions were used that contain (in mM) (1) 130 KCl, 1 $MgCl_2$, 10 BAPTA, 10 HEPES and 1 Na_2ATP (low Ca^{2+}, R_{min}); (2) 130 KCl, 1 $MgCl_2$, 3 $CaCl_2$, 4 BAPTA, 10 HEPES and 1 Na_2ATP (intermediate Ca^{2+}, 300 nM, according to a K_D of 107 nM for BAPTA (95)); and (3) 130 KCl, 1 $MgCl_2$, 10 $CaCl_2$, 10 HEPES and 1 Na_2ATP (high Ca^{2+}, R_{max}). To each solution, 100 μM Fura-2 was added. Intracellular fluorescence ratios were calculated according to Eq. 2. K is calculated as $K = 300$ nM $(R_{max}-R)/(R-R_{min})$.

Unlike this ratiometric method using the dual excitation dye Fura-2, Ca_i measurements using other dyes such as calcium green or Fluo-4 do not enable determination of absolute Ca_i levels that are mostly in the nanomolar range (47) (Chaps. 2-Ruangkittisakul et al. (50), 11-Kantor et al. (91), 12-De Curtis et al. (92) and 14-Biggs et al. (93)). In our hands, Fluo-4 is well suited for measuring relative dynamic changes of Ca_i and the pharmacological manipulations of such signals (47, 89, 96) (Chaps. 2-Ruangkittisakul et al. (50), 11-Kantor et al. (91) and 14-Biggs et al. (93)). Similar to reports by others (97), we found in the latter studies that focal injection into newborn rat brainstem slices of membrane-permeant Fluo-4 acetoxymethyl ester ('Fluo-4 AM') results in loading of both respiratory (and tonic) neurons and glial cells, whereas bath application of the same dye primarily stains astrocytic glia and microglia (see also Chaps. 2-Ruangkittisakul et al. (50) and 11-Kantor et al. (91)).

3.5.2. Optical pH_i Recording

For optical pH_i recording in DVN, we have used the fluorescent dye 2′,7′-bis-(2-carboxyethyl)-5-(and-6)-carboxyfluorescein (BCECF; Molecular Probes), which was added at a concentration of 50 μM to the patch electrode solution. The same photomultiplier-based system was used as described above for Ca_i measurements. The ratio of the fluorescence signals was measured at 540 nm (bandpass filtering 515–565 nm) in response to alternating excitation at 440 and 490 nm. Upon establishing the whole-cell configuration, steady-state loading of BCECF is achieved after <10 min as indicated by stable signals of the excitation fluorescence signals and a smooth ratio trace, representing pH_i. For calibration of BCECF signals, it is pivotal to determine whether diffusional barriers cause a lagging and blunting of pH changes in slices. A previous study from our group using pH-sensitive microelectrodes revealed that extracellular pH (pH_e) next to optically recorded DVN deviates notably from superfusate pH (60). Thus, pH of the solution used for calibration should be adjusted accordingly (60). Ratiometric fluorescence pH_i signals (see below) were calibrated using the 'nigericin method' (98). In brief, slices were exposed after the experiment to 4 μM of the K^+-H^+ ionophore nigericin in high-K^+ solution consisting of (in mM) 118 KCl, 3 NaCl, 1 $MgCl_2$, 1.5 $CaCl_2$, 25 HEPES and 10 D-glucose. The pH of this saline was varied between 6 and 8 with 1 N KOH. Presumably, cellular gradients of V_m or H^+-HCO_3^- are abolished under these conditions so that pH_i rapidly equilibrates with the pH of the calibration solutions (98).

3.5.3. Optical ΔΨ Recording

In a previous study, we analysed metabolism-related changes of Ca_i and ΔΨ in voltage-clamped CA1 pyramidal neurons of juvenile rat hippocampal slices (87). The same approach is feasible for recording ΔΨ simultaneously with V_m or I_m (plus Ca_i) in DVN. In the hippocampal neurons, measurements were done using either a photomultiplier (Luigs & Neumann) approach or an imaging system with a 12-bit CCD camera (TILL Photonics) attached to an upright fluorescence microscope (Sect. 2.7). The fluorescent ΔΨ indicator rhodamine-123 (10 μg/ml), plus eventually Fura-2, was administered to the cytosol via the whole-cell patch electrode. Ca_i signals were recorded and calibrated as described above. For rhodamine-123 measurements, the excitation wavelength was 485 nm while emission was measured at 530 nm. For simultaneous recording of Ca_i and ΔΨ, alternating excitation was done at 390 and 485 nm and emission was measured at 530 nm. For the photomultiplier system, a pinhole diaphragm was used to avoid disturbance from background illumination. Similar to hippocampal neurons, a patch electrode concentration of 10 μg/ml resulted in a stable rhodamine-123 fluorescence signal (RFS) within <5 min after establishing the whole-cell configuration in the DVN. Because determination of the absolute value (in mV) of ΔΨ is rather complex (99–102), experimentally evoked changes of RFS were normalised with reference to the initial stable baseline value. A depolarising change of ΔΨ was indicated by an increase in RFS in response to dequenching of the dye after release from mitochondria. In isolated mitochondria, the magnitude of the RFS has been shown to vary linearly with ΔΨ (99–102). The latter studies demonstrated that the resting value of $\Delta\Psi$ can vary considerably between neurons, even in the same slice and the same type of cells. The relative value of $\Delta\Psi$ and its dynamics can be estimated in a given neuron using two agents that interfere with mitochondrial properties. Bath application of 10 μg/ml of the antibiotics oligomycin induces a maximal hyperpolarisation of $\Delta\Psi$ by inhibiting mitochondrial ATP synthase, whereas the protonophore p-trifluoromethoxyphenylhyd razone (FCCP) depolarises $\Delta\Psi$ maximally by dissipating the mitochondrial transmembrane H^+ gradient (100–102).

3.5.4. Optical Cell-Death Recording

One of our previous slice studies was devoted to the optical analysis of acute death of DVN caused by severe metabolic perturbations (53). Specifically, cell death was assessed with propidium iodide, which upon loss of membrane integrity binds to nucleic acids and responds with increased fluorescence emission (103–105). This fluorescent dye was excited at 535 nm and emission was measured beyond 590 nm using the Omega Opticals XF34 filter set (excitation filter: 535/35 nm bandpass, dichroic mirror: 570 nm, emission filter: 590 nm longpass). Propidium iodide fluorescence was recorded using a TILL Photonics imaging system equipped with a CCD camera, mounted to an upright fluorescence microscope (Sect. 2.7). Addition to the superfusate of propidium iodide (2 μg/ml) resulted in ~5 min in bright red staining of the nuclei of individual cells in the DMNX, and build-up of propidium iodide fluorescence started to level off subsequently. For comparison of different slices, cell death was normalised to the total number of cells in the DMNX. This number was determined after the experiment in the recording chamber by permeabilising cells via 25 min bath application of 1% Triton X100 which led to intense staining of virtually every cell in the slice within ~15 min.

3.6. Gas- and Ion-Sensitive Microelectrodes

In several studies, mostly on respiratory active rat brainstem preparations, we have used microelectrodes sensitive to either O_2, CO_2, H^+, K^+ or Ca^{2+} (106–111). In fact, we succeeded to use double-barrelled ion-sensitive electrodes for measuring pH_i, Ca_i and also intracellular Na^+, K^+ or Cl^- in other neurons (and glia) of isolated mammalian nervous tissues (112–115). Details regarding use of such electrodes are described in those reports which focused on analysing dynamic changes of the microenvironment of anoxia-tolerant newborn rat medullary respiratory neurons during spontaneous and evoked activity and effects of anoxia, aglycaemia and complete blockade of metabolism ('in vitro ischemia') on these parameters (80, 81) (Chap. 2-Ruangkittisakul et al. (50)). Corresponding experiments were conducted in two of our studies on similarly anoxia-resistant juvenile rat DVN (22, 60). For recording PO_2, pH_e and the free extracellular K^+ concentration (K_e) during electrical stimulation and/or energy depletion, electrodes were positioned in the DMNX at a depth of 100–150 μm.

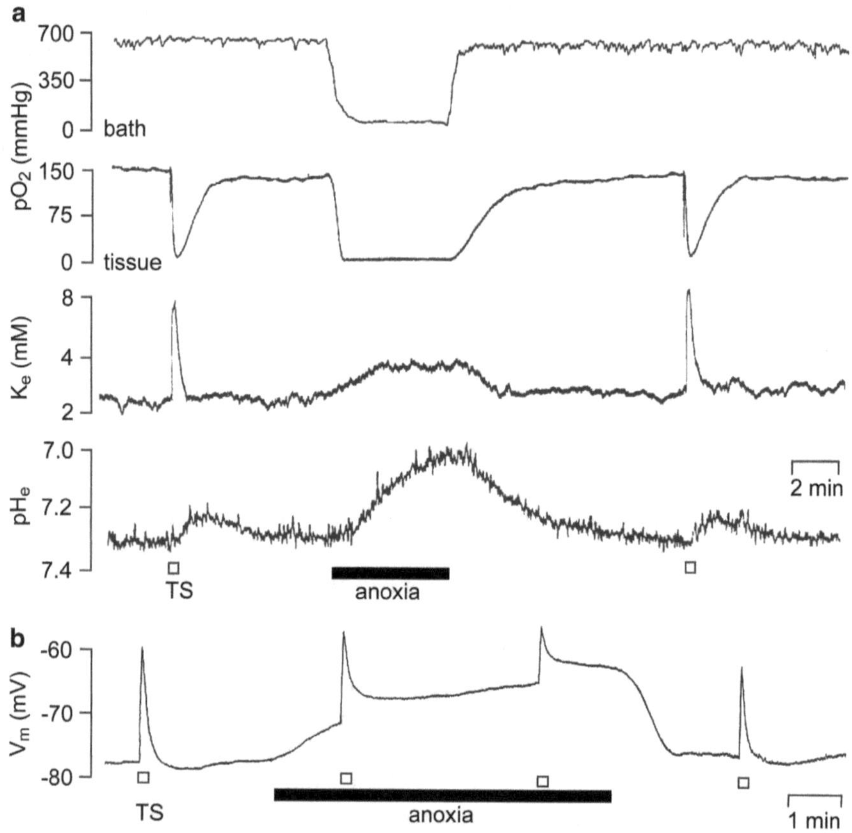

Fig. 7. *Effects of anoxia on extracellular K^+ (K_e), pH (pH_e), partial pressure of oxygen (PO_2) and V_m of glial cell in DMNX of rat brainstem slices.* (**a**) Substitution of 95% O_2 gassing with 95% N_2 gassing in carbogen of superfusate led to pronounced fall of PO_2 in recording chamber as measured with O_2-sensitive macroelectrode. This hypoxic solution evoked tissue anoxia as indicated by a fall of tissue PO_2 to nominally 0 mmHg as measured with O_2-sensitive microelectrode in the slice at a depth of 100–150 μm. Such hypoxic anoxia led to modest (<2 mM) increase of (K_e) and concomitant progressive fall of pH_e assessed with two double-barrelled ion-sensitive microelectrodes. Tetanic TS stimulation (50 V, 50 Hz, 10 s) before and after recovery from anoxia-evoked similar K_e and pH_e transients. (**b**) Modest anoxia-related K_e rise in vicinity of DVN was reflected by small depolarisation of whole-cell recorded glial cell whose V_m typically behaves like a K^+ electrode (see (114)) (With kind permission from (22)).

3.6.1. Microelectrode Recording of Tissue PO_2

PO_2 in the recording chamber was continuously monitored with a 'Clark-style macroelectrode' (Diamond Electro-Tech Inc., Ann Arbor, MI, USA), whereas tissue PO_2 was determined with O_2 microelectrodes obtained either from Diamond (outer tip diameter, 20–30 μm) or as a gift from H. Baumgärtl (Max-Planck-Institut fur Systemphysiologie, Dortmund, Germany; outer tip diameter 5–15 μm) (Fig. 7). Procedures for constructing these electrodes and their properties are described elsewhere (62, 110). The O_2-sensitive macroelectrode was calibrated in the recording chamber during superfusion with HEPES pH-buffered solution gassed with either 100% O_2 or room air (21% O_2) and during superfusion with

159 mM of the O_2-chelator Na_2SO_3 for determination of the 0 mmHg O_2 level ('anoxia'). Subsequent superfusion with standard superfusate resulted in a bath PO_2 value of ~700 mmHg in CO_2/HCO_3^- pH-buffered, carbogenated standard solution and a very similar value in HEPES pH-buffered saline gassed with 100% O_2 (Fig. 7). In hypoxic saline (95% or 100% N_2 in CO_2), PO_2 in the recording chamber ranged from 3 to 10 mmHg (Fig. 7). In contrast to the 20–30 μm tip diameter Diamond O_2 microelectrodes, some of the 5–15 μm non-commercial microelectrodes were unstable in CO_2/HCO_3^- pH-buffered saline also indicating imperfect 'membranisation' (62). Accordingly, measurements with these electrodes were mostly made in HEPES pH-buffered solution. Arrest of the superfusion flow reduced the signals of the O_2 macro- and microelectrodes by 7–15%. This 'stirring effect' might lead to underestimation of surface and tissue PO_2 because the electrode approaches the tissue surface and crosses a proposed 'unstirred layer' (62, 110).

3.6.2. Microelectrode Ion Recording

Procedures for the production of ion-sensitive microelectrodes and their properties have been described in detail elsewhere (22, 60, 107, 111, 112, 116, 117). Briefly, theta-style double-barrelled borosilicate microelectrodes are generated with a microelectrode puller (Sect. 2.4). For intracellular recording with sharp microelectrodes, which was difficult in DVN in our hands (but see (118)), their outer tip diameter needs to be <0.3 μm, whereas a good tip size for extracellular recording is 1–5 μm. K^+-sensitive microelectrodes contain a highly K^+ selective resin (Fluka 82500; WPI) in the silanised tip of one channel which is backfilled with 150 mM KCl. K^+-sensitive microelectrodes are calibrated before and after the experiments with solutions containing 1.5, 6 and 15 mM KCl and a constant background of 145 mM NaCl. The value for the slope for a tenfold change in the free K^+ concentration for these 'valinomycin'-type K^+ electrodes is typically 58 mV. Ca^{2+}-sensitive microelectrodes contain Fluka 21048 resin and are backfilled with 150 mM NaCl. They are calibrated in standard solution with different (buffered) Ca^{2+} content and have a mean slope of 29 mV/tenfold change in Ca^{2+}. pH-sensitive microelectrodes contain Fluka 95291 resin and are backfilled with 40 mM K_2HPO_4 + 23 mM NaOH + 15 mM NaCl (buffered to pH 7 with 1 N HCl). They are calibrated in solutions of different pH and have a mean slope of 58 mV/pH unit. Triple-barrelled pH-sensitive microelectrodes can be used, upon some modification, for measurement of partial pressure of tissue CO_2 as shown by us for respiratory active newborn rat en bloc brainstem preparations (106). The reference barrel of ion-sensitive microelectrodes used for recording either V_m or extracellular field potentials is filled with 3 mM KCl or standard solution, respectively.

3.7. Immunofluorescence Characterisation of Neurons

Simple immunofluorescence protocols do not require specific skills and equipment and can easily be performed in a laboratory that is primarily geared for electrophysiology on brain slices. The immunofluorescence protocols can be performed post hoc on brain slices that have been used for electrophysiology and, for example, contain cells that have been filled with dyes such as Lucifer yellow or tracers such as NEUROBIOTIN or biocytin (Sect. 3.3). These slices should be fixed in 4% paraformaldehyde (PFA) in phosphate-buffered saline (PBS) for 15–30 min at room temperature. Alternatively, if the protocol is not linked to previously functionally characterised cells, it is advisable to use 'perfuse-fixed' tissue for highest quality results. Perfuse-fixation follows the protocol for transcardial perfusion (Sect. 3.1), but after initial perfusion with standard solution (or PBS), the animal is perfused with 50 ml (mouse) or 100–300 ml (rat) 4% PFA in PBS. Subsequently, the brain is removed and 'post-fixed' in 4% PFA in PBS for 3 d at 4°C.

Protection of the tissue against freeze damage is achieved by incubation in PBS with 30% sucrose added. After transfer of the brain into this solution, it floats on top of the sucrose solution. Cryoprotection is complete once the solution has penetrated the brain and it has sunk to the bottom of the vial. This is usually achieved within 24 h. The tissue is then embedded in cryomatrix (e.g. O.C.T. compound; Sakura Finetek; www.sakuraus.com) and sectioned on a cryostat. Serial brainstem sections are kept in 'Eppendorf' caps (e.g. Fisher) filled with PBS until use in the immunofluorescence protocol. We obtain best results with 30 μm-thick free-floating cryostat brainstem sections.

For immunofluorescence visualisation of cholinergic cells, such as DVN, sections are incubated for 30 min in 'block' solution (10% rabbit serum, 1% BSA, 0.1% triton X-100 in PBS), followed by overnight incubation with a polyclonal goat antibody against choline acetyltransferase (ChAT) (1:1,500; Chemicon; www.millipore.com) in block solution at 4°C. After five 'washes' in PBS, sections are incubated in the dark for 2 h at room temperature in a rabbit anti-goat IgG CY-3 conjugate (1:500; Sigma; www.sigmaaldrich.com). Sections are washed again three times in PBS, mounted, dried until stuck firmly to the slide and then 'coverslipped' with VECTASHIELD (Vector Laboratories; www.vectorlabs.com), which is an aqueous mountant. An equivalent protocol is used to visualise catecholaminergic cells with a polyclonal rabbit antibody against tyrosine hydroxylase (anti-TH, 1:500, Santa Cruz; www.scbt.com). Block solution contains 10% sheep serum and the secondary antibody is a sheep anti-rabbit IgG CY-3 conjugate (1:500; Sigma). An even simpler protocol can be used to visualise glial cells. We use a CY-3 conjugated monoclonal mouse anti-GFAP antibody (1:500; Sigma). Since this antibody is directly conjugated, it can also be used in mice. Again, slices are incubated with this antibody in block solution (containing 10% goat serum) overnight at 4°C. Subsequently, the sections are washed three times in PBS and mounted with VECTASHIELD. Mounted sections were analysed under a Nikon upright microscope with epifluorescence to visualise fluorescence of Lucifer yellow, eYFP and CY-3.

For offline histological reconstruction, chemically fixed slices containing Lucifer yellow, NEUROBIOTIN or biocytin-filled neurons are incubated in 4% PFA (with 0.1 M PBS, pH 7.4) for 4–15 h at 4°C, followed by immersion in 30% sucrose solution (10 h). Thicker slices are cut in serial transversal sections (100 μm) using, for example, a 2800 Frigocut freezing microtome (Leica Microsystems, Wetzlar, Germany; www.leica-microsystems.com) and processed for avidin-biotin complex staining with horseradish peroxidase and diaminobenzidine. Sections are mounted on slides and eventually stained for Nissl substance with cresyl violet. Labelled cells are examined with a (fluorescence) light microscope and digitally photographed, for example, with a PL-A642 (PixeLINK, Ottawa, ON, Canada; www.pixelink.com).

3.8. Single-Cell RT-PCR

RT-PCR coupled with patch-clamp recordings is a powerful tool to link expression of specific proteins with electrical properties of individual cells (119). We have used this approach to confirm the molecular identity of K_{ATP} channels in DVN by matching single-channel properties with expression of the 6.2-type inward-rectifying K^+ channels (Kir6.2) and type-1 sulfonylurea receptors (SUR1) channel subunits (58) (Sect. 4.5). Similarly, we linked expression of the enzyme glucokinase with electrical responses observed in neurons of the vagal complex upon glucose removal from the superfusate (26) (Sect. 4.6). For single-cell RT-PCR, cytoplasm from individual cells is harvested into 6 μl of sterile-filtered pipette solution by suction at the end of a patch-clamp recording (Fig. 5). During suction, care is taken to maintain the 'gigaseal' between electrode and cell membrane. In order to synthesise first strand cDNA from the harvested cytoplasm, the 6 μl pipette solution is expelled with positive pressure into a 0.5-ml PCR tube containing 4 μl reverse transcription buffer with 5 μM random hexamer primers, 20 mM dithiothreitol, 0.5 mM (each) dNTPs, 20 U ribonuclease inhibitor (Promega; www.promega.com) and 100 U SuperScript II or III reverse transcriptase (Invitrogen). To ensure that all harvested cytoplasm is transferred into this solution, the electrode tip is crushed into the bottom of the tube.

First strand synthesis is carried out in a Thermomixer comfort (Eppendorf; www.eppendorf.com) for 1 h at 37°C. Subsequently, multiplex PCR amplification using appropriate primer pairs for desired products is carried out under a 'hot start protocol' with Taq polymerase in a Perkin Elmer Thermal Cycler 9700 (typically 30 cycles: 30 s at 94°C, 60 s at 55°C, 3 min at 72°C). A hot-start protocol proved essential to avoid formation of spurious products due to premature amplification. We use a specific hot start Taq polymerase (e.g. Promega) because this allows addition of the enzyme to a Master Mix, and thus we can reduce the amount of polymerase added to each reaction compared to protocols used by us earlier. This is also more cost-effective despite the higher cost of the heat-activated enzyme itself simply because previously the minimum amount of polymerase that could be added with an Eppendorf-style pipette was 0.5 μl, while only half of that is needed now. Furthermore, for each reaction vial, a new pipette tip had to be used and in addition to the 0.5 μl wanted some more polymerase stuck to the pipette contributing to the 'waste'. This is followed by a second PCR amplification using nested primers, that is, primers that bind within the sequence amplified in the first PCR, and 2 μl of the first PCR product as template. This second round of PCR amplification is not performed multiplex, but each primer pair is used in its own reaction mix. As negative controls for single-cell RT-PCR, reactions are routinely performed with H_2O instead of DNA sample, with solution from pipettes that are inserted into the slice without recording from a cell, and with samples that are not reverse transcribed. For positive controls, PCRs are performed on a 1:100 dilution of brainstem cDNA. Primer pairs are preferentially selected to bind to different exons. Should genomic DNA be amplified using such primer pairs, the PCR product would be larger than expected due to the inclusion of one or more introns.

3.9. Genotypying from Ear Biopsies with PCR

Genotyping of transgenic mice can easily be done from ear biopsies. Ear punching produces little trauma to animals (in contrast to the traditional tail biopsies) and can be done without anaesthesia. An ear puncher (Sect. 2.8) is used to excise a 1–2-mm-wide piece of ear tissue. This tissue is then digested in 50 μl proteinase K solution (0.5 mg/ml proteinase K in H_2O) for 40 min at 55°C, followed by 10 min at 95°C to inactivate the enzyme. A 20 mg/ml stock solution of proteinase K can be kept in a freezer and be used for several months. These incubations can be done in a thermal cycler. Three μl of the resulting suspension are then used in a 25-μl PCR reaction to amplify the transgene. It seems important to add 1.25 μl dimethyl sulfoxide to the PCR reaction to aid access to the genomic DNA. In our hands, reactions fail regularly if dimethyl sulfoxide is not added. We routinely run 40 PCR cycles for genotyping. Recently, for maximum convenience GoTaq Hot Start Green Master Mix (Promega) has been used. Use of a master mix minimises pipetting steps, inclusion of a 'Hot Start' polymerase, that needs to be activated by heating to 95°C, allows setting up of complete reactions at room temperature, and incorporation of green dye allows us to add the PCR reaction directly onto an electrophoresis gel for analysis without need of adding running buffer with dye. This protocol has proven highly reliable with several strains of transgenic mice (31, 120, 121).

3.10. Gel Electrophoresis

PCR samples are analysed by agarose gel electrophoresis (2% agarose in Tris-acetate-EDTA buffer; 110 V; approx. 20 min) using a 100-bp PCR ladder (Promega) as reference and visualised using ethidium bromide or SYBR Green. While slightly more expensive, the safer SYBR Green is strongly recommended over use of ethidium bromide. Images of gels are taken with a Kodak gel documentation system and stored on a PC for analysis and reproduction. PCR products are verified by direct sequencing (MWG Biotech AG, London, UK; www.mwg-biotech.com).

3.11. Data Analysis

Neuronal resting conductance (g_m) is measured upon application of hyperpolarising current pulses (0.05–0.5 nA, in current-clamp recording mode) or voltage pulses (−20 mV, in voltage-clamp mode) with a duration of 300–700 ms. V_h in voltage-clamp recording mode is typically −50 mV (DVN) corresponding to their mean resting V_m (8, 22). Steady-state or peak current-voltage (I/V) relationships can be determined in current-clamp mode, following injection of 300–700 ms rectangular DC current pulses or in voltage-clamp by membrane depolarisation or hyperpolarisation for 300–700 ms, usually in 10 mV steps over the desired voltage range. Alternatively, I/V relationships can be obtained from slow current or voltage ramps (39–41). If an IV relationship is required for passive (resting) membrane conductance only, it can be attempted to block all voltage-activated currents, or the cell can be held at a depolarised potential until all major voltage-gated conductances are inactivated and then apply a hyperpolarising voltage ramp (122). Holding the cell at −20 mV for a longer period of time, however, could potentially lead to a supraphysiological rise of Ca_i, which could still distort the IV relationship by activating Ca^{2+}-dependent K^+ channels. Different stimulation protocols are used to isolate different specific currents/conductances to be analysed. Examples for how to investigate the A-type K^+ current (Sect. 3.4) are shown in Fig. 3.

Electrophysiological signals are sampled into either a Macintosh PowerPC (Apple Computers Inc., Cupertino, CA, USA) or a PC equipped with either an internal PCI card (e.g. ICT-16) or an external USB-2 interface. The amplifier is controlled by software, such as ‘Pulse’ (now replaced by ‘Patchmaster’) when using a HEKA amplifier. In order to sample microfluorometrical signals at the same time, voltage signals from the photomultiplier are fed into spare A/D channels of the amplifier, and the X-chart/Fura software extension from HEKA is used to sample these data into the computer. Acquired data are analysed offline using, for example, ‘Pulsefit’ or ‘Patchmaster’ software (HEKA; www.heka.com). (For further detailed information about protocols for intracellular electrophysiological studies and (statistical) approaches for data analysis, see Chap. 3-Moore et al. (44)).

4. Results

Our interest in DVN was kindled by a study of Cowan and Martin (7) who reported that most of these neurons in mature rat brainstem slices are either hyperpolarised or modestly depolarised by 6–10 min periods of anoxia. At that time, research in the Ballanyi laboratory had focused (and still does) on analysing cellular processes involved in generation of rhythm in isolated newborn rat respiratory networks (50, 123, 124). One aspect of these studies is to explore mechanisms of the high tolerance to anoxia of newborn rat respiratory interneurons whose rostrocaudal location (in the ventrolateral medulla) partly overlaps with that of DVN (in the dorsomedial medulla). Also, these neonatal respiratory neurons are either hyperpolarised or modestly depolarised during sustained time periods (tested for up to 1 h) of anoxia (80, 81). Therefore, we hypothesised that underlying cellular mechanisms might be similar for both neuron types and that the larger (>20 μm soma diameter) and anatomically plus physiologically distinct DVN (Figs. 1 and 2) comprise a potent model for developing novel pharmaceutical strategies to protect neurons from damage by stroke-related 'complete ischemia', mimicked experimentally in vitro by combining anoxia with aglycaemia (82, 83).

Initially, we took a two-pronged approach to investigate DVN. On the one side, we followed the approach taken by Cowan and Martin (7) and investigated membrane properties and microenvironment in 'classical' 400-μm-thick brainstem slices from juvenile rats using sharp microelectrodes (22). On the other side, we applied whole-cell patch-clamp recording under visual control (8, 9) (Figs. 1, 3, and 5). This technique has been developed in the Konnerth laboratory (33, 35) and was originally adapted to DVN by Travagli and colleagues (6). Our latter two studies provided pharmacological evidence for functional K_{ATP} channels, which was later confirmed at a molecular level (58). One of our studies (22) comprises a comprehensive description of DVN and their environment under metabolic stress. These findings are discussed in the following two sections.

4.1. Membrane Properties and Microenvironment of DVN in 'Thick' Slices

Brainstem slices of 400 μm thickness were obtained from 2- to 6-week-old rats to record V_m and microenvironment of DVN during metabolic stress (22). DVN were identified via single-pulse electrical stimulation of their axons (Figs. 1 and 2), which elicited an antidromic action potential, whereas an excitatory postsynaptic potential was revealed upon stimulation of the tractus solitarius and an A-type K^+ current became evident upon termination of a hyperpolarising prepulse (Fig. 2). We showed next that repetitive (e.g. 50 Hz, 10 s) stimulation of the tractus solitarius leads to sustained depolarisation of DVN and adjacent glial cells (Fig. 2). As monitored with O_2- and ion-sensitive microelectrodes, this period of neuronal and glial excitation due to such tetanic stimulation (Fig. 2) is accompanied by a decrease of tissue PO_2 by 80–150 mmHg from a baseline of 90–200 mmHg. This fall of tissue PO_2 due to increased aerobic metabolism is accompanied by a pH_e fall from its baseline of 7.2 by 0.1–0.3 pH units, whereas K_e increases by up to 10 mM from a resting value of 3 mM (Fig. 7).

4.2. Responses to Metabolic Perturbations of DVN in Thick Slices

Next, we showed that hypoxic, N_2-gassed solution causes tissue anoxia in the DMNX (22). Such 'hypoxic anoxia' leads to a progressive decrease of pH_e (by ~0.25 pH units after 5 min) and a concomitant rise of K_e by <2 mM that is stable as tested during anoxia periods of up to 30 min (Fig. 7). The minor anoxic K^+ release by DVN is reflected by a sustained modest depolarisation by <10 mV of neighbouring glial cells whose membrane potential is typically almost exclusively determined by their K^+ conductance and thus closely follows changes of K_e in slices (114) (Fig. 7). Moreover, it was revealed that aglycaemia, after a delay of 5–15 min, also raises K_e by <2 mM and modestly depolarises glial cells as determined after 20 min of superfusion with glucose-free standard solution (Fig. 8), whereas pH_e increases by ~0.25 pH units during that time period. Additional withdrawal of O_2 at this point, to evoke in vitro ischemia, elevates K_e baseline by further ~10 mM and causes a >15 mV depolarisation of both DVN and glia (Fig. 8) with no accompanying change of pH_e.

Either fine-tipped microelectrodes or 'blind' whole-cell patch-clamp techniques (43) were used for V_m recording in DVN (and glia) in the 400-µm-thick slices (22). This showed that the spectrum of responses of DVN to anoxia is similar for both recording techniques and also similar to that in the study by Cowan and Martin (7). Specifically, in response to 5–20 min of anoxia, 38% of DVN develop a stable hyperpolarisation of ~15 mV (Fig. 8), 25% of cells respond with a modest and sustained depolarisation with a mean value of 9 mV and V_m plus tonic discharge are not changed in the remaining 37% of DVN. The similar distribution of anoxia responses in DVN for both V_m recording techniques indicates that diffusion of cellular constituents into the patch electrode or, conversely, of patch electrode content into the cell, does not impair the primary mechanism of metabolic sensing leading to hyperpolarisation-related functional inactivation of DVN. This finding is important because it indicates that dialysis of fluorescent dyes via the patch electrode is a feasible approach for optical recording of cellular signalling factors involved in such metabolic sensing (Sect. 4.7).

DVN that show a stable hyperpolarisation in response to anoxia are also hyperpolarised by aglycaemia. However, in contrast to rapid onset of the anoxic hyperpolarisation (typically during <3 min

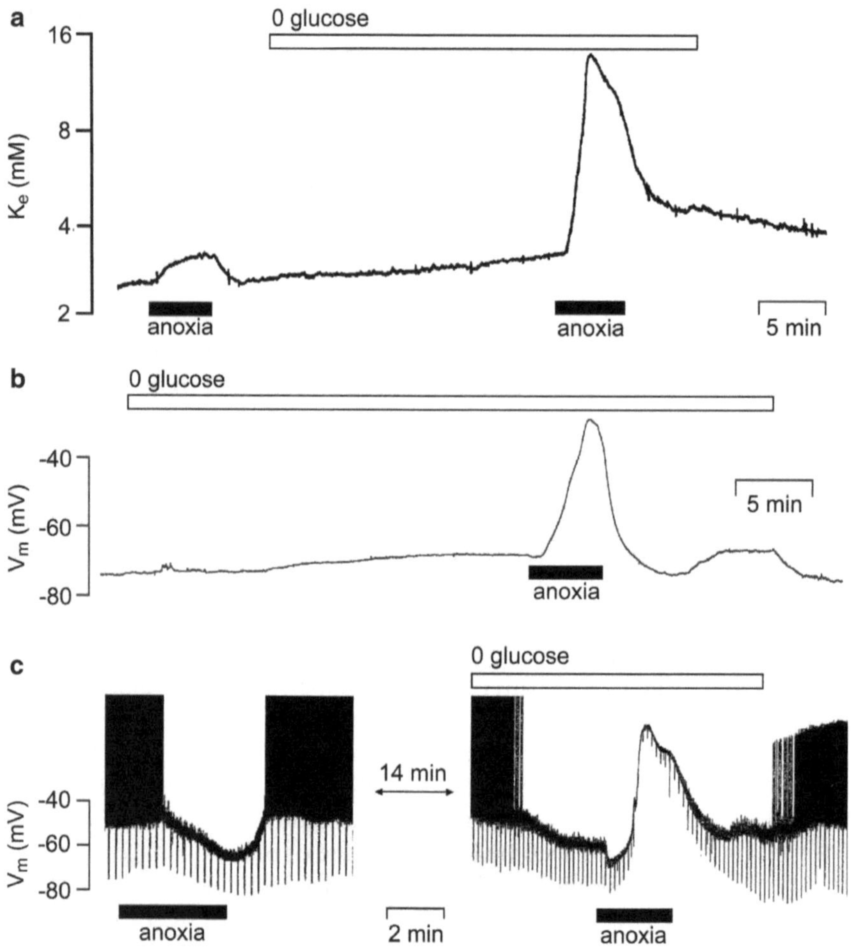

Fig. 8. *Effects of 'in vitro ischemia' on K_e and V_m of DVN and neighbouring glial cell in DMNX of rat brainstem slices.* (**a**) Tissue anoxia due to application of hypoxic superfusate evoked modest K_e rise. Following recovery, the brainstem slice was exposed to standard solution without glucose which led to delayed K_e rise by <2 mM. Anoxia after ~20 min in glucose-free solution, mimicking complete ischemia in vivo (82, 83), raised K_e by >10 mM. (**b**) glucose-free solution had no major effect on V_m of glial cell in DMNX, whereas in vitro ischemia depolarised V_m by >20 mV. (**c**) Hypoxic anoxia hyperpolarised DVN by >10 mV and concomitantly decreased its membrane conductance measured by injection of hyperpolarising DC current pulses at a regular interval. After recovery and pre-incubation in glucose-free solution for >10 min, in vitro ischemia caused an initial hyperpolarisation that turned into a depolarisation of >30 mV (With kind permission from (22)).

after change to hypoxic solution), hyperpolarisation in response to glucose depletion is often delayed by 12–60 min (Fig. 8). Anoxia during glucose depletion leads to a prominent depolarisation (Fig. 8) with a glutamate-dependent mechanism described below (Sect. 4.7.2). In extension to the 'thin' slice recordings described in the next section, we found that not only the anoxic but also the aglycaemic hyperpolarisation is countered by the K_{ATP} channel blockers tolbutamide and glibenclamide, whereas the K_{ATP} channel activator diazoxide induces a very similar stable hyperpolarisation in carbogenated standard solution (22).

4.3. Anoxia-Evoked Activation of K_{ATP} Currents in 'Thin' Slices

Our first analysis of membrane currents underlying anoxic hyperpolarisation of a subpopulation of DVN was based on recording from visually identified cells in 150-μm-thin brainstem slices from 16- to 22-day-old rats (8). Specifically, we showed that O_2 deprivation elicits an outward current with a reversal potential at approximately −80 mV. Increasing extracellular K^+ from 3 to 10 mM shifts the reversal potential by +23 mV, whereas an increase from 5 to 144 mM of pipette Cl^- has no effect in that regard. This established that the anoxic outward current is indeed carried by K^+ ions, and the finding that tolbutamide reverses the current provides compelling evidence for involvement of K_{ATP} channels (Fig. 9). This scenario was substantiated by the fact that the anoxic outward current in DVN is mimicked by diazoxide (Fig. 9) and occluded in the presence of diazoxide (8).

Prior to this study, questions remained as to whether (1) metabolism-related responses are preserved in superficial DVN despite the fact that some of their dendritic processes might be removed during sectioning, (2) anoxia can be achieved by hypoxic superfusate in thin slices rather than having to use a pharmacological blocker of mitochondrial function such as cyanide (CN^-) or azide, as is required in preparations of dissociated cells and (3) anoxic hyperpolarisation is only observed in rats older than 4 weeks. Our study verified that none of these concerns prevailed and that 'thin slice patch-clamp recording under visual control' (33, 35) (Fig. 1) is a valuable tool when studying responses of brainstem neurons to metabolic challenges. In fact, comparison of cellular responses observed in Trapp and Ballanyi (8) with those in the thick slices of Ballanyi et al. (22) revealed very little differences, thus paving the way for a host of studies utilising optical techniques as described below (Sect. 4.7).

4.4. Spontaneous K_{ATP} Current During Whole-Cell Recording

While application of patch-clamp in thin slices under visual control facilitates the use of the in vitro DVN brainstem model for investigating mechanisms of metabolic sensing, another finding of Trapp and Ballanyi (8) indicates a potential pitfall using whole-cell recordings. The observation of the reversible anoxic outward current indicates that the cellular machinery for O_2 sensing of DVN is principally kept intact in whole-cell recordings. However, in 15–20% of recordings from DVN 'spontaneous' activation of K_{ATP} channels occurred, mostly within few minutes after establishing the whole-cell configuration which could occlude the experimentally evoked anoxic outward current (8, 9).

Such 'washout' K_{ATP} currents have been described in various publications that investigate properties of isolated pancreatic β-cells using whole-cell recordings (125–127). These studies revealed a correlation between ATP concentration of the patch pipette solution and K_{ATP} current amplitude, whereas a related report found that K_{ATP} currents activate spontaneously even when these cells are

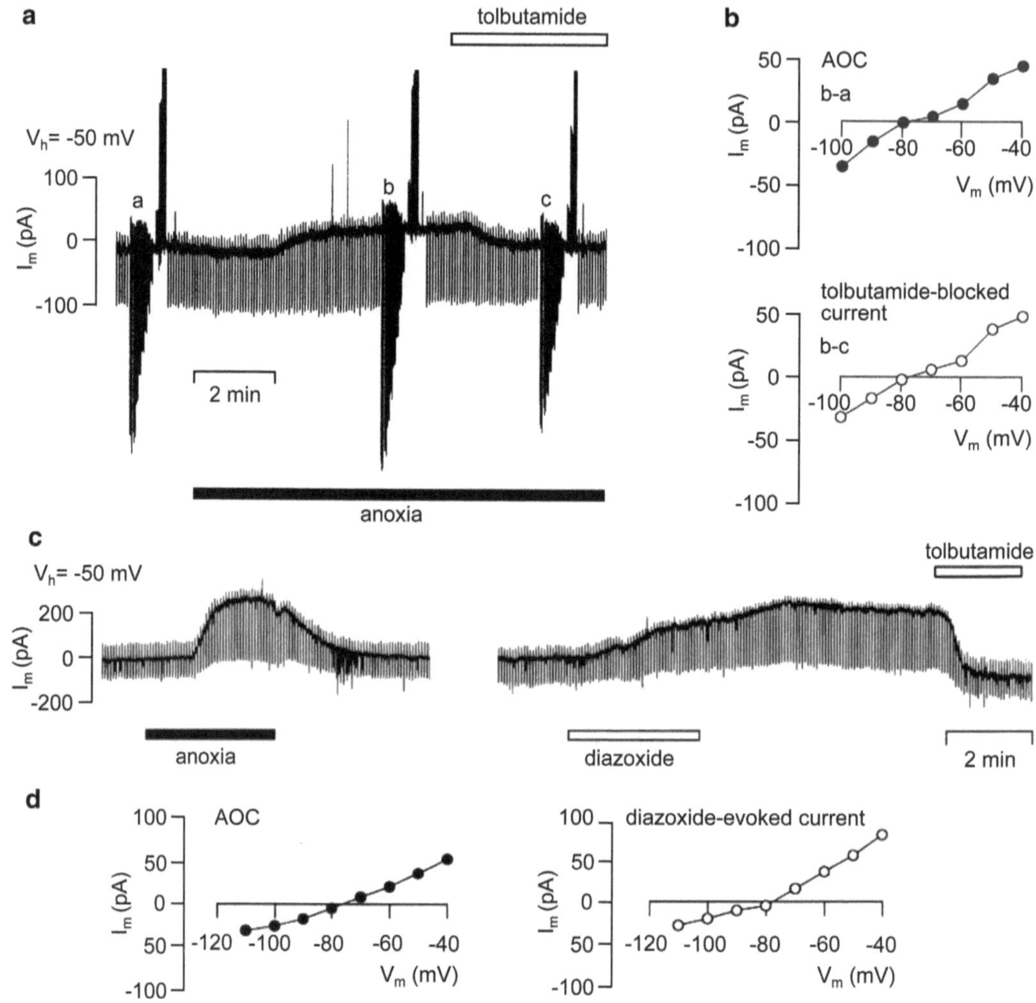

Fig. 9. *Pharmacological characterisation of anoxic outward current (AOC) in DVN of rat brainstem slices.* (**a**) In voltage-clamped rat DVN AOC and concomitant conductance increase in response to hyperpolarising voltage pulses were abolished by bath application of tolbutamide, a sulfonylurea-type K_{ATP} channel blocker. (**b**) Current/voltage (I/V) relationship in upper panel illustrates AOC obtained by subtracting currents in *b* from those in *a*, whereas (almost identical) tolbutamide-blocked current in lower panel was obtained by subtracting currents in *c* from those in *b*. (**c**) In different DVN AOC and conductance increase were similar to those elicited by bath application of the K_{ATP} channel activator diazoxide. Diazoxide-evoked outward current, which persisted after washout of drug from recording chamber, was blocked by tolbutamide. (**d**) In different DVN identical I/V relationships (subtracted as in **a**) were measured during anoxia and diazoxide (With kind permission from (8)).

dialysed with 10 mM or more ATP (128). Similar to the latter findings by Trube et al. (128), it seems unlikely that the sustained spontaneous outward current in DVN is caused by washout of cellular ATP. Firstly, patch electrodes in our above studies contained 2 mM ATP, which is in the estimated physiological cellular range (129). Secondly, neither increasing ATP content to 20 mM nor its removal from the pipette solution affected the probability of occurrence or delay of onset of the spontaneous outward current in one of our follow-up studies (55). Somewhat contrary to this, our more recent study on brainstem slices from 18- to 30-day-old Sprague-Dawley rats revealed some relation between patch pipette ATP content and occurrence of spontaneous K_{ATP} channel opening (26). When recording with ATP-free pipette solution, 9% of cells developed a spontaneous outward current. This probability was reduced to 4% and 3% with 1 mM and 2 mM ATP, respectively, and no spontaneous outward current was observed with 3 mM ATP in the pipette solution. Probably the most surprising finding was that only 9% of cells opened K_{ATP} channels spontaneously when recorded with ATP-free solution. This indicates that despite perfusion with pipette solution, mitochondrial metabolism determines K_{ATP} channel activity in these cells. This reiterates the findings from our earlier study (55), where 20 mM ATP in the patch pipette could not prevent opening of these channels in response to a metabolic challenge.

In our earlier studies (22, 55, 56), it seemed that spontaneous activation of K_{ATP} channels occurs more frequently when whole-cell recording was performed on DVN from brainstem slices kept for several hours in standard 'storage' solution compared to recordings from 'fresh' slices. In contrast, there were no obvious differences between early or late recordings regarding other basic properties of DVN such as the rate of their tonic firing or the probability of occurrence or the amplitude of the A-current (Fig. 2) during the early phase of whole-cell recordings, before development of the shunting spontaneous K_{ATP} channel outward current. Although all this has not been studied systematically, based on these observations, we would recommend the use of brainstem slices within <3 h after their preparation for a higher probability of a stable baseline of V_m or I_m during whole-cell recording. Interestingly however, this presumptive relation was not observed in our more recent studies (25, 26), where slices were routinely used for 5 h and more. One important difference between these studies and the previous ones was the use of transcardial perfusion of rats with ice-cold low solution (Sect. 3.1). Additionally, the percentage of neurons exhibiting a spontaneous outward current was 3% with 2 mM ATP in the pipette solution, compared to 15–20% reported by earlier (8, 9). This might indicate that transcardial perfusion and reduction of the Na^+ concentration during preparation might render the neurons in the slices more resilient later on, thus allowing intrinsic metabolism, rather than pipette solution, to determine K_{ATP} channel activity.

However, alternatively, the modified preparation technique might influence the selection of cells that remain 'perfectly' viable within the slice and thus determine which cells are available for patching. In this regard, it might be of significance that the mean whole-cell capacitance was ~40 pF in the early study by Trapp and

Ballanyi (8) but ~55 pF recently in Balfour et al. (26). This concurs with our subjective observation that slices from rats which underwent the transcardial perfusion showed more large cell bodies of healthy appearance, that is, opaque and not overtly rounded, in both DMNX and NTS, than those that did not.

4.5. Molecular and Single-Channel Characterisation of K^+ Currents in DVN

For determination of the molecular identity of K_{ATP} channels in rat DVN, we showed with single-cell antisense RNA amplification-PCR (58) (Sect. 3.8) that these channels are comprised of four SUR1 subunits which associate with four Kir6.2 channel subunits (Fig. 10). Thus, DVN express the same SUR1/Kir6.2-type K_{ATP} channels as most other neurons (130) and pancreatic β-cells that are innervated by a subpopulation of DVN (13, 61, 131–134).

Kir6.2 and SUR1 mRNA in DVN correlated with the presence of K^+ channels that had very similar single-channel properties as described for β-cell K_{ATP} channels (58). Specifically, patch-clamp recording from the soma membrane in the 'cell-attached' configuration (39–41) showed that 'chemical anoxia', induced by bath application of 1–2 mM CN^- (Sect. 4.3), blocks tonic spiking in DVN and activates single channels with a unitary conductance of 75 pS and a reversal potential ranging from −80 to −100 mV (Fig. 10). In juvenile rats, the probability for obtaining cell-attached patches with K_{ATP} channels was quite low though. Subsequently, we discovered that when using juvenile mice instead of rats, a larger proportion of patches contained functional K_{ATP} channels (55). These channels were studied not only in cell-attached but also in 'excised inside-out' configuration, where the intracellular surface of the patch is exposed to the superfusate (39–41). Thus, ATP sensitivity of K_{ATP} channels in DVN could be directly determined by superfusion of the patch with superfusate containing different ATP concentrations (Fig. 10). This revealed a half-maximal concentration of 5 μM for blockade of DVN K_{ATP} channels by 'intracellular' ATP, similar to what is observed for excised patches containing the recombinant Kir6.2/SUR1 channels expressed in Xenopus oocytes or in the 'COS' or 'HEK' cell lines (135–138). Our histological reconstruction of DVN filled with NEUROBIOTIN or biocytin via the patch electrode (Fig. 1) indicated that occurrence of anoxic K_{ATP}-channel-mediated outward currents does not depend on location of DVN within the DMNX (55). The morphology of murine DVN (Fig. 1) is apparently similar to that of DVN in mature rats (7) and guinea pigs (4, 5). Specifically, 2–5 dendrites with little arborisation originate from a bipolar-shaped soma with a diameter of 16–25 μm.

We further demonstrated with whole-cell recordings that kinetics of CN^--evoked activation of murine DVN K_{ATP} channels are not affected by the ATP content of the patch electrode (Sect. 4.4) or blockade of cellular ATP consumption via pharmacological inhibition of plasmalemmal Na^+/K^+ and Ca^{2+} ATPases with

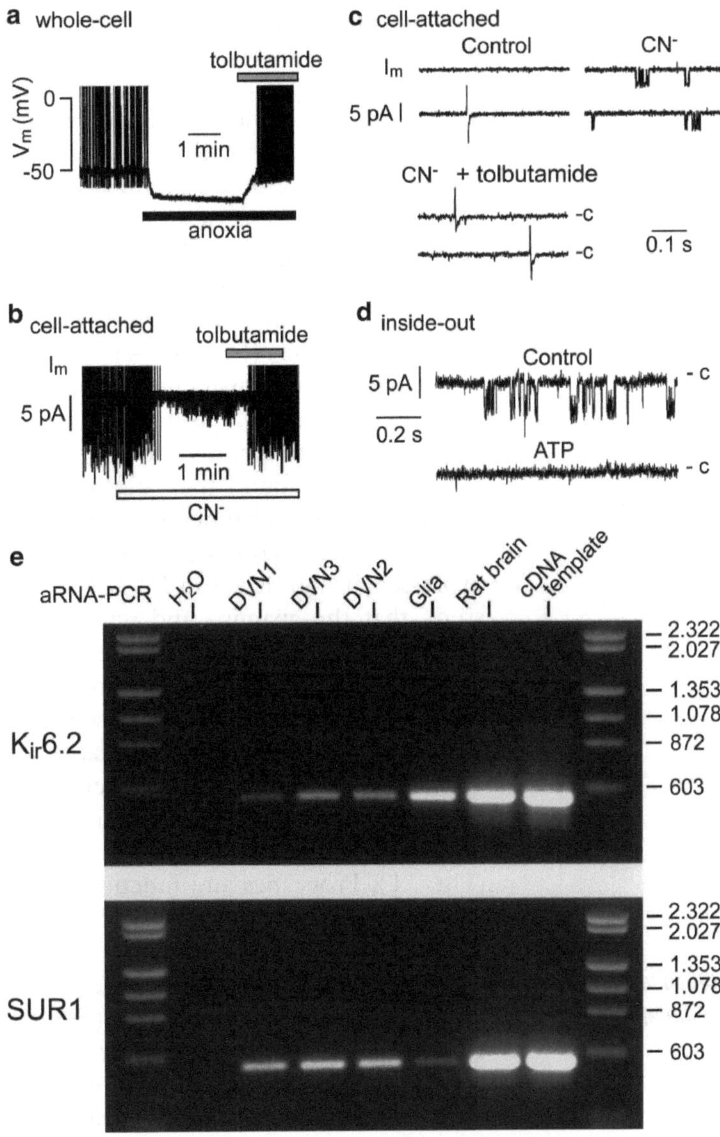

Fig. 10. *Molecular identity and biophysical properties of K_{ATP} channels in juvenile rodent DVN.* (**a**) In a rat DVN, hypoxic anoxia-evoked sustained hyperpolarisation and concomitant suppression of tonic spiking that were both reversed by tolbutamide (200 μM). (**b**) Anoxic hyperpolarisation of rat DVN is due to activation of single K_{ATP} channels as revealed for chemical anoxia due to bath application of 1 mM cyanide (CN^-). Sharp deflections on I_m trace in this 'cell-attached' patch recording from a rat DVN during control, CN^- plus tolbutamide, and wash are caused by tonic spiking. (**c**) Excerpts of I_m traces in **b** at higher time resolution. (**d**) In 'excised inside-out' patch from mouse DVN, K_{ATP} channel activity is abolished by addition of 20 μM ATP to superfusate with composition that mimics intracellular ion milieu. (**e**) Antisense RNA-PCR analysis of cytoplasm obtained during whole-cell recording reveals that 3 rat DVN (DVN1-3) coexpress mRNA for inwardly rectifying K^+ (K_{ir}) channel isoform, $K_{ir}6.2$, and SUR1. Thus, DVN express the same type of K_{ATP} channels as pancreatic ß-cells innervated by subpopulations of these neurons (With kind permission from (80); (**b**, **c** and **e**) with kind permission from (58); (**d**) from K. Ballanyi and J. Brockhaus, unpublished).

ouabain and vanadate, respectively (55). Similarly, reducing in vitro temperature from 30°C to 25°C had no effect on CN^- induced activation of K_{ATP} currents. Use of further pharmacological tools in that study excluded involvement of anoxia-evoked changes in cellular redox state for K_{ATP} channel activation in DVN. Based on these findings, we proposed that anoxic K_{ATP} channel activation in DVN is independent of a fall of ATP and a concomitant rise of ADP in the cytosol. We hypothesised that rather anoxia-induced relief of these channels from normal ongoing blockade by ATP is caused by signalling factor(s) related to impairment of mitochondrial function (55) (Sect. 4.7.4).

In parallel with analysing the molecular nature of K_{ATP} channels in rat and mouse DVN using PCR, we applied in situ hybridisation histochemistry to identify other types of K^+ channels in these cells. Specifically, we found that 'big' conductance Ca^{2+}- and voltage-activated (BK) K^+ channel α subunit mRNA and a 'small' conductance Ca^{2+}-activated (SK) K^+ channel subunit transcript, SK3, are expressed at high levels in DVN (57). We showed further in that study that the apamin- and scyllatoxin-sensitive SK3-type Ca^{2+}-activated K^+ channels contribute to spike after-hyperpolarisation and are thus important for the rate of tonic firing in DVN. In contrast, the iberiotoxin- and TEA^--sensitive BKα-type Ca^{2+}-activated K^+ channels shape action potential duration in DVN, but do not affect firing rate (57). More recently, we have shown (122) that DVN express a two-pore-domain acid-sensitive K^+ (TASK)-like conductance that is inhibited by activation of 5-HT_2 receptors (Fig. 11). Properties and functional roles of the latter type of K^+ channel are dealt with in the next section.

4.6. Glucose Sensing in the Vagal Complex

Our initial study in thick slices (22) had indicated that aglycaemia, like anoxia, hyperpolarises DVN (Fig. 8). This raised the question whether DVN might act as central glucose sensors. Glucosensing central neurons in mammals fall into two broad categories: those that increase action potential firing rate upon a rise in ambient glucose levels are called glucose-excited (GE) cells, and those that decrease firing rate when glucose levels rise are named glucose-inhibited (GI) (74). According to this nomenclature, DVN would fall into the category of GE cells. Aglycaemia silenced those cells and thus glucose excited them. However, these responses occurred under aglycaemia, rather than hypoglycaemia and required sometimes up to 1 h exposure of aglycaemia before they became apparent (Fig. 8) (Sect. 4.2). This is very different to the 'archetypal' GE cell type, pancreatic β-cells that respond almost instantaneously to changes in glucose levels, and the required changes are much less pronounced (139, 140). These cells 'sense' glucose via its impact on cellular metabolism, namely, by its effect on production of ATP from ADP. The ATP that is produced as a consequence of glucose metabolism then inhibits K_{ATP} channels in the plasma membrane of

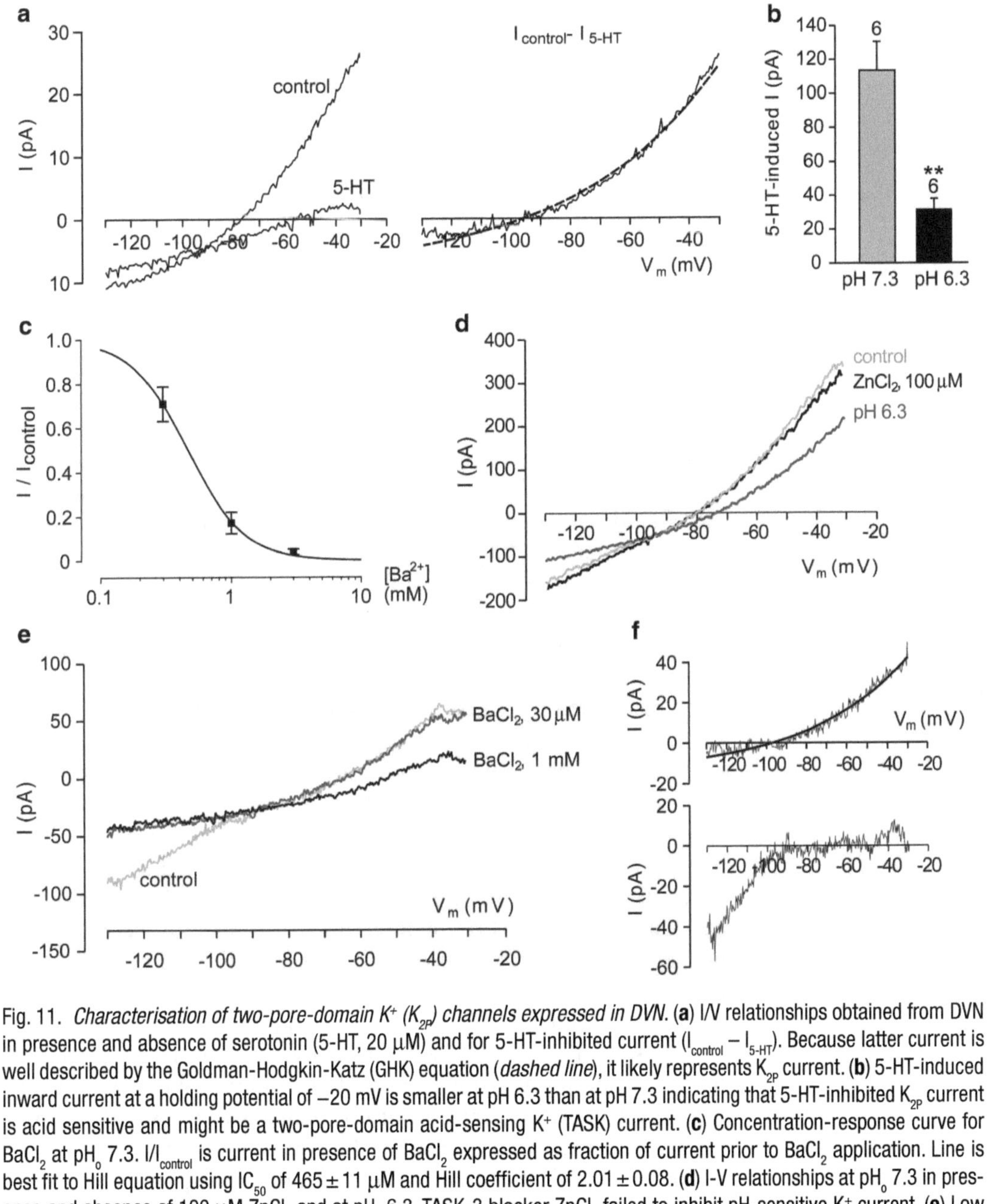

Fig. 11. *Characterisation of two-pore-domain K^+ (K_{2P}) channels expressed in DVN.* (**a**) I/V relationships obtained from DVN in presence and absence of serotonin (5-HT, 20 μM) and for 5-HT-inhibited current ($I_{control} - I_{5\text{-HT}}$). Because latter current is well described by the Goldman-Hodgkin-Katz (GHK) equation (*dashed line*), it likely represents K_{2P} current. (**b**) 5-HT-induced inward current at a holding potential of −20 mV is smaller at pH 6.3 than at pH 7.3 indicating that 5-HT-inhibited K_{2P} current is acid sensitive and might be a two-pore-domain acid-sensing K^+ (TASK) current. (**c**) Concentration-response curve for $BaCl_2$ at pH_o 7.3. $I/I_{control}$ is current in presence of $BaCl_2$ expressed as fraction of current prior to $BaCl_2$ application. Line is best fit to Hill equation using IC_{50} of 465 ± 11 μM and Hill coefficient of 2.01 ± 0.08. (**d**) I-V relationships at pH_o 7.3 in presence and absence of 100 μM $ZnCl_2$ and at pH_o 6.3. TASK-3 blocker $ZnCl_2$ failed to inhibit pH-sensitive K^+ current. (**e**) Low concentration (30 μM) of $BaCl_2$ inhibits primarily inward K^+ current. Increasing $BaCl_2$ to 1 mM inhibited mainly outward current. (**f**) Lower panel: I-V relationship of current inhibited by 30 μM $BaCl_2$, indicative of strongly rectifying K_{ir} channel. Upper panel: I-V relationship of current inhibited by increase of $BaCl_2$ from 30 μM to 1 mM. This current is described well by GHK equation (*solid line*), indicative of a K_{2P} channel (With kind permission from (122)).

the β-cell, causing depolarisation and subsequent release of insulin. Their exquisite sensitivity to changes in extracellular glucose concentration is brought about by use of two key components that distinguish β-cells from most other cells (141). Firstly, they express a low-affinity glucose transporter-2 (Glut-2) in their plasma membrane (142, 143). Due to this, intracellular glucose concentration in β-cells reflects the extracellular concentration. Secondly, β-cells express a low-affinity hexokinase (glucokinase) that phosphorylates glucose to glucose-6-phosphate as the first step of glycolysis. In contrast to ubiquitous hexokinase-1, that phosphorylates glucose with high affinity and matches glycolytic cellular energy demand via product inhibition, glucokinase has lower affinity to glucose, but is not inhibited by glucose-6-phosphate (140, 144, 145). Consequently, glucose-6-phosphate concentration is proportional to glucose availability, and thus ATP production reflects glucose availability and is not matched to metabolic demand. This means that in pancreatic β-cells ATP levels are a measure of glucose availability and K_{ATP} channel activity, which reflects ATP levels to determine V_m and thus their electrical activity.

Based on the latter findings in β-cells, we performed another set of studies analysing glucose sensitivity of neurons in the dorsal vagal complex (25, 26). In these studies, we not only examined the electrical response of DVN and NTS neurons to hypoglycaemia in detail but also used single-cell RT-PCR to interrogate the presence of the key components of the glucosensing machinery, namely, glucokinase and K_{ATP} channels (26) (Fig. 5). The findings in that work confirmed those from our earlier study (22) in demonstrating that a substantial fraction of DVN respond to prolonged aglycaemia with K_{ATP} channel activation. Additionally, this study reported that a small percentage of DVN and NTS neurons respond to aglycaemia much faster, that is, within 2–5 min (Fig. 5). These 'true' glucose-sensing cells belonged to either the GE type, showing K_{ATP}-channel-mediated hyperpolarisation, or the GI type, showing depolarisation upon brief aglycaemia. All remaining cells that did not alter V_m within 5 min of glucose removal were referred to as non-responding ('NR') neurons. These included both neurons that failed to change V_m during prolonged aglycaemia (>40 min) and those that hyperpolarised after >20 min. Most importantly, the 'fast' responding neurons expressed glucokinase, whereas NR cells did not. This demonstrates that GE neurons of the vagal complex utilise the same glucosensing machinery as pancreatic β-cells.

The second study (25) focused on GI responses observed previously in Balfour et al. (26). GI neurons have been described in various brain regions, but the nature of ion fluxes underlying depolarisation under aglycaemia remains controversial. Potential mediators include inhibition of the Na^+/K^+ pump, blockade of a K^+ conductance and closure of a particular type of Cl^- channels. The clearest evidence available is for inhibition of a specific K^+ channel. In a series of experiments of orexigenic neurons from lateral hypothalamus, Burdakov et al. (75) provided strong evidence for inhibition of TASK channels by raised glucose.

We had already shown (122) that DVN express a TASK-like conductance, whose closure mediates (at least in part) excitatory

5-HT effects, and that the pharmacological profile of this conductance excludes TASK-3 homomers thus implicating involvement of TASK-1-containing channels (Fig. 11). Consequently, when analysing GI responses of DVN, we examined this hypothesis. We discovered indeed that a fraction of GI cells depolarised during aglycaemia or chemical anoxia by means of inhibition of an outwardly rectifying K^+ channel (25). However, the volatile anaesthetic halothane failed to activate an outwardly rectifying K^+ current in those cells. Since activation by halothane is one hallmark feature of TASK channels, the GI response in DVN is unlikely mediated by such channels. Burdakov's group showed subsequently that the GI response of orexinergic hypothalamic neurons remains unaltered in TASK knockout animals (146, 147) thus suggesting involvement of a two-pore domain K^+ channel other than TASK1 or TASK3. However, while the GI response of orexin neurons is apparently always mediated by two-pore-domain K^+ channels, we found more variability in the vagal complex (25). Approximately 25% of DVN exhibited inhibition of an outwardly rectifying K^+ channel in response to metabolic inhibition by CN^-, azide or aglycaemia (Fig. 12). The remaining cells showed either an increase in conductance with a reversal potential positive of resting V_m or a small inward current without reversal potential within the voltage range tested (−120 to −20 mV). This pattern persisted whether low Cl^- or high Cl^- pipette solutions were used, thus suggesting that a Cl^- channel is not involved. Whether or not partial inhibition of the Na^+/K^+ pump was causally involved could not be resolved. In any case, these electrical responses were the same no matter how metabolic inhibition was achieved, that is, they were observed under chemical anoxia with CN^- or azide and also during hypoglycaemia (25) (Fig. 12). This is analogous to the finding that K_{ATP} channels are activated either by anoxia or aglycaemia (22) and suggests that in all cases metabolic inhibition triggers the response, rather than, for example, an extracellular glucose receptor.

Taken together, our studies on glucose and metabolic sensing in DVN suggest that mitochondrial function determines the metabolic sensitivity of the cell and that the presence or absence of functional K_{ATP} channels determines whether the cells respond with hyperpolarisation or (small) depolarisation to metabolic stress (Fig. 12). The subset of DVN that responds rapidly to changes in glucose availability is characterised by expression of glucokinase, and again the presence or absence of K_{ATP} channels determines whether the cell is of the GE or GI subtype (Fig. 12).

4.7. Optical Recording of Metabolism-Related Signalling Factors in DVN

We noted above that thin juvenile rodent brainstem slices can be used for whole-cell recording under visual control from DVN and that this recording mode leaves the cellular machinery for regulating K_{ATP} channels mostly intact. Thus, optical techniques are applicable for in vitro assessment of dynamic changes in cellular signalling factors associated with metabolic perturbations. At the

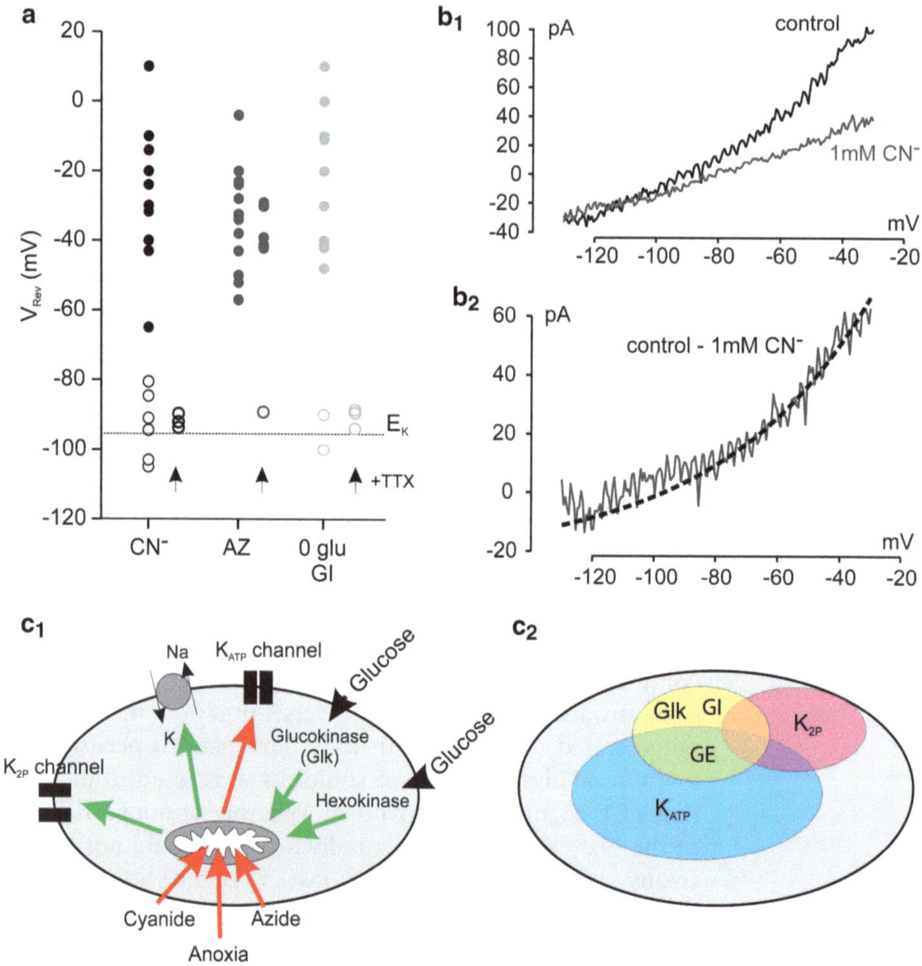

Fig. 12. *Inward currents of GI neurons of vagal complex*. (**a**) Reversal potential of inward current elicited by CN^-, azide (AZ) and aglycaemia (0 gluc GI) varied widely between individual DVN. In a subpopulation of these cells, a decrease in conductance with reversal potential close to the equilibrium potential for K^+ (E_K) was observed during metabolic inhibition (*open circles*). Other cells developed increase in conductance with wide range of reversal potentials (*filled circles*). Reversal potentials obtained in presence of 0.5 μM tetrodotoxin (TTX) to block voltage-activated Na^+ channels (39) are given in separate columns indicated by *arrows*. Cells that exhibited parallel, depolarising shift are not represented. **b**$_1$, DVN in which 1 mM CN^- inhibited current with reversal potential close to E_K. **b**$_2$, I/V relationships of current inhibited by CN^- were well fitted by GHK equation (*dashed line*). **c**$_1$, schematic pathways potentially involved in electrical responses to hypoglycemia or inhibition of mitochondrial function in DVN. *Green arrows* indicate stimulation and *red arrows* indicate inhibition; for example, mitochondrial function inhibits K_{ATP} channels, but stimulates K_{2P} channels. CN- inhibits mitochondrial function, thus activating K_{ATP} channels and inhibiting K_{2P} channels in some DVN **c**$_2$, schematic of different cell types (defined by metabolic responses) found within the vagal complex. About 40% of cells express K_{ATP} channels. A smaller fraction expresses K_{2P} channels, and a subgroup of those also express K_{ATP} channels. Another subgroup of cells express glucokinase (GLK), depending on their expression of K_{ATP} channels, they are GE cells (with K_{ATP} channels) or GI cells (without K_{ATP} channel) (Adapted with kind permission from (25)).

time when DVN were initially studied by us, we could not make use of multiphoton, confocal or CCD camera-based imaging technologies but instead used a photomultiplier-based optical approach.

4.7.1. Optical Recording of pH_i

In an initial study (60), we monitored pH_i in the soma region of DVN to investigate whether anoxia-related increase of cellular lactate directly activates their K_{ATP} channels as has been hypothesised for skeletal muscle fibres (148). For optical pH_i recording in that study, we dialysed juvenile rat DVN with the fluorescent dye BCECF, while we used pH-sensitive microelectrodes for monitoring pH_e in the vicinity of the whole-cell-recorded neurons. Firstly, we showed that tonic spiking of these neurons results in ongoing intracellular acidification by 0.05–0.2 pH units from a mean pH_i baseline of ~7.2 and that (additional) evoked bursts of activity have a comparable effect (Fig. 6). Similar to our study on activity-evoked decreases of pH_i in CA_1 neurons of hippocampal slices from juvenile rats (149), we presented evidence that acidosis of DVN in response to intense electrical activity or depolarisation in voltage clamp is due to activation of a vanadate-sensitive plasma membrane Ca^{2+}/H^+ ATPase which extrudes cellular Ca^{2+} in exchange for inward transport of protons. We showed further that similar intracellular acidification in response to bath-applied GABA is caused by HCO_3^- efflux through the $GABA_A$-receptor-coupled anion channel (Fig. 6), similar to GABA- and glycine-evoked acidification in neurons in respiratory areas of newborn rat brainstem slices (79) or expiratory neurons of adult cats in vivo (112).

While that study on DVN (60) presented some evidence with the optical approach that anoxia induces a notable fall of pH_i, this was investigated in detail in a follow-up report (52). In this work, bath application for several minutes of CN^-, rotenone or FCCP decreased pH_i by >0.4 pH units. Similarly, aglycaemia decreased pH_i in DVN with a normal resting pH (~7.2) with a delay of 7–16 min. This glucose-dependent acidification appeared to coincide with activation of the K_{ATP}-channel-mediated outward current, whereas K_{ATP} currents due to chemical anoxia or metabolic arrest could reach their maximum prior to a major pH_i change. We concluded from these observations (52) that the anoxic pH_i fall is due to enhanced anaerobic glycolysis and lactate formation, with often no obvious effect on K_{ATP}-channel activity in contrast to the scenario proposed for skeletal muscle fibres (148).

4.7.2. Optical Recording of Ca_i

We used Fura-2 to study whether metabolism-related K_{ATP} channel activation in DVN correlates with Ca_i changes (59). Firstly, we found that tonic activity of DVN determines Ca_i due to Ca^{2+} influx through voltage-activated Ca^{2+} channels. Consequently, blockade of tonic activity either by injection of hyperpolarising current or pharmacologically, by application of diazoxide, GABA or the voltage-activated Ca^{2+} and Na^+ channel blockers Ni^{2+} and tetrodotoxin, respectively, lowered Ca_i baseline by several tens of nanomolar. Contrary, increasing tonic firing rate by current-evoked depolarisation or with pharmacological tools, such as 5-HT (16), raised Ca_i to a comparable extent. Similarly, tolbutamide increased Ca_i in

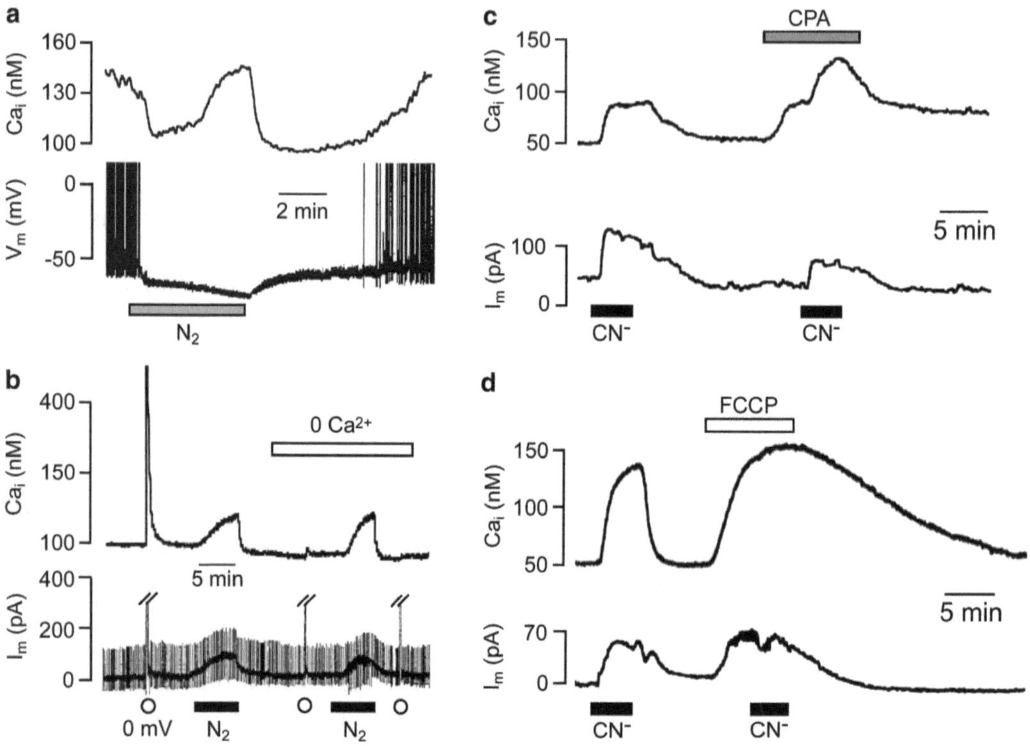

Fig. 13. *Activity- and metabolism-related changes of free cytosolic Ca^{2+} (Ca_i) changes in DVN of rat brainstem slices monitored with photomultiplier-based optical system.* (**a**) In DVN dialysed during whole-cell recording with Ca^{2+} sensitive fluorescent dye (Fura-2, 100 μM), hypoxic anoxia caused hyperpolarisation and initial fall of Ca_i due to blockade of spiking. Within <1 min, however, modest Ca_i increase occurred in association with blockade of aerobic metabolism. Similar modest anoxic Ca_i increase was revealed in voltage-clamped DVN (**b**). Recording in **b** also shows that superfusion of nominally Ca^{2+}-free superfusate abolished substantial Ca_i rise in response to membrane depolarisation from −50 to 0 mV, whereas anoxic Ca_i rise persisted. (**c**) CN^- (1 mM) led to AOC and similarly modest Ca_i rise as seen during hypoxic anoxia. Bath application of the blocker of endoplasmatic reticulum Ca^{2+} stores cyclopiazonic acid (CPA, 30 μM) evoked similar Ca_i rise as chemical (CN^-) or hypoxic anoxia, but did not induce AOC. Administration of CN^- in presence of CPA still evoked AOC plus additional increase of Ca_i. (**d**) The mitochondrial uncoupling agent and protonophore p-trifluoromethoxyphenylhydraz one (FCCP; 1 μM) had a very similar effect on both I_m and Ca_i compared to CN^-. However, CN^- in FCCP did not induce further effect ((**a**, **b**) With kind permission from (56); (**c**, **d**) K. Ballanyi and A. Kulik, unpublished).

current clamp because of an increase in firing rate indicating that some K_{ATP} channels are active under control conditions. Anoxia initially lowered Ca_i in DVN, if the evoked metabolic block of tonic firing occurred fast enough (Fig. 13). This was in particular the case with CN^--evoked chemical anoxia, which reached steady-state within <2 min, whereas hypoxic anoxia took up to 5 min to develop fully in our superfusion system (Sect. 2.3). The modest Ca_i fall due to arrest of spontaneous firing was countered by a delayed Ca_i rise of similar magnitude (Fig. 13). In voltage clamp, the anoxic increase in Ca_i baseline was typically <100 nM and remained stable for time periods >20 min (Figs. 13 and 14). This indicates that the high anoxia tolerance of DVN is reflected by only minor perturbation of Ca^{2+} homeostasis. This contrasts with a major Ca_i rise and

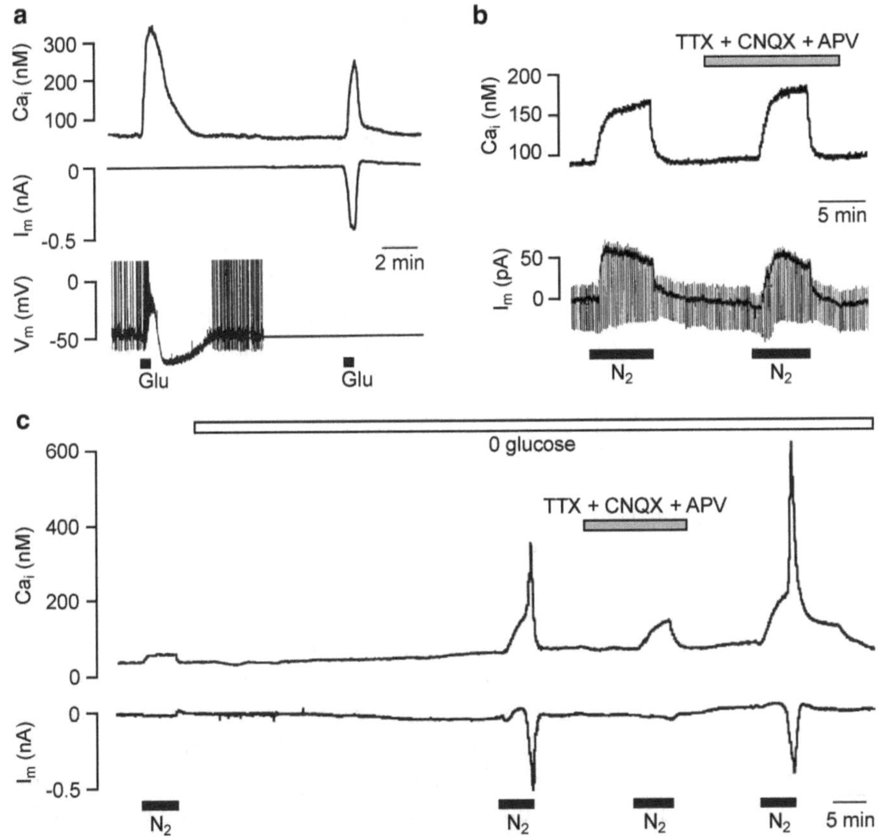

Fig. 14. *Involvement of glutamate in prominent ischemic membrane current and Ca_i rise in DVN.* (**a**) In a Fura-2-dialysed DVN bath application of 0.5 mM glutamate (Glu, 20 s) led to prominent depolarisation and concomitant Ca_i rise. After switching to voltage-clamp, the same GLU application evoked notably more modest Ca_i rise indicating major contribution of voltage-activated Ca^{2+} channels to this signal. (**b**) In different DVN, neither AOC nor concomitant modest Ca_i rise were affected by depressing (unclamped) spiking in DMNX with 0.5 μM TTX and blocking kainate/AMPA- and NMDA-type ionotropic glutamate receptors with 6-cyano-7-nitroquinoxaline-2,3-dione (CNQX, 50 μM) and (2R)-amino-5-phosphonopentanoate (APV, 100 μM), respectively. In contrast, this cocktail of neuronal blockers depressed progressive inward current and Ca_i rise in response to in vitro ischemia in different DVN (**c**) ((**a**, **b**, **c**) With kind permission from (56)).

impaired Ca^{2+} homeostasis in most other central neurons, for example, in neocortex, cerebellum or hippocampus (81–83).

We found that the minor anoxic Ca_i rise in whole-cell recorded DVN was very similar to that in intact cells loaded with membrane-permeant Fura-2 AM (59). This indicates that whole-cell recording does not substantially interfere with signalling pathways involved in the modest anoxic Ca_i rise and accompanying K_{ATP}-channel-mediated outward current. Both, anoxic outward current and concomitant Ca_i rise persisted after removal of Ca^{2+} from the superfusate, whereas more prominent Ca_i rises due to short stepwise depolarisation in voltage clamp were abolished (Fig. 13). These findings indicate that the anoxic Ca_i rise is due to Ca^{2+} release from intracellular stores. Specifically, release from mitochondria, which are Ca^{2+}-buffering cellular compartments, is likely the major mechanism because CN^- still evoked a Ca_i rise following depletion of endoplasmatic reticulum stores with cyclopiazonic acid (Fig. 13). While cyclopiazonic acid caused a release from endoplasmatic reticulum that was ~60% of that evoked by CN^-, the drug did not induce an outward current. In contrast, the K_{ATP}-channel-mediated outward current in response to FCCP was accompanied by a Ca_i rise similar to that evoked by CN^-, and addition of CN^- in the presence of FCCP affected neither I_m nor Ca_i (52) (Fig. 13).

Several findings from our latter study (59) and related work from our group (54, 56) suggest that the anoxic cytosolic Ca^{2+} increase is not linked to K_{ATP} channel outward current: (1) cyclopiazonic-acid-induced Ca_i rise is not accompanied by outward current; (2) in some DVN almost no Ca_i rise occurs during anoxia, despite sometimes prominent outward current; and (3) the anoxic outward current is already fully developed when Ca_i still rises. However, the majority of K_{ATP} channels might be located on distal dendrites of DVN, from which no optical signals were recorded in our studies, though our single-channel recordings showed that these channels are also found on the soma membrane (Sect. 4.5).

In most mammalian central neurons, anoxia-related 'excitotoxic' release of glutamate causes a major depolarisation and Ca_i rise that result in cell death (82, 83). Contrary, we showed (56) that activation of glutamate receptors does not contribute to the minor and stable rise in Ca_i baseline associated with anoxia in DVN (Fig. 14). In the same report, we found that anoxia after preincubation in glucose-free solution causes a major and progressive Ca_i rise in DVN. The prominent Ca_i increase in response to such 'in vitro ischemia' is caused by both vesicular and Ca^{2+}-independent release of glutamate acting on ionotropic kainate/AMPA- and NMDA-type of ionotropic receptors and can thus be depressed by blockers of these receptors (56, 82, 83) (Fig. 14).

4.7.3. Optical Recording of Cell Death

The ischemia-induced glutamate-dependent substantial Ca_i increase and the concomitant major inward current are irreversible unless carbogenated and glucose-containing standard solution is reapplied within <1 min following the onset of the inward current (56). We have developed a fluorometric method using propidium iodide for dynamic recording of acute ischemic death of DVN hypothesised in the latter report (Fig. 15). This work (53) revealed that some cells, which are located close to the surface and thus are probably damaged during the slicing procedure, die 'spontaneously' in standard solution within 1–7 h after generation of brainstem slices from juvenile rats. The number of dead cells does not increase during that time period following either anoxia for 30 min or aglycaemia for 1 h. Contrary, most DMNX cells die during 1–4 h after onset of in vitro ischemia due to simultaneous application of the blockers of aerobic and anaerobic metabolism, CN^- and

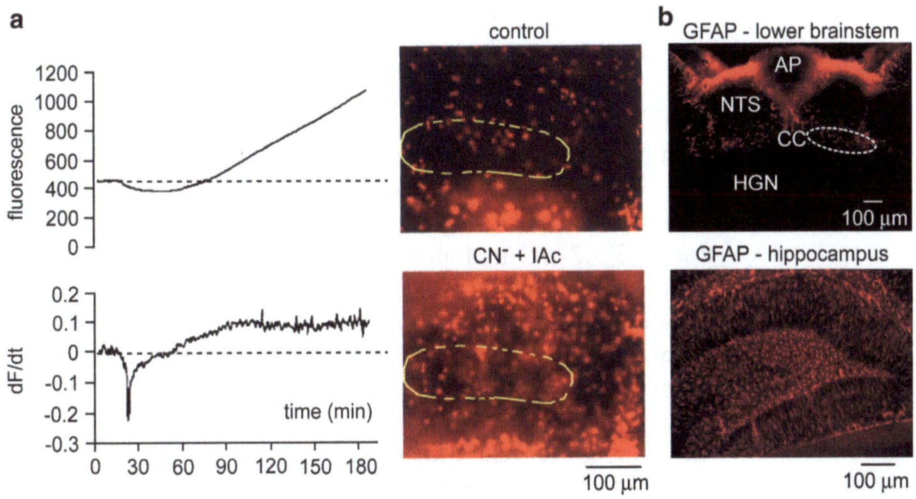

Fig. 15. *Dynamic fluorometric recording of ischemia-related cell death in DMNX.* (**a**) Chemical in vitro ischemia due to combined application of 1 mM CN^- and 5 mM of the glycolytic blocker iodoacetate (IAc) caused moderate initial decrease of the signal of fluorescent cell death marker propidium iodide (PI) imaged with a CCD camera-based optical system. However, within <1 h, this decreased fluorescence that likely indicates cell swelling, turned into steady increase of PI fluorescence intensity that continued for duration of experiment (>3 h). *Bottom* graph shows differentiated time course of changes in PI fluorescence. Micrographs illustrate PI-labelled cells before and after slice underwent in vitro ischemia, confirming that additional cell death generated dynamic increase in PI fluorescence plotted in **a**. (**b**) Photomicrographs showing glial fibrillary acidic protein (GFAP) immunoreactivity visualised with CY-3 conjugated anti-GFAP antibodies in sections of rat vagal complex (*top*) and hippocampus (*bottom*). Note the relatively low abundance of GFAP in NTS and DMNX (*dashed outline*) compared to hippocampus ((**a**) With kind permission from (53); (**b**) from S. Trapp, unpublished).

iodoacetate, respectively (53) (Fig. 15). This optical method is useful to study DVN as well as other brain slices but has the disadvantage that propidium iodide signals include both neurons and glial cells. However, glial cell density in the vagal complex is relatively low compared to, for example, hippocampus as shown with anti-GFAP antibodies (Fig. 15).

In summary of our optical measurements using Ca^{2+}- and H^+-sensitive fluorescent dyes, we did not obtain convincing evidence that metabolism-related changes in cytosolic Ca^{2+} or pH are causally involved in K_{ATP} channel activation in DVN. There are reports for other central neurons that anoxia causes outward current rather by activation of other types of K^+ channels, in particular Ca^{2+}-activated K^+ channels (150–152). We studied this in thin slices from juvenile rats using combined whole-cell recording and photomultiplier-based Ca_i measurements and found that CN^--evoked chemical anoxia blocks SK-type Ca^{2+}-activated K^+ channels, but does not affect BK-type Ca^{2+}-activated K^+ channels (54) (Sect. 4.5).

4.7.4. Optical Recording of $\Delta\Psi$ Changes

Our findings contribute notably to the understanding of various cellular properties of rodent DVN. However, we have not succeeded yet to answer one major question that is of primary interest for our research. Specifically, which cellular signalling pathway is involved in metabolism-related activation of their K_{ATP} channels? As described above, we hypothesise that processes associated with inhibition of mitochondrial functions are involved in anoxic activation of these channels. In an initial approach to study this, we dialysed DVN in thin slices from juvenile rats via the patch electrode with rhodamine-123, which is a fluorescent $\Delta\Psi$ indicator (99–101) (Sect. 3.5.3). Our unpublished data indicate that the mitochondrial uncoupler FCCP, and also CN^-- or rotenone-evoked chemical anoxia, lead to concomitant $\Delta\Psi$ depolarisation and K_{ATP} channel outward current, whereas oligomycin has no effect on either parameter (Fig. 16). Our simultaneous monitoring of $\Delta\Psi$ and Ca_i in whole-cell-recorded DVN revealed that anoxic $\Delta\Psi$ depolarisation kinetics correlate with those of K_{ATP} channel outward current, whereas the accompanying Ca_i rise is slower (Fig. 16). These findings, which are similar to those in intact hippocampal neurons loaded with both rhodamine-123 and Fura-2 via bath application (152), substantiate our hypothesis that anoxic K_{ATP} channel activation in rodent DVN is caused by a cellular process that is fast and correlates with depolarisation-related impairment of mitochondrial function and thus ATP production.

4.8. Properties of Tonically Active GLP-1 Neurons

GLP-1 neurons are a further population of neurons contained in the same brainstem slices as DVN, specifically 50–200 μm more dorsolaterally (Fig. 3). Unlike DVN, their main projections are ascending towards hypothalamic nuclei and other autonomic control areas (153–156). Due to the difficulty of identifying this cell population in vitro in slice preparations, nothing was known about their specific electrical properties until very recently (31). This study from our group was made possible by the creation of a mouse line that expresses Venus™, a YFP variant, under control of the preproglucagon (PPG) promoter (32). PPG is the precursor protein for glucagon, GLP-1 and GLP-2, and tissue-specific post-translational processing of PPG determines whether glucagon or GLP-1 and GLP-2 are produced by a specific cell population. Pancreatic α-cells produce glucagon, whereas enteroendocrine L-cells and some NTS neurons produce GLP-1 (157). We used these transgenic mice to study basic electrical and pharmacological properties of GLP-1 neurons (31). Firstly, we found that these neurons are spontaneously active in vitro similar to DVN. However, while DVN fire action potentials like a clockwork ('pacemaker'), activity of GLP-1 neurons is more variable (Fig. 17). Also like DVN, GLP-1 neurons express A-current and receive synaptic input from the tractus solitarius (Fig. 3). This input is primarily glutamatergic as excitatory postsynaptic currents that occur spontaneously or following stimulation of tractus solitarius are sensitive to antagonists of ionotropic glutamate receptors (Fig. 17). A first study designed to interrogate the role of these cells in control of

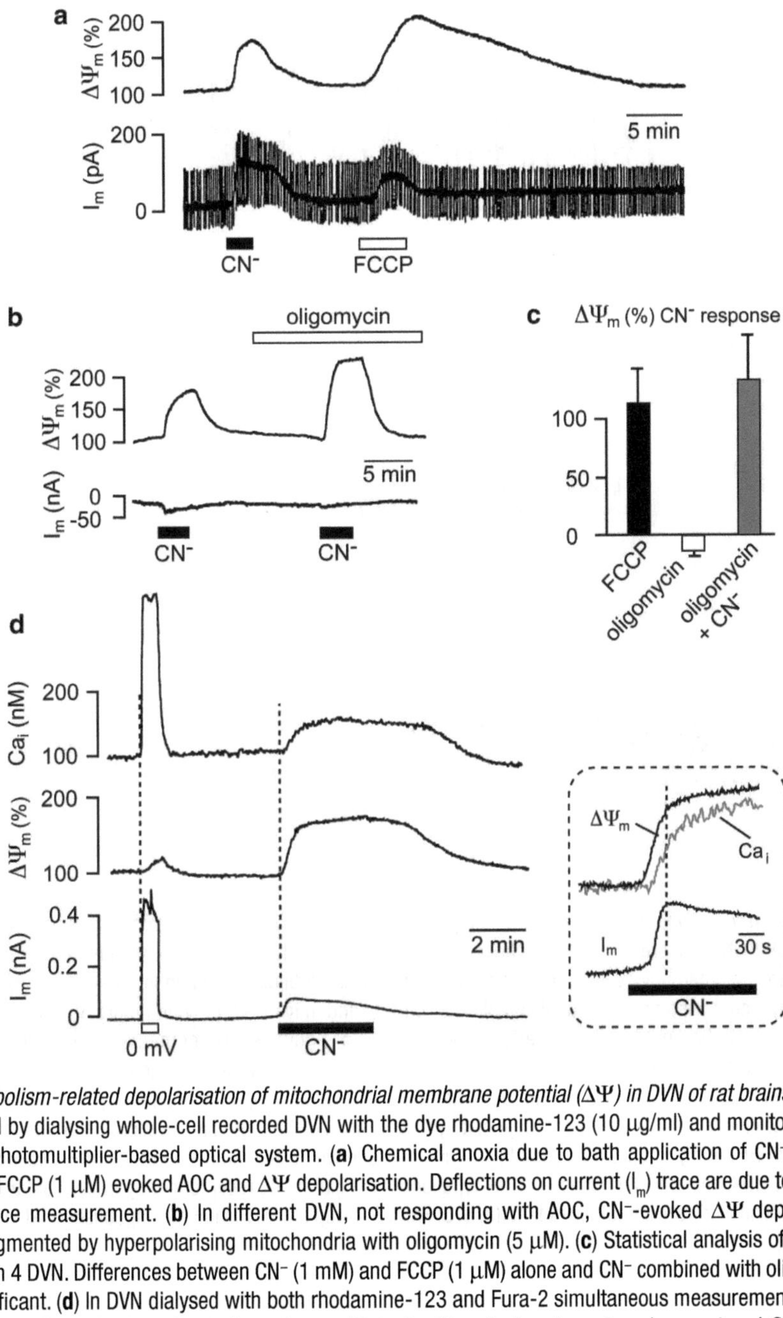

Fig. 16. *Metabolism-related depolarisation of mitochondrial membrane potential (ΔΨ) in DVN of rat brainstem slices.* ΔΨ was measured by dialysing whole-cell recorded DVN with the dye rhodamine-123 (10 μg/ml) and monitoring fluorescent signals with photomultiplier-based optical system. (**a**) Chemical anoxia due to bath application of CN^- (1 mM) or the protonophore FCCP (1 μM) evoked AOC and ΔΨ depolarisation. Deflections on current (I_m) trace are due to voltage pulses for conductance measurement. (**b**) In different DVN, not responding with AOC, CN^--evoked ΔΨ depolarisation was apparently augmented by hyperpolarising mitochondria with oligomycin (5 μM). (**c**) Statistical analysis of ΔΨ recordings in **a** and **b** from 4 DVN. Differences between CN^- (1 mM) and FCCP (1 μM) alone and CN^- combined with oligomycin (5 μM) were not significant. (**d**) In DVN dialysed with both rhodamine-123 and Fura-2 simultaneous measurement of ΔΨ and Ca_i was feasible. Depolarisation in voltage-clamp from −50 to 0 mV evoked major outward current and Ca_i rise, whereas associated ΔΨ depolarisation was modest. Subsequent chemical anoxia (1 mM CN^-) evoked a smaller outward current and Ca_i increase, whereas ΔΨ depolarization was more pronounced. Inset in *right* part shows that kinetics of anoxic ΔΨ depolarisation are similar to those of AOC whereas Ca_i rise develops more slowly ((**a–c**) From K. Ballanyi and A. Kulik, unpublished; (**d**) with kind permission from (80)).

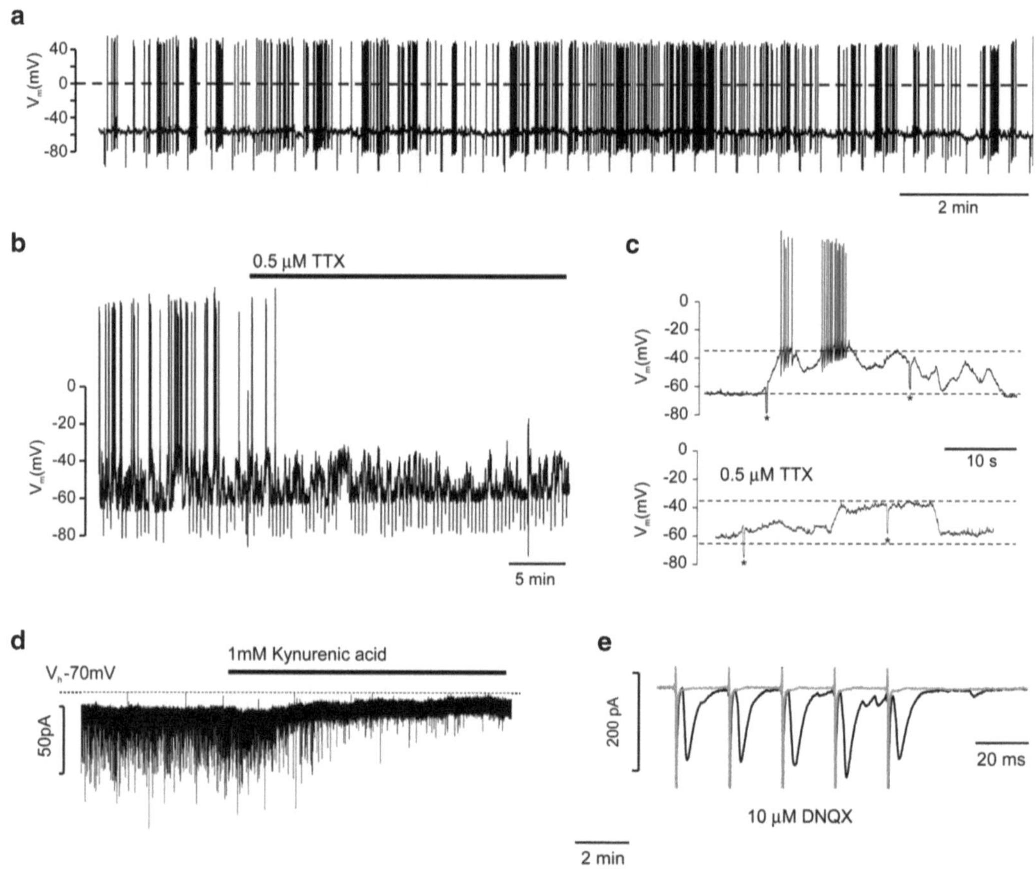

Fig. 17. *Basic electrical properties of GLP-1 producing PPG neurons of NTS in mouse brainstem slices*. (**a**) Current-clamp recording reveals that, compared to DVN, the firing pattern is more random with short periods of intense activity being followed by periods of sparse activity. (**b**, **c**) This firing behaviour is particularly prominent in ~10% of PPG neurons that burst fire. Here not only firing pattern, but also resting V_m varies randomly by ~20 mV and more. V_m fluctuations persist in TTX and thus seem to be an intrinsic property of these cells. (**c**) Shows short segments of recording in **b** at higher time resolution. Asterisks indicate the injection of a hyperpolarising current pulse to monitor membrane conductance. (**d**) Voltage-clamp recording from PPG neuron demonstrating that most of spontaneous synaptic activity is inhibited by 1 mM of the ionotropic glutamate receptor antagonist kynurenic acid. (**e**) Voltage-clamp recording demonstrating that also synaptic currents evoked by electrical TS stimulation (five consecutive pulses, 20 ms apart; control, *black trace*) are abolished by 10 μM DNQX, an antagonist of non-NMDA-type ionotropic glutamate receptors (*grey trace*). Fast transient up- and downward deflections indicate stimulus artefact ((**a**, **b**, **c**) From S. Trapp and K. Hisadome, unpublished; (**d**, **e**) from (31) with kind permission).

appetite revealed that NTS GLP-1 neurons are excited by the adiposity hormone leptin but that there is no direct response to the gut-derived satiety hormones PYY and GLP-1 or to the stomach-released hunger peptide ghrelin (31). Future studies will determine the exact circuitry these neurons are involved in and roles they play in regulating appetite and blood glucose levels.

5. Closing Remarks

This chapter should not only act as a guide how to optimise in vitro conditions for brainstem slices but also to demonstrate the versatility of brainstem slices in analysing intrinsic properties of identified neurons and their local connections in rodent brain. While we have shown here how various techniques can be used to study vagal preganglionic neurons and NTS neurons in brainstem slices, these techniques can easily be applied to other cell population in these slices or in slices of virtually any other part of the brain.

Acknowledgements

This work was supported by the Deutsche Forschungsgemeinschaft (ST, KB), Wilhelm Sander Stiftung (KB), Medical Research Council UK (ST), the Canadian Institutes of Health Research (KB) and the Alberta Heritage Foundation for Medical Research (KB).

References

1. Loewi O (1922) Ueber humorale Uebertragbarkeit der Herznervenwirkung. II Mitteilung. Pflugers Arch 193:201–213
2. Powley TL (2000) Vagal circuitry mediating cephalic-phase responses to food. Appetite 34:184–188
3. Groves DA, Brown VJ (2005) Vagal nerve stimulation: a review of its applications and potential mechanisms that mediate its clinical effects. Neurosci Biobehav Rev 29: 493–500
4. Yarom Y, Bracha O, Werman R (1985) Intracellular injection of acetylcholine blocks various potassium conductances in vagal motoneurons. Neuroscience 16:739–752
5. Yarom Y, Sugimori M, Llinas R (1985) Ionic currents and firing patterns of mammalian vagal motoneurons in vitro. Neuroscience 16:719–737
6. Travagli RA, Gillis RA, Rossiter CD, Vicini S (1991) Glutamate and GABA-mediated synaptic currents in neurons of the rat dorsal motor nucleus of the vagus. Am J Physiol 260:G531–G536
7. Cowan AI, Martin RL (1992) Ionic basis of membrane potential changes induced by anoxia in rat dorsal vagal motoneurones. J Physiol 455:89–109
8. Trapp S, Ballanyi K (1995) K_{ATP} channel mediation of anoxia-induced outward current in rat dorsal vagal neurons in vitro. J Physiol 487:37–50
9. Trapp S, Ballanyi K, Richter DW (1994) Spontaneous activation of KATP current in rat dorsal vagal neurones. Neuroreport 5: 1285–1288
10. Sah P (1995) Properties of channels mediating the apamin-insensitive afterhyperpolarization in vagal motoneurons. J Neurophysiol 74:1772–1776
11. Sah P (1995) Different calcium channels are coupled to potassium channels with distinct physiological roles in vagal neurons. Proc Biol Sci 260:105–111
12. Sah P, Davies P (2000) Calcium-activated potassium currents in mammalian neurons. Clin Exp Pharmacol Physiol 27:657–663
13. Loewy AD, Spyer KM (1990) Vagal preganglionic neurons. In: Loewy AD, Spyer KM (eds) Central regulation of autonomic functions. Oxford University Press, New York, pp 68–87
14. McTigue DM, Hermann GE, Rogers RC (1997) Effect of pancreatic polypeptide on rat dorsal vagal complex neurons. J Physiol 499:475–483
15. Mussa BM, Sartor DM, Verberne AJ (2010) Dorsal vagal preganglionic neurons: differential responses to CCK1 and 5-HT3 receptor stimulation. Auton Neurosci 156:36–43

16. Wang Y, Jones JF, Ramage AG, Jordan D (1995) Effects of 5-HT and 5-HT1A receptor agonists and antagonists on dorsal vagal preganglionic neurones in anaesthetized rats: an ionophoretic study. Br J Pharmacol 116:2291–2297
17. Holmes GM, Browning KN, Tong M, Qualls-Creekmore E, Travagli RA (2009) Vagally mediated effects of glucagon-like peptide 1: in vitro and in vivo gastric actions. J Physiol 587:4749–4759
18. Wan S, Browning KN, Coleman FH, Sutton G, Zheng H, Butler A, Berthoud HR, Travagli RA (2008) Presynaptic melanocortin-4 receptors on vagal afferent fibers modulate the excitability of rat nucleus tractus solitarius neurons. J Neurosci 28:4957–4966
19. Wan S, Browning KN, Travagli RA (2007) Glucagon-like peptide-1 modulates synaptic transmission to identified pancreas-projecting vagal motoneurons. Peptides 28:2184–2191
20. Travagli RA, Hermann GE, Browning KN, Rogers RC (2006) Brainstem circuits regulating gastric function. Annu Rev Physiol 68:279–305
21. Browning KN, Travagli RA (2009) Modulation of inhibitory neurotransmission in brainstem vagal circuits by NPY and PYY is controlled by cAMP levels. Neurogastroenterol Motil 21:1309–e1126
22. Ballanyi K, Doutheil J, Brockhaus J (1996) Membrane potentials and microenvironment of rat dorsal vagal cells in vitro during energy depletion. J Physiol 495:769–784
23. Cone RD, Cowley MA, Butler AA, Fan W, Marks DL, Low MJ (2001) The arcuate nucleus as a conduit for diverse signals relevant to energy homeostasis. Int J Obes Relat Metab Disord 25(suppl 5):S63–S67
24. Miki T, Seino S (2005) Roles of KATP channels as metabolic sensors in acute metabolic changes. J Mol Cell Cardiol 38:917–925
25. Balfour RH, Trapp S (2007) Ionic currents underlying the response of rat dorsal vagal neurones to hypoglycaemia and chemical anoxia. J Physiol 579:691–702
26. Balfour RH, Hansen AM, Trapp S (2006) Neuronal responses to transient hypoglycaemia in the dorsal vagal complex of the rat brainstem. J Physiol 570:469–484
27. Appleyard SM, Marks D, Kobayashi K, Okano H, Low MJ, Andresen MC (2007) Visceral afferents directly activate catecholamine neurons in the solitary tract nucleus. J Neurosci 27:13292–13302
28. Matsushita N, Okada H, Yasoshima Y, Takahashi K, Kiuchi K, Kobayashi K (2002) Dynamics of tyrosine hydroxylase promoter activity during midbrain dopaminergic neuron development. J Neurochem 82:295–304
29. Appleyard SM, Bailey TW, Doyle MW, Jin YH, Smart JL, Low MJ, Andresen MC (2005) Proopiomelanocortin neurons in nucleus tractus solitarius are activated by visceral afferents: regulation by cholecystokinin and opioids. J Neurosci 25:3578–3585
30. Cowley MA, Smart JL, Rubinstein M, Cerdan MG, Diano S, Horvath TL, Cone RD, Low MJ (2001) Leptin activates anorexigenic POMC neurons through a neural network in the arcuate nucleus. Nature 411:480–484
31. Hisadome K, Reimann F, Gribble FM, Trapp S (2010) Leptin directly depolarizes preproglucagon neurons in the nucleus tractus solitarius: electrical properties of glucagon-like Peptide 1 neurons. Diabetes 59:1890–1898
32. Reimann F, Habib AM, Tolhurst G, Parker HE, Rogers GJ, Gribble FM (2008) Glucose sensing in L cells: a primary cell study. Cell Metab 8:532–539
33. Edwards FA, Konnerth A, Sakmann B, Takahashi T (1989) A thin slice preparation for patch clamp recordings from neurones of the mammalian central nervous system. Pflugers Arch 414:600–612
34. Ballanyi K (1999) Isolated tissues: in vitro preparations. In: Windhorst U, Johansson H (eds) Modern techniques in neuroscience research. Springer, Heidelberg, pp 307–326
35. Konnerth A (1990) Patch-clamping in slices of mammalian CNS. Trends Neurosci 13: 321–323
36. Haas H, Büsselberg D (1992) Recording chambers-slices. In: Kettenmann H, Grantyn R (eds) Practical electrophysiological methods. Wiley-Liss, New York, pp 16–19
37. Wilson RJ, Straus C, Remmers JE (1999) Efficacy of a low volume recirculating superfusion chamber for long term administration of expensive drugs and dyes. J Neurosci Methods 87:175–184
38. Strupp M, Jund R, Schneider U, Grafe P (1991) Glucose availability and sensitivity to anoxia of isolated rat peroneal nerve. Am J Physiol 261:E389–E394
39. Hille B (2001) Ion channels of excitable membranes, 3rd edn. Sinauer Associates, Inc., Sunderland
40. Neher E, Sakmann B (eds) (2009) Single-channel recording, 2nd edn. Springer, New York/Dordrecht/Heidelberg/London
41. Walz W (2007) Patch-clamp analysis. In: Walz W (ed) Advanced techniques, 2nd edn. Humana, Totowa

42. Lowry JP, Griffin K, McHugh SB, Lowe AS, Tricklebank M, Sibson NR (2010) Real-time electrochemical monitoring of brain tissue oxygen: a surrogate for functional magnetic resonance imaging in rodents. Neuroimage 52:549–555
43. Blanton MG, Lo Turco JJ, Kriegstein AR (1989) Whole cell recording from neurons in slices of reptilian and mammalian cerebral cortex. J Neurosci Methods 30:203–210
44. Moore AR, Zhou W, Jakovcevski I, Zecevic N, Antic S (2012) Physiological properties of human fetal cortex in vitro. In Isolated Central Nervous System Circuits (Ed K Ballanyi), Neuromethods Series Vol. 73 (Ed W Walz). Springer Science+Business Media, LLC, New York, NY, pp 125–158
45. Dodt HU, Zieglgansberger W (1994) Infrared videomicroscopy: a new look at neuronal structure and function. Trends Neurosci 17:453–458
46. MacVicar BA (1984) Infrared video microscopy to visualize neurons in the in vitro brain slice preparation. J Neurosci Methods 12:133–139
47. Ruangkittisakul A, Okada Y, Oku Y, Koshiya N, Ballanyi K (2009) Fluorescence imaging of active respiratory networks. Respir Physiol Neurobiol 168:26–38
48. Yuste R, Konnerth A, Masters B (2006) Imaging in neuroscience and development, a laboratory manual. J Biomed Opt 11:19902
49. Neher E (1989) Combined fura-2 and patch-clamp measurements in rat peritoneal mast cells. In: Sellin LC, Libelius R, Theslaff S (eds) Neuromuscular junction. Elsevier, Science Publishers (Biomedical Division), Amsterdam, pp 65–75
50. Ruangkittisakul A, Panaitescu B, Secchia L, Bobocea N, Kantor C, Kuribayashi J, Iizuka M, Ballanyi K (2012) Isolated brainstem respiratory centers from perinatal rodents. In Isolated Central Nervous System Circuits (Ed K Ballanyi), Neuromethods Series Vol. 73 (Ed W Walz). Springer Science+Business Media, LLC, New York, NY, pp 61–124
51. Somjen GG (2002) Ion regulation in the brain: implications for pathophysiology. Neuroscientist 8:254–267
52. Raupach T, Ballanyi K (2004) Intracellular pH and K_{ATP} channel activity in dorsal vagal neurons of juvenile rats in situ during metabolic disturbances. Brain Res 1017: 137–145
53. Müller M, Ballanyi K (2003) Dynamic recording of cell death in the in vitro dorsal vagal nucleus of rats in response to metabolic arrest. J Neurophysiol 89:551–561
54. Kulik A, Brockhaus J, Pedarzani P, Ballanyi K (2002) Chemical anoxia activates ATP-sensitive and blocks Ca^{2+}-dependent K^+ channels in rat dorsal vagal neurons in situ. Neuroscience 110:541–554
55. Müller M, Brockhaus J, Ballanyi K (2002) ATP-independent anoxic activation of ATP-sensitive K^+ channels in dorsal vagal neurons of juvenile mice in situ. Neuroscience 109:313–328
56. Kulik A, Trapp S, Ballanyi K (2000) Ischemia but not anoxia evokes vesicular and Ca^{2+}-independent glutamate release in the dorsal vagal complex in vitro. J Neurophysiol 83: 2905–2915
57. Pedarzani P, Kulik A, Müller M, Ballanyi K, Stocker M (2000) Molecular determinants of Ca^{2+}-dependent K^+ channel function in rat dorsal vagal neurones. J Physiol 527:283–290
58. Karschin A, Brockhaus J, Ballanyi K (1998) K_{ATP} channel formation by the sulphonylurea receptors SUR1 with Kir6.2 subunits in rat dorsal vagal neurons in situ. J Physiol 509:339–346
59. Ballanyi K, Kulik A (1998) Intracellular Ca^{2+} during metabolic activation of K_{ATP} channels in spontaneously active dorsal vagal neurons in medullary slices. Eur J Neurosci 10:2574–2585
60. Trapp S, Lückermann M, Brooks PA, Ballanyi K (1996) Acidosis of rat dorsal vagal neurons in situ during spontaneous and evoked activity. J Physiol 496:695–710
61. Laughton WB, Powley TL (1987) Localization of efferent function in the dorsal motor nucleus of the vagus. Am J Physiol 252:R13–R25
62. Baumgärtl H (1987) Systematic investigations of needle electrode properties in polarographic measurements of local tissue P_{O2}. In: Ehrly AM, Hauss J, Huch R (eds) Clinical oxygen pressure measurement: Tissue oxygen pressure and transcutaneous oxygen pressure. Springer
63. Grote J, Zimmer K, Schubert R (1981) Effects of severe arterial hypocapnia on regional blood flow regulation, tissue P_{O2} and metabolism in the brain cortex of cats. Pflugers Arch 391:195–199
64. Abi-Saab WM, Maggs DG, Jones T, Jacob R, Srihari V, Thompson J, Kerr D, Leone P, Krystal JH, Spencer DD, During MJ, Sherwin RS (2002) Striking differences in glucose and lactate levels between brain extracellular fluid and plasma in conscious human subjects: effects of hyperglycemia and hypoglycemia. J Cereb Blood Flow Metab 22:271–279

65. De Vries MG, Arseneau LM, Lawson ME, Beverly JL (2003) Extracellular glucose in rat ventromedial hypothalamus during acute and recurrent hypoglycemia. Diabetes 52: 2767–2773
66. Fray AE, Forsyth RJ, Boutelle MG, Fillenz M (1996) The mechanisms controlling physiologically stimulated changes in rat brain glucose and lactate: a microdialysis study. J Physiol 496:49–57
67. Gramsbergen JB, Skjoth-Rasmussen J, Rasmussen C, Lambertsen KL (2004) On-line monitoring of striatum glucose and lactate in the endothelin-1 rat model of transient focal cerebral ischemia using microdialysis and flow-injection analysis with biosensors. J Neurosci Methods 140:93–101
68. Lowry JP, O'Neill RD, Boutelle MG, Fillenz M (1998) Continuous monitoring of extracellular glucose concentrations in the striatum of freely moving rats with an implanted glucose biosensor. J Neurochem 70:391–396
69. Silver IA, Erecinska M (1994) Extracellular glucose concentration in mammalian brain: continuous monitoring of changes during increased neuronal activity and upon limitation in oxygen supply in normo-, hypo-, and hyperglycemic animals. J Neurosci 14:5068–5076
70. Ashford ML, Boden PR, Treherne JM (1990) Tolbutamide excites rat glucoreceptive ventromedial hypothalamic neurones by indirect inhibition of ATP-K^+ channels. Br J Pharmacol 101:531–540
71. Ashford ML, Boden PR, Treherne JM (1990) Glucose-induced excitation of hypothalamic neurones is mediated by ATP-sensitive K^+ channels. Pflugers Arch 415:479–483
72. Minami T, Oomura Y, Sugimori M (1986) Electrophysiological properties and glucose responsiveness of guinea-pig ventromedial hypothalamic neurones in vitro. J Physiol 380:127–143
73. Wang R, Liu X, Hentges ST, Dunn-Meynell AA, Levin BE, Wang W, Routh VH (2004) The regulation of glucose-excited neurons in the hypothalamic arcuate nucleus by glucose and feeding-relevant peptides. Diabetes 53:1959–1965
74. Song Z, Levin BE, McArdle JJ, Bakhos N, Routh VH (2001) Convergence of pre- and postsynaptic influences on glucosensing neurons in the ventromedial hypothalamic nucleus. Diabetes 50:2673–2681
75. Burdakov D, Jensen LT, Alexopoulos H, Williams RH, Fearon IM, O'Kelly I, Gerasimenko O, Fugger L, Verkhratsky A (2006) Tandem-pore K^+ channels mediate inhibition of orexin neurons by glucose. Neuron 50:711–722
76. Dallaporta M, Perrin J, Orsini JC (2000) Involvement of adenosine triphosphate-sensitive K^+ channels in glucose-sensing in the rat solitary tract nucleus. Neurosci Lett 278:77–80
77. Dallaporta M, Himmi T, Perrin J, Orsini JC (1999) Solitary tract nucleus sensitivity to moderate changes in glucose level. Neuroreport 10:2657–2660
78. Kaila K (1994) Ionic basis of $GABA_A$ receptor channel function in the nervous system. Prog Neurobiol 42:489–537
79. Lückermann M, Trapp S, Ballanyi K (1997) GABA- and glycine-mediated fall of intracellular pH in rat medullary neurons in situ. J Neurophysiol 77:1844–1852
80. Ballanyi K (2004) Protective role of neuronal K_{ATP} channels in brain hypoxia. J Exp Biol 207:3201–3212
81. Ballanyi K (2004) Neuromodulation of the perinatal respiratory network. Curr Neuropharmacol 2:221–243
82. Haddad GG, Jiang C (1993) O_2 deprivation in the central nervous system: on mechanisms of neuronal response, differential sensitivity and injury. Prog Neurobiol 40:277–318
83. Lipton P (1999) Ischemic cell death in brain neurons. Physiol Rev 79:1431–1568
84. Lu VB, Biggs JE, Stebbing MJ, Balasubramanyan S, Todd KG, Lai AY, Colmers WF, Dawbarn D, Ballanyi K, Smith PA (2009) Brain-derived neurotrophic factor drives the changes in excitatory synaptic transmission in the rat superficial dorsal horn that follow sciatic nerve injury. J Physiol 587: 1013–1032
85. Metzger F, Klapproth N, Kulik A, Sendtner M, Ballanyi K (2005) Optical assessment of motoneuron function in a "twenty-four-hour" acute spinal cord slice model from fetal rats. J Neurosci Methods 141:309–320
86. Kulik A, Nishimaru H, Ballanyi K (2000) Role of bicarbonate and chloride in GABA- and glycine-induced depolarization and $(Ca^{2+})_i$ rise in fetal rat motoneurons in situ. J Neurosci 20:7905–7913
87. Schuchmann S, Lückermann M, Kulik A, Heinemann U, Ballanyi K (2000) Ca^{2+}- and metabolism-related changes of mitochondrial potential in voltage-clamped CA_1 pyramidal neurons in situ. J Neurophysiol 83:1710–1721
88. Kulik A, Haentzsch A, Lückermann M, Reichelt W, Ballanyi K (1999) Neuron-glia signaling via α_1 adrenoceptor-mediated Ca^{2+}

release in Bergmann glial cells in situ. J Neurosci 19:8401–8408
89. Ruangkittisakul A, Schwarzacher SW, Secchia L, Ma Y, Bobocea N, Poon BY, Funk GD, Ballanyi K (2008) Generation of eupnea and sighs by a spatiochemically organized inspiratory network. J Neurosci 28: 2447–2458
90. Nakamura T, Michel S, Block G, Colwell C (2012) Neural circuits underlying circadian oscillations in mammals: clocks in a dish. In Isolated Central Nervous System Circuits (Ed K Ballanyi), Neuromethods Series Vol. 73 (Ed W Walz). Springer Science+Business Media, LLC, New York, NY, pp 183–210
91. Kantor C, Panaitescu B, Kuribayashi J, Ruangkittisakul A, Jovanovic I, Leung V, Lee TF, MacTavish D, Jhamandas J, Cheung PY, Ballanyi K (2012) Early network oscillations in cortex and locus coeruleus of horizontal brain slices from newborn mammals. In Isolated Central Nervous System Circuits (Ed K Ballanyi), Neuromethods Series Vol. 73 (Ed W Walz). Springer Science+Business Media, LLC, New York, NY, pp 315–356
92. De Curtis M, Lilbrizzi L, Uva L, Gnatkovsky V (2012) Neuronal networks in the in vitro isolated guinea pig brain. In Isolated Central Nervous System Circuits (Ed K Ballanyi), Neuromethods Series Vol. 73 (Ed W Walz). Springer Science+Business Media, LLC, New York, NY, pp 357–383
93. Biggs J, Lu V, Kim H, Lai A, Todd K, Ballanyi K, Colmers W, Smith P (2012) Defined medium organotypic cultures of spinal cord put 'Pain in a Dish'. In Isolated Central Nervous System Circuits (Ed K Ballanyi), Neuromethods Series Vol. 73 (Ed W Walz). Springer Science+Business Media, LLC, New York, NY, pp 405–435
94. Grynkiewicz G, Poenie M, Tsien RY (1985) A new generation of Ca^{2+} indicators with greatly improved fluorescence properties. J Biol Chem 260:3440–3450
95. Tsien RY (1980) New calcium indicators and buffers with high selectivity against magnesium and protons: design, synthesis, and properties of prototype structures. Biochemistry 19:2396–2404
96. Ruangkittisakul A, Schwarzacher SW, Secchia L, Poon BY, Ma Y, Funk GD, Ballanyi K (2006) High sensitivity to neuromodulator-activated signaling pathways at physiological (K^+) of confocally imaged respiratory center neurons in on-line-calibrated newborn rat brainstem slices. J Neurosci 26:11870–11880
97. Homma R, Baker BJ, Jin L, Garaschuk O, Konnerth A, Cohen LB, Zecevic D (2009) Wide-field and two-photon imaging of brain activity with voltage- and calcium-sensitive dyes. Philos Trans R Soc Lond B Biol Sci 364:2453–2467
98. Thomas JA, Buchsbaum RN, Zimniak A, Racker E (1979) Intracellular pH measurements in Ehrlich ascites tumor cells utilizing spectroscopic probes generated in situ. Biochemistry 18:2210–2218
99. Duchen MR (1992) Ca^{2+}-dependent changes in the mitochondrial energetics in single dissociated mouse sensory neurons. Biochem J 283:41–50
100. Duchen M (1992) Fluorescence-monitoring cell chemistry in vivo. In: Monitoring neuronal activity-a practical approach. Stamford JA (ed) IRL Press at Oxford University Press, New York, pp. 231–260
101. Duchen MR (2004) Mitochondria in health and disease: perspectives on a new mitochondrial biology. Mol Aspects Med 25: 365–451
102. Emaus RK, Grunwald R, Lemasters JJ (1986) Rhodamine 123 as a probe of transmembrane potential in isolated rat-liver mitochondria: spectral and metabolic properties. Biochim Biophys Acta 850:436–448
103. Bevensee MO, Schwiening CJ, Boron WF (1995) Use of BCECF and propidium iodide to assess membrane integrity of acutely isolated CA_1 neurons from rat hippocampus. J Neurosci Methods 58:61–75
104. Laake JH, Haug FM, Wieloch T, Ottersen OP (1999) A simple in vitro model of ischemia based on hippocampal slice cultures and propidium iodide fluorescence. Brain Res Brain Res Protoc 4:173–184
105. Loo DT, Rillema JR (1998) Measurement of cell death. Methods Cell Biol 57:251–264
106. Voipio J, Ballanyi K (1997) Interstitial P_{CO2} and pH, and their role as chemostimulants in the isolated respiratory network of neonatal rats. J Physiol 499:527–542
107. Ballanyi K, Völker A, Richter DW (1996) Functional relevance of anaerobic metabolism in the isolated respiratory network of newborn rats. Pflugers Arch 432:741–748
108. Morawietz G, Ballanyi K, Kuwana S, Richter DW (1995) Oxygen supply and ion homeostasis of the respiratory network in the in vitro perfused brainstem of adult rats. Exp Brain Res 106:265–274
109. Völker A, Ballanyi K, Richter DW (1995) Anoxic disturbance of the isolated respiratory

network of neonatal rats. Exp Brain Res 103:9–19
110. Brockhaus J, Ballanyi K, Smith JC, Richter DW (1993) Microenvironment of respiratory neurons in the in vitro brainstem-spinal cord of neonatal rats. J Physiol 462:421–445
111. Ballanyi K, Kuwana S, Völker A, Morawietz G, Richter DW (1992) Developmental changes in the hypoxia tolerance of the in vitro respiratory network of rats. Neurosci Lett 148:141–144
112. Ballanyi K, Mückenhoff K, Bellingham MC, Okada Y, Scheid P, Richter DW (1994) Activity-related pH changes in respiratory neurones and glial cells of cats. Neuroreport 6:33–36
113. Ballanyi K, Kettenmann H (1990) Intracellular Na^+ activity in cultured mouse oligodendrocytes. J Neurosci Res 26:455–460
114. Ballanyi K, Grafe P, ten Bruggencate G (1987) Ion activities and potassium uptake mechanisms of glial cells in guinea-pig olfactory cortex slices. J Physiol 382:159–174
115. Ballanyi K, Grafe P (1985) An intracellular analysis of gamma-aminobutyric-acid-associated ion movements in rat sympathetic neurones. J Physiol 365:41–58
116. Ammann D (1986) Ion-selective microelectrodes. Springer Berlin, Heidelberg/New York/Tokyo
117. Grafe P, Rimpel J, Reddy MM, ten Bruggencate G (1982) Changes of intracellular sodium and potassium ion concentrations in frog spinal motoneurons induced by repetitive synaptic stimulation. Neuroscience 7:3213–3220
118. Cowan AI, Martin RL (1995) Simultaneous measurement of pH and membrane potential in rat dorsal vagal motoneurons during normoxia and hypoxia: a comparison in bicarbonate and HEPES buffers. J Neurophysiol 74:2713–2721
119. Sucher NJ, Deitcher DL (1995) PCR and patch-clamp analysis of single neurons. Neuron 14:1095–1100
120. Trapp S, Aller MI, Wisden W, Gourine AV (2008) A role for TASK-1 (KCNK3) channels in the chemosensory control of breathing. J Neurosci 28:8844–8850
121. Trapp S, Tucker S, Gourine AV (2011) Respiratory responses to hypercapnia and hypoxia in mice with genetic ablation of Kir5.1 (Kcnj16). Exp Physiol 96:451–459
122. Hopwood SE, Trapp S (2005) TASK-like K^+ channels mediate effects of 5-HT and extracellular pH in rat dorsal vagal neurones in vitro. J Physiol 568:145–154
123. Ballanyi K, Ruangkittisakul A (2009) Structure-function analysis of rhythmogenic inspiratory pre-Botzinger complex networks in "calibrated" newborn rat brainstem slices. Respir Physiol Neurobiol 168:158–178
124. Ballanyi K, Onimaru H, Homma I (1999) Respiratory network function in the isolated brainstem-spinal cord of newborn rats. Prog Neurobiol 59:583–634
125. Ashcroft FM, Kakei M, Gibson JS, Gray DW, Sutton R (1989) The ATP- and tolbutamide-sensitivity of the ATP-sensitive K-channel from human pancreatic B cells. Diabetologia 32:591–598
126. Kozlowski RZ, Hales CN, Ashford ML (1989) Dual effects of diazoxide on ATP-K^+ currents recorded from an insulin-secreting cell line. Br J Pharmacol 97:1039–1050
127. Niki I, Kelly RP, Ashcroft SJ, Ashcroft FM (1989) ATP-sensitive K-channels in HIT T15 beta-cells studied by patch- clamp methods, 86Rb efflux and glibenclamide binding. Pflugers Arch 415:47–55
128. Trube G, Rorsman P, Ohno Shosaku T (1986) Opposite effects of tolbutamide and diazoxide on the ATP-dependent K^+ channel in mouse pancreatic beta-cells. Pflugers Arch 407:493–499
129. Gribble FM, Loussouarn G, Tucker SJ, Zhao C, Nichols CG, Ashcroft FM (2000) A novel method for measurement of submembrane ATP concentration. J Biol Chem 275: 30046–30049
130. Karschin C, Ecke C, Ashcroft FM, Karschin A (1997) Overlapping distribution of K_{ATP} channel-forming Kir6.2 subunit and the sulfonylurea receptor SUR1 in rodent brain. FEBS Lett 401:59–64
131. Ashcroft FM, Gribble FM (1999) ATP-sensitive K^+ channels and insulin secretion: their role in health and disease. Diabetologia 42:903–919
132. Trapp S, Ashcroft FM (1997) A metabolic sensor in action: news from the ATP-sensitive K^+-channel. News Physiol Sci 12:255–263
133. Browning KN, Coleman FH, Travagli RA (2005) Effects of pancreatic polypeptide on pancreas-projecting rat dorsal motor nucleus of the vagus neurons. Am J Physiol Gastrointest Liver Physiol 289:G209–G219
134. Browning KN, Coleman FH, Travagli RA (2005) Characterization of pancreas-projecting rat dorsal motor nucleus of vagus neurons. Am J Physiol Gastrointest Liver Physiol 288:G950–G955
135. Gribble FM, Ashfield R, Ammala C, Ashcroft FM (1997) Properties of cloned ATP-sensitive

K^+ currents expressed in Xenopus oocytes. J Physiol 498:87–98

136. Tucker SJ, Gribble FM, Zhao C, Trapp S, Ashcroft FM (1997) Truncation of Kir6.2 produces ATP-sensitive K^+ channels in the absence of the sulphonylurea receptor. Nature 387:179–183
137. Bantel C, Maze M, Trapp S (2010) Noble gas xenon is a novel adenosine triphosphate-sensitive potassium channel opener. Anesthesiology 112:623–630
138. Inagaki N, Gonoi T, Clement JP 4th, Namba N, Inazawa J, Gonzalez G, Aguilar Bryan L, Seino S, Bryan J (1995) Reconstitution of IK_{ATP}: an inward rectifier subunit plus the sulfonylurea receptor. Science 270:1166–1170
139. Schuit FC, Huypens P, Heimberg H, Pipeleers DG (2001) Glucose sensing in pancreatic beta-cells: a model for the study of other glucose-regulated cells in gut, pancreas, and hypothalamus. Diabetes 50:1–11
140. Sakura H, Ashcroft SJ, Terauchi Y, Kadowaki T, Ashcroft FM (1998) Glucose modulation of ATP-sensitive K-currents in wild-type, homozygous and heterozygous glucokinase knock-out mice. Diabetologia 41:654–659
141. Efrat S, Tal M, Lodish HF (1994) The pancreatic beta-cell glucose sensor. Trends Biochem Sci 19:535–538
142. Guillam MT, Hummler E, Schaerer E, Yeh JI, Birnbaum MJ, Beermann F, Schmidt A, Deriaz N, Thorens B (1997) Early diabetes and abnormal postnatal pancreatic islet development in mice lacking Glut-2. Nat Genet 17:327–330
143. Thorens B (1992) Molecular and cellular physiology of GLUT-2, a high-K_m facilitated diffusion glucose transporter. Int Rev Cytol 137:209–238
144. Matschinsky FM (1996) Banting Lecture 1995. A lesson in metabolic regulation inspired by the glucokinase glucose sensor paradigm. Diabetes 45:223–241
145. Terauchi Y, Sakura H, Yasuda K, Iwamoto K, Takahashi N, Ito K, Kasai H, Suzuki H, Ueda O, Kamada N et al (1995) Pancreatic beta-cell-specific targeted disruption of glucokinase gene. Diabetes mellitus due to defective insulin secretion to glucose. J Biol Chem 270:30253–30256
146. Gonzalez JA, Jensen LT, Doyle SE, Miranda-Anaya M, Menaker M, Fugger L, Bayliss DA, Burdakov D (2009) Deletion of TASK1 and TASK3 channels disrupts intrinsic excitability but does not abolish glucose or pH responses of orexin/hypocretin neurons. Eur J Neurosci 30:57–64
147. Guyon A, Tardy MP, Rovere C, Nahon JL, Barhanin J, Lesage F (2009) Glucose inhibition persists in hypothalamic neurons lacking tandem-pore K^+ channels. J Neurosci 29: 2528–2533
148. Terzic A, Tung RT, Kurachi Y (1994) Nucleotide regulation of ATP sensitive potassium channels. Cardiovasc Res 28:746–753
149. Trapp S, Lückermann M, Kaila K, Ballanyi K (1996) Acidosis of hippocampal neurones mediated by a plasmalemmal Ca^{2+}/H^+ pump. Neuroreport 7:2000–2004
150. Erdemli G, Xu YZ, Krnjevic K (1998) Potassium conductance causing hyperpolarization of CA_1 hippocampal neurons during hypoxia. J Neurophysiol 80:2378–2390
151. Leblond J, Krnjevic K (1989) Hypoxic changes in hippocampal neurons. J Neurophysiol 62:1–14
152. Nowicky AV, Duchen MR (1998) Changes in $(Ca^{2+})_i$ and membrane currents during impaired mitochondrial metabolism in dissociated rat hippocampal neurons. J Physiol 507:131–145
153. Vrang N, Hansen M, Larsen PJ, Tang-Christensen M (2007) Characterization of brainstem preproglucagon projections to the paraventricular and dorsomedial hypothalamic nuclei. Brain Res 1149:118–126
154. Larsen PJ, Tang-Christensen M, Holst JJ, Orskov C (1997) Distribution of glucagon-like peptide-1 and other preproglucagon-derived peptides in the rat hypothalamus and brainstem. Neuroscience 77:257–270
155. Trapp S, Hisadome K (2011) Glucagon-like peptide 1 and the brain: central actions-central sources? Auton Neurosci 161:14–19
156. Llewellyn-Smith IJ, Reimann F, Gribble FM, Trapp S (2011) Preproglucagon neurons project widely to autonomic control areas in the mouse brain. Neuroscience 180:111–121
157. Holst JJ (2007) The physiology of glucagon-like peptide 1. Physiol Rev 87:1409–1439

Chapter 2

Anatomically "Calibrated" Isolated Respiratory Networks from Newborn Rodents

Araya Ruangkittisakul, Bogdan Panaitescu, Lucia Secchia, Nicoleta Bobocea, Chase Kantor, Junya Kuribayashi, Makito Iizuka, and Klaus Ballanyi

Abstract

En bloc and transversal slice preparations from perinatal rodents are established in vitro models for studying control of breathing by neurons and neighboring glial cells in the lower brainstem. These neural "respiratory networks in the dish" show features complementary with those in recent in vivo models that enable, for example, optogenetic manipulation for studying behavioral changes in awake animals or modulation of these circuits by higher brain regions during sleep. Contrary, the in vitro models allow respiratory network analysis at the (sub)cellular level. This is currently studied using powerful analytical tools such as quantitative pharmacology and patch-clamp recording of biophysical membrane properties of respiratory interneurons and/or motoneurons (plus neighboring glia). Increasingly, these approaches are combined with fluorescence imaging of both activity and morphology of individual cells or respiratory groups. Our recent work indicates that properties of the isolated respiratory networks depend critically on both their physical dimensions and the composition of superfusates used for their study. Based on this, we recommend to use anatomically "calibrated" rhythmic *en bloc* and slice preparations that we have developed and to study these models in superfusate that mimics in vivo conditions as closely as possible. We show here how these preparations are generated and present examples for pharmacological and "electrophysiological imaging" analyses that revealed novel properties of neuron–glial networks involved in respiratory control.

Key words: Breathing, Calcium, Fictive eupnea, Fictive sigh, Inspiratory, Multiphoton imaging, Pre-botzinger complex, Respiratory center

1. Introduction

Breathing movements are controlled by neurons that form "respiratory networks" with adjacent glial cells in the lower brainstem, the *medulla oblongata*. Since >200 years, findings from

Klaus Ballanyi (ed.), *Isolated Central Nervous System Circuits*, Neuromethods, vol. 73,
DOI 10.1007/978-1-62703-020-5_2,

in vivo experiments including brainstem transection indicate a rostrocaudal hierarchic organization of these neural respiratory networks that are located in three distinct areas. These bilaterally organized areas comprise the pontine, the dorsal and the ventral respiratory groups, with parts of the latter structure being referred to as the ventral respiratory column (VRC) (1–3). These three respiratory groups are located deep within the brain and are thus difficult to access. Consequently, there is a long-standing wish to isolate them for analyzing their properties. In 1984, Suzue (4) reported that an isolated newborn rat brainstem–spinal cord preparation (Fig. 1) generates inspiratory-related motor output. Since then, numerous studies aimed at elucidating respiratory network functions in this *en bloc* model (1, 2, 5, 6). In 1991, Smith and colleagues, involving our group (7), showed by serial brainstem section of this preparation that a limited VRC region, the pre-Bötzinger complex (preBötC), remains inspiratory active in a transversal brainstem slice, suggesting that it is responsible for generation of respiratory rhythm in mammals (Figs. 1 and 2). Since then, in vitro and in vivo studies established that the preBötC is crucial for normal breathing which greatly depends on active inspiration, whereas active expiration is less important (1–3, 8).

It is generally assumed that the preBötC comprises the only active respiratory network in the "classical" rhythmic slice which is a powerful model that enables application of various tools for analyses at both network and (sub)cellular levels. These tools include quantitative pharmacology, Ca^{2+} or voltage imaging, and patch-clamp analysis of membrane biophysics of preBötC neurons and preBötC-driven inspiratory active hypoglossal (XII) motoneurons (7, 9–15). Besides, "breathing" preBötC slices are studied increasingly from genetically engineered mice which either serve as models for respiratory diseases and/or contain respiratory neurons or glial cells that express fluorescent markers associated with cell-type-specific proteins such as transcription factors or neurotransmitter receptors (16–19).

The fact that various spinal and cranial ventral nerve roots in the brainstem–spinal cord preparation show inspiratory-related activity enables the analysis of different preBötC-driven inspiratory motor networks (5, 20–22). Such analyses can make use of most "electrophysiological imaging" approaches and genetic tools that are applicable to the slice model (5, 6, 15, 23, 24). Moreover, the *en bloc* model allows investigation of mutual interactions between inspiratory center neurons and another class of respiratory interneurons. This "parafacial respiratory group" (pFRG) was identified by voltage imaging in newborn rat *en bloc* medullas in the milestone study in 2003 by Onimaru and Homma (25). That report and previous work from that group reviewed in 1999 by Ballanyi and colleagues (5) hypothesized that a major rhythmogenic portion of the pFRG is located within the ventrolateral medulla several

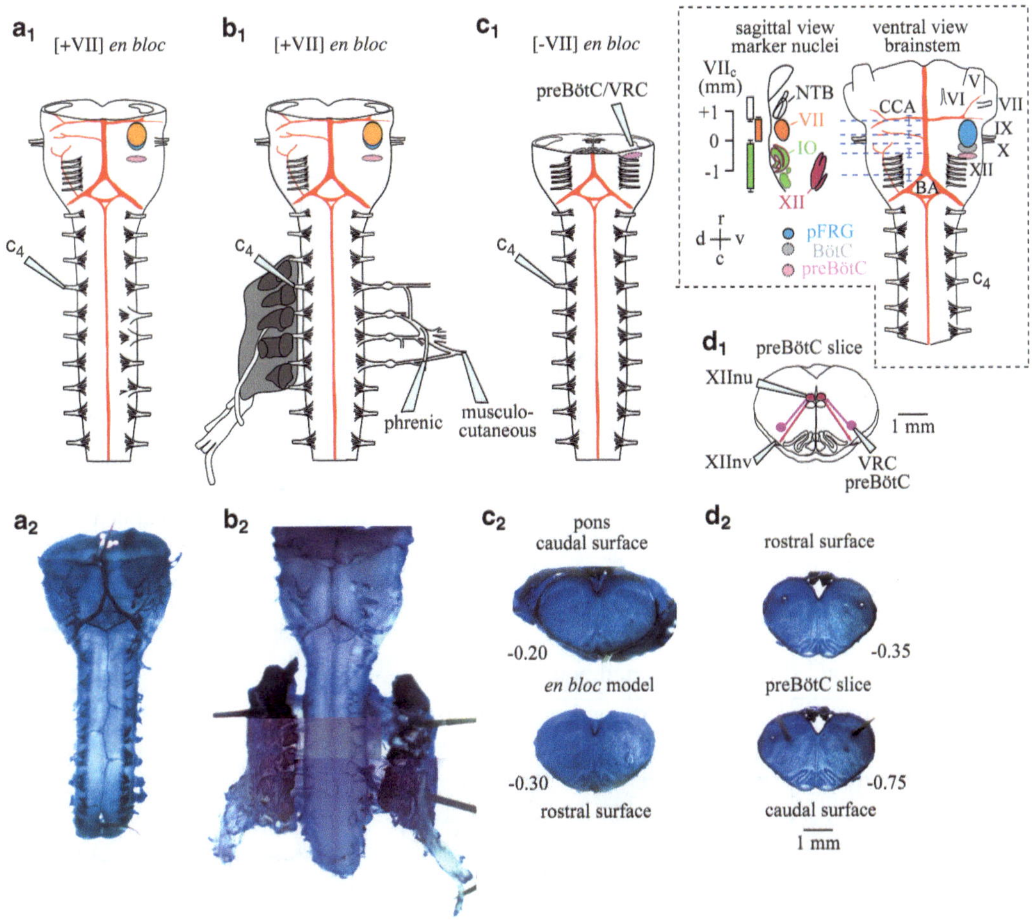

Fig. 1. Calibrated newborn rat "respiratory networks in the dish." The inset in the *dotted box* indicates in the ventral brainstem view the estimated rostrocaudal extents of inspiratory active pre-Bötzinger complex (preBötC) and expiratory pre-/postinspiratory active parafacial respiratory group (pFRG) neural networks that form a dual respiratory center. The region between these respiratory groups corresponds to the Bötzinger complex (BötC) which is primarily expiratory active in adults but not yet studied in newborns. The sagittal view in the inset indicates mean values plus standard deviations of rostrocaudal extents of some respiratory brainstem marker nuclei included in a newborn rat brainstem atlas for generating anatomically defined ("calibrated") inspiratory active transversal preBötC slices (see Fig. 2). These nuclei are the inspiratory active hypoglossal (XII) motor nucleus, the nucleus of trapezoid body (NTB), and particularly the "peri-inspiratory" active facial (VII) motor nucleus plus different subregions ("loops") of the inferior olive (IO). The respiratory groups and marker nuclei have a defined position relative to respiratory active nerve roots on the ventral brainstem surface. Specifically, these roots are from XII, trigeminal (V), abducens (VI), glossopharyngeal (IX), and vagal (X) nerves. Importantly, the most rostral XII root corresponds to the location and rostrocaudal extent of the preBötC. (**a**–**c**) Brainstem–spinal cord preparations. (**a_1**) Schema of the "Suzue preparation" (4), which is rostrally transected at the caudal cerebellar artery (*CCA*, see inset), whereas the cord is typically cut at the last cervical (C_8) or first thoracic (T_1) level. This is the "classical" *en bloc* model with complete *VII* nucleus ("[+VII] *en bloc*") for studying the dual respiratory center. (**b_1**) The *en bloc* model in **a_1** can be isolated with intact cervical and brachial plexuses for recording from cervical (e.g., C_4) roots plus phrenic, brachial, and/or musculocutaneous nerves. The drawing on the *left* is based on an actual preparation, while the drawing on the *right* is derived from (67). (**c_1**) shows a "calibrated" *en bloc* preparations with the manually cut rostral surface located –0.25 mm caudal to the caudal end of VII nucleus, VII_c (see inset and numbers next to stained preparations). This [–VII] *en bloc* model does not produce pFRG motor output and allows recording from the VRC/preBötC at the rostral surface and spinal (cervical) nerves. (**d_1**) Schema of preBötC slice with suction electrode positions for recording inspiratory-related rhythms. (**a_2**–**d_2**) show preparations according to **a_1**–**d_1** that were chemically fixed and thionin-stained after the experiment for better visualization of brainstem (surface) markers for histological analysis of their anatomical margins (For further details, see text or (33, 34, 40)).

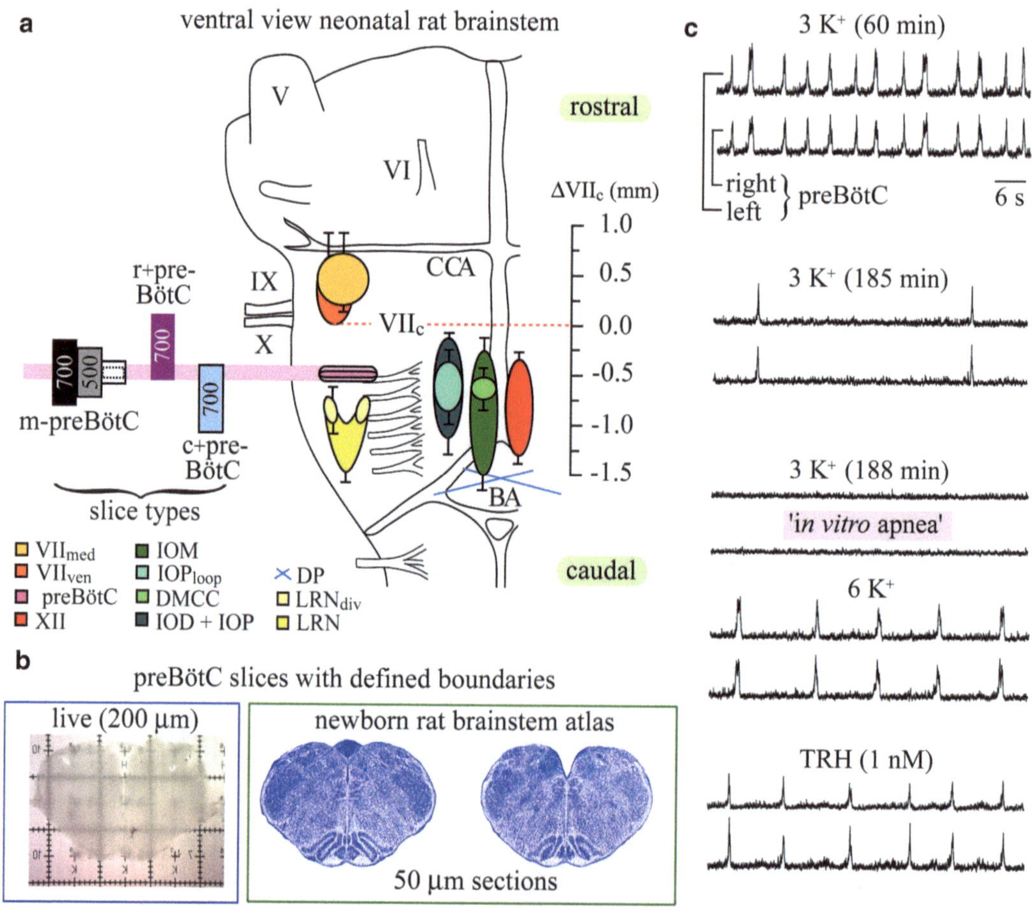

Fig. 2. Rhythm in physiological (3 mM) K+ of calibrated preBötC inspiratory center slices. (**a**) Rostrocaudal extents of respiratory brainstem marker nuclei such as subnuclei of the IO and the VII motor nucleus, projected onto the ventral brainstem surface, referred to VII_c (see Fig. 1) are constant (see small standard deviations of bars) in Sprague-Dawley and Wistar rats of postnatal days (P) 0–4. This allows generation with novel "online histology" of various types of preBötC slices with defined anatomical boundaries referred to VII_c. "m-preBötC" slices have the preBötC in the *middle*. r+preBötC or c+preBötC slices contain the <100-μm small preBötC kernel necessary for generation of rhythm plus additional rostral or caudal tissue, respectively. (**b**) shows in the *left* a 200-μm-thin living rhythmic m-preBötC slice after an experiment next to two sections from a newborn rat reference atlas used to determine the margins of rhythmic slices. (**c**), Rhythm recorded with two suction electrodes in solution with 3 mM K^+ and 1 mM Ca^{2+} ("3K/1Ca") in a 500-μm-thick m-preBötC (m-preBötC[500/−0.70]W-P3) slice was stable for ~3 h before burst rate dropped from >10 to <4 bursts/min. After rhythm stopped spontaneously shortly after 185 min ("in vitro apnea"), it was restored by raising K^+ to 6 mM or bath application of thyrotropin-releasing hormone (TRH) (details in (33, 40)). Abbreviations: *DMCC*, dorsomedial cell column of inferior olive (IO); *IOM*, medial IO; *IOD*, dorsal IO; *IOP*, principal IO; IOP_{loop}, lateral loop of IOP; *LRN*, lateral reticular nucleus; LRN_{div}, LRN divided into medial and lateral subnuclei; VII_{med}, medial subnucleus of VII; VII_{ven}, ventral subnucleus of VII; *DP*, pyramidal decussation. Other abbreviations in Fig. 1 (Modified after (40), with kind permission).

100 μm rostral to the preBötC in an area surrounding the caudal end of facial (VII) motor nucleus, VII_c (Figs. 1 and 2). This area also contains the retrotrapezoid nucleus that is involved in central respiratory chemosensitivity and may (partly) overlap with the pFRG (26, 27).

Recent findings, by others and us, indicate that important pFRG components are also located close to the rostral end of VII nucleus (28, 29). pFRG neurons drive, for example, lumbar motoneurons innervating expiratory abdominal muscles or facial motoneurons driving upper airway muscles (6, 8, 21, 22). These observations, along with the finding that opioids depress preBötC activity but not rhythmic bursting of pre-/postinspiratory active pFRG neurons, suggest that the inspiratory preBötC and the functionally expiratory pFRG constitute a pivotal dual respiratory center, at least in newborn mammals (2, 5, 23, 24).

Identification of the structure–function relationship for the dual respiratory center in the *en bloc* model is difficult because inspiratory and pre-/postinspiratory neurons are distributed widely along the rostrocaudal VRC axis, indicating that the preBötC and pFRG may partly overlap (5, 24, 30). Even in the more reduced inspiratory slice model, the structure–function relationship of the preBötC is not well elaborated. Between laboratories, preBötC slice thickness ranges from 200 μm to >1,000 μm, and superfusates used for their study differ notably, particularly regarding the contents of Ca^{2+} (0.8–2.5 mM) and K^{+} (3–11 mM) (4, 7, 10, 12, 18, 31–35). Because these cations greatly affect neuronal excitability, their extracellular concentrations in the mammalian brain are tightly regulated, specifically to 1.2 mM for Ca^{2+} and 3 mM for K^{+}, except for short activity-related transients (36–39). These differences in physical dimensions and/or ionic superfusate composition may explain some discrepant findings between groups, for example, regarding the relevance for rhythm of "intrinsic bursting" properties of respiratory neurons (2, 3, 7, 12–14, 25).

Here, we summarize factors that are important for identifying structure–function relationships for inspiratory versus expiratory networks in perinatal rodent *en bloc* and slice preparations. To achieve this, it was firstly necessary to determine anatomical brainstem marker regions in these models for generating "calibrated" respiratory networks in the dish with identified and reproducible rostrocaudal margins. It is one major aim of this chapter to summarize this "online histology" approach for obtaining calibrated *en bloc* and slice preparations from newborn Wistar and Sprague-Dawley rats and preBötC slices from C57BL/6 mice (33, 34, 40, 41). Based on this work and other recent studies from our group (18, 28, 35, 42–44), we also exemplify how physical dimensions and superfusate composition influence rhythmic activities generated by the preBötC and the dual respiratory center in the slice and *en bloc* models, respectively. Also, regarding previous findings on the "noncalibrated" in vitro models, we focus on our own publications reflecting our technical experiences with these preparations. Reference to work by others, for example, regarding interstitial brain levels of ions or glucose, is mostly represented by review articles from other leading groups using these respiratory models and from our own recent reviews.

2. Materials

Various materials used by our group for studying respiratory neurons in vitro are identical with those for analyzing dorsal vagal neurons in brainstem slices. Accordingly, complementary information is given in our chapter on these cells (Chap. 1-Trapp and Ballanyi (45)).

2.1. Animal Models

Respiratory functions are mostly studied in *en bloc* or slice models from perinatal rodents, traditionally mainly Wistar or Sprague-Dawley rats. Increasingly, mice from outbred (e.g., CD1, NMR, OF1) or inbred strains (e.g., C57BL/6) are being used due to ongoing development of transgenic models (16–19). Importantly, use of the *en bloc* and slice models from rats is restricted to newborn animals until about postnatal day (P) 7 (5). Rhythmic slices from some mouse strains such as CD1 or OF1, but likely not *en bloc* preparations, can be used from up to ~P14 animals unless a precocial strain is used (13, 14, 18, 32, 46, 47). This restriction is probably mostly due to inadequate supply of these nonperfused in vitro models with oxygen (O_2) as dealt with below (Sect. 5). In both species, functioning *en bloc* preparations and preBötC slices can be obtained from fetal animals, in which respiratory-related brainstem activities start ~1 week before birth (19, 46, 47). Only one *en bloc* preparation and normally also only one inspiratory active slice are obtained per animal, but see Sect. 4.2.1.

2.2. Solutions

In the following, we refer to compositions of (1) superfusates for optimal functions of the in vitro respiratory networks, (2) solutions for patch clamp and (ion-sensitive) microelectrodes, and (3) solutions for histology.

2.2.1. Superfusates

Commonly, the term "artificial cerebrospinal fluid" ('ACSF') is used for superfusate/perfusate in which isolated vertebrate nervous tissues are studied. However, firstly, the composition of cerebrospinal fluid differs substantially from that in the extracellular (interstitial) space of the brain because the latter fluid contains tonically present and/or phasically released neurotransmitters and other neuroactive agents such as endorphins or growth factors (36, 39, 48). Secondly, interstitial concentrations of these agents differ between brain regions and depend on patterns of spontaneous activity and/or developmental state in a given brain area (36, 39, 48). Accordingly, we use the term "superfusate" or "standard/control solution" instead of "ACSF".

Our recent findings pointed out that the ionic composition of the superfusate is pivotal for intrinsic activity of isolated perinatal rodent respiratory networks and their responses to various neuromodulators (Sect. 4.2.2). Based on this, we recommend for use with >600-μm-thick preBötC slices or *en bloc* preparations the so-called 3K/1Ca superfusate of the following ionic composition (in mM): 120 NaCl, 3 KCl, 1 $CaCl_2$, 2 $MgSO_4$, 1.25 NaH_2PO_4, and 26 $NaHCO_3$. The pH of this solution is adjusted to 7.4 by continuous gassing with "carbogen", which is a mixture of 95% O_2 and 5% carbon dioxide (CO_2). Such "carbogenation" is, for example, achieved with a gas dispersion tube that is also termed "microfilter candle" (Fisher Scientific, Canada; www.fisher.ca or ROBU, Hattert, Germany; www.robuglas.com).

Regarding the content of glucose (more precisely "D-glucose" or "dextrose") of standard superfusate, 20–30 mM is necessary for the *en bloc* model. For calibrated preBötC slices, 20 mM glucose is commonly used in superfusate, although they remain viable and rhythmic at levels as low as 1 mM (for details, see Sect. 4.2.2).

We found that inspiratory-related rhythm in 400-μm-thin preBötC slices in 3K/1Ca solution stops spontaneously within <2 h (in vitro apnea) but can be reactivated for time periods >7 h by raising K^+ to 7 mM (41, 42). In 7K/1Ca, the rate and amplitude of inspiratory rhythms in preBötC slices are appropriate for long-term studies and basically similar to rhythms in preBötC slices studied by others in 8–9 mM K^+ (and 1.5 mM Ca^{2+}) (2, 13, 14). Based on our recent unpublished findings, it may be appropriate for studies on "K^+-sensitive" cellular processes in thin preBötC slices (Sect. 4.2.2) to use 5K/1Ca solution at the start of recording and raise K^+ to 6 mM, and then eventually to 7 mM, when rhythm starts to slow down.

2.2.2. Electrode Solutions

The following "standard" patch electrode solutions are currently used for whole-cell patch-clamp recording (Sect. 3.2.2) in the Ballanyi laboratory: (1) standard ("low Cl^-") pipette solution contains (in mM) 120 K-gluconate, 1 NaCl, 1 $CaCl_2$, 1 $MgCl_2$, 11 K_4-BAPTA, 0.5 Na_2-ATP, and 10 HEPES (pH adjusted to 7.4 with KOH); (2) "high Cl^-" solution has the same composition, except that 120 mM KCl replaces 120 mM K-gluconate; and (3) "perforated patch" solution contains 5–10 μg/ml gramicidin prepared from two stock solutions. Stock I contains 2 mg gramicidin in 1 ml dimethyl sulfoxide and is freshly made every 4 h, whereas stock II contains 140 mM KCl and 10 mM HEPES; pH is adjusted to 7.4 with KOH. Stocks I and II are mixed at a ratio of 1:10, and this mixture is diluted 1:40 or 1:20 with one of the above patch solutions. Pipettes are prefilled with 2 μl of these solutions to avoid contamination of the tissue with gramicidin during cell searching. Rationales for particular compositions of patch electrode solutions are outlined in review articles (49, 50) and below (Sect. 3.2.2). Recording with intracellular sharp microelectrode is feasible for respiratory motoneurons but has not yet been reported for VRC interneurons in the respiratory networks in the dish. These electrodes with tip diameters <0.2 μm, resulting in a DC resistance of 30–200 MΩ, are typically filled with 3 M KCl, 1 mM K^+-acetate, or 1 M Mg^{2+}-acetate. Standard superfusate is convenient for filling suction electrodes for recording of extracellular population activity or electrical stimulation and also for filling reference barrels of extracellular double- or triple-barreled ion-sensitive microelectrodes (Sect. 3.2). Ion-sensitive barrels of the latter type of microelectrodes contain the following combination of ion-selective resin (in the silanized tip) and "backfilling" solution: (1) K^+ electrodes (Fluka 82500, from World Precision Instruments [WPI Sarasota, FL, USA; www.wpiinc.com]; 150 mM KCl), (2) Ca^{2+} electrodes (Fluka 21048, WPI; 150 mM KCl), and (3) pH electrodes (Fluka 95291, WPI; 40 mM K_2HPO_4 + 23 mM NaOH + 15 mM NaCl, buffered to pH 7 with 1N HCl).

2.2.3. Histological Solutions

For histological analysis using the dye thionin (Sigma-Aldrich; www.sigmaaldrich.com) (Sects. 2.9 and 3.1), brainstem tissue is fixed in 4% paraformaldehyde in phosphate buffer (1:2 mixture of 0.1 M NaH_2PO_4 + 0.1 M Na_2HPO_4 in H_2O, pH 7.2). The staining solution consists of 0.1 M Na^+-acetate trihydrate + 0.1 M acetic acid + 1% Thionin acetate dissolved in distilled water. The solution for embedding brainstems for facilitated sectioning consists of 6% agar-agar (Sigma-Aldrich) dissolved in boiling distilled water that is subsequently cooled down to ~40°C. A fixed brainstem is quickly dried (with filter paper or paper towel) and then placed in a small container (e.g., 10-ml beaker or small Petri dish) that is prefilled with agar solution. More agar solution is then added to cover the entire tissue block. The embedded block is subsequently cooled on an ice pad to eliminate air bubbles.

Whole-cell patch-clamp recorded respiratory neurons are stained with either 0.5% lucifer yellow (Sigma-Aldrich) or 0.2–0.5% biotin or biocytin (Sigma-Aldrich) added to the patch electrode solution. For histological reconstruction of these cells, we fix preparations either according to our studies in collaboration with the group of Dr. H. Onimaru (Showa University, Tokyo, Japan) (24, 51, 52) for >48 h at 4°C in "Lillie" solution (i.e., 10% formalin in phosphate buffer, pH 7.0). According to our previous collaboration with the group of Dr. S.W. Schwarzacher (University of Frankfurt, Germany) (53), tissue was fixed for 4–15 h at 4°C in 4% paraformaldehyde (in phosphate buffer, pH 7.4), followed by 10-h immersion in 30% sucrose solution. Further histological processing for neuron reconstruction is described below (Sect. 2.9).

2.3. Devices for Brainstem–Spinal Cord Dissection and Brainstem Slicing

Under anesthesia using isoflurane (Baxter Corp., Deerfield, IL, USA; www.baxter.com) (Sect. 3.1), the brainstem–spinal cord preparation is dissected with fine forceps (e.g., Dumont #5, straight tip [0.05 × 0.02 mm] or Dumont #5SF [0.025 × 0.005 mm]) and (iris) scissors (e.g., Vannas-style, straight tip angle [length 80 mm, 2.5-mm blade] or Vannas Tübingen [length 85 mm, 5-mm blade]) (supplied, e.g., by Fine Science Tools Inc., Vancouver, BC, Canada;

www.finescience.com). Also for generating rhythmic slices, the brainstem is firstly isolated and then transferred to the vise of a vibratome such as the VT1000S (Leica Biosystems, Nussloch, Germany; www.leica-microsystems.com) or the 5,000mz (Campden Instruments Ltd., Loughborough, UK; www.campdeninstruments.com). Use of a high-quality vibratome is recommended particularly when <250-µm-thin slices are planned to be used and/or recording from superficial visualized cells is intended (18, 35, 43, 48). The less expensive "McIlwain" tissue chopper (Warner Instruments, Hamden, CT, USA; www.warneronline.com) may cause more damage to superficial slice layers but can be used if recording is restricted to extracellular or "blind" patch-clamp techniques in thicker slices (Sect. 3.2).

There is no obvious need for slicing in ice-cold solution of particular composition, such as substitution of Na^+ with choline, or for storing preparations (in such solution) in a heated chamber for recovery from sectioning injury (48) (Chaps. 1-Trapp and Ballanyi (45), 3-Moore et al. (54), 8-Sanchez-Vives (55), 9-Broicher and Speckmann (56), 11-Kantor et al. (57) and 14-Biggs et al. (58)). In fact, respiratory rhythms can be recorded in both *en bloc* preparations and slices within <10 min after their generation, and these models remain active for extended time periods (Sects. 3.1.4 and 4.2).

2.4. Recording Chamber and Superfusion System

For experiments involving visualization of cells using an upright microscope (18, 35, 43, 48) (Sect. 2.8), we use a submerged-type acrylic chamber system with glass bottom, for example, from Warner Instruments (Fig. 3). A custom-made "submerged"-type acrylic recording chamber (Fig. 3) (see also Chap. 1-Trapp and Ballanyi (45)) is used by us for "blind" electrophysiological recording experiments, in which the preparation is mechanically fixed and electrodes are positioned under visual control using a stereomicroscope (Sect. 2.8) (Fig. 3). The latter type of chamber requires a cold-light source with flexible glass fiber guides, for example, from WPI or Zeiss (Oberkochen, Germany; www.zeiss.com). The recording chambers are continuously perfused with carbogenated standard solution (Sect. 2.2) that is prewarmed in a water bath (DC 10, Thermo Haake, Sigma-Aldrich) and administered via a peristaltic roller pump such as the Watson-Marlow Alitea (Sin-Can, Calgary, AB, Canada; www.sincaninc.com). The superfusate passes through a custom-made acrylic bubble trap to protect from perturbation of fluid flow in the recording chamber. The chamber is typically kept at 25–27°C via a temperature control system such as the TC-324B (Harvard Apparatus, Saint-Laurent, QC, Canada; www.sincaninc.com) (Fig. 3). Superfusate is removed from the recording chamber via a hypodermic needle (gauge 18 G1½") connected to either (a further channel of) a roller pump or a vacuum system, for example, an aquarium pump (Fig. 3). Tubings for the superfusion system should be impermeable to gases, in particular O_2 and

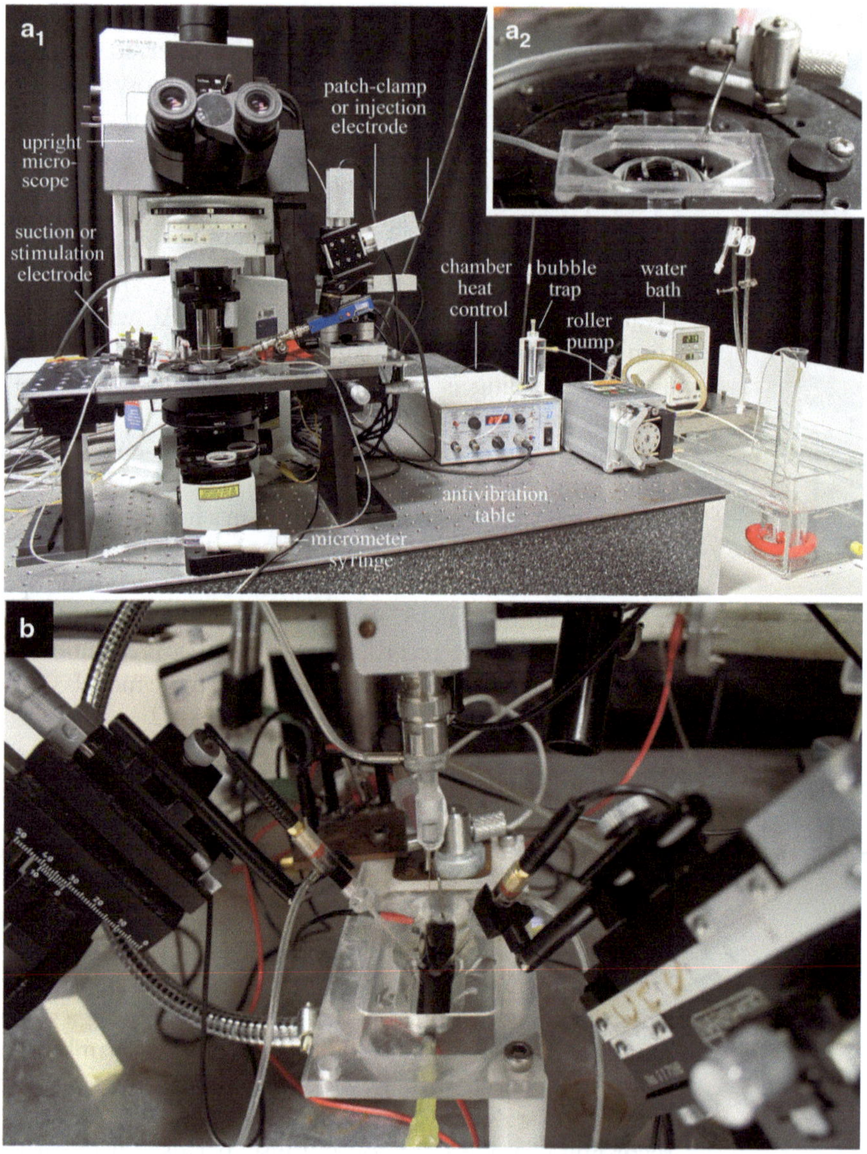

Fig. 3. Equipment for "electrophysiological imaging" in respiratory networks in the dish. (**a**) An upright microscope (**a$_1$**) and a "fixed" custom-made XY microscope stage are used for positioning and mechanical fixation of one calibrated preBötC slice or *en bloc* preparation with a "harp" to the glass bottom of a "Warner" submerged-type recording chamber (**a$_2$**), or with steel insect pins, if a layer of silicone ("Sylgard-184®") has been added to cover the bottom of the chamber. Superfusate enters the chamber via a Tygon® or steel tube and is removed at the opposite end via a hypodermic cannula connected to a vacuum system (e.g., an aquarium pump). Superfusate is prewarmed in a "Haake" water bath while the temperature of superfusate in the recording chamber is kept constant by a "Warner" temperature control system. Superfusate is transported via a "Watson-Marlow" peristaltic ("roller") pump and passes an acrylic "bubble trap" before entering the recording chamber. Cells in slices are visualized using a x20 objective which is part of a FV300 Olympus confocal laser scanning imaging system. Fluorescent dye is injected via a broken patch electrode positioned with a "Siskiyou" step motor also used for patch recording using a HEKA "patch clamp" (HEKA headstage shown here) or npi "single-electrode voltage-clamp" amplifier. Neuronal population activity is extracellularly recorded using suction electrodes positioned via a Narishige micromanipulator and connected to a "WPI" or "AM-Systems" differential amplifier. The entire recording system is positioned on a "Kinetic Systems" antivibration table. (**b**) Recording chamber and mechanical arrangement

CO_2 with which the standard superfusate is carbogenated (Sect. 2.2). This ensures that the partial pressure of O_2 (PO_2) and CO_2/HCO_3^--adjusted pH, respectively, do not change during superfusate transport to the recording chamber. Stainless steel or Tygon® tubing (e.g., Fisher Scientific, Canada; www.fishersci.ca) is a convenient option in that regard.

For long-term viability and (spontaneous) activity of respiratory networks in the dish, flow rates for superfusate administration should be at least 5 ml/min for recording chambers with a volume of ~1 ml. Larger chambers may be necessary if multiple electrodes are to be positioned, for example, for recording of bath and tissue PO_2 as well as ion activities combined with suction electrode and intracellular recording. These larger chambers require a higher flow rate for a similarly effective exchange rate of superfusate compared to smaller models. Importantly, flow rates <5 ml/min in a ~1-ml chamber may cause interstitial accumulation of (inhibitory) neuromodulators such as adenosine (48). A small recording chamber may be necessary, in combination with recirculation of superfusate, for particular experiments such as those involving expensive drugs. In our hands, normal rhythmic activity in both *en bloc* and slice preparations persists for several hours during such recirculation. A sophisticated recording chamber in that regard is referred to in a different chapter (Chap. 1-Trapp and Ballanyi (45)). Though not studied systematically by us, it seems that rhythm, in particular, in calibrated preBötC slices is faster and more robust when the meniscus of the fluid is <1 mm above the slice surface. However, the lower the fluid level, the more sensitive it is to perturbations of superfusate flow. This may result in movement of the slice toward the meniscus because of surface tension and would thus impair supply of superficial slice areas with carbogenated superfusate and disturb recording as well.

For imaging experiments or patch-clamp recording from superficial visualized cells (18, 35, 43, 48), a single brainstem slice or *en bloc* preparation is transferred to the recording chamber (Fig. 3). The rhythmic slice is held in place by a nylon net fixed to a platinum frame (48) (Fig. 4 in Chap. 1-Trapp and Ballanyi (45)).

Fig. 3. (continued) for combining suction electrode (and eventually ion- and/or gas-sensitive microelectrode recording) with "blind" patch-clamp recording from the respiratory networks in the dish. Preparations are positioned and fixed in the silicone-covered submerged-type custom-made recording chamber using a stereomicroscope that can be combined with (or replaced by) a video camera. Superfusate system including inlet and outlet is identical with that in **a**. Intracellular recording is done using a "Siskiyou" (see **a**) or "Sutter" step motor while "Märzhäuser" (*left*) or "Narishige" (*right*) micromanipulators are stable enough for >10-h recording, for example, of neuronal population activity. Suction and patch electrodes are conveniently placed in "ALA Scientific" or "Warner" holders. The latter models, shown here on the *left* and *right* sides, are not bulky, thus enabling positioning of up to eight electrodes in this type of chamber, in particular when using the slim Narishige manipulators. While extracellular electrodes are positioned at an angle, the patch electrode (connected here to the headstage of an npi amplifier) advances perpendicularly through the tissue using digital readouts for the Siskiyou or Sutter motors for identifying the recording depth.

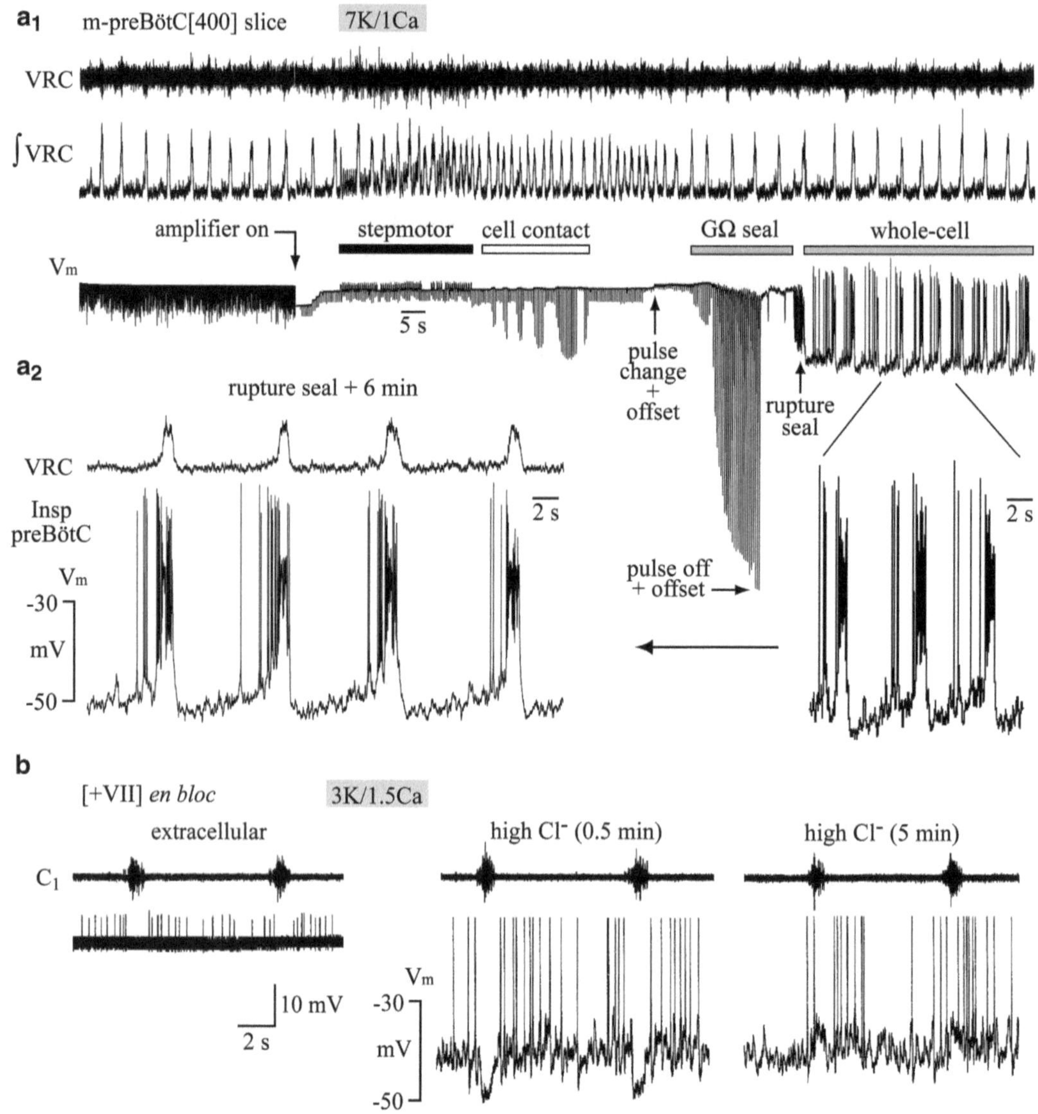

Fig. 4. "Blind" whole-cell recording from respiratory neurons. (**a**) Procedures involved in establishing a membrane potential (V_m) recording in "whole-cell" patch-clamp configuration. (**a**$_1$) In a m-preBötC[400/-0.75]W-P3 slice, inspiratory-related bursting is recorded with a suction electrode from the ventrolateral slice aspect containing the ventral respiratory column (*VRC*) with the preBötC extending rostrocaudally between 100 and 300 μm below the surface (see Figs. 1 and 2). Compared to the *upper VRC trace* showing the amplified (×10k) differentially recorded and band-pass filtered (0.3–3 kHz) raw signal, the integrated (τ: 50 ms) signal reveals the inspiratory bursts much better. After turning the intracellular npi amplifier on (see *arrow*), the patch electrode is positioned in the contralateral *VRC* and forwarded through the tissue using a Siskiyou step motor (see Fig. 3). Close to the cell membrane, the amplitude of the electrode voltage change in response to a repetitively applied rectangular DC current (-100 pA, 20 ms) increases and shows some flickering ("oscillations"). At this stage, the electrode is moved up and down by 1–5 μm until the oscillations during the test pulse are maximal and stable. Then, the amplitude of the test pulse is reduced to 10%, and the positive pressure (20–30 mmHg) applied to the electrode is removed. In most cases, a gigaohm seal establishes spontaneously, while in the other cases slight suction is applied to facilitate this. Once a seal of at least 1 GΩ is established, the pulse is switched off and the patch is ruptured by sudden suction, eventually combined with application of a short current pulse. In this example, the whole-cell configuration is established for an inspiratory ("Insp") preBötC neuron. The trace in **a**$_2$ from the same neuron shows that the recording

Platinum is mostly used because the frame needs to be both heavy and inert, but tantalum is a cheaper alternative, and both are available at Goodfellow (Huntington, United Kingdom; www.goodfellow.com). In recording chambers covered with silicone (Sylgard-184® ["Kwik-Gard"], WPI) (Fig. 3), a rhythmic slice can also be fixed with insect pins ("minutien pins," stainless steel, length 10 mm, 0.02-mm tip/0.2-mm base diameters; Fine Science Tools) inserted in its (dorso)lateral aspects, while suction electrode(s) provides further mechanical stability. Depending on the (angle of the) cold-light source used for illuminating the recording chamber, carbon powder may be added to the Sylgard-184® for better visualization of details of the preparation and electrode positioning. Also, *en bloc* preparations are fixed in Sylgard-184®-covered chambers with the ventral side up by placing a minutien pin at the rostral end of the brainstem and caudal end of the spinal cord. Further stabilization is achieved by suction electrodes used for recording nerve activities (Sect. 2.5).

2.5. Microelectrodes for Recording and Stimulation

Patch electrodes with outer tip diameters of 1.5–3 μm are used for "whole-cell" recording (7, 9, 42, 51–53, 59, 60) of membrane potential (V_m) or membrane currents of respiratory neurons for up to several hours (Sects. 3.2.2, 4.1.2, and 4.2). Fine-tipped microelectrodes with outer tip diameter <0.2 μm can be used for intracellular recording from most respiratory motoneurons, whereas stable recordings from preBötC or pFRG interneurons do not seem to be feasible with such "sharp" microelectrodes. Suction electrodes are a pivotal tool for recording spontaneous extracellular neural population activity from the *en bloc* and slice models (Sects. 3.2.1 and 4.1) (5, 18, 40). These electrodes have an outer tip diameter of 50–250 μm depending on the diameters of respiratory active nerve rootlets to which they are mostly attached besides positioning them in active slice regions (Fig. 3). They are pulled from borosilicate glass capillaries (GC-150TF-10; 1.5-mm outer diameter×1.17-mm inner diameter, Harvard Apparatus). The electrode tip can be grinded down ("beveled") to a desired size with sandpaper or

Fig. 4. (continued) stabilized within some minutes after the rupture of the seal. It is difficult to determine resting V_m in this cell because of its rhythmic fluctuations. Specifically, V_m is slightly more negative than −50 mV during a hyperpolarization after termination of the inspiratory burst and subsequently shows a progressive depolarization until action potential ("spike") discharge occurs in the preinspiratory phase preceding the onset of the inspiratory *VRC* population burst that is accompanied by a synaptic drive potential and further spike discharge in the neuron. Compared to the absence of respiratory-related neuronal activity in the patch electrode recording in **a** shortly before establishing the GΩ seal, recording in **b** showed inspiratory-related inhibition of tonic spiking, indicating that the electrode was close to a "tonic expiratory" neuron in a newborn rat *en bloc* preparation with complete VII nucleus ("[+VII] *en bloc*", see Fig. 1). Within 30 s after the rupture of the seal, the cell showed a very similar discharge pattern and a notable inspiratory-related hyperpolarization. After 5 min of whole-cell recording, this hyperpolarization transformed into a depolarization due to the fact that the patch electrode solution contained, among other components, 144 mM Cl^- ("high Cl^-") which changed the reversal potential for this anion from a value that was originally lower than V_m to a substantially less negative value (details in (59)).

whetstone ("True hard Arkansas whetstone," Fine Science Tools) and fire polished using a commercial lighter and blowing air through the electrode to avoid closure by melting. To prevent tissue damage particularly for recording under visual control using a short distance high-magnification objective (18, 35, 43, 48), electrodes are beveled at an angle (30–45°) to produce a cut tip surface that is parallel to the slice surface. Alternatively, a straight tip electrode is heated and bent down ~45° at its shank to obtain a parallel tip surface and allow maneuvering under the objective. The electrodes are inserted into a patch electrode holder system (ESP-M15-P + MHH-25, Warner Instrument Corp.) (Fig. 3) with an outlet for connecting a suction tube that is attached to a 5-ml standard medical syringe for pressure control via Tygon tubing. Fine-tuning of suction is achieved with a manually tunable syringe (e.g., Gilmont micrometer syringe; www.vwrsp.com) (Fig. 3).

Electrical stimulation, for example, of afferent spinal fiber tracts in the *en bloc* model (37) or the contralateral preBötC in rhythmic slices (Sect. 4.2.4), is feasible with suction electrodes used for extracellular recording. Alternatively, concentric 50–200 μm outer diameter platinum/iridium electrodes can be used such as the PTM3CC02INS or related models from WPI. Under visual control (18, 35, 43, 48) (Sects. 2.8, 3.3, and 4.2.4), single axons or local neural circuits can also be stimulated using a patch electrode filled with standard superfusate after breaking its tip to an outer diameter of 5–10 μm. A broken patch electrode or a triple-barreled microelectrode (3B120F-4 borosilicate glass with 1.2/0.68 mm outer/inner diameter; WPI) can also be used for injection of fluorescent dyes, for example, for Ca^{2+} imaging (Sects. 3.3.2, 3.3.3, and 4.2.4) or pressure application of drugs, respectively (15, 33, 40, 43, 61). These electrodes and double- or triple-barreled ion-sensitive microelectrodes (Sect. 3.2.4) consist of (thin-walled) borosilicate glass capillaries from, for example, Warner Instruments or WPI. The latter company also offers O_2-sensitive (micro)electrodes and compatible amperometers.

Patch electrodes, sharp microelectrodes, and ion-sensitive microelectrodes are fabricated using microelectrode pullers such as the P-1000 (Sutter Instrument Company, Novato, CA, USA; www.sutter.com) or the DMZ (Dagan Corporation, Minneapolis, MN, USA; www.dagan.com). The PC-10 (Narishige International Inc., East Meadow, NY, USA; www.usa.narishige-group.com) is a less expensive alternative but is not designed for pulling sharp (ion-sensitive) microelectrodes.

2.6. Micromanipulators and Antivibration Tables

Patch or sharp microelectrodes are positioned with either an MP-285 (Sutter Instrument Company) or an MX7600 step motor plus digital display of recording depth (Siskiyou Corporation, Grants Pass, OR, USA; www.siskiyou.com) (Fig. 3). For positioning extracellular or stimulation electrodes, less precise and consequently less expensive, yet reliable, micromanipulators can be used such as the Märzhäuser MM33 or the Narishige MMN-1 (both distributed by ALA Scientific Instruments, Farmingdale, NY, USA; www.alascience.com) (Fig. 3). For positioning electrodes on the XY stage of an upright microscope (Sect. 2.8), YOU-2 (Narishige) or Kleindiek Nanotechnik (Reutlingen, Germany; www.nanotechnik.com) manipulators are well suited. Various companies offer manual or motorized XY stages such as Siskiyou Design Inc. or Luigs-Neumann (Ratingen, Germany; www.Luigs-neumann.com).

If the laboratory floor is sensitive to vibration, intracellular recording requires use of an antivibration table, for example, from Newport (Irvine, CA, USA; www.newport.com) or Kinetic Systems (Boston, MA, USA; www.kineticsystems.com) (Fig. 3). These companies also provide Faraday cages that are necessary in a laboratory environment with electrical noise. The Faraday cage is mounted on the antivibration table and thus electrically shields the microscope, experimental chamber, and manipulators positioned on this table. A simple and less expensive (but often sufficient) system uses a stone or solid metal plate resting on rubber or silicone dampers inside a metal (tool) cabinet serving as Faraday cage.

2.7. Amplifiers, Stimulators, and Computer-Based Data Acquisition

For intracellular electrophysiological recording from respiratory neurons, either a patch-clamp amplifier from the "EPC family" (HEKA Lambrecht, Germany, www.heka.com) or a "sample-and-hold" SEC amplifier (npi electronic, Tamm, Germany; www.npielectronic.com) can be used. The latter company also provides high-impedance amplifiers for ion-sensitive microelectrode recording (Sect. 3.2.4). Suction electrode recording of neuronal population activity with amplitudes of 10–500 μV (mostly <100 μV) is done using an AC-coupled differential amplifier which has a gain of at least 1–10 k and is equipped with band-pass-filtering option (optimally 0.3–3 kHz). Examples are the Model-1700 from AM-Systems (Sequim, WA, USA; www.am-systems.com) or the DAM50 from WPI. For better resolution of <100 μV small signals, it is recommended to integrate the output of this amplifier, for example, by using the MA-821/RSP "Moving Averager" (CWE; www.cwe-inc.com) at a time constant of 20–50 ms (Fig. 4). Electrical signals are sampled into a computer via an interface and acquisition software at rates of 1–4 kHz for suction electrode recording and 4–20 kHz for V_m or membrane current recording. Convenient hardware/software packages are available from HEKA, npi, or Axon Instruments (Molecular Devices; Sunnyvale, CA, USA; www.moleculardevices.com). Commonly used multichannel systems for acquiring electrophysiological data are "PowerLab" (ADInstruments, Colorado Springs, CO, USA; www.adinstruments.com) and "Spike2" (Cambridge Electronic Design Ltd., Cambridge, UK; www.ced.co.uk). In most of these systems, software options are implemented for single-pulse or repetitive ("tetanic" or "train") electrical stimulation. For such stimulation, the output of the interface is connected with an electrical isolation unit, for example, from WPI, AM-Systems, or A.M.P.I. (Jerusalem, Israel; www.ampi.co.il), while the output of that unit is connected with the stimulation electrode.

2.8. Microscopes, Imaging Devices, and Fluorescent Dyes

For optic control or mechanic fixation of a preparation and subsequent microelectrode positioning, a stereomicroscope is used such as the Stemi 2000 (Zeiss), SZXZ (Olympus, Center Valley, PA, USA; www.olympusamerica.com), or the ELZ (SCIENSCOPE, Chino, CA, USA; www.scienscope.com). Visualization of slice areas and microelectrode positioning is facilitated by attaching to these microscopes a simple video camera, for example, from WPI. Alternatively, the stereomicroscope is substituted by a video system equipped with zoom optics. The video signal is displayed on a separate monitor or, using a frame grabber, as an inset on the computer monitor used for controlling the (intracellular) recording (Chap. 3-Moore et al. (54)). For recording under visual control from superficial neurons in slices (18, 35, 43, 48), an upright microscope is needed such as the Zeiss Axioskop 2 FS (or successor) and the Olympus BX61WI (Fig. 3). Such microscopes require a water immersion lens (usually 40× or 60× magnification) with a working distance >1 mm to facilitate pipette positioning. For approaching individual cells, the microscopes should be equipped with differential interference contrast (DIC) optics which makes them appear 3D-like (62) (Chaps. 1-Trapp and Ballanyi (45), 4-Fish et al. (63), 5-Nakamura et al. (64) and 9-Broicher and Speckmann (56)).

Cells located up to 80 μm below the slice surface can be visualized for targeted patch-clamp recording by attaching to the microscope a charge-coupled device (CCD) camera with sensitivity into the near-infrared (IR) spectrum and displaying the acquired IR-DIC image on a monitor. Examples for suitable CCD cameras are "ImageM" or "Orca-R2" (Hamamatsu, Bridgewater, NJ, USA; www.hamamatsu.com) or "CoolSNAP" (Röper, Ottobrunn, Germany; www.roperscientific.de). For patch-clamping cells that are tagged with a fluorescent label such as green fluorescent protein, an epifluorescence system is required. Details of such systems based on CCD camera or photomultiplier imaging are described elsewhere (62) (Chaps. 1-Trapp and Ballanyi (45), 3-Moore et al. (54), 4-Fish et al. (63), 5-Nakamura et al. (64) and 9-Broicher and Speckmann (56)).

For our studies on isolated respiratory networks, we used so far two optical imaging approaches. Fast voltage imaging with a temporal resolution in the order of tens of milliseconds was done in collaboration with the Onimaru group (24). In this study, changes in neuronal activity, which were proportional to changes in fluorescence intensity of the voltage-sensitive fluorescent dye

Di-2-ANEPEQ (Invitrogen, Carlsbad, CA, USA; www.invitrogen.com), were monitored with a CCD camera-based imaging system (MiCAM01, Brain Vision Inc., Tsukuba, Japan; www.scimedia.com) attached to an upright BX50WIF-2 microscope (Olympus). The optical path contained a 150-W tungsten halogen lamp, a 510–550 nm excitation filter, a dichroic mirror, and a 590 nm emission filter (U-MWIG2 mirror unit from Olympus). The imaged area was 4.20 × 3.25 mm (resolution 180 × 120 pixels). Differential images were processed with a 3 × 3 pixel software-spatial filter (BV Analyze, BrainVision Inc.).

Our laboratory was the first to adapt confocal and multiphoton ("2-photon") microscopy to image both activity and morphology of mammalian respiratory neurons (15, 33, 40). For this, we loaded these neurons (and adjacent glial cells) with the membrane-permeant acetoxymethyl (AM) form of the fluorescent Ca^{2+}-sensitive dyes Fluo-4 (TEFLabs, Austin, TX, USA; www.teflabs.com or Invitrogen) or Fluo-8L (ATT Bioquest, Sunnyvale, CA, USA; www.attbio.com). For this, Fluo-4-AM is firstly dissolved in dimethyl sulfoxide containing 20% pluronic acid to 5 mM and then diluted with standard superfusate (Sect. 2.2) to 0.5 mM for tissue injection or to 10 μM for loading via incubation (recently, we use mostly Fluo-8L AM for this). For the latter procedures, a slice is transferred to a 1.8-ml CryoTube (Nunc A/S, Roskilde, Denmark; www.nuncbrand.com) with a Tygon® tube inserted through the lid used for carbogenation of standard superfusate to which the dye is added. Direct contact of the Tygon® tube inlet with the solution is avoided to prevent excessive perturbation of the fluid which causes bubble formation due to the pluronic acid content and may also mechanically stress the slice. The CryoTube is inserted in a heating unit (Isotemp-125D, Fisher), and the dye is loaded for 30 min at 30°C. After entering the cells, the AM component of the Ca^{2+}-sensitive dye is cleaved by esterases rendering dye molecules membrane impermeant and activated for monitoring cytosolic Ca^{2+} dynamics (15, 23, 43, 62).

Activity-related increases in Fluo-4 (or Fluo-8L) fluorescence intensity (Sects. 3.3.2 and 4.2.4) are monitored using either a confocal laser scan microscope and software (20× XLUM-PlanFl-NA-0.95 objective; Olympus FV300, FluoView software) (Fig. 3) or a FV300 connected with a Ti:Sa laser (10 W Mira/Verdi, Coherent, Santa Clara, California, USA; www.coherent.com) for multiphoton imaging (15, 18, 23, 33, 40, 43, 61, 62, 65). The latter microscopes and software are also used by us for sulforhodamine 101 imaging that we have adapted to further improve the resolution of morphological features of preBötC neurons, XII motoneurons, and (some) adjacent glial cells (Sects. 3.3.3 and 4.2.4).

2.9. Histological Procedures

For generating brainstem atlases and anatomically "calibrated" *en bloc* preparations and preBötC slices (Figs. 1 and 2), thionin-stained respiratory marker nuclei were mapped in 50–100-μm-thin serial brainstem sections (Sect. 3.1). For this, isolated brainstems were fixed for >1 h, typically overnight until several months, in 4% paraformaldehyde in phosphate buffer. Agar-embedded neonatal brainstems (Sect. 2.2.3) were cut with the Leica vibratome (Sect. 2.3) to produce a series of 50-μm-thin slices which were then mounted on glass slides and thionin stained for analysis of marker structures. Specifically, slices were "washed" in phosphate buffer for 2 min, then immersed for 45–60 s in thionin solution (Sect. 2.2.3), and finally washed in 50% ethanol (4 min) and phosphate buffer (2 min). After experiments, rostrally sectioned *en bloc* preparations, preBötC slices, and also 100-μm-thin pre- and postrhythmic slice sections are fixed for at least 15 min (typically overnight) and stained, using the above-described procedure, but exposed to thionin for 105, 90, and 45 s, respectively. For histological analysis, these sections are transferred on a "hanging drop" glass slide (Fisher Scientific) to an upright microscope (Zeiss Standard 16, magnification ×32) (Sect. 2.8) and photographed (PL-A642-1.3Megapixel, PixeLINK, Ottawa, ON, Canada; www.pixelink.com). Stained preBötC slices or *en bloc* preparations are transferred in phosphate buffer to a Petri dish and photographed (PL-A686-6.6-Megapixel, PixeLINK) under a stereo microscope (Zeiss-SR15, magnification ×32) (Sect. 2.8) (Figs. 1 and 2).

For morphological reconstruction of respiratory neurons that were filled during electrophysiological recording with lucifer yellow, biotin or biocytin via the whole-cell patch electrode (Sect. 2.2.3), two techniques can be used. According to the approach of the Onimaru group (24, 51, 52), transverse 100 μm slices of brainstem–spinal cords fixed in Lillie solution (Sect. 2.2.3) are cut with a (Leica) vibratome (Sect. 2.3) and stained in distilled water with 1% neutral red (Sigma-Aldrich) for visualizing marker regions. Morphology and anatomical location of labeled neurons are reconstructed via a camera lucida attached to a fluorescence microscope (BH-2; Olympus) (Sect. 2.8) and comparing marker structures with our newborn rat brainstem atlas, respectively (24). For histological analysis of whole-cell recorded neurons in (>400 μm thick) preBötC slices (53), slices are immersed in 4% paraformaldehyde (in 0.1-M phosphate buffer, pH 7.4) for 4–15 h at 4°C, followed by immersion in 30% sucrose solution (10 h). Slices are then cut in serial transversal sections (100 μm) using a Frigocut (Leica, Germany) and processed for "ABC" staining with horseradish peroxidase and diaminobenzidine (both Sigma-Aldrich). Sections are mounted on slides, and cells are examined with a light microscope (for biotin/biocytin staining) or a fluorescence microscope (for lucifer yellow staining). For live-cell labeling of XII motoneurons and (primarily) neurons in the preBötC area, slices are incubated with 0.01% (165 μM) sulforhodamine 101 (in the dark) at 20–23°C for 1–1.5 h in carbogenated standard solution and then analyzed using the multiphoton/confocal imaging stations (Sect. 2.8).

2.10. Agents

Most agents for our studies on respiratory networks in the dish are obtained from Sigma-Aldrich except salts for superfusate, Na^+-acetate trihydrate, or acetic acid (Fisher Scientific). Other agents, mostly neurotransmitters and corresponding (ant)agonists, are mostly available at Tocris Bioscience (Ellisville, Missouri, USA; www.tocris.com), MP Biomedicals (www.mpbio.com), TEFLabs Inc. (Austin, Texas, USA; www.teflabs.com), or Alomone Labs Ltd. (Jerusalem, Israel; www.alomone.com). Fluorescence dyes are purchased from Invitrogen, ATT Bioquest and TEFLabs Inc.

3. Methods

In this section, techniques are described for generating anatomically defined *en bloc* and slice preparations in which respiratory activity is preserved for up to >1 day, while some basic procedures for isolating newborn rodent brainstems or their histological analysis have been described above (Sect. 2.2). Reference is also made here to microelectrode-based extracellular recording of neuronal population plus intracellular activities or dynamic changes of extracellular ions, PO_2 or PCO_2. Finally, morphological, voltage, and Ca^{2+} imaging at the (sub)cellular level plus spatiotemporal optical activity patterns in respiratory neuron populations are dealt with.

3.1. Generation of Calibrated Respiratory Networks in the Dish

Methods for generating transversal preBötC slices vary notably between laboratories, for example, regarding brainstem sectioning in rostral-to-caudal versus caudal-to-rostral direction or their rostrocaudal thickness ranging from 200 μm to >1,000 μm (18, 34). Moreover, previous studies did not provide histological proof of slice margins, and minimal margins sufficient and necessary for rhythm in these slices were also not determined until we addressed these issues (40) (Sect. 4.2.1) (Figs. 1 and 2). In the following, we (1) describe procedures for obtaining brainstem atlases for newborn rats and mice (Sect. 3.1.1), (2) outline how these atlases are used for identifying "respiratory marker" structures for "online histology" during live brainstem sectioning for generating preBötC slices with defined and reproducible anatomical boundaries (Sect. 3.1.1), and (3) describe how "calibrated" *en bloc* preparations are generated based on procedures outlined under (1) and (2) in concert with our analysis of ventral brainstem surface landmarks (Sect. 3.1.3). Experiments were performed in accordance with guidelines by the Canadian Council on Animal Care and were approved by the Health Sciences Animal Policy and Welfare Committee of University of Alberta.

3.1.1. Newborn Rat and Mouse Brainstem Atlases

For generating calibrated preBötC slices from commonly used P0-4 Wistar and Sprague-Dawley rats (33) and P0-7 C57BL/6 mice (41), we firstly showed that the rostrocaudal extents of anatomical marker structures associated with the yet solely functionally defined preBötC do not change during P0-4 for rats, whereas growth was evident between P0 and P6 mice. The brainstem marker structures include the VII motor nucleus, the lateral reticular nucleus, and in particular the "loops" (in coronal sections) of different subnuclei of the inferior olive, which are most pronounced in the rostrocaudal axis at the center of the preBötC (Figs. 1 and 2) (see also *Supplemental Web Material* including brainstem atlases in (33, 34, 41)). Less valuable information is obtained from dorsal marker regions such as the XII nucleus or the obex (i.e., the opening of the fourth ventricle) because their association with ventral structures depends on the cutting angle. However, this particularity can be used for quantifying such a "tilt" by comparing ventral and dorsal marker areas. Similarly, discrepancies between ventral marker structures on each brainstem side provide information on the extent of a possible lateral tilt. Motoneurons of the *nucleus ambiguus* do not aggregate to a detectable marker in live thin transversal brainstem sections, at least in our hands, and are therefore not a useful marker structure for our online histology approach (Sect. 3.1.2).

We analyzed the rostrocaudal extents of marker structures in 50-μm-thin thionin-stained transversal serial sections from agar-embedded paraformaldehyde-fixed brainstems (Sect. 2.9). The rostrocaudal position of each section was defined by its distance (in mm) from the caudal end of VII nucleus, introduced (Sect. 1) as "VII_c," with regions caudal to VII_c assigned negative values (Figs. 1 and 2). Also, serially sectioned 100-μm-thin slices from live brainstems kept in standard superfusate were stained with thionin and photographed. This analysis revealed that the rostrocaudal extents of marker nuclei (Fig. 2) and also the distance between the rostral margin of VII nucleus and the caudal end of the medial inferior olivary nucleus were identical. This was regardless of whether comparisons were made between rat strains (Wistar vs. Sprague-Dawley), age groups (P0 vs. P2 and P4), or live versus fixed sections. The constancy of these marker nuclei, both in absolute dimension and position relative to each other, allowed construction from a single representative 2 Wistar rat brainstem of a "reference atlas" comprised of 50-μm-thin serial transections. This atlas is applicable to P0-P4 Wistar and Sprague-Dawley rats (see *Supplemental Web Material* in (33)). Anatomical relations of the marker nuclei with respiratory brainstem regions, specifically the VRC including the pFRG, the preBötC, and the area corresponding to the Bötzinger complex (BötC) in adults, and some atlas sections are shown in Figs. 1 and 2. We reported (41) that, compared to P0-4 rats, the rostrocaudal marker structure extents are smaller (in particular at

P0) in inbred C57BL/6 mice that are often used for genetic engineering (16–19). For C57BL/6 mice at P6, marker structures also apparently differ with those in another developing mouse brain atlas based on 120-μm sections from a single brainstem of an unspecified mouse strain (66). The 50 μm resolution in our mouse atlas (41) provides details on brainstem marker structures that are necessary for generating calibrated slices.

Three conclusions can be drawn from our findings: (1) Neither our rat atlas nor the mouse atlas by Paxinos et al. (66) can be used for generating calibrated preBötC slices from C57BL/6 mice. Consequently, research groups that routinely use a mouse strain other than C57BL/6 (or a rat strain other than Wistar or Sprague-Dawley) should generate a reference atlas for that strain. (2) Our histological findings suggest that rhythmic slices of identical thickness from newborn rats versus mice contain a different portion of the VRC and other brainstem structures that may differentially modulate preBötC activity. Specifically, we found that the distance between the rostral margin of VII nucleus and the caudal margin of the inferior olive is ~1.5-fold larger in P0-4 rats versus P0 mice. This suggests that, at a thickness of 500–800 μm, P0-2 (C57BL/6) mouse slices used by numerous groups (2, 10, 13, 14, 18, 31, 32, 41) contain similar brainstem structures as 750–1,200-μm-thick newborn rat slices. (3) A >1,000-μm-thick newborn rat slice and a >650-μm-thick P0 mouse slice would likely include the entire preBötC plus a major portion of the (caudal aspect of the) pFRG, if the preBötC would be located close to their caudal margin (see Fig. 1 in (18)).

3.1.2. Calibrated preBötC Slices

Calibrated preBötC slices are generated according to procedures described in our original report (33). For this, animals are anesthetized with 100–200 μl isoflurane added to a small beaker with lid (Sect. 2.3) until the paw withdrawal reflex disappears. They are then decapitated at a level close to the shoulder, which ultimately provides a spinal cord length up to the first thoracic segments. Next, the brain plus rostral segments of the cervical spinal cord are isolated at 18–20°C in standard superfusate (Sect. 2.2). After removal of the cerebellum, the brainstem is transected between the fifth and sixth cranial nerves and just rostral to the C_1 spinal cord segment. For obtaining complete 200 μm or thinner slices as described below, it is pivotal to remove ("unzip") at that stage with a fine forceps (Sect. 2.3) the arachnoid layer, which is the component of the meningeal membranes that loosely ensheathe the brainstem. The brainstem block is then fixed with a thin layer of commercial grade cyanoacrylate glue rostral side down to a metal plate that is subsequently transferred to a Leica vibratome (Sect. 2.3). The plate is fixed in the slicing chamber that is afterward filled with standard superfusate (18–20°C) that is continuously carbogenated via a small gas dispersion tube (Sect. 2.2).

The brainstem block is sectioned, initially in 200–400 µm steps, in caudal-to-rostral direction starting at the caudal end of the inferior olive (Fig. 2). Each section is inspected using a microscope (Sect. 2.8) for comparing marker structures in the living tissue with those in the reference brainstem atlas (Sect. 3.1.1). Varying the focus plane and/or the condenser aperture reveals the medullary marker nuclei, particularly diverse subnuclei of the inferior olive (Figs. 1 and 2). Using such online histology, section thickness is reduced to 100 µm when approaching the desired level for generating a rhythmic slice. At that stage, it is particularly important to adjust the Leica vibratome (Sect. 2.3) to cut close to both the slowest speed of forward blade movement (0.05 mm/s, which takes >1 min per single slice section) and the highest frequency of lateral blade movement (~100 Hz). This is to avoid partial disintegration of the slice and to minimize damage of cells in superficial layers of the rhythmic slice. After cutting the preBötC-containing slice, 100–200-µm-thin "postrhythmic" sections are cut and photographed for determining the rostral border of the rhythmic slice (Figs. 1 and 2). For more detailed "off-line" analysis of slice margins, pre- and postrhythmic sections are fixed and stained with thionin (Sects. 2.9 and 3.1.1) (Figs. 1 and 2). For estimating the boundaries of the prospective rhythmic slice, it needs to be considered that the slicing procedure disintegrates a tissue layer of several tens of micrometer thickness. The predicted margins of calibrated preBötC slices vary by about ±50 µm due to the above-mentioned reasons and related to the fact that the extent and position of marker structures vary slightly between individual brainstems.

In our initial study (33), we generated calibrated slices that would likely contain in their middle the preBötC whose rostrocaudal thickness has originally been estimated to be ~200 µm, with its center located ~0.5 mm caudal to VII_c (7). According to terminology introduced in our study (33), a 400-µm-thick slice from a P4 Wistar rat with the preBötC in the middle and the caudal boundary located 0.8 mm caudal to VII_c is labeled "m-preBötC[400/−0.80]W-P4 slice" (Fig. 2). Corresponding m-preBötC slices from C57BL/6 mice were generated in a further study from our group (41). We also used (175–700 µm thick) newborn rat m-preBötC slices in one section of our study on identification of necessary and sufficient margins of preBötC slices for inspiratory rhythm (40) (Sect. 4.2.1) (Fig. 2). For the latter analysis, we developed two additional types of 700-µm-thick slices, in which the preBötC was located close to one slice boundary or was even partially exposed to the superfusate. In addition to the preBötC, these slices thus contained a notable amount of either rostral or caudal brainstem tissue. They were consequently labeled "r+preBötC[700]" and "c+preBötC[700]" slices, respectively, with the additional attributes of distance from VII_c of caudal or rostral slice surface, respectively, plus rat strain and age (see above) (Fig. 2) (Sect. 4.2.1).

3.1.3. Calibrated Brainstem–Spinal Cords

Different types of perinatal rodent brainstem–spinal cord preparations are available for studying the preBötC, the pFRG, their mutual interactions and/or properties of cranial versus spinal motor networks that are driven by either constituent of the dual respiratory center (Sects. 1 and 4.2.1) (Fig. 1). Our group uses primarily newborn Wistar or Sprague-Dawley rats for this research. For generating the *en bloc* models, animals are anesthetized with isoflurane until disappearance of the paw withdrawal reflex (Sects. 2.3 and 3.1.2). Following decerebration, the neuraxis is isolated at 18–20°C in standard superfusate, which contains 30 mM glucose instead of 20 mM typically used by us for calibrated slices (Sects. 2.2, 3.1.2, and 4.2.2). As described for generating rhythmic slices (Sect. 3.1.2), it is important to remove with fine forceps the arachnoid layer from the brainstem and spinal cord. If not removed, the arachnoid layer may act as a diffusion barrier and would also hamper microelectrode insertion into the ventral brainstem or spinal cord aspects.

We use three types of *en bloc* models. The first type, in which the preBötC interacts with the pFRG, is anatomically closely related to the preparation originally described by Suzue (4) (see also (5, 6)) (Fig. 1). For its generation, the isolated brainstem is transected rostrally, just rostral to the caudal cerebellar artery, and thus contains the complete VII motor nucleus ("*en bloc* [+VII] model"). In most studies, the spinal cord is partially exposed via laminectomy and then cut at the last cervical (C_8) or first thoracic (T_1) level. If the lumbar motor output of the pFRG is to be studied, the complete spinal cord is isolated (21, 22, 28). For investigating, for example, whether stimulatory neuroactive agents, such as caffeine, serotonin, or the ionotropic glutamate receptor agonist N-methyl-D-aspartate, induce hyperexcitability primarily in respiratory or rather nonrespiratory motoneurons, "classical" Suzue-type *en bloc* [+VII] preparations are isolated with the cervical and brachial plexuses attached for recording from the fourth cervical root (C_4), the contralateral phrenic, and the musculocutaneous (or brachial) nerves (Figs. 1 and 5). Inspiratory active cervical roots (C_1–C_8) contain axons from both inspiratory and nonrespiratory motoneurons, whereas the phrenic nerve contains solely axons from inspiratory (phrenic) motoneurons located in spinal segments C_3 to C_6, while the musculocutaneous nerve contains only nonrespiratory axons (5, 67, 68). In this model, it is possible to evoke fictive locomotion by combined bath application of serotonin and N-methyl-D-aspartate (for references, see (28)) (see also Chap. 15-Mandadi et al. (69)). Figure 5 shows that these agents induce, following an initial period of tonic discharge, (bilaterally alternating) fictive locomotor activity in several ventral nerve roots, whereas phrenic nerve activity is not much perturbed.

Finally, we use calibrated *en bloc* preparations (Fig. 1) in which manual brainstem transection with a razor blade aims to expose the

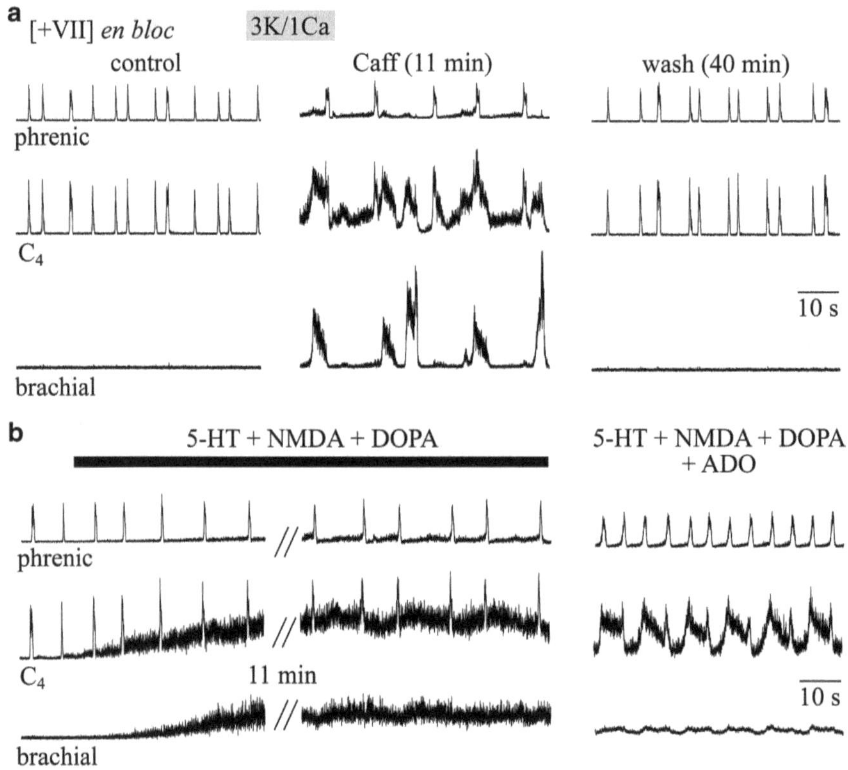

Fig. 5. Differences in fictive inspiratory and locomotor bursting between cervical nerve roots and phrenic plus brachial nerves in the *en bloc* model. (**a**) Suction electrode recording from phrenic nerve in the Suzue-type brainstem–spinal cord preparation with complete VII nucleus ("[+VII] *en bloc*") revealed inspiratory-related bursting very similar to that in the most commonly recorded C_4 root, whereas no activity occurred in brachial nerve not containing axons from respiratory motoneurons. Both, brachial nerve and C_4 root showed reversible (synchronous) massive bursting 11 min after start of application of caffeine (10 mM), whereas inspiratory phrenic activity was only slightly perturbed. (**b**) After recovery from caffeine, the same preparation was exposed to a "neuromodulator cocktail" comprised of serotonin (*5-HT*, 10 μM), N-methyl-D-aspartate (*NMDA*, 5 μM), and dopamine (*DOPA*, 100 μM). This induced massive tonic activity in C_4 root and brachial nerve, whereas phrenic inspiratory rhythm was barely affected even when slow rhythmic bursting occurred in particular in brachial nerve as indication of chemically evoked fictive locomotion. After several applications of different concentrations of neuromodulators (not shown), a very regular fictive locomotor pattern was revealed in the presence of 10 μM *5-HT*, 5 μM *NMDA*, 500 μM *DOPA*, and 1 mM adenosine (*ADO*). These findings show firstly that C_4 recording of inspiratory activity can be contaminated by nonrespiratory bursting that is much less pronounced in phrenic nerve. Secondly, this demonstrated that this preparation is principally suitable of analyzing cervical respiratory and locomotor networks simultaneously.

preBötC more or less to the cut surface. For newborn rats, we found that the extension of the preBötC between −0.4 and −0.6 mm caudal to VII_c is identical with that area on the ventral brainstem surface, from which the most rostral XII nerve root originates (34) (Fig. 1). In this transected *en bloc* model, the pFRG does not seem to be active as indicated by absence of pre-/postinspiratory activity in the two most rostral lumbar nerve roots containing axons from motoneurons that drive abdominal expiratory muscles (8, 22, 28, 34). For these reasons, transection should be done slightly rostral to the most rostral XII root, if an *en bloc* [-VII] preparation is to be studied in which the preBötC is not subject to a major influence from the pFRG (Sect. 4.2.1). The spinal cord is either transected at the midcervical level or completely dissected (which takes ~15 min longer). As with the other models, suction electrodes can be used for simultaneous recording of inspiratory-related rhythms from ventral spinal nerve roots. Normally, the lumbar roots in these transected *en bloc* [-VII] preparations are not active or show small-amplitude inspiratory activity (28, 34). Moreover, suction electrodes can be positioned in the ventrolateral aspect of the cut medullary surface containing the VRC with the more- or less-exposed preBötC, and also in the XII motor nucleus, which is though rather small at this rostral level (Fig. 1). Figure 1 also shows that thionin staining of the (chemically fixed) transected *en bloc* model after the experiments enables determination of its rostral boundary by comparison with the reference brainstem atlas. For some studies, it may be intended to use *en bloc* preparations in which a defined amount of tissue is preserved rostral to the preBötC, for example, the area corresponding to the BötC in adults (1–3) (Sect. 1). We determined for newborn rats that the vagal nerve is located on average 0.06 mm caudal to VII_c (34). This is the area in which the caudal aspect of the pFRG may overlap with the BötC (Fig. 1).

The location of the caudal cerebellar artery matches well with that of the rostral border of VII nucleus, that is, 0.74 mm versus 0.76 mm rostral to VII_c, respectively. The latter surface landmark may delineate an important part of the rostral aspect of the pFRG that may also be relevant for respiratory control, at least in newborns (28, 29, 34) (Sect. 1). For preparations transected between VII_c and the rostral end of the inferior olive, there are no marker nuclei to help determine the rostral boundary. In these cases, thionin-stained sagittal sections are generated from the transected and fixed rostral brainstem aspect for identifying the distance of VII_c from the caudal boundary of the rostral brainstem block. Based on this, the rostral boundary of the transected *en bloc* model can be calculated. For further information on anatomical properties of the calibrated *en bloc* model, see the original report (34).

3.1.4. Influence of Temperature and Preparation Time on Respiratory Rhythms

For slices studied in 3K/1Ca, and *en bloc* preparations in general, the closest possible range compared to in vivo of the temperature in the recording chamber is 25–27°C (Sect. 2.4). At >30°C, inspiratory-related bursting ceases irreversibly within <1 h in the *en bloc* [+VII] preparation as indicated by our unpublished results, while rhythm may also not be stable at temperatures <22°C. Similarly, we found that rhythm of calibrated preBötC slices in 3K/1Ca is "washed out" spontaneously within <0.5 h upon increasing the temperature in the recording chamber from 27°C to 37°C. However, as tested by us in m-preBötC slices of <700 μm thickness, bursting is reactivated by increasing K^+ to 7 mM and remains stable for >5 h. Similarly, mouse preBötC slices can be studied in 8K/1.5Ca at physiological temperature and even during "in vitro fever" (32). We use standard superfusate at 18–20°C for brainstem-spinal cord dissection and slice generation. We did not find any obvious advantage of using ice-cold standard solution or modified solution that may improve the viability of slices or *en bloc* models from other brain regions (see also other chapters of this book and Sects. 2.2 and 5).

When a temperature of 25–27°C is maintained in the recording chamber, cervical inspiratory-related rhythms in the *en bloc* model in 3K/1Ca solution last often >10 h, regardless of whether preparation time was 15 min or up to 2 h. Contrary, pFRG bursting, pFRG-driven cranial plus spinal (motor) activity, and cranial preBötC-driven motor activities may depend substantially on preparation time (Sect. 5.2). The average time for generating a calibrated slice by an experienced experimenter ranges between 35 and 45 min, while noncalibrated slices can be obtained within 15–25 min, mainly depending on how fast slicing is done (Sect. 2.3). Preparation time is not critical for the capability of the slices to generate inspiratory rhythm in high K^+, while the "longevity" of rhythms in 3K/1Ca decreases notably when it takes >1 h to obtain a slice. Importantly, no recovery time from the isolation/cutting procedures is necessary for either the *en bloc* or slice models.

Within <10 min after transferring a preparation to the recording chamber and start of the superfusion pump, respiratory activities are present in either 3K/1Ca or high-K^+ solution while their amplitudes may still develop during the first 30 min of recording. Only when <300-μm-thin slices are used or slices are generated in solution with <10 mM glucose, it may take up to 30 min until robust rhythm develops (in 3K/1Ca).

3.2. Electrophysiological Recording and Stimulation

Suction electrode recording is the most widely used electrophysiological approach for studying respiratory networks in the dish (Sect. 2.5) (Figs. 2, 4, 5, 6, 7, 8, 9, 10, and 11). Importantly, suction electrodes can be applied not only to proximal cut ends of ventral nerve roots but also to other inspiratory active structures such as the VRC region and XII motor nucleus in preBötC slices. Besides, they can be used for electrical stimulation (Sects. 2.5 and 4.2.4). Our group pioneered two decades ago the use of ion- and gas-sensitive microelectrodes for in vitro analyses of the ionic microenvironment of respiratory structures (37). Moreover, we were the first to adapt 10 years ago "blind" patch-clamp techniques for analyzing biophysical membrane properties of respiratory neurons (7, 9). Currently, whole-cell recording from respiratory neurons using either the blind approach for cells located deeper within brainstem tissues or visually targeted recording from (fluorescent protein-expressing) cells in superficial tissue layers is to an increasing

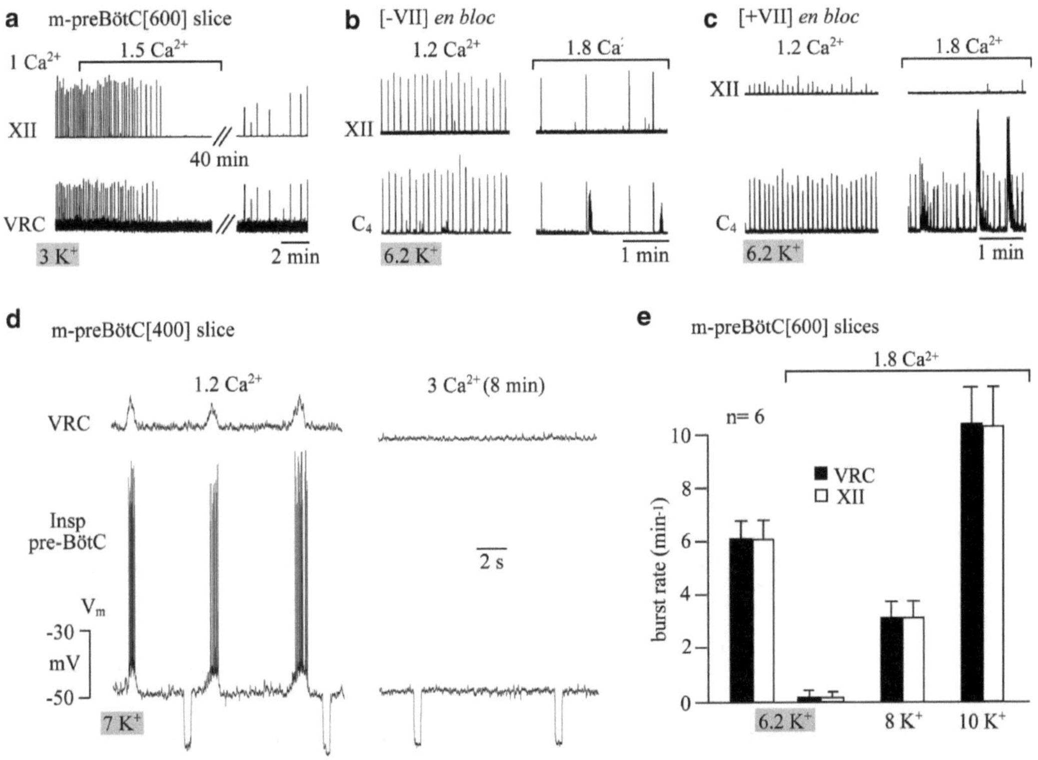

Fig. 6. Ca^{2+}/K^{+} antagonism of isolated preBötC bursting. (**a**) In an m-preBötC[600/−0.85]W-P1 slice, raising superfusate Ca^{2+} in 3 mM K^{+} from 1 to 1.5 mM abolished inspiratory bursting in both the VRC/preBötC area and *XII* nerve, and this effect reversed only partially even after >40 min following return to 1 mM Ca^{2+}. (**b**) In a brainstem–spinal cord preparation without VII nucleus transected close to the rostral preBötC boundary ("[−VII] *en bloc*"), raising Ca^{2+} from 1.2 to 1.8 mM in 6.2 mM K^{+} greatly depressed inspiratory-related C_4 and *XII* nerve bursting. (**c**) In a brainstem–spinal cord preparation with complete VII nucleus ("[+VII] *en bloc*"), raising Ca^{2+} from 1.2 to 1.8 mM in 6.2 mM K^{+} did not block C_4 rhythm but evoked nonrespiratory cervical discharge (compare Fig. 5). It may appear that high Ca^{2+} blocked *XII* bursting, but depression of this rhythm, which is typical for that model, started already before changing to high-Ca^{2+} solution. (**d**) In an m-preBötC[400/−0.70]W-P2 slice, raising Ca^{2+} from 1.2 to 3 mM in 7 mM K^{+} abolished both inspiratory-related VRC bursting and concomitant bursting in a preBötC inspiratory neuron. This blockade occurred without postsynaptic effect on V_m or membrane resistance measured by repetitive injection of negative DC current pulses causing a hyperpolarization of V_m. This indicates that the inhibitory Ca^{2+} effect on the isolated preBötC is primarily of presynaptic origin. (**e**) In 6 m-preBötC[600/−0.86] slices, strong depression of inspiratory-related *VRC* and *XII* bursting due to raising Ca^{2+} from 1.2 to 1.8 mM was countered by raising K^{+} from 6.2 to 8 and 10 mM (For details, see (18, 34, 42)) (**a–c, e** with kind permission from (34), **d** with kind permission from (42)).

extent combined with genetic or acute fluorescence labeling of respiratory neuron populations for identifying their genetic origin and for Ca^{2+} or voltage imaging (Sects. 3.3 and 4.2.4).

3.2.1. Suction Electrode Recording

Extracellular AC-coupled signals obtained with a suction electrode connected to a differential amplifier are typically in the low millivolt range, often as small as 10–100 μV (Sect. 2.7) (Fig. 4). These signals are thus very sensitive to electrical noise, which may require the use of a Faraday cage (Sect. 2.6), while it is recommended to record and display them as a moving average "integrated" trace for

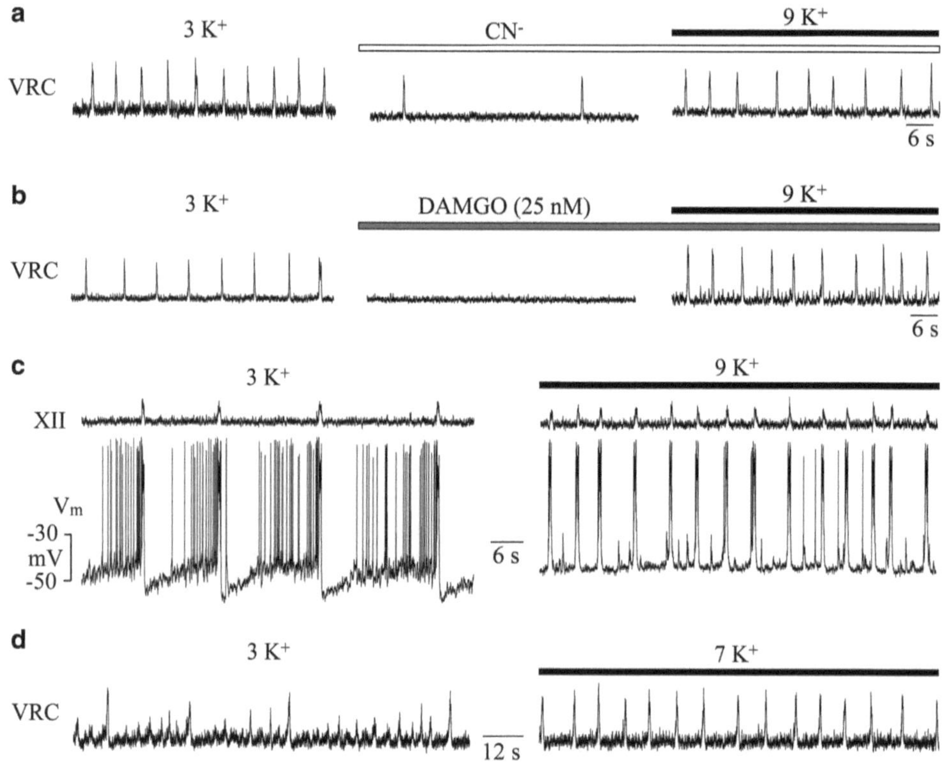

Fig. 7. Modulatory effects of superfusate K^+ on in vitro inspiratory activities. (**a**) In an m-preBötC[600/–0.85]W-P3 slice, substantial slowing of *VRC* bursting due to chemical anoxia evoked by bath application of cyanide (CN^-, 1 mM) was countered by raising superfusate K^+ from 3 to 9 mM. (**b**) In a different m-preBötC[600/–0.83]W-P3 slice, blockade of *VRC* rhythm by the μ-opioid receptor agonist [D-Ala2, N-MePhe4, Gly-ol]-enkephalin (*DAMGO*) was countered by increasing K^+ from 3 to 9 mM. (**c**) Raising K^+ from 3 to 9 mM in an m-preBötC[500/–0.80]W-P2 slice not only greatly accelerated the rate of rhythm but also changed the V_m trajectory in an inspiratory preBötC neuron notably. (**d**) Increasing K^+ from 3 to 7 mM reconfigured *VRC* bursting in a preBötC[500/–0.75]SD-P2 slice from a mixed pattern of fictive eupnea and fictive sigh bursts to a regular burst pattern.

a better signal-to-noise ratio (Sect. 2.7) (Fig. 4). Suction electrodes of apparently identical shape can differ notably in the capability of recording particularly small signals. Prior to starting a series of experiments, it is thus recommended to prepare a set of about ten such electrodes and record rhythmic bursting in a test *en bloc* preparation or preBötC slice for choosing the most responsive electrodes. These electrodes can be used for up to several months provided they are cleaned with distilled water after the experiment. For suction electrode recording from a nerve root, the initial (modest) suction may be relieved by disconnecting the syringe from the suction tube (Sect. 2.5) while the electrode is slightly pushed against the brainstem or spinal base of the nerve root. For recording from the cut brainstem surface, the electrode should only make contact with the superficial tissue layer, particularly for recording from the VRC/preBötC in <300-μm-thin slices when the inspiratory center is partly exposed to the recorded surface. Though,

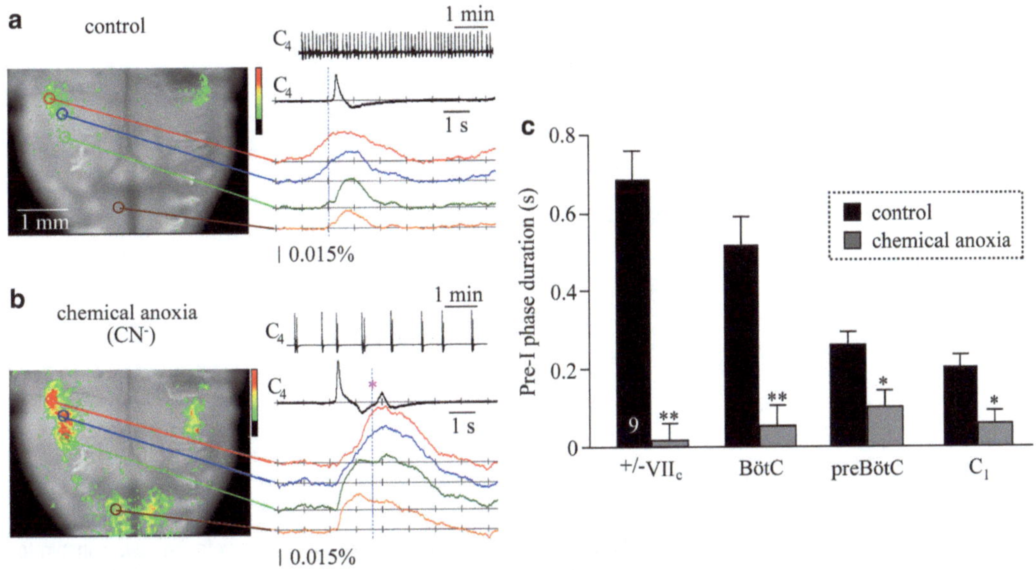

Fig. 8. Shortening by anoxia and prolongation by opioids of delay between onset of preinspiratory (pFRG) and inspiratory (preBötC) network bursting in newborn rat *en bloc* [+VII] medullas. (**a, b**) Simultaneous voltage imaging and electrophysiological cervical inspiratory burst recording during control (**a**) and anoxia (**b**). Averaged (n = 40) imaging pixels in the *left* image summarize respiratory activity immediately at the onset (indicated by *dotted line* in *right panel*) of averaged inspiratory-related integrated cervical root (C_4) burst that was plotted synchronous with kinetics of fluorescence changes in four regions of interest (ROIs). These ROIs were located within [in rostral to caudal direction] the area around VII_c, the *BötC*, the *preBötC*, and the C_1 spinal motoneuron area (for exact location of these areas, see Figs. 1 and 2). Note that preinspiratory activity is particularly evident in the VII_c and *BötC* areas. (**b**) Chemical anoxia due to bath application of CN^- (1 mM) slowed rhythm and evoked double bursts (compare integrated original C_4 recordings in **a**, **b**). Anoxia abolished the preinspiratory phase of optical pFRG activity as shown in the bar graphs of **c** based on measurements in nine preparations. Moreover, anoxia elicited in all regions a prominent second peak of activity during the postinspiratory phase after the augmented initial C_4 peak. Signals were sampled and averaged (n = 20) only upon occurrence of C_4 double bursts during the late phase of chemical anoxia. The peak was attenuated here by averaging due to a variation in the time of its onset ("jitter") (With kind permission from (24)).

recording from the cut surface of the VRC/preBötC area or XII motor nucleus can benefit from gentle suction using a micrometer syringe (Sect. 2.5), which often improves the signal-to-noise ratio greatly. In case of a "leaky" recording with steadily decreasing signal amplitude, it may be necessary to reestablish modest suction every 10–20 min. When recording from calibrated preBötC slices in 3K/1Ca, the amplitude of inspiratory-related XII nerve or nucleus signals may be small and disturbed by random nonrespiratory discharge contrary to pronounced inspiratory activities in elevated K^+. When recording from cranial nerves, such as the XII, VII, or trigeminal nerves in the *en bloc* [+VII] model (Sect. 3.1.3), signal amplitude may progressively and irreversibly decrease for unknown reasons as indicated by our unpublished results on VII and trigeminal bursting and our published findings on unstable XII rhythm (34) (Sects. 4.2.1 and 5). In particular, in the ventrolateral slice aspect, suction electrodes record tonic neuronal population

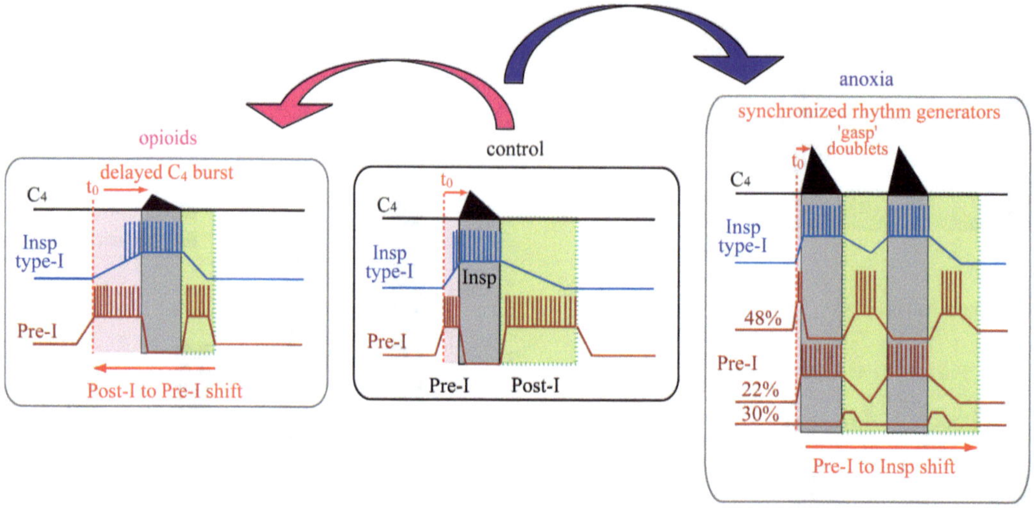

Fig. 9. Opioid versus anoxia effects on onset of pFRG and preBötC bursting. As shown in the box in the *center*, inspiratory motor output is recorded, for example, from C_4 roots. pFRG ("Pre-I") neurons provide preinspiratory drive to type I Insp neurons that may play an important role in triggering the onset of preBötC bursting. *Left box* shows that opioids prolong preinspiratory bursting in Pre-I neurons and concomitantly shorten their postinspiratory bursting compared to control (compare with *center box*). Assuming that pFRG activity starts the respiratory cycle, this effect may be due to delayed generation of inspiratory motor output. As shown in the *right box*, anoxia almost synchronizes the dual respiratory rhythm generator. Accordingly, the preinspiratory drive potential is shortened in type I Insp neurons, whereas Pre-I neurons show three different responses to sustained (10–20 min) anoxia. In 48% of cells, the bursting pattern shifts toward the postinspiratory phase with concomitant depression of the duration of preinspiratory activity and the strength of postinspiratory bursting. In 22% of Pre-I neurons, the activity pattern changes to inspiratory bursting, whereas the remaining 30% are "functionally inactivated" because of greatly depressed drive potentials (and concomitant hyperpolarizations). Not shown in this schema are hypothetic medullary neurons, which develop pronounced postinspiratory activity during anoxia as indicated by the voltage imaging data. The overall enhanced inspiratory/postinspiratory medullary activity during anoxia may be responsible for triggering inspiratory motor bursts, doublets or triplets. More details in (24) (With kind permission from (24)).

discharge in addition to phasic inspiratory-related bursting in the ventrolateral aspect. This becomes most evident in the integrated VRC/preBötC trace (Sect. 2.7) when inhibition of such activity, for example, with opioids, high superfusate Ca^{2+}, or blockers of voltage-activated Na^{+} or Ca^{2+} channels, occasionally decreases the baseline of this signal profoundly (18, 42) (Figs. 6 and 7).

3.2.2. Patch-Clamp Recording

Single-channel recording is the original domain of patch-clamp technology (49, 50) that is briefly dealt with elsewhere in this book (Chap. 1-Trapp and Ballanyi (45)). This approach is feasible in the perinatal rodent respiratory networks in the dish as documented by various papers, in particular from the groups of Mironov and Richter (31). Our group has so far only used whole-cell or "perforated" patch-clamp techniques for analyzing electrical properties of respiratory interneurons and motoneurons (7, 9, 42, 46, 51–53, 59, 60, 70–72) (Figs. 4, 6, and 7). Both types of recording modes can be stable for up to several hours (mostly 30–60 min), independent on whether cells are studied in deep tissue layers with blind patch-clamp or have been approached under visual control

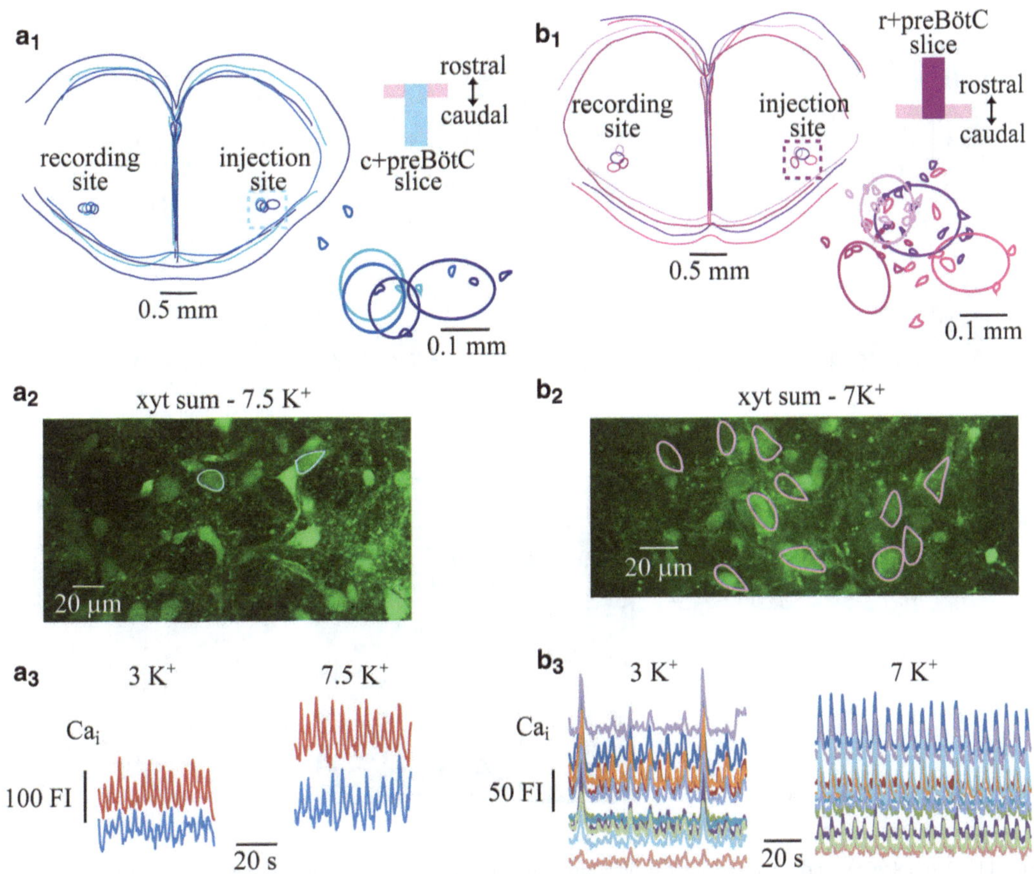

Fig. 10. Multiphoton/confocal Ca^{2+} imaging in the isolated newborn rat pre-Bötzinger complex (preBötC). (**a**) Location, morphology, and activity of active inspiratory neurons in superficial layers (30–80 µm) of slices with the preBötC exposed to the rostral border ("c + preBötC slices"). (**a**$_1$) Locations in four superimposed slices of inspiratory neurons loaded with membrane-permeant Ca^{2+}-sensitive dye (Fluo-4-AM) via pressure injection (25–50 mmHg, 10 min) into the preBötC correspond with positions in contralateral preBötC of suction electrodes for recording inspiratory rhythm with maximal amplitude. Overlapping dye injection sites outlined by *dotted square* are shown in the *lower right inset* at higher magnification. Presumptive inspiratory active neurons are marked by ROIs with equal shading indicating the same injection site (one per slice). (**a**$_2$) Examples for cellular morphology in one injection spot of **a**$_1$ with inspiratory neurons indicated by ROIs. Cells were located 0.43 mm caudal to VII_c, that is, 70 µm rostral to the preBötC center located 0.50 mm caudal to VII_c. (**a**$_3$) Inspiratory-related rises of cytosolic Ca^{2+} (Ca_i) shown by plotting Fluo-4-AM fluorescence intensity (FI) traces over time in the two neurons indicated by ROIs of **a**$_2$ were synchronous with electrophysiologically recorded population bursting in contralateral preBötC (not shown). Raising superfusate K^+ from 3 to 7.5 mM produced more robust oscillations with a concomitant increase in Ca_i baseline (see Movies 1 and 2). (**b**) Location (**b**$_1$), morphology (**b**$_2$), and activity (**b**$_3$) of inspiratory neurons in superficial layers (30–80 µm) of slices with the preBötC exposed to the caudal border ("r+preBötC slices"). (**b**$_1$) Overlapping dye injection sites outlined by *dotted square* are shown in *lower right inset* at higher magnification. (**b**$_2$) Examples for cellular morphology in one injection spot of **b**$_1$ with 11 inspiratory neurons indicated by ROIs. Cells were within the preBötC center at 58 µm caudal to VII_c. (**b**$_3$) Raising K^+ from 3 to 7 mM transformed inspiratory Ca_i rises from a eupnea sigh to regular eupnea pattern in the 11 neurons outlined in **b**$_2$ (see Movies 3 and 4) (With kind permission from (94)).

(18, 35, 43, 48), often using IR-DIC or fluorescence optical approaches (Sects. 2.8 and 3.3).

Whole-Cell Recording: This widely used recording mode has several advantages over sharp microelectrode recording (49, 50) (see also Chaps. 1-Trapp and Ballanyi (45), 3-Moore et al. (54),

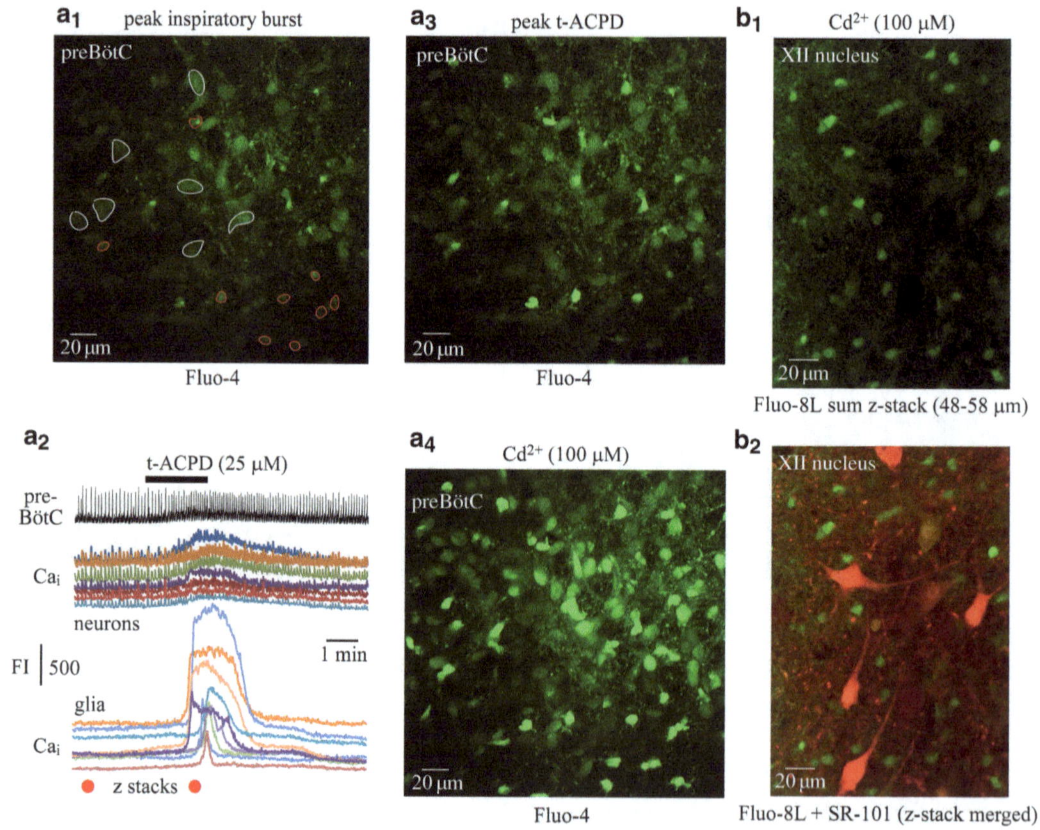

Fig. 11. Morphological and pharmacological discrimination of respiratory neurons and neighboring glia in calibrated preBötC slices. (**a**) In a r+preBötC[400/–0.65]W-P2 slice kept in 7 mM K^+, the presumptive inspiratory neurons, loaded via injection with *Fluo-4*, indicated by white ROIs (**a**$_1$) showed rhythmic Ca_i increases (image taken at the peak of a single inspiratory burst at the time indicated by the *left red dot* in **a**$_2$) that were synchronous with inspiratory population activity in contralateral preBötC (**a**$_2$). Notably smaller cells, presumably representing astrocytes, were sometimes not visible in standard solution, likely indicating a lower Ca_i baseline, and did not show inspiratory-related Ca_i rises. However, such cells (as exemplified by red ROIs in **a**$_1$) responded with a notable Ca_i increase to bath application of the metabotropic glutamate receptor agonist trans-1-aminocyclopentane-1,3-dicarboxylic acid (*t-ACPD*) which accelerated rhythm without a major accompanying Ca_i rise in inspiratory preBötC neurons (**a**$_2$). (**a**$_3$) shows an image at the peak of the response to *t-ACPD* (*right red dot* in **a**$_2$) which reveals that more small (astrocytic) cells are active at this time compared to during the inspiratory peak in control solution in **a**$_1$. (**a**$_4$) Even more presumptive astrocytes could be visualized during the initial 7–10 min of bath application of Cd^{2+} (100 μM) after recovery from *t-ACPD*. Several minutes later also neurons would show an increase in Fluo-4 fluorescence intensity that is due to interaction of the cation with the dye. (**b**) In an m-preBötC[400/–0.80]W-P1 slice kept in 3 K^+ and loaded with *Fluo-8L-AM* via incubation, presumptive astrocytes in XII nucleus became apparent after adding Cd^{2+}. (**b**$_1$) shows summated image from a z-stack (single steps 0.5 μm) between 48 and 58 μm below the slice surface. Contrary, XII motoneurons were only faintly labeled both in control solution and after adding Cd^{2+}. Bath application for 1.5 h at room temperature with 165 μM of the membrane-permeant fluorescent dye sulforhodamine-101 (*SR-101*) stained XII motoneurons (image not shown), whereas glial cells did not accumulate the dye as indicated by lack of appearance of small yellow cells upon merging the green Fluo-4 image in **b**$_1$ with the *SR-101* image (**b**$_2$) (see Movies 5 and 6).

4-Fish et al. (63), 5-Nakamura et al. (64), 6-Stachniak et al. (73), 8-Sanchez-Vives (55), 9-Broicher and Speckmann (56) and 14-Biggs et al. (58)). For example, the recorded cell can more easily be filled with fluorescent dyes for histological analysis and/or imaging (Sects. 2.2.3, 2.9, and 4.2.4). Also, the composition of the pipette solution can be modified to unmask specific membrane currents. In our study on the distribution of subtypes of voltage-activated Ca^{2+} channels in VRC neurons in the newborn rat *en bloc* model, K^+ in the pipette solution was substituted with Cs^+ and tetraethylammonium$^+$ (TEA^+) for depressing K^+ currents (51) (Sect. 4.1.2). In a more recent study (52), we showed in the same model, for example, that Ca^{2+}-activated K^+ channels are functional in respiratory neurons (Sect. 4.1.2). The extent of activation of these channels and various other cellular processes depend greatly on cytosolic Ca^{2+}, which shows notable fluctuations during activity in these cells (11, 15, 18, 33, 40, 43) (Sect. 4.2.4). Accordingly, it is important to add to the patch electrode solution a Ca^{2+} buffer that mimics endogenous buffering as closely as possible. While some groups prefer ethylene glycol tetraacetic acid (EGTA) in that regard (25, 32, 52), others including us (mostly) use 1,2-bis(o-aminophenoxy)ethane-N,N,N ,N -tetraacetic acid (BAPTA) (7, 9, 11, 31, 42, 53, 59, 60, 70, 71). Both agents are usually applied at a concentration range of 1–11 mM, optimally close to 1 mM to avoid excessive Ca^{2+} buffering in the cytosol. For this reason, some studies preferred not to use either buffer in the patch electrode solution (10–12). Besides, our patch pipette solutions contain Ca^{2+} and also Mg^{2+} as the second physiologically important divalent cation, while 4-(2-hydroxyethyl)-1-piperazine-ethanesulfonic acid (HEPES, mostly 10 mM) is used to buffer pH_i (74). Importantly though, the major relevant physiological intracellular pH buffer is comprised of (dissociated) intracellular CO_2 and HCO_3^- (24–28 mM) provided by the carbogenated standard solution (74). Finally, varying concentrations of Na_2ATP or K_2ATP (usually 1–2 mM) and eventually Na_2GTP (0.2–1 mM) are added to compensate for washout from the cytosol of these important metabolites and/or "second messengers" (49, 50).

According to our experience, the choice between EGTA or BAPTA and 1–11 mM of either agent does not have a major effect on resting V_m, input resistance (R_i), or bursting pattern of newborn rat respiratory neurons in vitro. For these reasons, 1 mM BAPTA is recommended because this presumably approximates the endogenous Ca^{2+}-buffering power (49, 50). Similar to the superfusate, these solutions have an osmolarity of 270–290 mOsmol, which is similar to the value of cellular and interstitial fluids (39). Important to note, long-term whole-cell recording from some types of neurons may require use of slightly hypotonic (85–90%) electrode solution to avoid damage that is presumably due to the pressure of the fluid column in the pipette (Chap. 6-Stachniak et al. (73)).

Perforated Patch Recording: As one disadvantage of whole-cell recording, the low electrical impedance contact between patch electrode and cell interior results in bidirectional diffusional exchange of solute. Therefore, perforated patch-clamp techniques should be applied if the (ionic) composition of the cytosol needs to be maintained. For this, the pipette solution contains ionophores such as nigericin or gramicidin, which incorporate into the plasma membrane (49, 50). These antibiotics form pores that are primarily permeable to K^+ and H^+ but (mostly) not to Na^+, Ca^{2+}, and most anions or larger uncharged molecules like proteins. After formation of a gigaohm (GΩ) seal (see below) as a pivotal precondition for a stable (perforated) patch recording, the electric "series resistance" (comprised of the electrode resistance plus the resistance between electrode and cytosol) decreases progressively from several hundred MΩ to 40–100 MΩ to reveal a V_m of at least −40 mV and overshooting (rhythmic) action potential discharge. Even at steady state, values for series resistance and R_i are though notably higher for perforated patch than whole-cell recording. This can be problematic for space clamp in (fast) voltage-clamp recordings (Chap. 3-Moore et al. (54)). Perforated patch recordings are also more time consuming because it may take 10–30 min from GΩ seal formation until a sufficient number of channels are integrated into the membrane to allow V_m or I_m recording (49, 50, 59).

A critical step is to figure out empirically the optimal concentration of the ionophore in the pipette solution (Sect. 2.2.2), considering that this concentration may change slightly even from day to day by unknown reasons. In studies referred to in Chap. 1-Trapp and Ballanyi (45), the group of Dr. S. Trapp (Imperial College, London, UK) used solubilized amphotericin B (Sigma-Aldrich) and added it to the pipette solution at a final concentration of ~137 μg/ml freshly every day. Contrary, we used a gramicidin-based solution (Sect. 2.2.2) for a perforated patch recording study on synaptic inhibition in the dual respiratory network in the newborn rat *en bloc* model (59) (Fig. 4) (Sect. 4.1.2).

In the latter study (59), the Cl^- content of the electrode solution was crucial for elucidating whether Cl^--dependent inhibitory postsynaptic potentials in newborn respiratory neurons are depolarizing as in immature cortical neurons (46, 48, 75) (Chap. 11-Kantor et al. (57)). Spontaneous or intended rupture of the perforated patch was evident from a positive change by ~10 mV in the apparent resting V_m likely due to a liquid junction and/or "Donnan" potential (51, 59) plus reversal of spontaneous inhibitory postsynaptic potentials within few minutes (Fig. 4). This indicated that intracellular Cl^- in newborn rat respiratory neurons is similarly low as in adult respiratory neurons (46). Consequently, most groups studying postnatal respiratory networks use now <10 mM Cl^- in whole-cell patch electrode solutions. This is only one example showing that diffusional exchange between patch electrode and the cytosol modifies basic biophysical properties of neurons in isolated brain tissues (for other examples, see Chaps. 1-Trapp and Ballanyi (45), 3-Moore et al. (54) and 7-McKay et al. (76)).

Approaching Respiratory Neurons for Patch-Clamp Recording: For blind patch recording from respiratory neurons in slices, the electrode is advanced perpendicularly using an MX7600 motor-driven micromanipulator with added display of recording depth, both from Siskiyou (Sect. 2.6), into the preBötC region that is not used for VRC suction electrode recording (Sect. 3.2.1). Inspiratory neurons (the main type of respiratory cells particularly in thin preBötC slices) are identified by their rhythmic extracellular discharge that is in phase with contralateral suction electrode-recorded VRC population bursting (Fig. 4). For blind patch recording in the *en bloc* model from >P1 old animals, it is important to remove the arachnoid and some portion of the pia that cover the brainstem surface to avoid electrode clogging (Sects. 3.1.2 and 3.1.3). Removal of pia is achieved, for example, via scratching using an unfilled patch pipette controlled by a joystick such as used for several types of hydraulic micromanipulators from Narishige. For patch recording under visual control, we use either the MX7600 or the similar motor-driven MP-2 (Sutter) (Sect. 2.6).

While advancing the electrode through the tissue, positive pressure (~30 mmHg) is applied to avoid clogging which is evident from an increase in its electric resistance (7, 9, 35, 48, 50). Independent on the type of patch recording (blind vs. visually targeted), augmentation by up to 300% of a voltage signal in response to repetitive injection at a rate of 25 Hz of negative DC current pulses (−1 nA, 10 ms) indicates that the electrode is close to a neuronal cell membrane (Fig. 4). The probability for a good GΩ seal seems to correlate in this situation with the occurrence during the current injection of electrode potential oscillations that resemble sine waves. At this instance, the patch electrode is moved upward and downward by 1–5 μm several times until the amplitude of electrode potential oscillations becomes maximal. At this point, the pulse is attenuated to −0.1 nA and positive pressure is then removed from the patch pipette (Fig. 4). In ~30% of cases, a GΩ seal develops spontaneously within <5 s as evident from an increase in the amplitude of the current-induced electrode potential change (Fig. 4). If the potential does not start to continuously increase upon removal of positive electrode pressure within this time period, negative pressure of up to −50 mmHg is applied to facilitate GΩ seal formation.

After formation of a seal >1 GΩ, the whole-cell configuration is established by abrupt and brief suction (~100 mmHg) (Fig. 4). If patch electrodes are not "perfect," for example, by corrosion of the filament of the electrode puller (Sect. 2.5), suction may have to be substituted or combined with a short pulse (500–1,000 nA, 1 ms) using the "ZAP" function of the intracellular amplifier/software, if implemented. Depending on the respiratory neuron class (5), resting V_m and R_i in our studies on the neonatal rat in vitro models range from −35 to −60 mV and from 150 to 700 MΩ, respectively (7, 9, 24, 42, 46, 51–53, 59, 60, 70–72) (Fig. 4, 6, and 7). V_m values in our published studies tend to be lower than those in reports from other groups (10, 12, 25). This is partly because we do not correct V_m for liquid-junction potentials with an estimated amplitude of 10–20 mV (49–51). We do not correct for these potentials because we did not find a significant change between resting V_m for different classes of (respiratory) neurons, regardless of whether the cells were recorded with high or low Cl^- solution (see above), which should theoretically change liquid-junction potential by at least 10 mV. Once V_m stabilizes within several minutes after establishing the whole-cell configuration, R_i is monitored via repetitive injection of hyperpolarizing DC current pulses (40–250 pA, 0.5–0.7 s) at intervals of 5–15 s (Fig. 6).

3.2.3. Electrical Stimulation

Electrical stimulation of neighboring neuronal areas or fiber tracts at regular intervals with either a single pulse or repetitive ("tetanic") pulse trains is a powerful tool for analyzing synaptic neuronal connectivity and roles of glial cells in neural network function. We have used this approach in a variety of respiratory and nonrespiratory slice and *en bloc* preparations (77–82). Findings using stimulation in the respiratory networks in the dish are referred to in the "Results" section (Sects. 4.1.1 and 4.2.4). Typical stimulation parameters using, for example, concentric bipolar stimulation electrodes (Sect. 2.5) are 5–90 V and 20–200 Hz for 0.02–10 s at single-pulse duration of 0.1–0.5 ms.

3.2.4. Gas- and Ion-Sensitive Microelectrodes

In several studies using the newborn rat *en bloc* model (37, 38, 83–85), we have applied microelectrodes that are sensitive to O_2, CO_2, H^+, K^+, or Ca^{2+} (86, 87). Contrary to our positive experience with other isolated mammalian tissues, such as astrocytes in guinea pig olfactory cortex slices (77), neurons in rat sympathetic ganglia (88), and expiratory neurons and neighboring glia in cats in vivo (89), intracellular recording does not seem to be feasible with ion-sensitive microelectrodes in isolated respiratory networks from perinatal rodents. Details regarding extracellular use of such electrodes by us are given in the above reports that focused on analyzing the extracellular microenvironment of newborn rat VRC neurons during spontaneous and evoked activity and on effects of oxygen and glucose depletion, that is "anoxia" and "aglycemia," respectively, and complete blockade of metabolism ("in vitro ischemia") (Sect. 4.1.1). Corresponding experiments were conducted in two of our studies on similarly anoxia-resistant dorsal vagal neurons in juvenile rat brainstem slices (Chap. 1-Trapp and Ballanyi (45); see also Chap. 12-De Curtis et al. (90)).

For recording dynamic changes of the above-mentioned gases or ions, the microelectrodes were positioned within the VRC in a tissue depth of 300–600 μm below the ventrolateral brainstem surface, corresponding to the main distribution within the transversal plane of newborn rat VRC neurons (5, 30, 37) (Figs. 1 and 2).

PO_2 in the recording chamber was continuously monitored with a "Clark-style" macroelectrode (Diamond Electro-Tech Inc.,

Ann Arbor, MI, USA), whereas tissue PO_2 was determined with O_2 microelectrodes obtained either from Diamond Electro-Tech (outer tip diameter, 20–30 µm) or as a gift from H. Baumgärtl (Max Planck Institut fur Systemphysiologie, Dortmund, Germany; outer tip diameter 5–15 µm). Procedures for constructing these electrodes and their properties are described elsewhere (87). The O_2-sensitive macroelectrode was calibrated in the recording chamber during superfusion with HEPES pH-buffered solution gassed with either 100% O_2 or air (21% O_2) and during superfusion with 159 mM of the O_2 chelator Na_2SO_3 for obtaining the 0 mmHg O_2 level ("anoxia"). Subsequent superfusion with standard superfusate resulted in a bath PO_2 value of ~700 mmHg in CO_2/HCO_3^- pH-buffered, carbogenated standard solution and a very similar value in HEPES pH-buffered saline gassed with 100% O_2. In hypoxic saline (95% N_2 plus 5% CO_2 or 100% N_2), PO_2 in the recording chamber ranged from 3 to 10 mmHg. Arrest of superfusate flow reduced the signals of the O_2-sensitive macro- and microelectrodes by 7–15%. This "stirring effect" may lead to underestimation of surface and tissue PO_2 because the electrode approaches the tissue surface and crosses a proposed "unstirred layer" (37, 85, 87).

Procedures for the production of ion-sensitive microelectrodes and their properties have been described elsewhere (37, 38, 80, 83–86). In brief, theta-style double-barreled borosilicate micropipettes are fabricated with a microelectrode puller (Sect. 2.5). A good tip size for extracellular recording is 1–5 µm because smaller tip diameters result in a reduced signal-to-noise ratio due to the high resistance (Sect. 2.2.2), while larger tip sizes may cause leakage of ion resin. K^+-sensitive microelectrodes are calibrated before and after experiments with solutions containing 1.5, 6, and 15 mM KCl and a constant background of 145 mM NaCl. The value for the slope for a tenfold change in the free K^+ concentration for these "valinomycin"-type K^+ electrodes is typically 56 mV. Ca^{2+}-sensitive microelectrodes are calibrated in standard solution with different (buffered) Ca^{2+} content and have a mean slope of 29 mV/tenfold change in Ca^{2+}. pH-sensitive microelectrodes are calibrated in solutions of different pH (Sect. 2.2.2) and have a mean slope of 58 mV/pH unit. Triple-barreled pH-sensitive microelectrodes have been used by us for simultaneous measurement of extracellular pH and partial pressure of tissue CO_2 in the VRC of the newborn rat *en bloc* model (85).

For gas- and ion-sensitive microelectrode measurements, it is important to determine the surface of the in vitro respiratory preparations. For ion-sensitive microelectrodes, touching the surface is reflected by a sudden perturbation of the voltage of the reference barrel. Lateral disposition during downward movement of O_2-sensitive electrodes is used to determine the point of entry into the tissue. Importantly, the reference barrel of ion-sensitive electrodes, which is filled with standard solution (Sect. 2.2.1), can be used to record extracellular respiratory neuronal activity (37).

3.3. Optical Recording

Both voltage- and Ca^{2+}-sensitive fluorescent dyes have been applied since ~10 years for monitoring the activity of populations of respiratory neurons in both *en bloc* and slice models. These studies by various groups including us are summarized in our recent review articles (15, 18, 23, 43). These articles also point out how we have adapted multiphoton microscopy for monitoring cellular morphology and activity based on rhythmic cytosolic Ca^{2+} rises (Sect. 3.3.2) (62) in preBötC neurons and neighboring glia with greatly improved spatial resolution compared to previous CCD camera imaging (e.g., in (11)) (Sect. 4.2.4). In the following, principles of such Ca^{2+} imaging are described after briefly referring to the methodology of voltage imaging using the newborn rat *en bloc* model based on our collaboration with the Onimaru group (24).

3.3.1. Voltage Imaging

In our voltage imaging study (23, 24), *en bloc* [+VII] preparations (Sect. 3.1.3) were immersed for 40–50 min in "Suzue-type" (4) solution with 50 μg/ml of the fluorescent voltage-sensitive dye Di-2-ANEPEQ. They were then placed ventral side up in an acrylic chamber that was mounted on an Olympus BX50WIF-2 fluorescence microscope (Sect. 2.8). After focusing the objective on the ventral brainstem surface, spontaneous activity in the VRC including both the preBötC and the pFRG was optically monitored as a change in Di-2-ANEPEQ fluorescence intensity with a MiCAM01 CCD camera (Sect. 2.8). Optical recording, triggered by electrophysiological cervical inspiratory-related burst activity, was done with an acquisition time of 20 ms per image resulting in a sampling rate of 50 Hz. The sampling period (6.8 s) included a time period of 1.7 s that preceded the steepest point of the cervical burst, which constituted the trigger threshold. Because of the extremely low signal-to-noise ratio associated with voltage imaging (15), 40–50 images and corresponding cervical bursts needed to be averaged and were digitally stored. Fluorescence changes were expressed as ratios of fluorescence intensity that were compared with a reference image.

Differential images were processed with a 3×3 pixel software-spatial filter (Sect. 2.8) and displayed in pseudocolor with red corresponding to a fluorescence decrease indicating depolarization. For representation of the time course of fluorescence changes, averaged fluorescence intensity was assessed and inverted in a 50-pixel region of interest covering an area of 92×108 μm. It should be noted that optical signals presumably reflect primarily depolarizations that are associated with respiratory drive potentials but may also include a glial component (15).

3.3.2. Multiphoton/Confocal Ca^{2+} Imaging

Bursting in newborn rat VRC neurons is accompanied by Ca^{2+} influx through specific sets of voltage-activated Ca^{2+} channels (51, 52). Consequently, their activity can be visualized with optical imaging of changes in the free cytosolic [Ca^{2+}] as established since several decades for neurons in various brain regions, both in vitro and in vivo (43, 62, 65). For our multiphoton/confocal Ca^{2+} imaging approach in preBötC interneurons and XII motoneurons plus neighboring glia of calibrated brainstem slices, 0.5 mM Fluo-4-AM is backfilled into a broken patch pipette with an outer diameter of 5–10 μm (Sect. 2.8). The dye is then pressure-injected (25–50 mmHg) for 10 min (65) into the VRC/preBötC area or XII nucleus at 50 μm depth. This leads to staining of cells that are located between the surface and a depth of at least 100 μm, which is the resolution limit for our multiphoton system. While imaging, inspiratory-related neuronal population activity is monitored with a suction electrode from the contralateral VRC/preBötC (Sect. 2.5). Fluorescence signals are measured using either an Olympus FluoView-300 confocal laser scanning microscope (Fig. 3) or a custom-made multiphoton system using a further FluoView-300 system connected to a Ti:Sa IR laser (Sect. 2.8) Details of the extents of stained areas and the quality of cellular imaging are given below (Sect. 4.2.4). The Ca^{2+} dye-loaded areas are mostly monitored using a 20× objective with 2–3× digital zoom at reduced settings for y-axis scanning. Compared to full frame acquisition (512 × 512 pixels), such "clipped mode" imaging samples an area of 512 × 100–220 pixels and provides scanning rates of 1.25–1.43 Hz which are sufficient to detect 70–100% of the peak of inspiratory-related Ca^{2+} rises. The morphology of cell bodies and primary dendrites is best revealed by combining "z-stack" image series (single z-steps 0.1–0.5 μm) with 3D animation using FluoView software (Sect. 2.8). Fluo-4 or Fluo-8L fluorescence emission is weak at low cytosolic Ca^{2+} levels (62, 65). Therefore, to improve morphological characterization of cells, slices are exposed at the end of experiments to glutamate (0.25–1 mM, 5–10 min) for generation of z-stacks. Glutamate-induced fluorescence signals are 3–10 times larger than those associated with spontaneous Ca^{2+} rises or Ca^{2+} increases evoked by other neuromodulators such as substance-P.

Another method of loading cells in preBötC slices with Ca^{2+} dye is via injection into the midline (11). We do not use that approach because retrograde transport of the dye from the injection spot to the preBötC takes several hours compared to 10 min for direct injection into the preBötC or 30 min for incubation of slices with Ca^{2+} dye. For the latter procedure, a preBötC slice is kept for 30 min at 30°C in carbogenated standard superfusate containing 10 μM Fluo-8L-AM (Sect. 2.8). In our hands, bath application of Fluo-8L-AM (or Fluo-4-AM) primarily loads small cells, likely astrocytic glia, whereas others find that a sufficient number of (respiratory active) neurons are loaded as well (for references, see (15, 23)) (Sect. 4.2.4).

3.3.3. Morphological Characterization of preBötC Cells in Calibrated Slices

We outlined above (Sect. 2.9) how whole-cell recorded neonatal rodent respiratory neurons can be histologically identified using lucifer yellow or neurobiotin/biocytin in the patch electrode. While this approach is performed on chemically fixed and subsequently sectioned *en bloc* or slice preparations, we recently developed a method for acutely labeling XII motoneurons and (primarily) neurons in the preBötC area of living slices that were incubated with sulforhodamine-101 (Sect. 2.9). Note that this technique likely cannot be used to differentially label neurons versus glial cells in the brainstem slices contrary to the feasibility of this approach in newborn rat hippocampal or neocortical slices (Chap. 11-Kantor et al. (57)).

3.4. Data Analysis

Inspiratory-related neuronal population activity in preBötC slices in 3K/1Ca solution is continuously recorded at a sampling rate of 1–4 kHz. Burst rate is averaged every 20 min over 2-min time windows. The "longevity" of rhythm in 3 mM K^+ solution is defined as the time from start of recording until the time when a burst is not followed by a further burst within a time period of 10 min. Single-burst duration is defined using ClampFit software (Molecular Devices Corporation, Chicago, Il, USA) as the time interval from when the signal increases above and decreases below a threshold set at 10% of the peak amplitude for that burst. Pharmacologically reactivated rhythms are described by averaging burst rates over a 2-min time period at steady state. Values are means ± SEM except for histological analyses (means ± SD). Significance ($P < 0.05$) is assessed by Student's t-test or one-way ANOVA using SigmaPlot (Systat Software Inc., Point Richmond, Ca, USA) or Prism (GraphPad Software Inc., La Jolla, CA, USA; www.graphpad.com).

For analysis of Ca^{2+}-imaging experiments, firstly, FluoView software is used for drawing regions of interests (ROIs) around cell somata to obtain plots of fluorescence intensity over time (Sect. 2.8). Fluorescence intensity analysis data are then transferred to Excel of the Microsoft Office suite for further computation and plotting of graphs. 3D reconstruction is obtained from z-stack series (Sect. 2.8). As typical 3D settings, we use an x-y plane with rotation in x and y axes to give a 3D impression of cell morphology and x-z plane with y axis rotation to reveal the depth of fluorescence signal. For generating figures for publication, fluorescence intensities are plotted in graphs using both Excel and Illustrator (Adobe; www.adobe.com).

4. Results

Here, we summarize our findings using the (calibrated) in vitro respiratory models. Reference to work by other groups can be found in the reference lists of our original papers and review articles that are primarily referred to in the following.

4.1. Findings on Noncalibrated En Bloc Preparations and preBötC Slices

Our previous studies using the "classical" Suzue-type (*en bloc* [+VII]) brainstem–spinal cord preparation (Sect. 3.1.3) and preBötC slices from newborn rats aimed primarily at analyzing the extracellular microenvironment of VRC neurons (only in the *en bloc* model) and elucidating intrinsic biophysical membrane properties of VRC neurons and their response to (endogenous) neuromodulators. This work, which has been reviewed in detail in several articles (5, 46, 72), is summarized in the following.

4.1.1. Microenvironment of VRC Neurons in the En Bloc Model

One major aspect of our research is devoted to the analysis of cellular mechanisms of the extreme tolerance to anoxia of neural respiratory networks in newborn mammals as established by previous in vivo work (for citations, see (5, 46, 72)). For this, we blocked aerobic metabolism either by exposing the preparations to hypoxic superfusate or chemically via bath application of cyanide (1 mM). Our measurements with O_2-sensitive microelectrodes revealed that hypoxic superfusate causes anoxia within the VRC, whereas tissue PO_2 increases to almost superfusate levels in response to cyanide (37, 38, 83). Contrary, our ion-sensitive microelectrode and whole-cell V_m recordings revealed very similar responses to both approaches for blocking aerobic metabolism. Specifically, extracellular pH shows (often after a <2-min period of alkalinization) a progressive acidosis that can exceed >0.3 pH units at the end of a >10 min anoxia period (37, 38, 84, 85). Within <3 min into anoxia, extracellular K^+ rises by <2 mM and remains stable at this level for at least 30 min, whereas extracellular Ca^{2+} does not change during this time period (37, 38, 83, 84). In ~30% of cases, inspiratory-related cervical bursting accelerates by up to 100% during the initial 1–2 min of anoxia. Rhythm transiently stops during this period in other ~30% of preparations, while bursting is not affected in the remaining cases. After >3 min into anoxia, inspiratory rhythm slows down progressively. At the end of a 30 min anoxia period, in vitro inspiratory burst rate typically falls about tenfold, from 5–10 bursts/min in control to <0.5 bursts/min. This is very similar to slowed, but persisting breathing in newborn mammals in vivo at reduced temperature (for references, see (5, 46, 72)).

Slowing of in vitro inspiratory rhythm in response to hypoxic or chemical anoxia is accompanied by a persistent hyperpolarization in >50% of inspiratory VRC neurons, a major portion of which becomes "functionally inactivated" (70). There is pharmacological evidence that ATP-sensitive K^+ channels are involved in this anoxic hyperpolarization and thus possibly contribute to slowing of rhythm, which may be an adaptive mechanism for reduction of energy consumption (5, 46, 70, 72, 84). Arrest of metabolism, by combined exposure to cyanide and the glycolytic inhibitor iodoacetate, blocks inspiratory rhythm irreversibly within <5 min after start of application. Such in vitro apnea is followed within several minutes by a rise of extracellular K^+ of >10 mM that is mirrored by a fall of extracellular Ca^{2+} by >0.5 mM (84). Perturbation of extracellular ions and pH within the VRC in response to in vitro ischemia is accompanied by irreversible depolarization of VRC neurons (5). All this showed that sustained anoxia does not cause a major disruption of ionic homeostasis in the VRC contrary to its rapid deregulation in response to metabolic arrest. This indicates that stimulation of anaerobic metabolism during anoxia ("Pasteur effect") is able to maintain basic cellular functions in newborn rat brainstems (5, 80, 84).

Our above studies also revealed that the >2-mm-thick *en bloc* preparation has an anoxic core in carbogenated standard solution (37, 38, 83, 84). The same reports also indicated that cells in depths of 300–800 μm below the ventral surface including spontaneously active neurons within the most densely aggregated area of the VRC in the transversal area 300–600 μm (Figs. 1 and 2) are normoxic and that cells in depths <300 μm are even hyperoxic. The tissue PO_2 gradient in the ventrolateral aspect of isolated brainstems is reflected by gradients for PCO_2 and pH which is ~0.5 pH units more acidic in the anoxic core compared to the surface. This pronounced extracellular acidosis in control solution is likely the consequence of enhanced anaerobic metabolism in the anoxic core leading to lactate formation and associated H^+ production (37, 38, 83, 84). Along the VRC column in the transversal plane, there is also a gradient for extracellular K^+, which increases from ~4 mM in 300 μm depth to ~5.5 mM in 600 μm depth (37). Contrary, extracellular Ca^{2+} is stable throughout the *en bloc* preparation at 1.5 mM corresponding to the superfusate Ca^{2+} content used as standard at that time (84).

Upon tetanic electrical stimulation (Sects. 2.5 and 3.2.3) in control solution, PO_2 in the VRC decreases by up to 50 mmHg with a concomitant increase of both extracellular K^+ and pH by maximally 5 mM and 0.07 pH units, respectively (37, 38). We did not measure extracellular Ca^{2+} during tetanic stimulation but found that it remains unaltered during spontaneous inspiratory activity, which does also not affect extracellular pH (37, 38, 84). Contrary, extracellular K^+ in the VRC increases during each inspiratory burst by up to 100 μM, while PO_2 falls by maximally 0.4 mmHg (37).

In the *en bloc* model, use of O_2-, CO_2-, and H^+-sensitive microelectrodes enabled us to study central chemosensitivity which means adaptation of breathing to altered PCO_2 levels by brainstem sites that are intrinsically sensitive to the resulting change of medullary pH (26, 27, 85) (see also Chap. 13-Day and Wilson (91)). We found that the pH gradient in the VRC is reflected by a PCO_2 gradient with a value of 77 mmHg in 300 μm depth versus 95 mmHg at 600 μm. Despite the fact that both pH (see above) and PCO_2 in the VRC differ notably from in vivo values (see also Chap. 13-Day and Wilson (91)), the *en bloc* model is capable of responding to experimentally evoked changes of either parameter.

We also showed that acidosis counters the depression of inspiratory rhythm by opioids (5). Our findings from such experiments indicate that extracellular H^+ is the primary stimulatory factor in central chemosensitivity which may often mask a less evident direct effect of CO_2 (85). In addition, the latter work and our further studies using these electrodes (38, 48) provided evidence that respiratory chemosensitive structures are either located several hundred micrometers within the tissue, and not close to the ventral surface as has been proposed, and/or that chemosensitive structures close to the surface do respond with a delay of several minutes (for references, see (5, 26, 46, 48, 85)).

From these findings, we hypothesized that (1) VRC neurons in the newborn rat *en bloc* model are under aerobic conditions which is reflected by baselines of extracellular K^+ and Ca^{2+} that are close to physiological levels (36, 39), (2) anoxic slowing of inspiratory-related rhythm in newborn mammals represents an adaptive mechanism for energy conservation and seems to involve hyperpolarization of VRC neurons via ATP-sensitive K^+ channels, (3) tolerance to anoxia of respiratory networks in neonatal mammals may be due to their capability to enhance their utilization of anaerobic glycolysis, and (4) preBötC slices of <1,000 μm thickness should not have an anoxic core, at least under our superfusion conditions (Sects. 2.4 and 3.1).

4.1.2. Properties of Newborn Rat VRC Neurons

In various studies using whole-cell recording in both the *en bloc* and preBötC slice models, we established that resting V_m and R_i (Sect. 3.2.2) range between −35 to −60 mV and 150–750 MΩ, respectively, depending on the specific class of VRC neurons (Figs. 4, 6, and 7). In the en bloc model, these cell classes comprise several types each of pre-/postinspiratory active pFRG, inspiratory, and "tonic" expiratory neurons (5, 6, 24).

Using primarily the *en bloc* [+VII] model (Sect. 3.1.3), we elucidated the following intrinsic biophysical membrane properties of newborn rat VRC neurons: (1) Following initial identification of the type of VRC neurons and subsequent blockade of their rhythmic activity by the voltage-activated Na^+ channel blocker tetrodotoxin, use of specific (snail peptide) blockers revealed that these cells are equipped with distinct sets of L-, T-, N-, R-, and P/Q-type voltage-activated Ca^{2+} channels (51). Our follow-up study on analyzing the functional relevance of voltage-activated Ca^{2+} channels (52) unraveled that blockade of a single type of such Ca^{2+} channels (or Ca^{2+}-activated K^+ channels) does not block inspiratory rhythm despite specific cellular effects on individual VRC neurons. (2) As stated above (Sect. 4.1.1), ATP-sensitive K^+ channels seem to be involved in persistent postsynaptic hyperpolarization of VRC neurons during anoxia (5, 46, 70, 72). B-type γ-aminobutyric acid ($GABA_B$) receptors are coupled in a subpopulation of these neurons to postsynaptic K^+ channels whose activation by the $GABA_B$ receptor agonist baclofen leads to sustained hyperpolarization, whereas there is no evidence for endogenous activation of these channels during in vitro inspiratory rhythm (59). (3) The latter study provided evidence that $GABA_A$ and glycine receptor-coupled anion channels hyperpolarize VRC neurons rhythmically during different phases of the in vitro respiratory cycle (Fig. 4), contrary to depolarizing actions of these major inhibitory neurotransmitters on neonatal hippocampal and cortical neurons (46, 48, 75) (see also Chap. 11-Kantor et al. (57)). (4) Hyperexcitability of spinal inspiratory motoneurons and some VRC interneurons in response to the $GABA_A$ receptor antagonist bicuculline (60) is depressed by adenosine acting on A_1 receptors, although neither adenosine nor A_1 receptor agonists have a major postsynaptic effect on these cells in control solution. However, adenosine acts through A_1 receptors to depress nonrespiratory activity and stabilize irregular inspiratory rhythms, while there is no evidence for functional A_2 receptors in newborn rat VRC neurons (60). (5) Agents raising cellular cyclic adenosine monophosphate (cAMP) such as forskolin or isobutylmethylxanthine reverse depression of both cellular bursting in VRC neurons and inspiratory cervical motor output by either prostaglandins or agonists of μ-type opioid receptors. Depending on the inspiratory VRC neuron type, the latter agents can cause either a postsynaptic hyperpolarization or presynaptic inhibition (71) (Sect. 4.2.3). (6) Using preBötC slices (in 8–9-mM K^+), we found that tonic excitation of XII motoneurons is mediated by postsynaptic 5-HT_2-type serotonin receptors (53). The same study also revealed that the latter receptors are also activated in preBötC neurons during excitatory serotonin effects on inspiratory rhythm, whereas other serotonin effects are mediated by 5-HT_{1A} receptors (53).

4.2. Findings on Calibrated En Bloc Preparations and preBötC Slices

Here, we summarize our recent findings on calibrated in vitro models from newborn rats and mice (Sect. 3.1). Specifically, we refer to (1) the structure–function analysis of the dual respiratory center via anatomical and pharmacological dissection, (2) influence of superfusate composition on isolated preBötC, (3) cellular mechanisms of preBötC depression by opioids and anoxia, and (4) "pharmoptic" approaches for preBötC neuron–glia discrimination.

4.2.1. Structure–Function Analysis of the Dual Respiratory Center

Studies using the *en bloc* [+VII] model (Sect. 3.1.3) established that the preBötC and pFRG constitute a dual inspiratory-expiratory center, at least in newborns (2, 5, 6, 8, 26) (Sect. 1). Both components of this dual center can be isolated by transecting this model. Onimaru's group found 5 years ago that a rostral brainstem "slab," generated by transection of the *en bloc* model 0.1–0.2 mm caudal to VII_c, shows rhythmic VII nerve activity presumably representing cranial pFRG motor output (6, 92). The caudal boundary of this rhythmic pFRG block neighbors the rostral boundary of ~350-μm-thick preBötC slices that we developed in our milestone study on the discovery of the inspiratory center about two decades ago (7). In the latter study, we estimated based on consecutive sectioning that the newborn rat preBötC extends ~200 μm in rostrocaudal direction. Specifically, the brainstem aspect of the *en bloc* model was sectioned in either rostral-to-caudal or caudal-to-rostral direction until cervical or cranial inspiratory motor output, respectively, was irreversibly blocked. Since then, numerous studies make use of preBötC slices, however, with notable differences between laboratories in regard to their thickness (200–1,100 μm) and anatomical margins that were not determined histologically (for references, see (18, 33)).

We dealt with the latter issues by developing online histology for generating calibrated preBötC slices with anatomically defined and reproducible boundaries (33) (Sect. 3.1.2). In a follow-up study (40), we used "m-preBötC" slices with centered preBötC (Sect. 3.1.2) to determine the minimal thickness that is sufficient for inspiratory-related rhythm in "physiological" 3K/1Ca solution (Sect. 3.1) (Figs. 1 and 2). Specifically, we found that the center of the newborn rat preBötC is located exactly 0.5 mm caudal to VII_c and that the minimal slice thickness for synchronous bursting in the ipsilateral and contralateral preBötC area is 200 μm. In that study, we also developed 700-μm-thick "r+preBötC" and "c+preBötC slices" that contain the exposed (and partially sectioned) preBötC plus either rostrally or caudally adjacent brainstem tissue, respectively (Sect. 3.1.2) (Fig. 2). Using the latter slice types, we firstly found that <100 μm preBötC tissue is necessary for rhythm generation in 3K/1Ca if rostrally or caudally neighboring brainstem areas are present to provide (tonic) excitatory drive to rhythmogenic neurons. Secondly, we demonstrated that the preBötC has a "spatio-chemical organization." Specifically, in 3K/1Ca, c+preBötC slices generate regular fast inspiratory-related "fictive eupnea" bursts. Contrary, r+preBötC slices show a sequence of fictive eupnea bursts that are interrupted by occurrence of a single, larger amplitude "fictive sigh" which is often followed by transient post-sigh depression of one or several subsequent fictive eupnea bursts.

We also showed that upon onset of in vitro apnea after several hours of recording in 3K/1Ca (33, 40), bath application of various neuromodulators can restore rhythm with a fictive eupnea burst pattern (Fig. 2). These agents are thyrotropin-releasing hormone (0.1–100 nM), the cAMP-dependent phosphodiesterase-4 blocker rolipram (1–5 μM), the methylxanthine theophylline (0.25–10 mM), the group I metabotropic glutamate receptor agonist 3,5-dihydroxyphenylglycine (at 1–10 μM), and also 9 mM K^+ (15, 18, 33, 40). Contrary, substance-P (or the agonist GR73632 for neurokinin-1 receptors that are normally activated by substance-P), 3,5-dihydroxyphenylglycine (at 10–50 μM), and also 6–8-mM K^+ induce mostly a mixed fictive eupnea-sigh burst pattern. These different patterns of suction electrode-recorded inspiratory population activity were reflected by corresponding patterns of small (eupnea-like) and large (sigh-like) cytosolic Ca^{2+} rises in preBötC neurons (Sect. 4.2.4) (Fig. 7). The latter observations indicate that structures rostral to the preBötC are capable of inducing sighs in preBötC networks which can reconfigure between different burst patterns under the influence of particular neuromodulators.

Our latter published findings substantiate hypothesis that the preBötC comprises a very thin kernel of likely only several hundred rhythmogenic inspiratory neurons (2, 7). However, our most recent data indicate that the preBötC, nevertheless, represents a small distributed neural network as proposed previously based on both in vivo and in vitro findings (for references, see (18)). Specifically, we isolated two neighboring inspiratory active 700-μm-thick slices from a single newborn rat brainstem (18). This is achieved by slicing firstly at 1.2 mm caudal to VII_c, then at 0.5 mm caudal to VII_c, and finally 0.2 mm rostral to VII_c (Sect. 3.1.1). The caudal one of these "sandwich" preBötC slices shows rhythmic activity with ~1-s single burst duration in both ventrolateral aspects of the rostral margin. In the rostral slice, recording from the ventrolateral area of the caudal margin reveals similar bursting but at lower rate (i.e., <3 vs. >5 bursts/min). Moreover, the bursts in the rostral slice with a single-burst duration of ~1 s are often partly obscured by an even slower rhythm with a single-burst duration of several seconds. The short-burst rhythms in both aspects of the sandwich slices seem, indeed, to represent inspiratory-related bursting because they are blocked by the μ-opioid receptor agonist [D-Ala^2,N-Me-Phe^4,Gly^5-ol]-enkephalin (DAMGO) [compare Fig. 7]. Contrary, slow bursting in the rostral slice is either not affected, or even stimulated, by DAMGO and may thus be generated by pFRG networks. In fact, findings from various in vivo and in vitro studies revealed that the preBötC is inhibited by opioids, whereas the pFRG is resistant, thus substantiating the >20-year-old hypothesis by the group of Onimaru and Homma on a dual respiratory center (for references, see (2, 5, 6)).

Recently, we showed using the *en bloc* [+VII] model that cyanide-evoked chemical anoxia is a further pharmacological tool, in addition to opioids, for elucidating the structure–function relationship between the preBötC and the pFRG (24). Specifically, we found, using combined suction electrode and whole-cell recording, that opioids increase the delay between the onsets of preinspiratory pFRG and inspiratory preBötC bursting. Contrary, we showed in that study (24) that synchronization of both rhythm generators by anoxia seems to result in generation of repetitive cervical bursts and enhanced postinspiratory activity (Figs. 8 and 9). These electrophysiological changes were reflected by optical (neuronal) population signals obtained by voltage imaging from three different brainstem regions containing the preBötC, the area of the BötC in adults, and the caudal aspect of the pFRG (Fig. 8). All this indicates that similarly severe frequency depression of preBötC network bursting by opioids and anoxia (Fig. 7) is due to fundamentally different cellular mechanisms that are referred to below in more detail (Fig. 9) (Sect. 4.2.3). In a related study (28), we found that the change of respiratory burst patterns in the brainstem aspect of the *en bloc* [+VII] model upon anoxia revealed by voltage imaging is reflected by a similar transformation of pre-/postinspiratory into inspiratory/postinspiratory bursting in pFRG-driven lumbar nerves. That study (28) reported further that such pre-/postinspiratory lumbar bursting in control solution, and also repetitive cervical nerve bursts during evoked anoxia, is blocked following transection of the *en bloc* [+VII] preparation ~0.2 mm caudal to VII_c. In a related report (34), we identified landmarks on the ventral brainstem surface that enable generation of the *en bloc* model with an anatomically defined rostral section level (Sect. 3.1.3). We used this approach to demonstrate that the critical level, at which brainstem transection abolishes pFRG-mediated lumbar nerve bursting, is located close to the rostral end of VII nucleus as also indicated by findings of Giraudin and colleagues (29). In our latter study (34), we also reported that inspiratory XII nerve bursting disappears spontaneously in the *en bloc* [+VII] model within ~30 min after start of recording in control solution, whereas such activity is stable for several hours in *en bloc* [−VII] preparations.

In summary, we showed on the one hand that both preBötC-driven inspiratory and pFRG-driven expiratory activities in calibrated newborn rat brainstem models depend greatly on both their physical dimensions and the relative location of the preBötC and pFRG within the preparation depending on transection levels. On the other hand, we added experimentally evoked anoxia as a further tool besides opioids for pharmacological dissection of these networks.

4.2.2. Influence of Superfusate Composition on Isolated preBötC

In this section, we outline that superfusate K^+, Ca^{2+}, and glucose greatly influence preBötC (slice) rhythms. We point out that (close to) physiological superfusate concentrations of these constituents do not provide long-term in vitro inspiratory rhythm while, depending on the slice type, it is beneficial to use, at least, values that are as close as possible to physiological levels.

Superfusate K^+: In our paper on the preBötC discovery (7), we noted that "slices >500 µm thick generate rhythmic motor output with a cycle frequency that was 40 to 70% of that in the *en bloc* brainstem-spinal cord before the slice was cut. This frequency could be increased by changing the K^+ concentration in the slice bathing solution from 3 to 9 mM *[…]*. Thinner slices did not generate motor output at 3 mM K^+ but generated rhythmic output at K^+ concentrations of 9–11 mM." These statements indicated, on the one hand, a modulatory effect on preBötC networks of superfusate K^+, which does not fully equilibrate with extracellular K^+ in *en bloc* preparations (Sect. 4.1.1), but most likely in slices of <1 mm thickness. This established the view that preBötC slices are principally only active in superfusate with a higher than physiological K^+ level, which is 3 mM (33, 34, 36, 39, 48). Specifically, it was argued that the ("tonic") excitatory drive from neighboring neurons for (intrinsic) bursting of rhythmogenic neurons is too low in preBötC slices and that elevated K^+ depolarizes these neurons and thus compensates for that attenuated drive (2). As a result, most groups use 8 or 9 mM K^+ even when studying thick preBötC slices despite the fact that such K^+ levels are an established in vitro epilepsy model using cortical brain slices (for references, see (18, 33, 34) and Chap. 11-Kantor et al. (57)).

Our initial study (33) showed that calibrated preBötC slices of ≥500 μm thickness (Sect. 3.1) are capable of inspiratory rhythm in 3K/1Ca for several hours (Fig. 2). Our recent unpublished observations indicate that raising K^+ from 3 to 9 mM does not principally change rhythmic Ca^{2+} rises in preBötC neurons except that a few more cells may become active in the same ventrolateral slice region while the rate of their inspiratory Ca^{2+} oscillations increases modestly (Fig. 10). This indicates that preBötC bursting in slices in elevated K^+ differs from seizure-like K^+-evoked hyperexcitability in cortical brain slices (see above). Our group also demonstrated that bursting in the preBötC area of slices is also not perturbed by blockers of $GABA_A$ and glycine receptors, such as bicuculline and strychnine, respectively, which are further common epilepsy models for cortical brain slices (for references, see (18, 33, 34) and Chap. 11-Kantor et al. (57)). Contrary, the latter agents cause large-amplitude seizure-like rhythmic bursting in XII motoneurons of slices and in spinal motoneurons of the *en bloc* model (18, 59, 60).

Despite the fact that preBötC slices are principally functional in physiological K^+, elevated superfusate K^+ is needed if long-term (>2 h) recording is intended in <500-μm-thick preBötC slices that are used to minimize modulation of rhythmogenic inspiratory networks by rostrocaudally neighboring brainstem tissue (40, 42) (Sect. 4.2.1). Above (Sect. 2.2), we stated that a convenient strategy may be to perform such experiments initially in 5K/1Ca solution and gradually increase K^+ in steps of 1 mM once rhythm starts to slow down. Ultimately, in 7K/1Ca solution, preBötC-driven rhythms in the ventrolateral slice area and concomitant activities XII nerve and/or nucleus are stable for >6 h at a rate of 5–15 bursts/min (42). Although this is slightly slower compared to bursting in 9 mM K^+ in 1 mM Ca^{2+}, it is still fast enough for pharmacological and patch-clamp analyses (42, 44). For Ca^{2+} imaging, such burst rates in 7K/1Ca are even advantageous because recovery of inspiratory-related Ca^{2+} rises takes sometimes >5 s, at least in our hands using either Fluo-4 or Fluo-8L (Fig. 10). Thus, preBötC rhythm at >15 bursts/min would probably interfere with Ca^{2+} recovery (Sect. 4.2.4) (15, 24, 33, 40, 43).

While elevated K^+ is needed for long-term rhythm in slices, this manipulation affects the sensitivity of the preBötC to some neuromodulators. For example, we showed that DAMGO abolishes preBötC slice rhythm in 3K/1Ca at 10–25 nM (33). Our recent findings indicate that this blockade is reversed by increasing K^+ to 9 mM in the presence of such low DAMGO doses (Fig. 7) and that DAMGO needs to be elevated to almost micromolar concentrations to depress rhythm again in 9K/1Ca (see Fig. 11 in (18)). Similarly, pronounced slowing of preBötC slice rhythm by chemical anoxia is antagonized by raising K^+ from 3 to 9 mM in 1 mM Ca^{2+} (Fig. 7) (see also Fig. 11 in (18)). Moreover, we found that V_m oscillations in inspiratory preBötC neurons are much affected by superfusate K^+. In 3K/1Ca, a large population of inspiratory neurons show a ramp-like depolarization before the onset of the inspiratory (synaptic) drive potential and concomitant action potential discharge (Fig. 7). This "inspiratory burst" is often followed by a pronounced hyperpolarization that turns into a progressive preinspiratory depolarization (Fig. 7). Raising K^+ only modestly depolarizes resting V_m which becomes rather "flat" between inspiratory bursts (Fig. 7) (see also Fig. 12 in (18)).

Superfusate Ca^{2+}: Measurements with Ca^{2+}-sensitive microelectrodes in various brain regions in vivo indicated a most common physiological extracellular value of 1.1–1.2 mM for this cation (for references, see (18, 33, 34, 36, 39)). Despite this, most in vitro respiratory studies use superfusate Ca^{2+} levels that vary, mostly between 1.5 and 2.4 mM, and sometimes as low as 0.8 mM (for references, see (4, 10, 18, 25, 33, 34)). Recently, we investigated whether extracellular Ca^{2+} influences preBötC networks (34). Specifically, we found that robust rhythm of calibrated newborn rat preBötC slices in 3K/1Ca is strongly depressed by raising Ca^{2+} to 1.5 mM and that this depression is countered by increasing also K^+ to at least 7 mM (Fig. 6). This substantiates the hypothesis that inspiratory network activity is determined by a "Ca^{2+}/K^+ antagonism." This term has been introduced several decades ago by Leusen based on in vivo findings by his group using injection of both these ions into respiratory brainstem areas (for references, see (18, 33, 34)). We showed further in this study (34) that Ca^{2+} elevation from 1 to 1.5 mM also strongly depresses preBötC bursting in the newborn rat *en bloc* [–VII] model with rostrally exposed preBötC. Contrary, we found in the same study (34) that >1.5 mM Ca^{2+} is necessary to depress inspiratory rhythm in 3 K in *en bloc* preparations with more brainstem tissue rostral to the preBötC (Fig. 6) likely because K^+ is ~2 mM higher in the VRC area (Sect. 4.1.1). Our unpublished findings indicate that slice rhythms are slightly more robust and faster in 3 mM K^+ and 0.75 mM Ca^{2+}. In some preparations, however, seizure-like activity develops in inspiratory motor networks at 0.75 mM Ca^{2+}, and such hyperexcitability becomes more pronounced when Ca^{2+} is lowered further. In a recent study (42), we presented evidence that blockade of bursting in the isolated preBötC by raised Ca^{2+} is due to presynaptic inhibition. This view is based on an apparent lack of a postsynaptic effect of high Ca^{2+} on V_m or R_i in inspiratory preBötC neurons and neighboring tonically active neurons that may provide excitatory drive to the rhythmogenic inspiratory cells (Fig. 6). In summary, our latter findings indicate that superfusate Ca^{2+} levels above physiological values inhibit the preBötC and thus necessitate the use of 8–9 mM K^+ to counter this inhibition. At close-to-physiological Ca^{2+} levels of 1–1.2 mM, rhythm in (thick) slices occurs in 3 K for time periods of up to several hours and can be maintained for many hours after raising K^+ to maximally 7 mM.

Glucose: Extracellular brain glucose levels appear to range between 0.1 and 1 mM, depending somehow on the type of analytical tool and/or the brain region tested (for references see Chap. 1-Trapp and Ballanyi (45)). Contrary, 30 mM glucose was used in control solution in all our published studies on the *en bloc* model and non-calibrated preBötC slices and also in the vast majority, if not all, of studies by other groups using these models, for example, (4, 10–12, 22, 25, 29, 30, 32). In our published work on calibrated preBötC slices, 20 mM glucose was sufficient for robust rhythm over time periods of at least 8 h (in raised K^+). As indicated by our yet unpublished findings, in vitro inspiratory activity depends notably on superfusate glucose content. Specifically, we showed that (1) preBötC slices generated at 18–20°C in superfusate with 30 mM glucose show a slower rhythm (<5 bursts/min) that stops earlier in 3K/1Ca compared to slices generated under the same conditions and using the same solution except that glucose was 5 mM; (2) in 7K/1Ca, preBötC slices of up to 700 μm thickness generate stable rhythm in the ventrolateral aspect for >5 h in 1 mM glucose, but inspiratory XII activity seems to be more robust in 5–10 mM; and (3) for sustained viability of the *en bloc* model, superfusate glucose needs to be at least 20 mM while 30 mM are typically used. This is based on both our recent unpublished data and our previous findings on the *en bloc* model (84). Specifically, the latter study showed that 10 mM glucose is not sufficient for long-term inspiratory rhythm and that the typically highly anoxia-tolerant preparations (Sect. 4.1.1) respond with irreversible blockade of rhythm and major perturbation of ion homeostasis to anoxia in 10 mM glucose, particularly when preparations were obtained from P2-3 compared to P0-1 rats.

4.2.3. Cellular Mechanisms of preBötC Depression by Opioids and Anoxia

The above findings (Sect. 4.2.1) indicate that opioids and anoxia are powerful pharmacological tools for dissecting the individual components of the dual respiratory center. Currently, we study cellular mechanisms by which opioids depress preBötC networks and the clinically used methylxanthines caffeine and theophylline counter this depression. Our preliminary data indicate that opioids primarily inhibit preBötC neurons via presynaptic mechanisms, similar to inhibitory effects of high Ca^{2+} (Sect. 4.2.2) (Fig. 6), and that also methylxanthines exert their respirogenic action mainly at presynaptic sites (18, 42, 93). Our recent studies show that methylxanthines need to be applied at low millimolar doses for reversing in vitro inspiratory depression by opioids (44, 93). Based on the necessity for using such high doses of caffeine and theophylline, we hypothesized that the stimulatory action of caffeine and theophylline on inspiratory rhythms is due to blockade of cAMP-degrading phosphodiesterase-4 in rhythmogenic preBötC neurons (44) to which cAMP is presumably a major stimulus (5, 18, 31, 33, 44, 46, 71). Our related preliminary results indicate that low millimolar methylxanthine doses, applied in control solution, cause seizure-like hyperexcitability in spinal inspiratory motor networks (93) (Fig. 5). This is in line with the established clinical epileptogenic side effect of these agents (for references, see (44)).

Our recent study using the *en bloc* [+VII] model indicated that, similar to opioids, also anoxia does not hyperpolarize a major number of identified preBötC neurons (24). This contrasts with our previous work on the uncalibrated *en bloc* model and our recent unpublished data on calibrated preBötC slices, indicating that ~50% of inspiratory (preBötC) neurons are persistently hyperpolarized by anoxia with a notable concomitant fall of R_i as indication of a metabolically induced increase in K^+ conductance (5, 46, 70, 72). The above discrepancy in anoxia response may partly stem from the use of different models, that is, uncalibrated versus calibrated *en bloc* or *en bloc* versus slice. Contrary, comparison between anoxia and opioid responses points to a different cellular mechanism of a seemingly very similar frequency depression of inspiratory motor output by anoxia and opioids. This hypothesis is supported by our published findings based on voltage imaging in the *en bloc* model (Sect. 4.2.1) (Figs. 8 and 9) ((24) and unpublished Ca^{2+} imaging data obtained in the preBötC of calibrated slices). The latter findings show that DAMGO-evoked depression of preBötC inspiratory-related neuronal activity is accompanied by a fall of cytosolic Ca^{2+} in these neurons while Ca^{2+} in adjacent presumptive astrocytes does not change. Contrary, depression of preBötC neuronal bursting by anoxia is accompanied by a notable Ca^{2+} rise in inspiratory neurons and a concomitant, often more pronounced, Ca^{2+} increase in neighboring glia (see Fig. 4 in (15)).

4.2.4. Pharmoptic Approaches for preBötC Neuron–Glia Discrimination

In a recent "Perspective" article, we noted that glial cells may contribute to control of inspiratory preBötC neuron networks, similar to their recently established important roles in other behaviors (27). However, our published imaging data do not provide straightforward criteria for discrimination of glial from neuronal responses. For example, our voltage imaging study revealed different reconfiguring preBötC/pFRG activities in the *en bloc* [+VII] model in response to opioids and anoxia (24) (Figs. 8 and 9).

Importantly, current CCD camera-based voltage imaging in the in vitro respiratory models has principally a temporal resolution in the millisecond range, whereas spatial resolution is rather limited (15, 23, 24, 43). Consequently, voltage imaging data, as shown in Fig. 8, represent, on the one hand, the merged activity of inspiratory, pre-/postinspiratory, and tonic expiratory neurons (5). On the other hand, these presumably neuronal signals may be "contaminated" by responses from glial cells (15).

Compared to voltage imaging, the kinetics of (recovery from) respiratory-related Ca^{2+} rises are in the range of 1–5 s. This is, at least partly, due to buffering and binding properties of the currently available fluorescent dyes (15, 23, 62). This disadvantage is outweighed by spatial resolution at the (sub)cellular level (see below). Focal injection of Fluo-4-AM, as our preferred method of loading cells in the preBötC of calibrated slices with fluorescent Ca^{2+} dye (15, 23, 33, 40, 43, 94) (Sects. 2.8 and 3.3.2), loads both neurons and glia. Contrary, bath application of Fluo-4-AM or Fluo-8L-AM seems to primarily load glia in our hands. Consequently, our injection approach does not discriminate between preBötC neurons and glial cells unless additional techniques are applied, such as patch-clamp recording from at least one cell during population imaging, or imaging combined with genetic fluorescent protein marking of identified neuron or glial populations (15–17, 43). In the following, we present our preliminary data on combined pharmacologic and optic ("pharmoptic") approaches to discriminate neurons from glia. Also, in this section, we summarize some of our recent Ca^{2+} imaging data in preBötC slices, in extension to those regarding effects of anoxia and opioids outlined at the end of the previous section.

Our group was the first to adapt multiphoton/confocal microscopy techniques (62, 65) to respiratory network analysis in mammals. In a series of papers, (15, 18, 23, 33, 40, 43, 94), we unraveled both the activity and morphology of inspiratory active, histologically identified preBötC neurons in calibrated newborn rat brainstem slices (Sect. 3.3.2). Within <20 min upon focal injection of Fluo-4-AM into the preBötC or XII nucleus, typically 5–15 cells in a given focal imaging plane show inspiratory-related Ca^{2+} rises in a circular area of 150–300 µm diameter (15, 94) (Fig. 10). Presumptive inspiratory preBötC neurons in transversal newborn rat slices can be visualized into tissue depths of up to 80–100 µm (15, 33, 40). Thus, 50–100 active Fluo-4-labeled preBötC cells can theoretically be studied in a total of 4–5 imaging planes in each preBötC aspect or XII nucleus using our custom-made multiphoton system (Sect. 2.8). However, for optimal spatial resolution, corresponding imaging with our confocal or custom-made multiphoton systems (Fig. 3) is restricted to imaging planes of maximally 50–60 µm below the slice surface (Sect. 2.8). Either imaging system provides greatly increased spatial resolution compared to previous CCD camera-based imaging (11). This is particularly evident when 3D animations are made based on z-stack imaging series in slices exposed to glutamate (Sect. 3.3.2) (see figures in (15, 18, 23, 33, 40, 43, 94)). Our multiphoton imaging data indicate that presumptive inspiratory preBötC neurons have a soma diameter of ~12–25 μm and 2 to >5 primary dendrites, whereas astrocytic glial cells appear mostly round and seem to have soma diameters of <10 μm with no indication of successful labeling of their normally widespread but very fine processes (Figs. 10 and 11). Our published Ca^{2+} imaging studies were done in r+preBötC slices from cells located within the caudal 50 μm of the caudal preBötC margin. Our preliminary data (94) indicate that inspiratory active neurons in the most rostral preBötC aspect of c+preBötC slices are less abundant and more sparsely distributed in the transversal plane (Fig. 10). It is unlikely that fewer cells take up Fluo-4-AM in the rostral preBötC aspect because bath application of glutamate (0.1–1 mM) reveals a prominent Ca^{2+} rise in many cells with neuron-like morphology in both the rostral and caudal preBötC aspect (Fig. 10) (see also Fig. 4 in (15, 94)).

As a finding of major importance for understanding preBötC properties, we demonstrated that individual presumptive inspiratory neurons visualized with multiphoton/confocal Ca^{2+} imaging are capable in 3K/1Ca of reconfiguring their activity pattern between a fictive eupnea and a mixed eupnea-sigh burst pattern (see Fig. 7 in (40)) (Sect. 4.2.1). Specifically, after onset of in vitro apnea in 3K/1Ca, we used mostly thyrotropin-releasing hormone (0.1–100 nM), the phosphodiesterase-4 blocker rolipram (1–5 μM), or 7–9 mM K^{+} (Fig. 7) to evoke regular eupnea-like Ca^{2+} rises in preBötC neurons, whereas we applied the neurokinin-1 receptor agonist GR73632 to evoke a eupnea-sigh burst pattern (15, 18, 33, 40) (Sect. 4.2.1). More recently, we have found that phenylephrine (5–50 μM) promotes a prominent Ca^{2+} increase in neurons corresponding to fictive sigh signals in control solution (see Fig. 4 in (15)). However, stimulation of the amplitude of inspiratory bursts by phenylephrine is not accompanied by a major rise of Ca^{2+} baseline in active preBötC neurons, whereas presumptive glial cells respond often with a major Ca^{2+} increase. A modest rise of Ca^{2+} baseline is also typical for neurons in response to GR73632 while this agent has a minor, if at all, effect on presumptive glia. Moreover, thyrotropin-releasing hormone does not evoke a major Ca^{2+} rise in any preBötC cell type (see Fig. 4 in (15)).

Another of our recent studies (61) used our Ca^{2+} imaging approach to complement a wide range of other data from the group of Dr. G. D. Funk (Edmonton) for elucidating a potential contribution of preBötC glia to stimulation of inspiratory slice rhythm by adenosine triphosphate (ATP). Specifically, we found in that collaboration (61) that ATP evokes a Ca^{2+} rise in the preBötC area primarily in cells with a soma diameter of <10 μm, whereas Ca^{2+}

was not notably changed in larger rhythmically active neighboring cells, likely representing neurons. Prior to this study, we observed that further pharmacological agents besides chemical anoxia and phenylephrine evoke a less prominent Ca^{2+} rise in presumptive inspiratory active neurons versus astrocytes that are noninspiratory active, at least regarding cytosolic Ca^{2+} signaling. Specifically, the endoplasmic reticulum Ca^{2+} ATPase inhibitor cyclopiazonic acid and the metabotropic glutamate receptor agonist 1-aminocyclopentane-1,3-dicarboxylic acid (t-ACPD) apparently evoke a notably more prominent Ca^{2+} rise in glia (Fig. 11), which occurs often in an oscillatory pattern in the case of t-ACPD.

We have currently developed a further assay for highlighting glial cells in newborn rat brainstem slices. Specifically, bath application of Fluo-4-AM or Fluo-8L-AM can be combined with (or followed by) bath application of Cd^{2+} (Fig. 11). Cd^{2+} seems to enter brainstem glial cells more easily than neurons, in particular during the first 10 min of exposure, and subsequently interacts with Fluo to increase fluorescence intensity. Prelabeling with Fluo-4 or Fluo-8L can be used, for example, to target glial cells for patch-clamp recording in preBötC slices (11). We currently use the novel glial "Fluo-Cd^{2+} staining" approach after pharmacological analysis in preBötC slices to highlight the 3D organization of astrocyte networks. In addition, by using the fluorescent cell marker sulforhodamine-101, we can label XII motoneurons (in red) while using the combined Fluo-Cd^{2+} approach for labeling neighboring presumptive astrocytes (in green) (Fig. 11). Our most recent findings indicate that we can label newborn rat astrocytes in hippocampal and cortical slices after electrophysiological experiments with a green fluorescent antibody against glial fibrillary acidic protein (Chap. 11-Kantor et al. (57)). We have just started to use this approach to identify, after the electrophysiological imaging experiments, astrocytes in the preBötC and XII nucleus of calibrated slices.

Recently, we have started to use tetanic electrical stimulation (Sect. 3.2.3) in the calibrated slices for identifying properties of preBötC neurons and glia. Preliminary findings indicate that stimulation in the preBötC (or XII) area in slices loaded with Fluo-4 or Fluo-8L via bath application evokes a wave of Ca^{2+} rises in small cells. This Ca^{2+} wave spreads in all directions to a distance of sometimes >200 μm from the starting point of the signal at the electrode tip (Fig. 12). Similar tetanic stimulation of the preBötC following loading of both glia and neurons via injection of dye reveals that also (inspiratory active) neurons respond with a notable Ca^{2+} rise.

By combining electrical stimulation with suction electrode recording, we demonstrated that evoked activity in one preBötC aspect can depress activity in the ipsilateral preBötC aspect (Fig. 12). Moreover, we found that stimulation in the midline of the slice

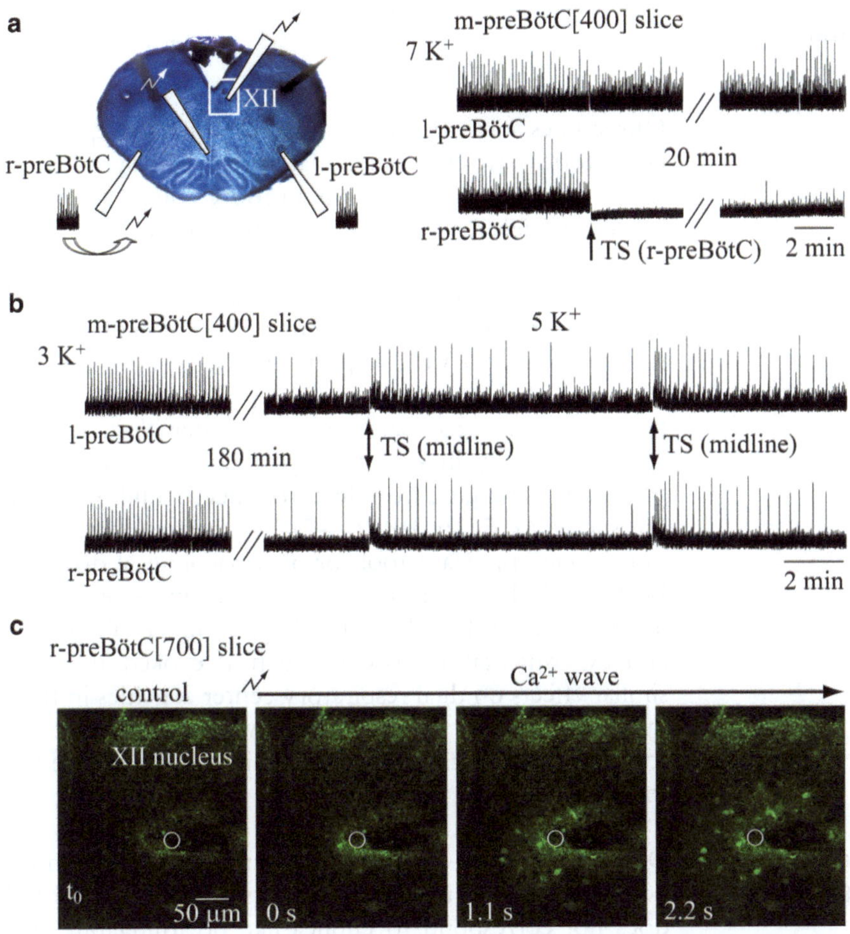

Fig. 12. Effects of electrical stimulation on inspiratory rhythm and glial Ca^{2+} in calibrated newborn rat preBötC slices. (**a**) In an m-preBötC[400/–0.75]W-P2 slice kept in 7 mM K^+ application of repetitive electrical pulses (50 V, 100 Hz, 100 pulses with 100-μs single-pulse duration) via the suction electrode in the *right* ("r"), preBötC aspect (see schema) caused long-term blockade of inspiratory rhythm in this region, whereas bursting in contralateral (*left*, "l") preBötC was notably less perturbed. (**b**) In an m-preBötC[400/–0.80]W-P1 slice, initially robust inspiratory rhythm in 3 K^+ stopped after 90 min. Raising K^+ to 5 mM reactivated inspiratory bursting in both preBötC aspect. Tetanic electrical stimulation (60 V, 100 Hz, 100 pulses with 250-μs single-pulse duration) in the slice midline (see schema in **a**) accelerated rhythm in both preBötC aspects. (**c**) In the XII nucleus (see schema in **a**) of a 700-μm-thick caudal "sandwich preBötC slice" (see Sect. 4.2.1) loaded via incubation with Fluo-4 local tetanic stimulation (70 V, 100 Hz, 100 pulses with 250-μs single-pulse duration) evoked a wave of cytosolic Ca_i rises in presumptive astrocytes that progressed at a speed of ~25 μm/s.

between both preBötC aspects can activate bouts of inspiratory rhythm in 5K/1Ca solution which, by itself, is not sufficient to restore rhythm after onset of in vitro apnea in 3K/1Ca (Fig. 12). These preliminary findings indicate that electrical stimulation is a new powerful tool for identifying neuron–glia interactions within the preBötC as we have demonstrated successfully for neural circuits in cerebellar slices (82).

5. Limitations

Our studies summarized above and numerous reports from other laboratories have established that both newborn rodent brainstem–spinal cord preparations and preBötC slices are highly potent in vitro models for studying respiratory control. Examples for outstanding results are the discovery of the preBötC and the pFRG and the fact that both these rhythmogenic neural networks remain active in a transversal slice and after transection of the *en bloc* model slightly caudal to VII_c, respectively (Sect. 4.2.1). The functional relevance of the preBötC and the pFRG and also the hypothesis that they comprise a dual respiratory center (Sect. 4.2.1) have been substantiated by in vivo findings that used a wide spectrum of preparations ranging from fetal rodents to adult mammals of various species (2, 5, 16, 19, 26, 47). For that research, opioids turned out as one important tool for pharmacological dissection of preBötC- and pFRG-driven activities, both in the *en bloc* model and in vivo (Sect. 4.2.1). In fact, a large number of neuromodulators and experimental approaches, such as evoked anoxia, have very similar effects on dual respiratory center activities in that model as in vivo (5, 46, 72). Despite these advantages, both types of respiratory networks in the dish have limitations as pointed out in the following.

5.1. Limitations of the En Bloc Model

Despite the above-described ultimately important findings, it has been argued that in vitro respiratory activity in the *en bloc* model does not correspond to eupnea-like breathing but rather represents (anoxia-related) gasping (95). However, in recent review articles, we summarized findings from our group and others showing that respiratory-related rhythms in the newborn rat *en bloc* model are very similar to in vivo activity in deafferented newborn rats in vivo at reduced temperature (5, 46, 72). Although our studies revealed an anoxic core in this preparation, this core does not seem to influence the respiratory networks that are located in more superficial layers where (respiratory) neurons operate in this model under aerobic (and even hyperoxic) conditions for cells close to the ventral surface (Sect. 4.1.1) (5, 37, 46, 72, 83). We note in the latter articles that the fact that respiratory networks in the *en bloc* model respond to experimentally evoked anoxia like intact newborn mammals indicates that these structures are not hypoxic or anoxic in control solution. This view is supported by the findings from our group and others summarized in our above review articles (5, 46, 72) that inspiratory rhythms in the *en bloc* model are not affected by removal via sectioning of the dorsal brainstem aspect which very likely eliminates the anoxic core. As long as superfusate glucose content is at least 20 mM (Sect. 2.2), anaerobic metabolism appears to be sufficient for maintaining cellular/axonal functions even in the anoxic core for several hours.

Inspiratory-related (cervical) spinal motor output persists in all types of calibrated *en bloc* models for >8 h (Sect. 3.2.1), whereas cranial inspiratory activity seems to deteriorate over time. Specifically, XII bursting disappears in *en bloc* [+VII] preparations within <1 h of recording according to our published study (34). Similarly, respiratory VII or trigeminal nerve bursting seems to stop during that time period based on our unpublished observations. Arrest of inspiratory XII bursting is likely not related to the anoxic core because evoked chemical anoxia of the entire preparation can restore such bursting. In addition, XII bursting is not abolished in *en bloc* [–VII] preparations during evoked anoxia as indicated by our preliminary findings. In our hands, also pre-/postinspiratory activity in lumbar and VII nerve roots disappears after 15–60 min in >50% of complete brainstem–spinal cord preparations, whereas such activity is typically stable for several hours in other laboratories, for example, that of Dr. Onimaru. At least in newborn mice, a further limitation of using the *en bloc* model with a major portion of tissue attached is related to the fact that the pons in this species exerts a tonic depressing action on the preBötC (96).

5.2. Limitations of Calibrated preBötC Slices

Our finding that preBötC slices are capable of rhythm in 3K/1Ca led to various important discoveries: (1) spatio-chemical organization of rhythmogenic inspiratory networks (Sect. 4.2.1) (40), (2) Ca^{2+}/K^{+} antagonism of bursting of (isolated) preBötC networks (34), (3) high sensitivity to some neuromodulators that is in various cases greatly attenuated by commonly used high K^{+} solutions, for example, during inspiratory depression by opioids and anoxia (18, 33). While we found that supraphysiological superfusate Ca^{2+} levels depress bursting of the isolated preBötC, use of 1 mM superfusate Ca^{2+} is not sufficient for countering occurrence of in vitro apnea in 3 mM K^{+} that occurs within <2 h in 400-μm-thin slices and within ~4 h in 700-μm-thick slices (33, 34, 40). While we found that use of close-to-physiological superfusate glucose concentrations (1–5 mM) during generation of preBötC slices results in more robust slice rhythm in 3K/1Ca, this does not delay onset of in vitro apnea. Consequently, particularly <700-μm-thick slices need to be studied in elevated K^{+} (7 mM) with glucose levels between 1 and 20 mM, and we recommend to start recording in 5K/1Ca (Sects. 2.2 and 4.2.2).

5.3. Limitations Common to En Bloc and Slice Models

One major limitation of both *en bloc* or slice models is the restriction of the time window of their use to the perinatal period, specifically to ~1 week old in rats for both models and to ~2 weeks for slices in mice, at least in some strains (Sect. 2.1). One major reason for this restriction may be related to the fact that the minimal neural substrate capable of respiratory rhythm likely expands rostrocaudally during the early postnatal time period. Consequently, thicker transversal preBötC slices would have to be generated, if older animals are intended to be used. These thicker slices from older animals would have a larger anoxic core that may result in inactivation and/or damage of preBötC (and pFRG) neurons. Similar to other neural structures, respiratory groups seem to lose, during the first postnatal week, their extreme tolerance to depletion of oxygen (O_2) that is evident at birth (5, 46, 72, 83). As discussed above (Sect. 4.1.1) an anoxic core is already present in the newborn rodent *en bloc* model (37, 83). However, this core does not seem to include essential cellular elements of the dual respiratory center except for apparently anoxia-tolerant axons that connect both bilaterally organized VRC aspects (Sects. 4.1.1 and 5.1). The anoxic core would extend notably after the first postnatal week in parallel with an increase in brainstem size. To avoid this limitation while preserving some of the advantages of an in vitro approach, respiratory networks can be studied in older animals in an internally perfused brainstem model that is also applicable to newborn rodents (for references, see (46, 48) and Chap. 13-Day and Wilson (91)).

As a further restriction, *en bloc* preparations in general and preBötC slices in 3K/1Ca need to be studied at a temperature that is ~10°C lower than in vivo (Sect. 3.1.4). This results in slowing of respiratory-related bursting and may also distort some (synaptic) interactions within the respiratory networks and possibly affect their responses to some neuromodulators as well. Our unpublished observations indicate that the robustness of inspiratory-related rhythm in 3K/1Ca is more pronounced in preBötC slices from Wistar versus Sprague-Dawley rats and that onset of in vitro *apnea* also varies between litters from the same rat strain. This indicates that properties of the isolated preBötC differ between rodent species and also depend on factors such as nutrition or stress which may even differ between litters in a given species and strain. Another limitation of either model is related to the fact that slices with partially exposed preBötC in close-to-physiological 3K/1Ca solution differ in their inspiratory-related burst patterns, depending on whether the preBötC is neighbored by rostral or caudal tissue (Sect. 4.2.1) (40). Modulation by adjacent brainstem tissue may also affect the pFRG upon isolation in a slice, but this has not been studied yet. The latter considerations indicate that it may be advantageous to keep the rostrocaudal microenvironment of the preBötC (or pFRG) kernel intact, as is the case in the *en bloc* [+VII] model. While analysis of neuronal properties in both networks is possible using blind whole-cell recording techniques (Sect. 4.1.2), imaging at cellular resolution has yet been reported only for respiratory neurons located <100 μm below the (slice) surface based on previously used (custom-made) multiphoton microscopes (Sect. 4.2.4) (15, 33, 40). The latest generation of such microscopes is presumably capable of dynamic (Ca^{2+}) imaging deeper within living brainstem tissues, both in slices and in the *en bloc* model, but this still needs to be demonstrated. Alternatively, "GRIN" lenses or micro-optrodes may be used which combine microelectrode penetration of tissue with the capability for cellular imaging (for references, see (15, 23, 43, 62)).

A further obstacle for "respiratory imaging" (15) is the reduced capability for long-term analysis of in vitro respiratory networks which is mostly related to the toxicity of this approach. For voltage imaging, most of the currently used dyes seem to depress respiratory network activity, and extended time periods of imaging are required because of the necessity for averaging (62), normally for ~50 cycles with the system we used (Sects. 2.8, 3.3.1, and 4.2.4) (15, 23, 24). Regarding our currently used multiphoton system, continuous Ca^{2+} imaging at a sampling rate of ~1 frame/s is limited to 1–2 h. Beyond that time period at that sampling rate or after correspondingly shorter time for higher scanning rates, both neurons and glial cells show a progressive increase in resting fluorescence intensity along with decreased resolution of morphological details. The cells stop showing spontaneous Ca^{2+} rises (in the case of neurons) or responses to experimental manipulations. On average, the total time for Ca^{2+} imaging in healthy cells using the confocal system seems to be slightly shorter than for the multiphoton imaging system. It needs to be considered that multiphoton imaging may reduce some phototoxicity because only cells in the focal plane are excited by the potentially harmful excitation laser light. However, the necessity for using an IR laser for this imaging approach comes at the expense of a potential caloric cell damage, particularly in more superficial tissue layers (15, 23, 24, 43, 62). Possibly, these problems will be solved, or at least improved, by the new generation of voltage and multiphoton imaging systems and/or other (genetically-engineered) dyes.

6. Perspective

Despite these limitations, the isolated calibrated respiratory networks from perinatal rodents will certainly provide further results of ultimate importance for understanding the neural control of breathing. We conclude with resounding a hypothetical scenario from our review article on "respiratory imaging" (15) that would likely answer, with advanced imaging techniques, most relevant questions in neural control of breathing: "Subpopulations of respiratory interneurons and output cells will be identified via specific patterns of expressed fluorescent proteins and acutely fluorescence-tagged ion channels, receptors and/or transporters. Populations of these identified cells will then be loaded with cell-permeant fluorescent dyes for, e.g., simultaneous imaging of dynamic changes of signaling factors such as Ca^{2+} or pH in the cytosol and/or subcellular compartments. Subsequently, one or a few of these cells will be whole-cell recorded with additional functional or morphological dyes in the patch electrode solution for a correlation analysis of dynamic changes of subcellular activities with respiratory-related membrane potential oscillations or underlying membrane currents. Further improvement of computerized data processing and the spatiotemporal resolution of fluorescence microscopy will enable simultaneous 4-dimensional imaging of all these events for a thorough structure-function relationship of respiratory centers and motor circuits."

Acknowledgments

Work contributing to this study was supported by the Alberta Heritage Foundation for Medical Research (AHFMR), Alberta Innovates Health Solutions-Hotchkiss Brain Institute (AIHS-HBI), the Canada Foundation for Innovation (CFI-ASRIP), and operating plus training grants from the Canadian Institutes of Health Research (CIHR).

References

1. Bianchi AL, Denavit-Saubié M, Champagnat J (1995) Central control of breathing in mammals: neuronal circuitry, membrane properties, and neurotransmitters. Physiol Rev 75:1–45
2. Feldman JL, Del Negro CA (2006) Looking for inspiration: new perspectives on respiratory rhythm. Nat Rev Neurosci 7:232–242
3. Smith JC, Abdala AP, Rybak IA, Paton JF (2009) Structural and functional architecture of respiratory networks in the mammalian brainstem. Philos Trans R Soc Lond B Biol Sci 364:2577–2587
4. Suzue T (1984) Respiratory rhythm generation in the *in vitro* brain stem-spinal cord preparation of the neonatal rat. J Physiol 354:173–183
5. Ballanyi K, Onimaru H, Homma I (1999) Respiratory network function in the isolated brainstem-spinal cord of newborn rats. Prog Neurobiol 59:583–634
6. Ballanyi K, Onimaru H (2009) Respiratory network analysis, isolated respiratory center functions. In: Binder M, Hirokawa N, Windhorst U (eds) Encyclopedia of neuroscience. Springer, Heidelberg/New York/Tokyo, pp 3453–3458
7. Smith JC, Ellenberger HH, Ballanyi K, Richter DW, Feldman JL (1991) Pre-Bötzinger complex: a brainstem region that may generate respiratory rhythm in mammals. Science 254:726–729
8. Molkov YI, Abdala AP, Bacak BJ, Smith JC, Paton JF, Rybak IA (2010) Late-expiratory activity: emergence and interactions with the respiratory CpG. J Neurophysiol 104: 2713–2729
9. Smith JC, Ballanyi K, Richter DW (1992) Whole-cell patch-clamp recordings from respiratory neurons in neonatal rat brainstem *in vitro*. Neurosci Lett 134:153–156
10. Rekling JC, Champagnat J, Denavit-Saubi M (1996) Electroresponsive properties and membrane potential trajectories of three types of inspiratory neurons in the newborn mouse brain stem in vitro. J Neurophysiol 75:795–810
11. Koshiya N, Smith JC (1999) Neuronal pacemaker for breathing visualized *in vitro*. Nature 400:360–363
12. Del Negro CA, Morgado-Valle C, Feldman JL (2002) Respiratory rhythm: an emergent network property? Neuron 34:821–830

13. Ramirez JM, Viemari JC (2005) Determinants of inspiratory activity. Respir Physiol Neurobiol 147:145–157
14. Doi A, Ramirez JM (2008) Neuromodulation and the orchestration of the respiratory rhythm. Respir Physiol Neurobiol 164:96–104
15. Ruangkittisakul A, Okada Y, Oku Y, Koshiya N, Ballanyi K (2009) Fluorescence imaging of active respiratory networks. Respir Physiol Neurobiol 168:26–38
16. Gaultier C, Gallego J (2008) Neural control of breathing: insights from genetic mouse models. J Appl Physiol 104:1522–1530
17. Gray PA (2008) Transcription factors and the genetic organization of brain stem respiratory neurons. J Appl Physiol 104:1513–1521
18. Ballanyi K, Ruangkittisakul A (2009) Structure-function analysis of rhythmogenic inspiratory pre-Botzinger complex networks in "calibrated" newborn rat brainstem slices. Respir Physiol Neurobiol 168:158–178
19. Champagnat J, Morin-Surun MP, Bouvier J, Thoby-Brisson M, Fortin G (2011) Prenatal development of central rhythm generation. Respir Physiol Neurobiol 178:146–155
20. Iizuka M (2001) Respiratory activity in glossopharyngeal, vagus and accessory nerves and pharyngeal constrictors in newborn rat in vitro. J Physiol 532:535–548
21. Iizuka M (2004) Rostrocaudal distribution of spinal respiratory motor activity in an in vitro neonatal rat preparation. Neurosci Res 50:263–269
22. Iizuka M (2011) Respiration-related control of abdominal motoneurons. Respir Physiol Neurobiol 179(1):80–88
23. Ballanyi K, Onimaru H, Keller B (2009) Respiratory network analysis, functional imaging. In: Binder M, Hirokawa N, Windhorst U (eds) Encyclopedia of neuroscience. Springer, Heidelberg/New York/Tokyo, pp 3447–3453
24. Ballanyi K, Ruangkittisakul A, Onimaru H (2009) Opioids prolong and anoxia shortens delay between onset of pre-inspiratory (pFRG) and inspiratory (preBötC) network bursting in newborn rat brainstems. Eur J Physiol (Pflügers Archiv) 458:571–587
25. Onimaru H, Homma I (2003) A novel functional neuron group for respiratory rhythm generation in the ventral medulla. J Neurosci 23:1478–1486
26. Guyenet PG, Mulkey DK (2010) Retrotrapezoid nucleus and parafacial respiratory group. Respir Physiol Neurobiol 173:244–255
27. Ballanyi K, Panaitescu B, Ruangkittisakul A (2010) Control of breathing by nerve glue. Sci Signal 3:pe41
28. Taccola G, Secchia L, Ballanyi K (2007) Anoxic augmentation of lumbar respiratory bursts and block of lumbar locomotion in newborn rat brainstem-spinal cord. J Physiol 585:507–524
29. Giraudin A, Cabirol-Pol MJ, Simmers J, Morin D (2008) Intercostal and abdominal respiratory motoneurons in the neonatal rat spinal cord: spatiotemporal organization and responses to limb afferent stimulation. J Neurophysiol 99:2626–2640
30. Smith JC, Greer JJ, Liu GS, Feldman JL (1990) Neural mechanisms generating respiratory pattern in mammalian brain stem-spinal cord in vitro. I. Spatiotemporal patterns of motor and medullary neuron activity. J Neurophysiol 64:1149–1169
31. Mironov SL, Langohr K, Richter DW (1999) A_1 adenosine receptors modulate respiratory activity of the neonatal mouse via the cAMP-mediated signaling pathway. J Neurophysiol 81:247–255
32. Tryba AK, Ramirez JM (2004) Hyperthermia modulates respiratory pacemaker bursting properties. J Neurophysiol 92:2844–2852
33. Ruangkittisakul A, Schwarzacher SW, Secchia L, Poon BY, Ma Y, Funk GD, Ballanyi K (2006) High sensitivity to neuromodulator-activated signaling pathways at physiological [K^+] of confocally imaged respiratory center neurons in on-line-calibrated newborn rat brainstem slices. J Neurosci 26:11870–11880
34. Ruangkittisakul A, Secchia L, Bornes TD, Palathinkal DM, Ballanyi K (2007) Dependence on extracellular Ca^{2+}/K^+ antagonism of inspiratory centre rhythms in slices and *en bloc* preparations of newborn rat brainstem. J Physiol 584:489–508
35. Ballanyi K, Ruangkittisakul A (2009) Brain slices. In: Binder M, Hirokawa N, Windhorst U (eds) Encyclopedia of neuroscience. Springer, Heidelberg/New York/Tokyo, pp 483–490
36. Hansen AJ (1985) Effect of anoxia on ion distribution in the brain. Physiol Rev 65:101–148
37. Brockhaus J, Ballanyi K, Smith JC, Richter DW (1993) Microenvironment of respiratory neurons in the in vitro brainstem-spinal cord of neonatal rats. J Physiol 462:421–445
38. Völker A, Ballanyi K, Richter DW (1995) Anoxic disturbance of the isolated respiratory network of neonatal rats. Exp Brain Res 103:9–19
39. Somjen GG (2002) Ion regulation in the brain: implications for pathophysiology. Neuroscientist 8:254–267
40. Ruangkittisakul A, Schwarzacher SW, Secchia L, Ma Y, Bobocea N, Poon BY, Funk GD, Ballanyi K (2008) Generation of eupnea and sighs by a spatiochemically organized inspiratory network. J Neurosci 28:2447–2458

41. Ruangkittisakul A, Panaitescu B, Ballanyi K (2011) K^+ and Ca^{2+} dependence of inspiratory-related rhythm in novel "calibrated" mouse brainstem slices. Respir Physiol Neurobiol 175:37–48
42. Panaitescu B, Ruangkittisakul A, Ballanyi K (2009) Silencing by raised extracellular Ca^{2+} of pre-Bötzinger complex neurons in newborn rat brainstem slices without change of membrane potential or input resistance. Neurosci Lett 456:25–29
43. Ruangkittisakul A, Ballanyi K (2009) Neuron-glia-imaging. In: Binder M, Hirokawa N, Windhorst U (eds) Encyclopedia of neuroscience. Springer, Heidelberg/New York/Tokyo, pp 2756–2764
44. Ruangkittisakul A, Ballanyi K (2010) Methylxanthine reversal of opioid-evoked inspiratory depression via phosphodiesterase-4 blockade. Respir Physiol Neurobiol 172:94–105
45. Trapp S, Ballanyi K (2012) Autonomic nervous system in vitro: studying tonically active neurons controlling vagal outflow in rodent brainstem slices. In Isolated Central Nervous System Circuits (Ed K Ballanyi), Neuromethods Series Vol. 73 (Ed W Walz). Springer Science+Business Media, LLC, New York, NY, pp 1–59
46. Ballanyi K (2004) Neuromodulation of the perinatal respiratory network. Curr Neuropharmacol 2:221–243
47. Greer JJ, Funk GD, Ballanyi K (2006) Preparing for the first breath: prenatal maturation of respiratory neural control. J Physiol 570:437–444
48. Ballanyi K (1999) Isolated tissues: in vitro preparations. In: Windhorst U, Johansson H (eds) Modern techniques in neuroscience research. Springer, Heidelberg, pp 307–326
49. Neher E, Sakmann B (2009) Single-channel recording, 2nd edn. Springer, New York/Dordrecht/Heidelberg/London
50. Walz W (2007) Patch-clamp analysis: advanced techniques, 2nd edn. Humana, Totowa
51. Onimaru H, Ballanyi K, Richter DW (1996) Calcium-dependent responses in neurons of the isolated respiratory network of newborn rats. J Physiol 491:677–695
52. Onimaru H, Ballanyi K, Homma I (2003) Contribution of Ca^{2+}-dependent conductances to membrane potential fluctuations of medullary respiratory neurons of newborn rats *in vitro*. J Physiol 552:727–741
53. Schwarzacher S, Pestean A, Günther S, Ballanyi K (2002) Serotonergic modulation of respiratory motoneurons and interneurons in brainstem slices of perinatal rats. Neuroscience 115:1247–1259
54. Moore AR, Zhou WL, Jakovcevski I, Zecevic N, Antic SD (2012) Physiological properties of human fetal cortex in vitro. In Isolated Central Nervous System Circuits (Ed K Ballanyi), Neuromethods Series Vol. 73 (Ed W Walz). Springer Science+Business Media, LLC, New York, NY, pp 125–158
55. Sanchez-Vives MV (2012) Spontaneous rhythmic activity in the adult cerebral cortex in vitro. In Isolated Central Nervous System Circuits (Ed K Ballanyi), Neuromethods Series Vol. 73 (Ed W Walz). Springer Science+Business Media, LLC, New York, NY, pp 263–284
56. Broicher T, Speckmann EJ (2012) Living human brain slices: network analysis using voltage sensitive dyes. In Isolated Central Nervous System Circuits (Ed K Ballanyi), Neuromethods Series Vol. 73 (Ed W Walz). Springer Science+Business Media, LLC, New York, NY, pp 285–300
57. Kantor C, Panaitescu B, Kuribayashi J, Ruangkittisakul A, Jovanovic I, Leung V, Lee TF, MacTavish D, Jhamandas JH, Cheung PY, Ballanyi K (2012) Spontaneous neural network oscillations in hippocampus, cortex and *locus coeruleus* of newborn rat and piglet brain slices. In Isolated Central Nervous System Circuits (Ed K Ballanyi), Neuromethods Series Vol. 73 (Ed W Walz). Springer Science+Business Media, LLC, New York, NY, pp 315–356
58. Biggs JE, Lu VB, Kim H, Lai A, Todd KG, Ballanyi K, Colmers WF, Smith PA (2012) Defined medium organotypic cultures of spinal cord put 'pain in a dish'. In Isolated Central Nervous System Circuits (Ed K Ballanyi), Neuromethods Series Vol. 73 (Ed W Walz). Springer Science+Business Media, LLC, New York, NY, pp 405–435
59. Brockhaus J, Ballanyi K (1998) Synaptic inhibition in the isolated respiratory network of neonatal rats. Eur J Neurosci 10:3823–3839
60. Brockhaus J, Ballanyi K (2000) Anticonvulsant adenosine A_1 receptor-mediated adenosine action on neuronal networks in the brainstem-spinal cord of newborn rats. Neuroscience 96:359–371
61. Huxtable AG, Zwicker JD, Alvares TS, Ruangkittisakul A, Fang X, Hahn LB, Posse de Chaves E, Baker GB, Ballanyi K, Funk GD (2010) Glia contribute to the purinergic modulation of inspiratory rhythm generating networks. J Neurosci 30:3947–3958
62. Yuste R, Konnerth A, Masters B (2006) Imaging in neuroscience and development, a laboratory manual. J Biomed Opt 11:19902
63. Fish KN, Gonzales-Burgos G, Zaitsev AV, Lewis DA (2012) Histological characterization of physiologically determined fast spiking interneurons in slices of the primate dorsolateral prefrontal cortex. In Isolated Central Nervous System Circuits (Ed K Ballanyi), Neuromethods

Series Vol. 73 (Ed W Walz). Springer Science+Business Media, LLC, New York, NY, pp 159–181
64. Nakamura TJ, Michel S, Block GD, Colwell CS (2012) Neural circuits underlying circadian oscillations in mammals: clocks in a dish. In Isolated Central Nervous System Circuits (Ed K Ballanyi), Neuromethods Series Vol. 73 (Ed W Walz). Springer Science+Business Media, LLC, New York, NY, pp 183–210
65. Stosiek C, Garaschuk O, Holthoff K, Konnerth A (2003) *In vivo* two-photon calcium imaging of neuronal networks. Proc Natl Acad Sci USA 100:7319–7324
66. Paxinos G, Halliday G, Watson C, Koutcherov Y, Wang H (2007) Atlas of the developing mouse brain. Elsevier, Amsterdam/Boston/Heidelberg
67. Greene EC (1935) The anatomy of the rat. Braintree Scientific, Philadelphia
68. Cho YR, Pawela CP, Li R, Kao D, Schulte ML, Runquist ML, Yan JG, Matloub HS, Jaradeh SS, Hudetz AG, Hyde JS (2007) Refining the sensory and motor ratunculus of the rat upper extremity using fMRI and direct nerve stimulation. Magn Reson Med 58:901–909
69. Mandadi S, Nakanishi ST, Han P, Humphreys J, Whela PJ (2012) The use of rodent isolated spinal cord preparations to examine motor output. In Isolated Central Nervous System Circuits (Ed K Ballanyi), Neuromethods Series Vol. 73 (Ed W Walz). Springer Science+Business Media, LLC, New York, NY, pp 437–464
70. Ballanyi K, Völker A, Richter DW (1994) Anoxia induced functional inactivation of neonatal respiratory neurons *in vitro*. Neuroreport 6:165–168
71. Ballanyi K, Lalley PM, Hoch B, Richter DW (1997) cAMP dependent reversal of opioid and prostaglandin mediated depression of the isolated respiratory network in newborn rats. J Physiol 504:127–134
72. Ballanyi K (2004) Protective role of neuronal K_{ATP} channels in brain hypoxia. J Exp Biol 207:3201–3212
73. Stachniak TS, Sudbury JR, Trudel E, Choe KY, Ciura S, Bourque CW (2012) Osmoregulatory circuits in slices and *en bloc* preparations of rodent hypothalamus. In Isolated Central Nervous System Circuits (Ed K Ballanyi), Neuromethods Series Vol. 73 (Ed W Walz). Springer Science+Business Media, LLC, New York, NY, pp 211–231
74. Trapp S, Lückermann M, Brooks PS, Ballanyi K (1996) Acidosis of dorsal vagal neurons in situ during spontaneous and evoked activity. J Physiol 496:695–710
75. Ben-Ari Y, Gaiarsa JL, Tyzio R, Khazipov R (2007) GABA: a pioneer transmitter that excites immature neurons and generates primitive oscillations. Physiol Rev 87:1215–1284
76. McKay BE, Tadayonnejad R, Anderson DM, Engbers JDT, Fernandez FR, Iftinca M, Turner RW (2012) Establishing in vivo like activity in rat cerebellar cells maintained in vitro. In Isolated Central Nervous System Circuits (Ed K Ballanyi), Neuromethods Series Vol. 73 (Ed W Walz). Springer Science+Business Media, LLC, New York, NY, pp 233–262
77. Ballanyi K, Grafe P, ten Bruggencate G (1987) Ion activities and potassium uptake mechanisms of glial cells in guinea-pig olfactory cortex slices. J Physiol 382:159–174
78. Konnerth A, Keller BU, Ballanyi K, Yaari Y (1990) Voltage sensitivity of NMDA-receptor mediated postsynaptic currents. Exp Brain Res 81:209–212
79. Ballanyi K, Branchereau P, Champagnat J, Fortin G, Velluti J (1993) Extracellular potassium, glial and neuronal potentials in the solitary complex of rat brainstem slices. Brain Res 607:99–107
80. Ballanyi K, Doutheil J, Brockhaus J (1996) Membrane potentials and microenvironment of rat dorsal vagal cells *in vitro* during energy depletion. J Physiol 495:769–784
81. Ballanyi K, Kaila K (1998) Activity-evoked changes in intracellular pH. In: Kaila K, Ransom BR (eds) pH and brain function. Wiley, New York, pp 283–300
82. Kulik A, Haentzsch A, Lückermann M, Reichelt W, Ballanyi K (1999) Neuron-glia signaling via α_1 adrenoceptor-mediated Ca^{2+} release in Bergmann glia *in situ*. J Neurosci 19:8401–8408
83. Ballanyi K, Kuwana S, Völker A, Morawietz G, Richter DW (1992) Developmental changes in the hypoxia tolerance of the in vitro respiratory network of rats. Neurosci Lett 148:141–144
84. Ballanyi K, Völker A, Richter DW (1996) Functional relevance of anaerobic metabolism in the isolated respiratory network of newborn rats. Eur J Physiol (Pflügers Archiv) 432:741–748
85. Voipio J, Ballanyi K (1997) Interstitial P_{CO2} and pH, and their role as chemostimulants in the isolated respiratory network of neonatal rats. J Physiol 499:527–542
86. Ammann D (1986) Ion-selective microelectrodes. Springer, Berlin/Heidelberg/New York/Tokyo
87. Baumgärtl H (1987) Systematic investigations of needle electrode properties in polarographic measurements of local tissue $P0_2$. In: Ehrly AM, Hauss J, Kolde G (eds) Clinical oxygen pressure

measurement. Springer, Berlin/Heidelberg/New York/London/Paris/Tokyo, pp 17–42

88. Ballanyi K, Grafe P (1985) An intracellular analysis of γ-aminobutyric-acid-associated ion movements in rat sympathetic neurones. J Physiol 365:41–58
89. Ballanyi K, Mückenhoff K, Bellingham MC, Okada Y, Scheid P, Richter DW (1994) Activity-related pH changes in respiratory neurones and glial cells of cats. Neuroreport 6:33–36
90. De Curtis M, Lilbrizzi L, Uva L, Gnatkovsky V (2012) Neuronal networks in the in vitro isolated Guinea Pig Brain. In Isolated Central Nervous System Circuits (Ed K Ballanyi), Neuromethods Series Vol. 73 (Ed W Walz). Springer Science+Business Media, LLC, New York, NY, pp 357–383
91. Day TA, Wilson RJA (2012) The study of respiratory chemoreflexes using a novel dual-perfused rodent preparation. In Isolated Central Nervous System Circuits (Ed K Ballanyi), Neuromethods Series Vol. 73 (Ed W Walz). Springer Science+Business Media, LLC, New York, NY, pp 385–403
92. Onimaru H, Kumagawa Y, Homma I (2006) Respiration-related rhythmic activity in the rostral medulla of newborn rats. J Neurophysiol 9:55–61
93. Ruangkittisakul A, Panaitescu B, Kuribayashi J, Ballanyi K (2010) Caffeine reversal of opioid-evoked and endogenous inspiratory depression in perinatal rat en bloc medullas and slices. Adv Exp Med Biol 669:123–127
94. Bobocea N, Ruangkittisakul A, Ballanyi K (2010) Multiphoton/confocal Ca^{2+}–imaging of inspiratory pre-Bötzinger complex neurons at the rostral or caudal surface of newborn rat brainstem slices. Adv Exp Med Biol 669: 81–85
95. St. John WM (2009) *Noeud vital* for breathing in the brainstem: gasping-yes, eupnoea-doubtful. Philos Trans R Soc Lond B Biol Sci 364:2625–2633
96. Zanella S, Roux JC, Viemari JC, Hilaire G (2006) Possible modulation of the mouse respiratory rhythm generator by A1/C1 neurones. Respir Physiol Neurobiol 153:126–138

Chapter 3

Physiological Properties of Human Fetal Cortex In Vitro

Anna R. Moore, Wen-Liang Zhou, Igor Jakovcevski, Nada Zecevic, and Srdjan D. Antic

Abstract

Emergence of neuronal membrane excitability (e.g., voltage-gated ion conductances) and spontaneous electrical activity (e.g., spontaneous firing of action potentials) is an imperative for the normal development and maturation of brain circuits. Understanding the interplay between electrical activity and normal brain development may prove critical for prevention and therapy of devastating neurological and psychiatric diseases. Due to the limited availability of human fetal tissue, our current understanding of functional (physiological) maturation of the human cerebral cortex has been limited to *in vivo* and *in vitro* animal models. Although invaluable for the generation of basic insights into this process, animal models fall short in providing an accurate model of human cortical development. The discrepancy between animal and human models of brain development is largely due to evolutionary changes such as differences in the time course, number of cellular divisions, and cellular composition of cortical layers, making it difficult to ascertain the step by step physiological maturation in human. Here, we provide detailed methodology on how to handle human fetal tissue, generate acute brain slices, and perform whole-cell patch-clamp recordings of biophysical membrane properties in human cortical neurons in their natural surroundings, that is, preserved neighboring neurons, glia, and extracellular matrix.

Key words: Action potential, Cerebral cortex, Human, Membrane excitability, Patch-clamp, Subplate neuron

1. Introduction

The development of neuronal excitability as expressed by action potential firing is essential for the establishment of neuronal electrical activity, which plays a vital role in maturation of individual neurons and development of functional cerebral cortical synaptic circuits (1–3). Understanding when a developing neuron becomes excitable is important for determining its role in normal cerebral cortical development and how disruptions to that natural process

Klaus Ballanyi (ed.), *Isolated Central Nervous System Circuits*, Neuromethods, vol. 73,
DOI 10.1007/978-1-62703-020-5_3, © Springer Science+Business Media New York 2012

can lead to long-term neurological deficits. One of the major techniques used for measuring cell excitability is the patch-clamp technique (4), which allows direct and reliable access to the electrical properties of an individual cell in any given cortical region or cortical layer (5, 6). Analysis of cerebral cortical development has previously been limited to recordings in animal models (e.g., rodents, cat, and monkey) or human neurons dissociated from brain tissue and grown in culture (7–14). Although valuable for determining general trends in neuronal differentiation, studies of animal models cannot provide an accurate analysis of the physiological maturation in the human cerebral cortex. The differences between the rodent and human brain development are significant. More specifically, rodent neuron progenitors divide 11 times, while in the human cortex, the number of cell divisions exceeds 30 (15). A greater number of cell divisions leads to a drastically larger cortical volume (~1,000-fold) and significantly greater diversity in cell type expression in the human cerebral cortex compared to rodent (15, 16). This is especially evident when comparing cortical layers 2/3 between the two species. Besides these remarkable differences in total cell count, cell density, and structure of the neuropil (morphology and geometry of dendritic and axonal arbors), there is an obvious difference in temporal dynamics of cortical development. Young neurons in the rodent brain mature over a period of just a few days, whereas in human, maturation occurs over a period of several months. This expanded period of development may reveal some critical time points and events, which are either absent in rodents or difficult to discern due to temporal and spatial overlap. More complex human cortical development may, at least partly, explain why rodents often fail to develop the same neurological deficits seen in human (15).

Due to restricted availability and technical problems pertaining to experimental work with acute human (fetal) brain tissue (17, 18) (see also Chap. 9-Broicher & Speckmann (19)), detailed physiological characterization of the developing human cortex has not been performed thus far. This chapter outlines some basic techniques and challenges associated with using human fetal tissue in terms of measuring neuronal excitability and spontaneous electrical activity in an isolated brain circuit. The clear advantage of using whole-cell patch-clamp techniques to explore membrane potential (V_m) and membrane current (I_m) responses of young neurons is to provide details about maturation of individual cellular elements, which cannot be detected using, for example, electroencephalography or functional magnetic resonance imaging from preterm babies (20–22). By using whole-cell patch-clamp techniques in human cells within semi-intact cortical tissue (preserved extracellular matrix, preserved neighboring glial cells, neurons, and synaptic contacts, all of which are absent from primary cell cultures), we are able to detect the development of basic membrane excitability (e.g., insertion of functional voltage-gated ion channels into the membrane and formation of the axon initial segment). We can also elucidate the early presence of spontaneous electrical neuronal activity and networking such as spontaneous background synaptic inputs and sustained depolarizations crowned with action potentials, similar to cortical activity states ("UP" states) that occur in slow wave sleep (23, 24). Successful utilization of acute human fetal brain slices and patch-clamp recordings can open a door to a set of powerful experimental techniques. These techniques include intracellular injections of anatomical markers; Ca^{2+}-sensitive, voltage-sensitive, or pH-sensitive fluorescent dyes; a plethora of pharmacological agents; exogenous mRNAs; and various vectors for cDNA. All this will further our understanding of cellular and molecular processes that govern cerebral cortical development in humans. The focus of this chapter is to provide a detailed overview of how to prepare, handle, and maintain human fetal acute brain slices for electrophysiological recordings and analysis. Together with practical advice, we include actual examples of basic electrophysiological measurements and immunolabeling performed in human fetal brain tissue.

2. Materials

2.1. Electrophysiological Setup

Details of standard electrophysiological plus optical methods and equipment have been provided elsewhere (Chaps. 1-Trapp & Ballanyi (25) and 2-Ruangkittisakul et al. (26)) (27, 28). Here, we just list the actual equipment our group uses for whole-cell patch-clamp recordings in human fetal brain tissue: (1) an anti-vibration table (Minus K Technology, Inglewood, CA, USA; www.minusk.com) with a massive baseplate to reduce mechanical vibrations during recordings; the microscope X-Y platform and fixed stage (Siskiyou Design Inc., Grants Pass, OR, USA; www.siskiyou.com) are mounted on the baseplate of the anti-vibration table; (2) a custom-made Faraday cage is placed around the setup to prevent ambient electrical noise from interfering with the patch-clamp recordings; (3) a BX51WI upright microscope (Olympus, Center Valley, PA, USA; www.olympusamerica.com) with a 5× objective and 40× water immersion objective is connected to an infrared (IR) charge-coupled device (CCD) camera (IR-1000, Dage-MTI, Michigan City, IN, USA; www.dagemti.com) through an optical zoom that provides additional magnification (0.75×–5.25×), Thales-Optem Zoom (70×L); (4) MPC-200 motor-driven remote control micromanipulators with movement along three axes (XYZ) and the fourth (x-y-combined) diagonal axis (Sutter Instrument Company, Novato, CA, USA; www.sutter.com); (5) a multibarrel perfusion system (VC-6MCS Perfusion Valve Control System, Warner Instruments, Hamden, CT, USA; www.warneronline.com); (6) an Axon Instruments Multiclamp 700B amplifier for recording (intracellular) electrical signals that are digitized (Digidata Series 1322A, Molecular Devices Corp., Sunnyvale, CA, USA; www.moleculardevices.com) and stored on a computer using the included pClamp 9.2 acquisition software while off-line data analysis is performed with Clampfit 9.2 (Molecular Devices); (7) WinTV computer card (Hauppauge Computer Works Inc., NY, USA; www.hauppauge.com) is used for digitization of "IR video feed" (taking still pictures); and (8) dual TC344B automated temperature control heater (Warner Instruments) for maintaining the fluid in the recording chamber at 32–33°C.

2.2. Fabrication of Patch Pipettes

The following equipment, materials, and procedures are used for the fabrication of patch pipettes: (1) a horizontal micropipette puller (P-97 Flaming/Brown; Sutter Instrument Company) for pulling the pipettes in several steps to attain the desired tip size and electrical resistance, (2) borosilicate glass (length 10 cm, external/internal diameter 1.5/0.86 mm) with internal glass filament (Warner Instruments), and (3) prior to use, the glass capillaries are thoroughly washed to remove dust and contaminants. In our experience, this procedure reduces the chance of pipette clogging during patching. For washing, glass capillaries (450 pieces) are loaded into a 500-ml glass beaker filled with 70% ethanol. The beaker is next placed on an electric heater and the heat gradually increased until the ethanol boiling point is reached. Caution is necessary because rising the temperature too quickly may cause ethanol to burst out of the beaker. It is important that capillaries stay in the vertical position inside the beaker during heating, so that bubbles of boiling ethanol permeate through the capillary lumen on their way towards the surface. Twenty to thirty minutes of washing in boiled ethanol is usually sufficient to clean the lumen of the capillaries. Next, the glass is cooled down at room temperature, taken out of ethanol, rinsed in acetone, and laid in a clean covered box to dry overnight. Prior to placing wet capillaries into the box, we line the bottom of the box with a fresh aluminum foil and we fold the same foil to create a cover, which prevents dust particles from getting in touch with clean capillaries.

2.3. Solutions

The electrolyte solution used for bathing and perfusion of brain slices is traditionally called artificial cerebrospinal fluid (ACSF). ACSF was designed with the intention to mimic the ionic concentration of the cerebral spinal fluid (CSF). However, the concentration of Ca^{2+} in standard electrophysiological ACSF (2–2.5 mM) is about twofold greater than in natural CSF (see below), partly because it is believed that calcium ions (Ca^{2+}) stabilize the neuronal membranes and improve the quality of (patch-clamp) recordings.

Table 1
Solutions

Standard ACSF	"Low" Ca^{2+} ACSF	Intracellular solution	Cs^+ intracellular solution
(mM)	(mM)	(mM)	(mM)
125 NaCl	125 NaCl	135 K-gluconate	70 Cs_2SO_4
2.3 KCl	2.3 KCl	10 HEPES	10 HEPES
26 $NaHCO_3$	26 $NaHCO_3$	2 $MgCl_2$	6 EGTA
2 $MgSO_4$	2 $MgSO_4$	3 ATP-Na_2	0.1 $CaCl_2$
1.26 KH_2PO_4	1.26 KH_2PO_4	0.3 GTP-Na_2	4 ATP-Mg_2
2 $CaCl_2$	1.2 $CaCl_2$	10 phosphocreatine	0.3 GTP-Na_2
10 glucose	10 glucose		10 phosphocreatine

Besides Ca^{2+} concentration, there are other discrepancies between standard ACSF and naturally occurring CSF (29) (see also Chaps. 1-Trapp & Ballanyi (25), 2-Ruangkittisakul et al. (26), 8-Sanchez-Vives (30) and 11-Kantor et al. (31)). Nevertheless, brain slice electrophysiologists, including our group, continue to use the term "ACSF" because (1) of the large body of previously published literature that had adopted this expression, (2) the composition of ACSF is similar between different research groups, which allows direct comparison of data coming from different laboratories, and (3) the attribute "artificial" indicates that discrepancies exist between this experimental solution and natural CSF.

In our experiments, standard ACSF is made on the day of recordings from cold stock and cold double deionized water (ddH_2O) used to dilute the stock solution. This stock solution contains the same electrolytes as the final solution, though at a 5× concentration, with the exception of glucose and $CaCl_2$ to avoid, respectively, growth of microorganisms or Ca^{2+} precipitation in the stock bottle. D-glucose is added to the final ACSF as a dry substance. $CaCl_2$ is the last ingredient added to the final ACSF. $CaCl_2$ is stored as a 1-M stock solution in the refrigerator, together with the ACSF 5× stock solution and deionized water. Cold temperature (+4°C to +8°C) of these three components ($CaCl_2$, 5× stock ACSF, and H_2O) prevents the precipitation of $CaCl_2$ during mixing of final ACSF. Complete ionic composition of the final standard ACSF can be found in Table 1. In standard ACSF with 2 mM $CaCl_2$, human subplate neurons (32–34) exhibit spontaneous electrical activity (35). The outbursts of action potential firing are separated by prolonged periods of silence. To intensify spontaneous electrical activity in the cerebral cortex, an ACSF with close-to-physiological Ca^{2+} concentration (1.2 mM) is used in the perfusion system, as described in (Chaps. 2-Ruangkittisakul et al. (26), 8-Sanchez-Vives (30) and 11-Kantor et al. (31)) (36). We term this solution "Low Ca^{2+} ACSF." The preparation of Low Ca^{2+} ACSF is identical to standard ACSF, except that a smaller volume of $CaCl_2$ stock is added.

Patch electrode solutions aim at mimicking the intracellular ionic concentration of the patched cell, unless particular (unphysiological) agents are used to manipulate, for example, ion channel activities or cellular signaling pathways (see also Chap. 1-Trapp & Ballanyi (25)). In control conditions, we use K^+-gluconate-based solution enriched with adenosine triphosphate (ATP) and guanosine triphosphate (GTP) whose exact ionic composition (Table 1) is identical to that of intracellular solution used for patching animal neurons (37). To pharmacologically isolate transmembrane Na^+ currents, K^+ channels are blocked with alkali metal cesium ions (Cs^+). For these experiments, K^+-gluconate in the internal solution is substituted by Cs^+ sulfate (Cs_2SO_4, 70 mM). We term this solution "Cs^+ intracellular solution" (Table 1). The pH of extracellular and intracellular solutions is measured with a standard pH meter and adjusted with NaOH for extracellular solutions, KOH for normal intracellular solution, and CsOH for Cs^+-based intracellular solutions to a pH value of 7.3. A freezing point osmometer (μOsmete, Precision Systems Inc., Natick, MA, USA; www.precisionsystemsinc.com) is used to measure the osmolality of the solutions. Both standard ACSF and Low Ca^{2+} ACSF should have a measured osmolality of 305–310 mOsm/kg, and both intracellular solutions should be in the range between 290 and 295 mOsm/kg.

2.4. Human Fetal Brain Tissue

All tissue for our studies is obtained from the Albert Einstein College of Medicine Tissue Repository (The Bronx, NY, USA). Tissue is collected with proper parental consent and approval of the ethics committees at both universities. Fetal age is estimated based on weeks after ovulation, crown-rump length, and anatomical landmarks (38). Aside from fetal age and sex, no other information is provided. The tissue is transferred in a plastic collection specimen vial containing ice-cold "Hank's Balanced Salt Solution" (5.33 mM KCl, 0.441 mM KH_2PO_4, 4.17 mM $NaHCO_3$, 137.93 mM NaCl, and 0.338 mM Na_2HPO_4) with 5.5 mM D-glucose and antibiotic/antimycotic (10 μl/ml). The container is kept in an ice cooler, and the content is continuously carbogenated with a portable 5% CO_2/95% O_2 containing tank fitted with a gas dispenser (porous stone).

For brain dissection, all instrumentation is sterilized prior to use and includes (1) an ice bucket/tray (approximately 20″ × 12″ × 4″), (2) a polystyrene tissue culture dish (150 × 25 mm), (3) small tweezers (approximate size # 3), (4) a scalpel #10, (5) small scissors, (6) a 10–15-cm-long flat anatomical knife, (7) a beaker to discard solution, and (8) a beaker with ice-cold ACSF to collect the tissue chunk for acute brain slices.

Brain slices are cut by hand using a sterile razor blade (Gillette – super stainless, super inoxydable). Other commercially available razor blade brands can serve the same purpose. Tissue is cut on a custom-made frozen 7 cm × 7 cm acrylic block of 1.25-cm thickness. Brain slices are transferred with a blunt end glass Pasteur pipette into a holding chamber. The slice-holding (incubation) chamber is a 250-ml beaker with mesh platform approximately 2/3 distance between the bottom of the beaker and surface of ACSF (see diagram in Fig. 6.19.1 in (39)) (see also Fig. 4 in Chap. 1-Trapp & Ballanyi (25)). The holding chamber is filled with standard ACSF and continuously carbogenated first at 37°C for 40 min and then at room temperature.

3. Human Fetal Brain Slice Preparation and Recording

3.1. Tissue Collection

Tissue, typically one brain hemisphere, is obtained within 15 min postmortem and placed into a specimen container containing ice-cold Hank's Balanced Salt Solution (Sect. 2.3). It is kept on ice and carbogenated with a portable 5% CO_2/95% O_2 tank for the total duration of the ~2.5 h lasting transfer from New York to Farmington (Connecticut). Upon arrival in our laboratory, the tissue is placed in a −20°C freezer for 5–10 min to harden for facilitating the dissection. For dissection, it is placed on ice in a Petri dish, and meninges are carefully removed using tweezers and scissors. The dorsal and ventral portion of the occipital cortex (~10–20 mm thick) is cut out with a flat anatomical knife and then submerged in ice-cold (it contains ice pieces) ACSF for 5 min.

Maintaining the fetal brain tissue at low temperature, near 0°C, is an essential step for improving the quality of electrophysiological recordings in cortical slices. At room temperature, the human fetal brain is extremely soft. In our experience, the generation of viable slices is mechanically very difficult at room temperature. In the field of *in vitro* electrophysiology, there is a common belief that cooling down brains prior to dissection, and especially during the actual slicing step, significantly reduces tissue damage and improves the quality of thin slices. Note that neuron protective solutions (e.g., containing a low Na^+, low Ca^{2+}, and high Mg^{2+} and, eventually, glutamate receptor antagonists) (see also Chaps. 1-Trapp & Ballanyi (25) and 11-Kantor et al. (31)) are not being employed in our laboratory during the dissection step. Rather, brain slices are cut in the same standard ACSF as used for electrophysiological recordings (Table 1).

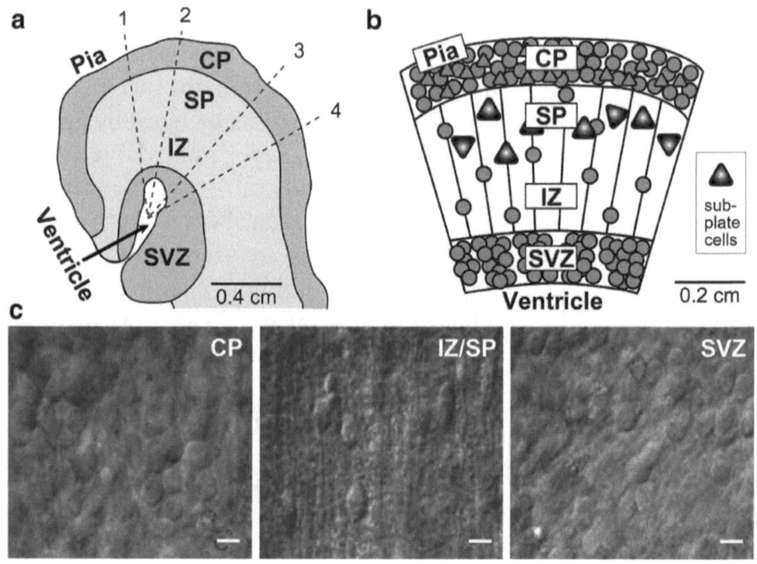

Fig. 1. *Human fetal brain tissue – acute brain slice preparation.* (**a**) Schematic diagram of the human fetal telencephalon including the cortical plate (CP), subplate (SP), intermediate zone (IZ), subventricular zone (SVZ), and the lateral ventricle (Ventricle). The *dashed lines* (1–4) illustrate the angles through which slices can be cut to include all areas of interest. (**b**) Diagram illustrating the architecture of the acute human fetal brain slice including the dense cellular regions SVZ and CP and the more sparse areas, the IZ and SP. In layer SP, one often finds large cells dubbed "SP" neurons (33). (**c**) Live cells in the acute human fetal brain slice residing in the CP (*left*), IZ/SP (*middle*), and SVZ (*right*), as seen during patch-clamp recordings. Microphotographs obtained using the same camera used for infrared differential interference contrast (IR-DIC) video microscopy. Scale bar 10 μm. Unpublished data.

3.2. Acute Brain Slices

After 5 min in ice-cold ACSF, the brain tissue block (~25 × 12 mm cortical slab) is placed on the frozen plastic block (Sect. 2.4) placed inside a polystyrene tissue culture dish (150 × 25 mm style). The bottom of this dish is in direct contact with ice during the procedure. In addition, both the cortical slab and acrylic block are surrounded by chunks of frozen ACSF during the entire slice-cutting procedure. Prior to slicing, the cortical slab is oriented in such way that it lies on the acrylic block with its ventricular (ependymal) surface facing down, while the pial surface is facing up. Cutting slices (in either sagittal or coronal plane) is performed by holding the razor blade in free hand. A decisive, but not too fast, cut is made through the pia towards the ventricular surface, severing all cortical layers in between. We seek the shortest distance between pia and ventricular surface (Fig. 1a). Depending on the dorsolateral location of the pial entering point (where blade enters the pia), our slices, made from one ~25 × 12 mm cortical slab, constitute a continuum of planes between the sagittal plane and horizontal plane or between the coronal plane and horizontal plane. The exact thickness of the human cortical slice is impossible to control by blunt (freehand, no vibratome) cutting. Nevertheless, we aim to cut <1-mm-thick slices. More often than not, the main cortical slab is slanted with the pial surface protruding slightly over the surface of the ventricular base. For generation of slices of relatively uniform thickness, the plane of the razor blade must follow (in parallel) the plane of the slanted surface. To fulfill this goal, quite often the razor blade has to be held at a 45° angle against the acrylic block to provide that the cutting edge of the razor blade is following an oblique (45°), rather than a perpendicular (90°), trajectory. After sectioning, slices are placed into a holding chamber with carbogenated ACSF, incubated at 37°C for 30 min, and then stored at room temperature (Sect. 2.4) prior to recordings.

Despite the set of conditions that can be considered suboptimal for any animal slice preparation (long postmortem delay, soft tissue, freehand cutting), we are still able to create translucent, healthy human brain slices containing the subventricular zone (SVZ), intermediate zone (IZ), subplate (SP), and cortical plate (CP) (Fig. 1b). At 40× magnification, IR differential interference contrast (IR-DIC) microscopy of the acute brain slices easily resolves individual cells in each anatomical region. Healthy cells have robust and smooth cellular membranes (Fig. 1c), whereas unhealthy and apparently dead cells are characterized by sharp, high-contrast edges, porous plasma membranes, and by prominent or swollen nuclei. SVZ cells are properly round and densely packed together. They do not show obvious cytoplasmic protrusions (processes) under regular IR-DIC video microscopy (Fig. 1c). On thin histological sections and upon histological staining, the SVZ displays radial cellular bands composed of chains of migrating neurons (38, 40). These radial SVZ bands are difficult, but not impossible, to spot in acute brain slices using IR-DIC microscopy and a water immersion 40× objective. The IZ, on the other hand, is identified by the presence of long radial fibers extending abundantly from the ventricular to pial surfaces (Fig. 1c). Young migrating cells of the IZ (putative pyramidal neuron progenitors and interneuron progenitors) are small and round, or small and fusiform, with one or two main processes extending at either pole. Some migrating IZ cells could also be seen in juxtaposition to a thick radial fiber. Located between the IZ and CP is a large region identified as the (cortical) subplate layer (SP). In the SP zone, we find specialized cells, so-called SP neurons (32–34), which are some of the firstly born neurons in the developing cerebral cortex. SP neurons can be identified as large cells, typically double the size of migrating neurons, ovate or pyramidal-like in shape, with multiple primary processes emerging from the cell body. The CP layer is characterized by a dense cell population in fetal cortex, similar in density to the SVZ. Note that these two densely packed cellular regions flank the fetal cortical slice (Fig. 1b). CP cells can be distinguished from SVZ cells because they are not as tightly packed together and typically have a more defined elongated cell shape.

Microscopic appearances (at 5× or 40× magnification) of the SVZ and CP are somewhat similar, and inexperienced experimenters may have a hard time discriminating between these cortical layers. Should there be any confusion between the pial and ventricular edge in the isolated brain slice, here we offer additional anatomical orienting points: Note that radial fibers extend (fan-out) from the SVZ, with the distance between them gradually increasing as fibers transverse the IZ and approach the pia. For this reason, the pial edge is notably and invariably longer than the ventricular edge in any given slice (we had encountered so far). In other words, acute slices obtained from human fetal cortex have a trapezoid shape with a shorter (often concave) ventricular edge and a more lengthy (often convex) pial edge (Fig. 1b).

3.3. Whole-Cell Patch-Clamp Recordings

Acute tissue slices of the human fetal occipital cortex are placed in the recording chamber and gravity-perfused with carbogenated ACSF at a rate of 200 ml/h. Fast perfusion rates and continuous flow of solution increase the longevity of the human brain slices in the recording chamber. Individual cells are selected using an upright microscope equipped with IR-DIC video features (Sect. 2.1). Cells are identified based on anatomical positioning within the slice (anatomical layers SVZ, IZ, SP, CP), size and shape of the cell body, and the number and orientation of primary processes emerging from the cell body.

Patch pipettes, when filled with normal intracellular solution (Sect. 2.2), have a DC resistance of 7–12 MΩ. If the pipette mouth is too big (as indicated by a DC resistance <7 MΩ), then too much stress is imposed on the fetal cell, and this shortens the cell's life. If the pipette mouth is too small, this results in poor quality current- or voltage-clamp recordings (Sect. 4.3). We have determined empirically that 7–12-MΩ patch pipettes represent a good trade-off between the longevity and quality of recordings in human fetal cerebral cortex. Note that cells in different anatomical regions vary in size, with larger soma diameters seen in the SP and CP. Thus, pipette resistance and tip size should complement the dimensions of the selected cell. Additionally, one can decide how to fabricate recording pipettes based on the experimental aim. We use 10–12-MΩ patch pipettes for longer recording sessions (details below).

The maneuvering of the pipette through the brain slice (approach) and formation of a "gigaohm seal" are attained in voltage-clamp mode of the amplifier (Sect. 2.1) in a fashion identical to that developed for animal neurons in cortical brain slices (41). Here, we discuss four points that are somewhat different from the majority of animal experiments: (1) In animal experiments, positive pressure (in the order of 3–6 kPa) is applied to the back of the pipette to prevent pipette contamination with particles or tissue as it travels to and below the surface of the slice. Because cellular membranes are softer and more delicate in human fetal tissue, it is recommended to use smaller pressure (~1.5 kPa) to improve the success rate of stable whole-cell recordings. (2) In animal experiments, as the pipette is brought to the cell surface and nearly touches the neuronal cell membrane, an outward flow of electrode solution in response to the positive pressure creates a visible dimpling of the cell surface. At this point, positive pressure is released and slight negative pressure (suction) is applied to seal the membrane to the pipette tip. Here, the resistance of contact increases with concomitant substantial reduction in the current, creating what is referred to as a gigaohm seal. In the case of human fetal neurons, it is both difficult, because the membrane is apparently fragile, and also unnecessary to form a dimple prior to gigaohm seal formation. It is sufficient to detect a tiny reflection of the pipette tip on the surface of the human cell, prior to reversing the pressure gradient and pursuing the formation of tight seal. (3) Although strong suction can be used to establish the whole-cell configuration in a human fetal neuron, it is recommended to combine modest negative pressure (e.g., −2 kPa) and a short "ZAP" (duration = 100 μs) which is a voltage pulse capable of disrupting the membrane under the patch implemented in several patch-clamp amplifier systems such as Multiclamp 700A or B (Molecular Devices) and Patchmaster (HEKA Instruments, Lambrecht/Pfalz, Germany; www.heka.com). (4) Unlike adult animal neurons, where membrane "break-in" is manifested with a huge increase in amplitude and duration of the electrical transient, in human fetal neurons, the membrane rupture under the patch is manifested as a tiny increase in amplitude and virtually no change in the duration (half-width) of the capacitative transient. For practical advice on adjusting the bridge balance and capacitance neutralizations in whole-cell recordings, see, for example, the manual (42) by Boris Barbour, Ecole Normale Superieure, Paris, France. Barbour's manual is used by students of the Microelectrodes Techniques Course in Plymouth, UK. It is available at http://www.biologie.ens.fr/perso/barbour.

3.4. Passive Membrane Properties

3.4.1. Resting Membrane Potential

In experiments performed in our laboratory on pyramidal neurons of the rodent cerebral cortex, using the same equipment and solutions as described above, resting V_m varies between −58 and −68 mV (not corrected for the liquid junction potential) (43). V_m in human fetal neurons, on the other hand, varies in a wider range, specifically between −30 and −60 mV. This trend for a positive shift in V_m obtained in the human data set may be interpreted in two ways. Firstly, a positively shifted V_m may be the consequence of neuronal immaturity. V_m of cortical neurons becomes more negative during the course of development (44, 45). Secondly, such a "positive" V_m may be the sign of neuronal injury due to prolonged hypoxic conditions that characterize the storage, handling, and transportation of the postmortem tissue. Being unable to experimentally resolve this issue and given the scarcity of the human brain tissue, we employ the following strategy when recording from human cells. Upon changing from voltage-clamp to current-clamp recording mode and documenting the cell's intrinsic V_m, we next apply negative DC current ("holding current") in the range of −3 to −15 pA. In the majority of cells, this modest amount of current is sufficient to shift V_m to the target value of approximately −60 mV because the membrane input resistance (R_i) is typically in the gigaohm range (6). Holding V_m at −60 mV is beneficial for our experimental goals. For example, at this potential, a greater fraction of voltage-gated Na^+ channels escapes inactivation and becomes available for the generation of an action potential (a critical experimental test of membrane excitability). Also, at −60 mV, the flux of Ca^{2+} ions through voltage-gated Ca^{2+} channels is reduced, which suppresses potentially harmful Ca^{2+}-sensitive intracellular reactions and thus improves the longevity of cellular recordings. Finally, at −60 mV, the driving force for ion fluxes (e.g., those associated with glutamatergic synaptic inputs) is increased, which improves their detection and analysis.

3.4.2. Input Resistance

R_i corresponds to the electrical resistance which DC current encounters as it tries to find the shortest path from the tip of the electrode (through cytoplasm and pores in the lipid membrane) to the ground wire. This parameter includes the series resistance, the sum of pipette and access resistance, axial (cytoplasmic) resistance, and membrane resistance (see Fig. 4 in Barbour's manual (42)). When measuring neuronal R_i, it is important to keep the access resistance small. The access resistance can be reduced by fabricating electrodes with large pipette tips, maintaining a clean pipette tip during approach and "punching" a large hole in the plasma membrane during the whole-cell "break-in" (resilient suction). We measured R_i of human fetal cortical cells in current-clamp mode. After adjusting the bridge balance and pipette capacitance, a series of current steps, usually in the narrow range −20 to +20 pA (step size 2 or 4 pA), is applied to the cell. The most accurate way for calculating R_i is to plot the change in voltage (V) versus the amplitude of applied current (I) and use the slope of the linear fit (Fig. 2a). However, this type of analysis is quite cumbersome and impractical to perform during the actual experiment. In order to make a quick estimate of neuronal R_i directly from the oscilloscope trace, the steady-state displacement of V_m is divided by the intensity of the current step [$R_i = \Delta V_m / I_{injected}$].

Systematic measurements of R_i are particularly useful in developmental neuroscience because three major cellular processes affect this parameter. Firstly, the formation of gap junctions (i.e., electrical synapses) between cortical cells tends to reduce R_i. Cells with significantly lower R_i than their counterparts in the same cortical layer (SVZ, IZ, or CP) are likely to be electrically interconnected

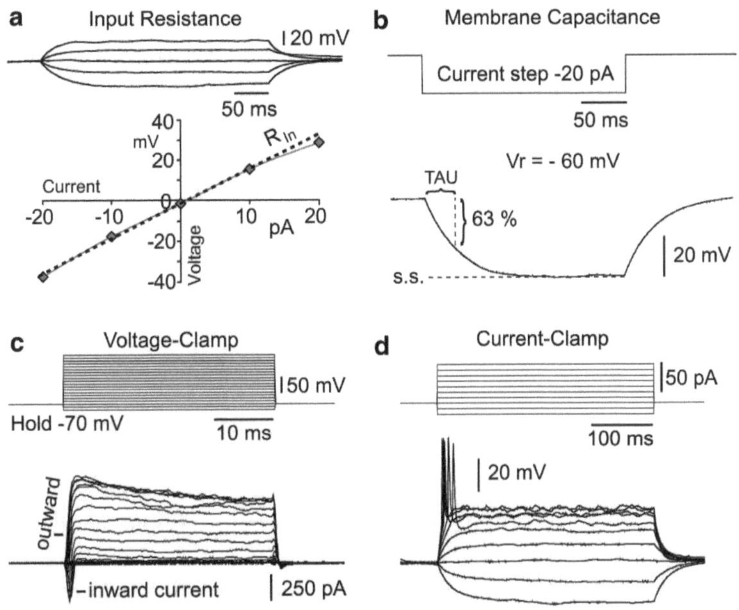

Fig. 2. *Passive and active membrane properties of fetal human neurons.* (**a**) To determine input resistance (R_i) in current-clamp whole-cell recording mode, each cell was exposed to a series of current pulses from −20 to +20 pA for 350 ms. The amplitude of the membrane potential (V_m) change (at steady state) was plotted versus the amplitude of the injected current. The slope of the linear regression fit through the I-V plot is the measure of R_i. In this example, $R_i = 1.85$ GΩ. (**b**) To determine membrane capacitance (C_m), we used negative current pulses (range: −40 to −20 pA), duration 350 ms. C_m was calculated by dividing the time constant tau (τ) by R_i. τ is the time needed for the V_m transient to reach 63% of its steady-state value (s.s.). (**c**) Transmembrane currents measured in voltage-clamp recording mode of human fetal neuron (*top*). V_m was held at −70 mV and then stepped in 10 mV increments from −90 to +100 mV for 50 ms (*bottom*). (**d**) To determine if fetal human cells were able to generate action potentials, we held them in current-clamp at −60 mV using a DC negative current (range: −3 to −15 pA) and then applied a series of current steps from −20 to +40 pA for a duration of 350 ms (*top*). V_m responses (*bottom*) indicate that some human fetal cells fire an action potential in response to direct current stimulation. Unpublished data.

through gap junctions (46–48). Secondly, the growth of dendrites and axons increases the neuronal membrane surface area, thus providing greater number of electrical paths for injected current to reach the ground wire (see definition of R_i above). Thirdly, during the process of differentiation, voltage-gated channels are "trafficked" to the membrane. An insertion of voltage-gated ion channels into the plasma membrane increases the number of electrically conductive pores (making the membrane leaky) and consequently the neuronal R_i drops notably (47, 49). In summary, determination of R_i may hold an important clue on how far a neuron has advanced in the process of differentiation.

3.4.3. Membrane Capacitance

The plasmalemmal membrane acts as an insulator between the interior and exterior electrolytes of the cell and comprises also a significant electrical capacitor. The term membrane capacitance expresses the ability of the cell membrane to store electrical charge; it is directly proportional to the surface area of the cell. The greater the surface area, the larger the capacitance, as more charge will be stored along the membrane. In the study of human brain development, total cell capacitance can be used as a measure of the neuronal process growth. Membrane capacitance (C_m) can be measured from the membrane voltage response to a negative current pulse (Fig. 2b). The first step is to determine the membrane time constant Tau (τ), where $\tau = C_m \, x \, R_i$. Tau is equal to the time period required to charge the membrane to 63% of its steady-state value. The second step is to determine R_i as described above. To calculate C_m, τ is divided by *Rin* ($C_m = \tau/R_i$). C_m of human fetal cortical neurons varies between 10 and 25 pF (6).

3.5. Active Membrane Properties

3.5.1. Voltage-Gated Ion Currents

Developmental onset of electrical excitability is arguably the most important cellular and molecular process in neuronal differentiation. This event unequivocally denotes cells of the neuronal lineage because neurons are the only cellular elements in the brain capable of generating Na^+ action potentials. Na^+ action potentials depend on the synthesis, trafficking, and insertion of voltage-gated Na^+ and K^+ channels. Traditionally, the presence of voltage-gated channels is detected, and their biophysical properties are studied with voltage-clamp techniques. To properly clamp the voltage of young human neurons, especially in experiments aiming at analyzing the (in)activation kinetics of transmembrane ion currents, one should use pipette tips as large as the membrane can tolerate. In our experience, human fetal neurons can be successfully whole-cell-recorded with 7–10-MΩ patch electrodes (Sect. 3.3). In a typical experiment (Fig. 2c), a human neuron is held at −70 mV, and a series of voltage steps is applied to reveal voltage-sensitive currents present at a given gestational age and in a particular cortical layer (6). It is recommendable to initiate the voltage-clamp protocol with voltage pulses more negative than the holding potential. Stepping from the holding potential to more hyperpolarized V_m values (e.g., −90 mV) results in closure of voltage-gated Na^+ channels and removal of their residual inactivation. Young human neurons express two predominant types of voltage-gated currents, selective for Na^+ and K^+ ions, manifested by brief inward and sustained outward current transients, respectively (Fig. 2c). In our experience, every human fetal cortical cell, regardless of its location in the SVZ, IZ, or CP, displays a certain amplitude of voltage-gated K^+ current, but in respect to voltage-gated Na^+ currents, these cells vary greatly. The majority of SVZ, IZ, and CP cells display Na^+ current of only 0–100 pA, whereas a smaller fraction of cells (presumably more mature neurons), located exclusively in the CP and SP, can occasionally have larger peak Na^+ current amplitudes ranging from 500 to 2,000 pA (6).

3.5.2. Isolated Voltage-Gated Na^+ Currents

Mammalian voltage-gated Na^+ channels constitute a family of molecularly distinct subtypes (isoforms). The most frequent mRNA isoforms, Nav1.2 and Nav1.6, are detected by immunolabeling and single-cell analysis, while Nav1.3 and Nav1.7 are moderately expressed in embryonic and neonatal neurons (50, 51). Furthermore, it has been suggested that in the same cell type and the same individual cell, one Nav isoform (e.g., Nav1.2) is replaced by another (e.g., Nav1.6) as a natural part of the differentiation process (52). Each Nav isoform has a unique characteristic activation and inactivation dynamics (53, 54). Ideally, one could study the activation and inactivation dynamics of the isolated Na^+ current and determine the molecular composition of membrane proteins.

One strategy for isolating the voltage-gated Na^+ current is "TTX subtraction." In this procedure, an identical series of voltage steps is applied to the same cell before (control) and after bath application of the selective Na^+ channel blocker tetrodotoxin (TTX, 0.5–1 μM) (Fig. 3b). Current traces corresponding to same

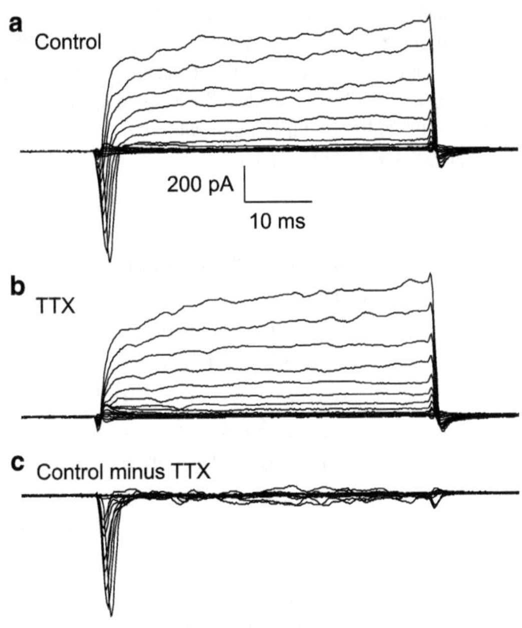

Fig. 3. *Pharmacological isolation of voltage-gated Na^+ current in human neurons.* (**a**) Membrane currents (I_m) measured with voltage-clamp of human fetal cells bathed in standard artificial cerebrospinal fluid (ACSF; "control"). Each trace may contain both inward (putative Na^+) and outward (putative K^+) voltage-gated ion currents. (**b**) Following bath application of 500 nM of the voltage-gated Na^+ channel blocker tetrodotoxin (TTX) for 5 min, the same voltage steps are applied to the cell resulting in only outward (putative K^+) current. (**c**) Resulting isolated voltage-gated Na^+ current following subtraction of traces "**b**" from traces "**a**" performed using Clampfit 9.2 software by Molecular Devices (see text for details). Unpublished data.

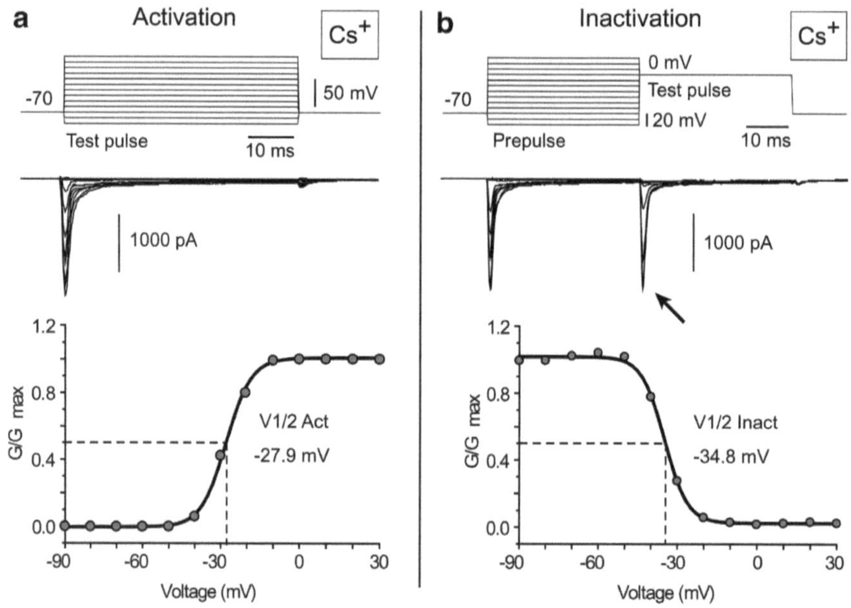

Fig. 4. *Characterization of voltage-gated I_{Na} in the human fetal cortex.* (**a**) To block K^+ currents, CP cells were whole-cell recorded with Cs^+-based intracellular solution (Table 3.1). Neurons were held at −70 mV in voltage-clamp, and voltage was stepped from −90 to +30 mV in 10 mV steps for 50 ms time periods, each (*top*) revealing isolated voltage-gated Na^+ current (I_{Na}, *middle*). Relative peak size of Na^+ conductance (G) versus test pulse potential (*bottom*) constitutes the "steady-state activation curve" (or the "m" curve) of the Hodgkin-Huxley model (82). (**b**) Voltage-clamp experiment to measure steady-state I_{Na} inactivation. Cs^+-treated cells were held at −70 mV, stepped from −90 to +30 mV for 50 ms followed by a 50 ms step at 0 mV (*top*), revealing the isolated test pulse current (*middle, arrow*). Relative peak size of test pulse Na^+ conductance versus the prepulse potential forms the "steady-stated inactivation curve" (or "h" curve) of the Hodgkin-Huxley model (*bottom*). Unpublished data.

voltage step are next subtracted off-line to reveal the isolated Na^+ current ("Control minus TTX") (Fig. 3c). An alternative strategy for isolating voltage-gated Na^+ currents in human fetal cells is based on whole-cell-recordings with Cs^+-based intracellular solution because Cs^+ ions block the pores of voltage-gated K^+ channels (Sect. 2.3). Cs^+ intracellular solution (Table 1) must be allowed to diffuse into the cell for at least 5 min prior to data acquisition. We have empirically determined that in human fetal neurons from tissue of gestational weeks 16–24, 5–8-min cellular dialysis with Cs^+ is sufficient for adequate blockade of K^+ channels (Fig. 4). To measure the activation dynamics of voltage-gated Na^+ current, human cells were loaded with Cs^+ and exposed to a series of voltage steps ranging from −90 to +30 mV (duration 50 ms) from a holding potential of −70 mV (Fig. 4a). For assessment of channel inactivation dynamics, the voltage protocol is slightly different. To determine the voltage dependence of inactivation, cells are exposed to a series of prepulse potentials ranging from −90 to +30 mV (duration 50 ms) and subsequently clamped to 0 mV for 50 ms (Fig. 4b). The activation and inactivation plots (Fig. 4) are generated using a standard procedure (53). This analysis is based on equation $I_m = G_m (V_m - E_{rev})$. Where I_m is the measured ion current, G_m is the membrane conductance, V_m is membrane command potential, and E_{rev} is the reversal potential for Na^+. The normalized conductances (G/G_{max}) are fitted with Boltzmann equation $y = A1 - A2/(1 + e(v - v_{1/2})/k) + A2$. The characteristic values termed voltage of half-activation ($V_{1/2}$Act) and voltage of half-inactivation ($V_{1/2}$Inact) represent membrane potentials (voltages) at which 50% of the available pool of Na^+ channels is activated or inactivated, respectively. Recall that different Nav isoforms have different $V_{1/2}$Act and $V_{1/2}$Inact values (51).

Although we have determined the actual activation and inactivation properties of neurons in human fetal CP, this information was not sufficient to identify plasmalemmal Na^+ channel α-subunits (6). Our conclusions are in agreement with recent studies demonstrating that the Na^+ current properties of one cell type alone cannot explain complex patterns of Na^+ channel α-subunit gene expression (55, 56). Numerous factors may contribute to the voltage dependence of Na^+ channel activation and inactivation, such as the association of auxiliary β-subunits, cell-specific factors, or protein trafficking (57–59). A combination of experimental approaches involving electrophysiological recordings, pharmacological manipulations, immunological labeling, and biochemistry would probably have to be put together in the same tissue sample in order to elucidate the exact subunit composition of voltage-gated Na^+ channels during development.

3.5.3. Action Potential Firing

The hallmark of a neuronal progeny is its ability to fire an action potential. Action potentials are principle carriers of neuronal activity in the cerebral cortex in both the mature and developing brain. Neuronal electrical activity (i.e., action potential firing) is important for neuronal proliferation, migration, neurites extension, synapse formation, and establishment of functional circuits (2, 60, 61). To test whether fetal neurons in the human cerebral cortex can generate action potentials, a series of command pulses ranging from −20 pA to +40 pA (duration 350 ms) is injected into the cells (Fig. 2d). In this experimental paradigm, negative current pulses are used to test the R_i and to determine whether action potentials are triggered upon cessation of the hyperpolarizing current pulse corresponding to "rebound firing." The selection of positive current intensities should be such to satisfy a wide range of potential voltage thresholds, so as not to miss action potentials in weakly excitable cells. For practical purposes and direct cell-to-cell comparison, one should use a fixed current stimulation protocol. In this project, we have set the maximal current pulse at +40 pA. At this current drive, majority of human SVZ and CP cells depolarize by more than 50 mV from resting V_m.

Already at the 16th gestational week, the human fetal CP is endowed with young neurons with enough voltage-gated Na^+ current to generate an action potential upon direct current injection (6). Suprathreshold current intensities have been used to show that despite sustained (>600 ms) suprathreshold stimulation, most young neurons are typically able to fire only a single spike (Fig. 2d). Repetitive action potential firing is a feature of mature cortical neurons (62) and can only occasionally be found in the fetal CP (6). The apparent lack of repetitive action potential firing suggests that the majority of human fetal cortical neurons are in a physiologically immature state. After the 20th gestational week, a small percentage of neurons, located randomly in the SP and CP, are indeed able to generate repetitive action potential firing. Thus, they appear to be more differentiated than their counterparts in the same cortical layer (6). To summarize, in the same cortical layer, one can find neurons of very different maturity levels, residing side-by-side (see Fig. 7 in (6)). Currently, it is not clear what cellular mechanisms govern the patchy and inhomogeneous differentiation time course of human CP neurons.

3.5.4. Action Potential Properties

In mature pyramidal and non-pyramidal neurons of the rodent cerebral cortex, the voltage-gated Na^+ current peak value regularly exceeds 3,000 pA. With this amount of available Na^+ current, every current-evoked action potential is a truly "all-or-none" phenomenon. Conversely, developing human neurons exhibit a large spectrum of the peak current amplitudes, specifically from 0 to 1,500 pA (see Fig. 3 in (6)). Visual inspection of current-clamp traces reveals "spiky" voltage transients of various shapes and sizes (Fig. 5). Since both voltage- and current-clamp recordings comprise an uninterrupted continuum of shapes and amplitudes, a definite determination on the presence or absence of a regenerative potential is not trivial, as it relies heavily on subjective opinion. Ideally, one would like to have some type of quantitative measure on how active, how regenerative, and how nonlinear is the voltage response in a given neuron. For this purpose, we use the first derivative (dV_m/dt) of the V_m waveform. This type of analysis is performed off-line. Previously recorded waveforms (Fig. 5, grey traces) are passed through a derivative function constructed in Clampfit 9.2 (Molecular Devices), resulting in a second waveform (Fig. 5, black traces). The actual algorithm incorporates the following steps: (1) open voltage trace in the Clampfit window; (2) open Analyze→Arithmetic tab; (3) type in an expression: t(x) = diff (tx), where x is the selected sweep number; and (4) select time window to analyze, for example, between Cursors 1 and 2 (Fig. 5a, b, vertical dashed lines).

Sequential voltage- and current-clamp recordings obtained in the same CP cell (6) enable to perform a correlation analysis between dV/dt of the voltage waveform and the peak Na^+ current amplitude. We have performed such analysis on 107 CP neurons in

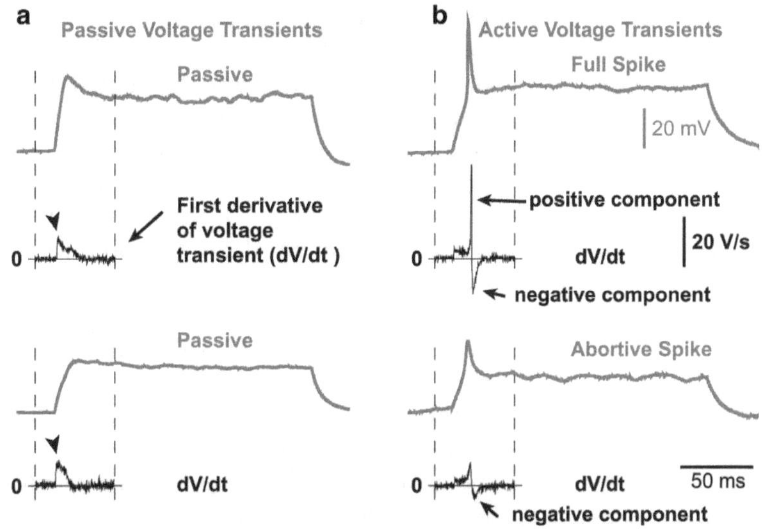

Fig. 5. *Quantitative assessment of membrane excitability.* Examples of the four V_m transients characteristic of human fetal CP neurons in response to current-evoked depolarization. Each V_m transient (*grey* trace) is accompanied by its first derivative (dV_m/dt, *black* trace). *Dashed lines* denote time interval of interest used to calculate dV_m/dt in Clampfit 9.2. Passive voltage responses (**a**) produce notable positive components in dV/dt (*arrowheads*). Cells with active (regenerative) membrane responses have notable negative components of dV_m/dt (**b**, *arrows*). *Vertical dashed lines* indicate positions of cursors "1" and "2" (Clampfit 9.2) during off-line data analysis. The amplitudes of negative components (dV_m/dt) are very strongly correlated with the peak amplitude of I_{Na} (6). The positive component of full-Spike – dV/dt is truncated for clarity. Unpublished data.

6 fetal human brain slices from different tissue samples. Although the positive component of dV/dt (Fig. 5b) is due to action potential rise (rapid activation of Na^+ channels), we found a better correlation between the negative components of dV/dt and peak *I*Na (see supplementary Fig. 5 in (6)). Using dual (sequential) voltage-clamp and current-clamp recordings in the same cell, we empirically determined that zero or small amplitude of the negative dV/dt component (in the range 0 to –2 *V/s*) signifies a passive response (Fig. 5a), whereas nonlinear membrane responses (spikes) produce large negative deflections in dV/dt, in the range of –2 to –18 *V/s* (Fig. 5b).

3.6. Cellular Morphology

To study the geometry of neuronal processes (dendrites and axons) in the human tissue, the fluorescent dye sulforhodamine (50 μM, MW 606) is added to the patch pipette solution. It takes at least 10–15 min of diffusion from the patch pipette into the cytosol for sufficient dye filling of the cellular processes. Labeled cells are visualized with epifluorescence microscopy, and images are captured on the electrophysiology recording station using an IR-CCD video camera. The quality of fluorescent images is compromised by having the dye-filled patch pipette attached in whole-cell configuration.

Fluorescent light emanating from concentrated dye inside the pipette is scattered through brain tissue, thus producing large background fluorescence. With high background fluorescence, the thin cellular processes are not showing in microphotographs (Fig. 6a2). It is thus recommended to detach the recording pipette prior to taking fluorescent microphotographs (Fig. 6b). To preserve the integrity of the cell body, the recording pipette is removed using an identical procedure for pulling an outside-out patch (41). This maneuver is done in voltage-clamp mode with the holding potential set to mimic resting V_m. A "seal-test" voltage pulse is applied at 20 Hz to monitor the resistance of the membrane patch. By engaging two axes of the motorized micromanipulator at the same time (Sect. 2.1), the patch pipette is carefully pulled away from the cell body, until the seal test indicates reformation of a gigaohm seal. At this point, it is safe to speed up the motorized micromanipulator and completely remove the patch pipette from the recording chamber. Two potential problems may cause a failure of this "pipette removal" procedure. Firstly, the somatic membrane

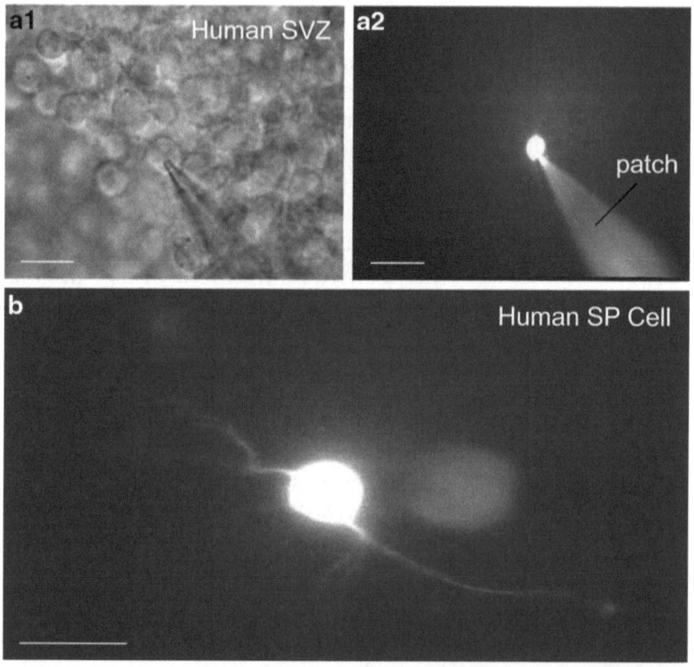

Fig. 6. *Morphology of cortical cells in human fetal tissue.* (**a1**) IR-DIC microphotograph of an acute human fetal brain slice showing a 10-MΩ patch pipette (filled with the fluorescent dye sulforhodamine) attached to a SVZ cell. Scale bar, 10 μm. (**a2**) Same visual field, same cell, same magnification (scale bar, 10 μm), except the image of the sulforhodamine-injected cell was captured using epifluorescence. Fluorescence from the sulforhodamine-loaded patch pipette is interfering with imaging of thin dendritic branches. (**b**) Example of a human SP cell upon removal of the patch pipette used for intracellular injection of sulforhodamine. Unpublished data.

may overextend and rupture in the course of the process before an outside-out patch has been formed, thus spilling the fluorescent content into the surrounding tissue. Secondly, human fetal cells are not firmly embedded in the tissue, so quite often the entire cell body travels with the pipette. Forceful extraction of the soma ruptures primary processes and destroys the morphology of the cell under investigation. Because it is difficult to remove the patch pipette while keeping the human fetal cell in the slice, images should be taken both before and after pipette extraction (Fig. 6a2, b).

3.7. Spontaneous Activity

In the developing mammalian brain, spontaneous electrical activity is essential for proper neuronal differentiation, migration, synaptogenesis, neurotransmitter specification, and synaptic plasticity (2, 63, 64). Spontaneous activity begins to emerge early in cortical development and appears to coincide with the neuron's ability to generate Na^+ action potentials (13). A relatively large number of physical parameters (e.g., electromagnetic radiation, temperature), chemical compounds (e.g., toxins), and biological factors (e.g., microorganisms), to which pregnant women are potentially exposed in the first and second trimester of gestation, have the potential to interrupt network electrical activity and in this way fatally affect the number and layering of neurons. Some subtle cortical changes at early fetal age could manifest later in adulthood (after puberty), often in the shape of devastating mental disease (65–67). A technology to record spontaneous electrical activity in human fetal neurons may facilitate our understanding of basic biological process in development. Moreover, such technology may provide an excellent opportunity to analyze how particular physical, chemical, and biological agents interfere with it. In human fetal cortical slices, one can attain recording sessions longer than 20 min with small-diameter patch pipettes of 10–12-MΩ resistance. Negative DC current should be applied to adjust the cell's resting V_m to approximately −60 mV (Sect. 3.4.1). In slices perfused with 2 mM Ca^{2+}, quite often >1-min-long time periods pass without occurrence of spontaneous depolarization or indication of strong background synaptic activity (Fig. 7, silent period). Patience is required because a random surge of spontaneous electrical activity (plateau depolarization accompanied by action potential burst) has occurred in a significant number of human SP neurons during 5 min of whole-cell recording (35).

3.8. Evoked Spontaneous Activity

Network activity can be induced experimentally in human fetal cortical slices. One of the most commonly used experimental techniques for triggering spontaneous activity in isolated neural circuits is reduction of the Ca^{2+} concentration in the perfusate from conventionally used values of around 2 mM to physiological levels of 1–1.2 mM (31, 68). Possible mechanisms by which physiological extracellular Ca^{2+} levels stimulate or, conversely, artificially raise

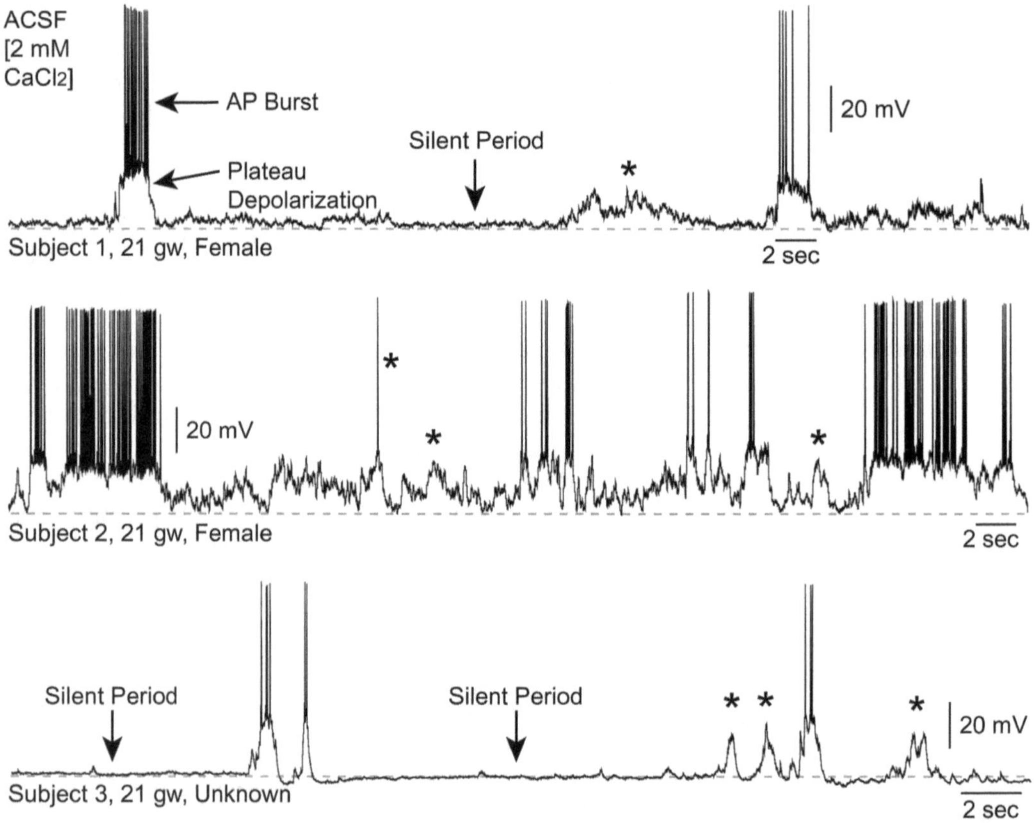

Fig. 7. *Generation of spontaneous electrical activity by human subplate neurons in vitro.* Spontaneously active SP neurons from three human subjects in brain slices perfused with standard ACSF containing 2 mM $CaCl_2$. *Dashed lines* indicate holding potentials (−60, −62, and −63 mV, respectively). Asterisks indicate UP states with one or no action potentials (APs). These data were acquired by performing uninterrupted 5-min recording sessions in 49 neurons obtained from five human subjects. UP states were detected in 25 out of 49 SP neurons. (Adapted from (35) with kind permission).

Ca^{2+} may depress in spontaneous electrical activity are discussed in Chaps. 2-Ruangkittisakul et al. (26), 8-Sanchez-Vives (30), and 11-Kantor et al. (31). Here briefly, lowering extracellular Ca^{2+} abolishes inhibitory synaptic transmission, as well as inhibitory intrinsic processes such as afterhyperpolarization (69). Secondly, under natural conditions, Ca^{2+} ions constitute a major portion of the membrane surface charge, which exerts powerful biophysical effects on V_m and ion channel gating (53), such that higher extracellular concentrations have the net effect of decreasing neuronal excitability (36). Thirdly, in low extracellular Ca^{2+}, K^+ ions accumulate in the extracellular space and further depolarize neurons (70, 71).

Just like animal brain slices, human fetal slices also increase the amount of spontaneous electrical activity upon switching from standard ACSF ($[Ca^{2+}] = 2.0$ mM) (Fig. 8a) to "low" calcium

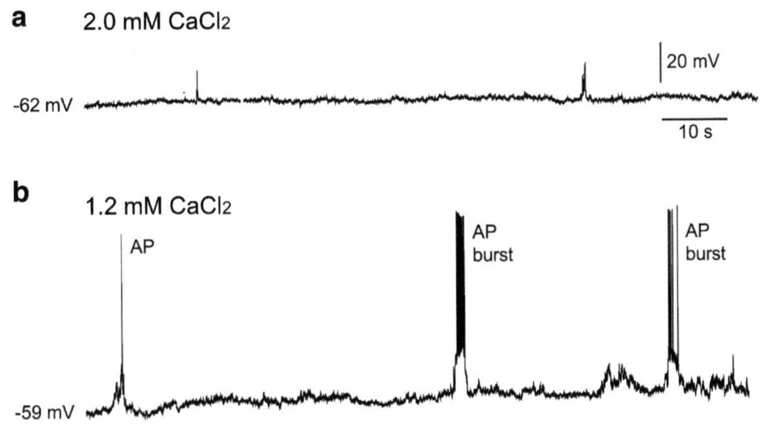

Fig. 8. *Physiological level of Ca^{2+} promotes spontaneous electrical activity.* (**a**) Spontaneous activity recorded in an SP neuron (23 gestational weeks) at its natural resting V_m. (For this approach, neurons were whole-cell recorded with 10–12-MΩ electrodes to assure longer-lasting recording sessions.) In the presence of standard ACSF containing 2 mM Ca^{2+}, this human SP cell displays weak spontaneous activity. (**b**) Lowering Ca^{2+} in the ACSF to a close-to-physiological level (1.2 mM) caused an increase in spontaneous electrical activity. Similar increases in the level of spontaneous activity were detected in ten further human SP neurons. Unpublished data.

ACSF ($[Ca^{2+}]=1.2$ mM) (Fig. 8b). Not having to wait for large depolarizations to occur spontaneously (see silent periods in previous section) has great practical value for a large number of potential experimental designs, intended to explore cellular mechanisms of early cortical rhythms.

3.9. Immuno-histochemistry

Immunohistochemistry is an essential tool in modern neurobiology (72). Its widespread use owns to a good compatibility with other neurobiological approaches, including physiological recordings. Here, we describe immunolabeling of axon initial segments in the human fetal slice. In this practical example, the molecular marker of our choice is βIV spectrin, a protein localized to the axon initial segment (AIS) and nodes of Ranvier (73).

3.9.1. Significance of AIS

The AIS of (cortical pyramidal) neurons is a distinct cellular compartment which hosts high densities of cell adhesion molecules, modulatory proteins, and a unique extracellular matrix. This complex structure is thought to support heavy clustering of voltage-gated Na^+ and K^+ channels. The AIS is the major integration point for afferent synaptic activity and the actual site of action potential initiation in cortical pyramidal neurons (74). Since action potentials are the main output signal of each cortical cell, it is likely that development of (spontaneous and evoked) cortical network activity in early stages of fetal brain formation critically depends on the assembly of molecularly and physiologically mature AIS. The first study to combine electrophysiological recordings and immunolabeling in the developing human fetal brain indicates that arousal of neuronal excitability (i.e., the ability to generate action potential) coincides with formation of AISs in the SP and CP (6). This experimental approach may be useful for further analysis of cellular and molecular mechanisms involved in the development of spontaneous electrical activity, in the first and second trimester of human gestation.

3.9.2. Immunolabeling

In the experimental design described below, immunolabeling is performed in the same slice in which electrical recording has been carried out. This strategy is useful because it assures that electrical recordings and immunolabelings were performed in the same cortical area of interest, while it maximizes the use of human postmortem material. All solutions for immunolabeling procedures were made in 0.1-M phosphate buffer. The acute slice is transferred from the recording chamber into 4% paraformaldehyde for 1 h to overnight chemical fixation, depending on the type of antibody to be used. In the case of overnight fixation, the fixed slice is stored at +4°C. For immunodetection of βIV spectrin, for example, the preferred fixation period is 1 h at room temperature. Following the fixation step, a human slice is cryoprotected by 30% sucrose for 24 h at +4°C. Next, the slices are embedded in Tissue-Tek OCT mounting medium and resectioned on a microtome (or cryostat) to 20 μm sections which are mounted onto gelatin-coated glass coverslips, permeabilized, and blocked with 10% normal goat serum in 0.3% Triton X-100. βIV spectrin polyclonal antibodies are applied overnight at 1:300 dilution (75). Sections are rinsed 3 × 8 min in phosphate buffer with 10% normal goat serum and 0.3% Triton X-100, and then secondary antibodies are applied at room temperature for 1 h (Alexa Fluor 488 anti-rabbit secondary antibody (Invitrogen, Carlsbad, CA)). Sections are rinsed 3 × 10 min with phosphate buffer solution and "coverslipped" with glycerol and *p*-phenylenediamine (Sigma).

At 19 gestational weeks, the number of βIV spectrin–positive AIS in the human CP is on average 6.8 ± 0.5 per visual field (220 × 160 μm (Fig. 9, top)), and its average length is 16.2 ± 0.8 μm (6). These numbers are in stark contrast to mature AIS in the human cortex (76) and our own measurements in the juvenile rat cortex (postnatal day 36). Under identical tissue processing conditions, we found that the density of AIS is more than fivefold higher in rat cortex (30.3 ± 1.7 per visual field). Our measurements also show that the average AIS length in the postnatal day 36 rat is nearly 3 times greater (49 ± 0.6 μm) than in the human fetus (16.2 ± 0.8 μm) (Fig. 9, bottom).

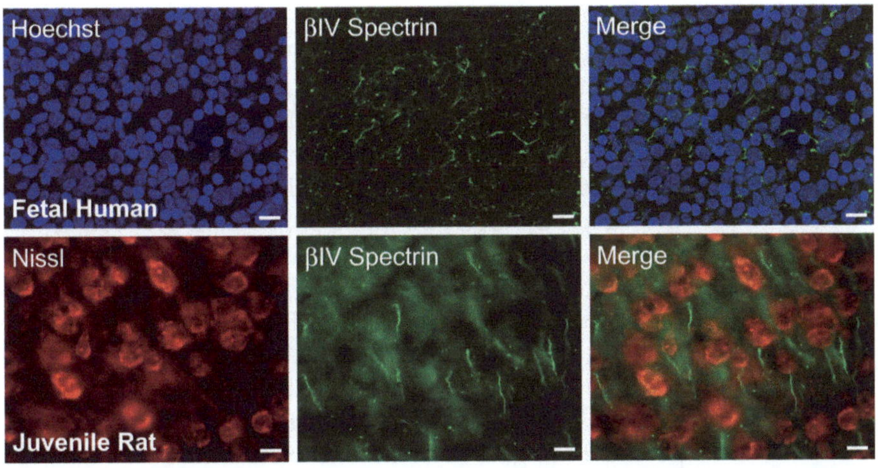

Fig. 9. *Immunolabeling of the human fetal cerebral cortex and juvenile rat. Top row*: Following electrophysiological recording, human fetal brain slices were fixed in 4% paraformaldehyde (1 h), resectioned on a cryostat to slices 20 μm thickness, and then treated with antibodies against βIV spectrin (*green*), a protein localized to axon initial segments. The pia is located in the direction of *upper right* corner of the photograph. *Bottom row*: An identical immunolabeling procedure is used in the prefrontal cortex of the juvenile rat (P 36) to identify βIV spectrin (*green*) positive segments. Hoechst (*blue*), nuclear stain. NeuroTrace (*red*), Nissl stain. Scale bar 10 μm. Pia is located in the direction of *upper left* corner. Unpublished data.

4. Potential Problems with Electrophysiological Recordings in Human Fetal Tissue

4.1. Viability of Human Fetal Brain Slices

One important question that needs to be addressed before experimental data are used to draw conclusions is whether, and to what extent, individual cellular elements in postmortem human material have been preserved. This concern is valid given the significant postmortem delay, comprised of three distinct time periods: (1) time spent in tissue repository, (2) time spent during transfer of tissue from repository to our laboratory, and (3) time spent during preparation of acute slices. Attempts should be made to keep each of these three phases as short as possible. Besides shortening these time periods, human material meant for physiological recordings should continuously be kept in carbogenated solution on ice. If there is a metabolic need for oxygen and nutrients in this state, the human tissue possesses some excellent features. Unlike adult animal (e.g., rodent) brain tissue, young human brain tissue has a high content of water, low cellular density, and soft consistency. These features facilitate diffusion of metabolic gases and nutrients deep into the tissue during transportation. When mature neurons are exposed to metabolic stress (e.g., hypoxia, hypoglycemia), they begin to lose membrane excitability through the downregulation of voltage-gated Na^+ channels (77). Unlike in mature brains, embryonic and fetal neurons are able to withstand metabolic stress with little change in the number of expressed voltage-gated Na^+ channels within the first 24 h (78, 79). The ability of a young human neuron to maintain its excitable membrane properties despite metabolic stress is critical for the success of experiments described in this chapter. It is probable that the mature neurons are more susceptible to the hypoxic/ischemic environment associated with transportation and tissue preparation procedures, leading to their underrepresentation in the data set. In other words, due to accelerated cell death of the most mature population of neurons, the experiments in acute human fetal slice may underestimate the number of differentiated neurons based on any of the following criteria: (1) R_i analysis, (2) extension of primary and secondary processes, (3) expression of voltage-gated ion channels, (4) action potential firing, (5) formation of AIS, and (6) formation of synaptic connections.

4.2. Limited Cell Death

Due to the postmortem time delay of 15 min and long transport time of ~2 h, followed by an additional 30 min needed for making the acute slices, it is predicted that a majority of cells would be unable to withstand such extreme conditions. But as mentioned previously, young neurons are resilient to hypoxic situations and metabolic stress. The amount of cell death present in fetal brain slices following extraction is tested using the in situ TUNEL (terminal deoxyribonucleotidyl transferase (TDT)-mediated dUTP-digoxigenin nick end labeling) assay (80). The TUNEL in situ method is able to detect early DNA fragmentation in apoptotic cells (81). Based on "Neurolucida" drawings illustrating the TUNEL(+) nuclei (80), the majority of cell death is limited to the proliferative regions of the ventricular zone and SVZ with a maximum apoptotic index of 4.7 ± 0.7 TUNEL(+) nuclei per $10^6\ \mu m^3$ (Fig. 10). Overall, the majority of cells present in the acute human fetal brain slice are still viable, which can also be inferred from visual appearance of cells under IR-DIC visualization and whole-cell-recordings.

4.3. Sectioning

We have tried different methods of cutting human fetal tissue. The biggest problem is that this tissue is of a "fluid-gel" consistency, unmyelinated, and void of turgor. Placing the tissue directly onto a mounting block with cyanoacrylate adhesive (superglue) and sectioning with a vibratome, as is traditionally done with acute animal brain slices, is very difficult. Using an agarose mold to support the tissue for cutting is also problematic. We tried placing the tissue in an agar mold and sectioning it on a vibratome. Although this process produced nicely shaped brain slices, the rapid and repeated changes in temperature were unfavorable to the cell's electrophysiological response. The most propitious way is to keep the tissue cold at all times during sectioning. Cold ACSF preserves cell viability and provides the tissue with some stiffness

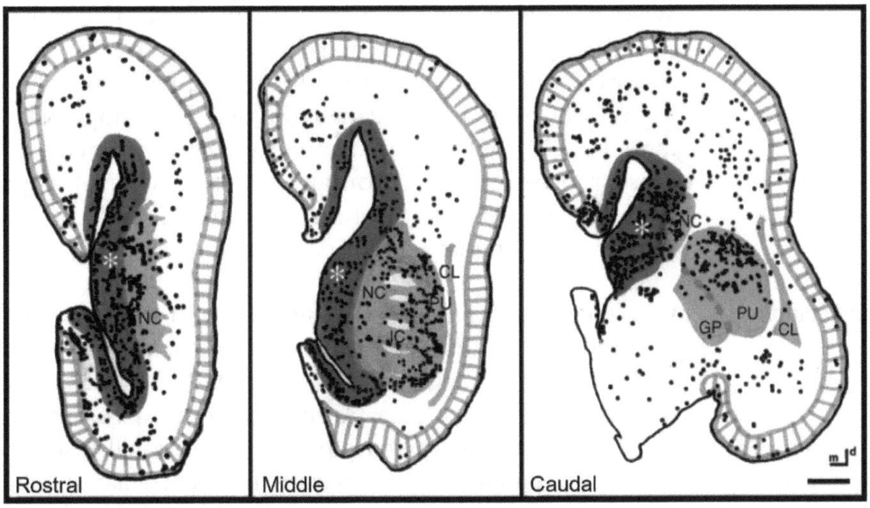

Fig. 10. *Programmed cell death in acute fetal brain slices.* "Neurolucida" drawings illustrating the distribution of "TUNEL(+)" nuclei (*black dots*) at 17 gestational weeks. Layer I and the CP are presented as a striped area, while the proliferative zones, VZ, SVZ, and ganglionic eminence, are *dark grey* and marked with asterisks. TUNEL(+) nuclei were concentrated in the proliferative zones and basal ganglia and less abundant in the surrounding subcortical white matter. Scale bar 3 mm. *CL* claustrum, *IC* internal capsule, *NC* nucleus caudatus, *PU* putamen. (Adapted from (80) with kind permission).

and rigidity against the cutting blade. Profound cooling of tissue blocks prior to cutting, and keeping the tissue surrounded by ice crystals during the cutting, has yielded good quality slices proven with physiological measurements.

Because we are cutting slices with a razor blade by hand without support, they can sometimes be up to 1 mm thick and uneven. As a rule, the two marginal zones (CP and SVZ) end up being thicker than the middle zones (containing SP and IZ), most likely due to differences in cell density. Although some slices are 1 mm thick and have an uneven appearance, the fetal tissue is relatively transparent at this age, and we are routinely able to visualize individual cells with IR-DIC microscopy (Fig. 1). Cutting tissue slices by hand is not easy, but this skill can be quickly improved with practice. Each section of fetal human brain tissue has a slightly different texture, some being more pliable than others. It is important to understand the anatomy of the tissue block prior to cutting (refer to Fig. 1). More importantly, one must be prepared to adjust the cutting angle and cutting speed, so that while the tissue is bending and stretching in front of the blade, one can change the blade angle to include all anatomical layers of the developing cortex.

4.4. Data Selection Criteria

To insure that all cells in the data collection are viable and capable of providing physiologically meaningful conclusions, the following set of elimination criteria has been adopted. Prior to patching, we select cells with a defined shape and smooth plasma membrane, avoiding cells which appear to have high visual contrast, rugged edges, saggy, and punctuated (rough) membrane. The starting electrical seal must be in the gigaohm range. Upon "break-in," V_m must be more negative than −30 mV. The bridge balance value must be <30 MΩ for the duration of the experiment. C_m and R_i must remain stable throughout the course of experiment. By following this set of criteria, we can effectively minimize artifacts and contamination of experimental data.

4.5. "Bad" Recordings

Whole-cell-recording from small cells with high-resistance patch pipettes (10–12 MΩ) may introduce artifacts. If these recording artifacts are not recognized and eliminated, they can severely affect the interpretation of data. Arguably, the most common problem with high-resistance patch pipettes is occasional resealing of the membrane patch. Rapid electrical transients like somatic action potentials are reduced in amplitude and broader in shape, eventually up to the point that they are missed (Fig. 11a1). One may erroneously conclude that such a cell is not capable of action potential firing. In order to avoid artifacts and collection of artifactual data, one should systematically check the bridge balance and abandon cells when the series resistance exceeds 30 MΩ. In rare occasions, the experimenter may succeed to apply negative pressure to the back of the recording pipette to break into the cell again. If successful, the access resistance drops significantly and the action potential amplitude and shape are restored (Fig. 11a2). Similar to current-clamp errors (see above), the recording artifacts can also compromise voltage-clamp data. Figure 11b exemplifies a space-clamp problem. In theory and in practice, the voltage-clamp technique is fully applicable only to spherical cells. In nonspherical cells, such as neurons, V_m is not effectively clamped distal to the patch electrode. This means that some parts of the neuron, in particular remote dendrites, are experiencing a V_m that may differ notably from the holding potential. If voltage-gated Na^+ currents show maxima at different time points (Fig. 11b), this is a clear sign of voltage-clamp error, and such data must not be used for construction of activation and inactivation curves or calculations of time constants of the Hodgkin-Huxley model (82, 83).

When recording spontaneous electrical activity in human fetal slices, sometimes the cell will enter a period of sustained depolarization accompanied by bursts of action potentials (Fig. 12a). In the majority of such cells, sustained depolarizations are interrupted by periods of quiescence when V_m is close to its resting level (Fig. 7, silent period). In the absence of at least one spontaneous return to the resting V_m, it can be suspected that the sustained depolarizing state was not natural. In other words, if a cell fails to repolarize and remains in a depolarized state until the end of the recording session (Fig. 12a), this recording should be considered inappropriate for

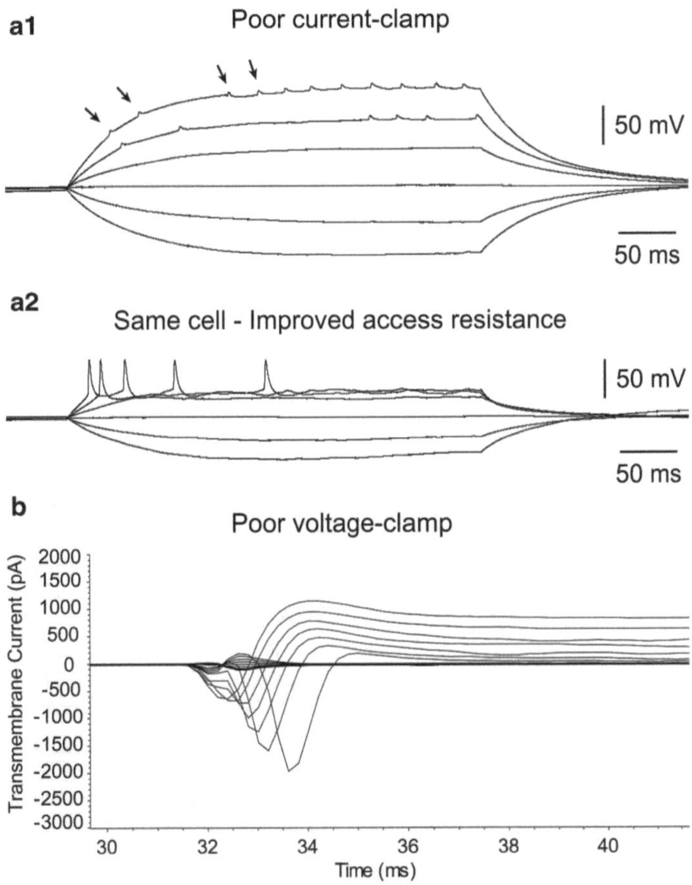

Fig. 11. *Criteria for discarding experimental data.* (**a1**) Whole-cell current-clamp recording of the V_m response to a series of hyperpolarizing and depolarizing current pulses. The electrical contact between patch pipette and intracellular space has been compromised, most likely due to a partial closure (sealing) of the membrane opening between pipette tip and cell interior. The two signs of inadequate recording conditions are (1) large bridge balance (≥50 MΩ) and (2) distorted shape of action potentials. Action potentials (*arrows*) appeared to be heavily filtered (low-pass). (**a2**) Same cell and same recording pipette, except negative pressure was applied to break the patch of membrane under the pipette (increase the hole in the membrane) and thus decrease access resistance. (**b**) I_{Na} shows maxima at different time points. This is an indication of a space-clamp problem. Unpublished data.

inclusion in the data set. Another example of a bad patch electrode recording is shown in Fig. 12b. In this case, the identifying signs of a compromised seal between electrode tip and cellular membrane are sudden change in action potential amplitude followed by spike-less depolarized state, reminiscent of a so-called depolarization block (84, 85). Such events should not be mistaken for an episode of spontaneous electrical activity.

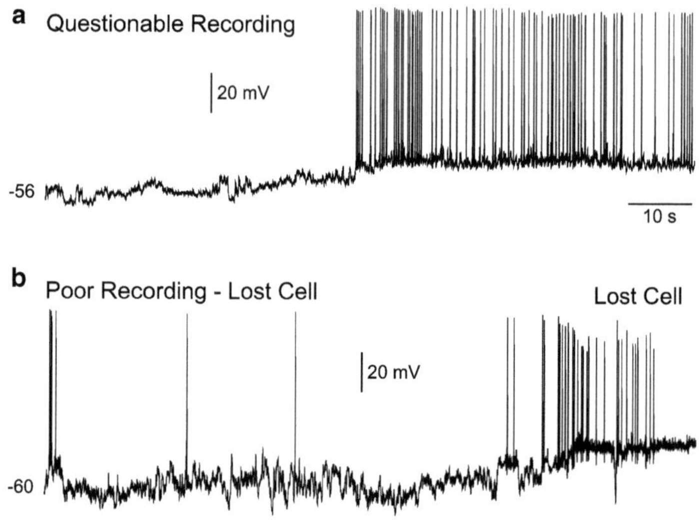

Fig. 12. *Problems associated with long recordings of spontaneous activity.* Because of the rapid V_m changes during episodes of intense spontaneous activity, it is often difficult to ascertain whether or not meaningful experimental data have been acquired (i.e., whether recordings have been carried out in healthy or compromised cells). (**a**) This trace illustrates a cell becoming electrically active, firing a continuous string of action potentials, and remaining in a sustained depolarized state. (**b**) Example of a cell experiencing an episode of unstable (noisy) resting V_m, followed by severe reduction in action potential amplitude, and finally entering a state which can be characterized as a "depolarization block." Both "**a**" and "**b**" obtained after the introduction of 1.2 mM Ca^{2+} ACSF. Unpublished data.

5. Conclusions

This chapter provides an overview of the techniques to detect and analyze electrical membrane properties and spontaneous electrical activity in cortical cells of the human fetal tissue. The methods described here can be used for identification of neuronal progeny as well as for determining the stage of neuronal differentiation. These methods could become essential tools for studying cellular and molecular events in the course of cortical development, as well as for testing the influence of chemical and biological agents (toxins, microorganism) on membrane excitability. We have provided some experience-based insights and practical advice on how to overcome the technical issues associated with handling human fetal tissue, maintaining cell viability during transportation, cutting slices, and patching small and fragile cells. We believe that the set of techniques demonstrated here is useful for future studies aiming at answering questions about cellular and molecular determinants of human cortical development in health and disease.

References

1. Zhang LI, Poo MM (2001) Electrical activity and development of neural circuits. Nat Neurosci 4:1207–1214
2. Spitzer NC (2006) Electrical activity in early neuronal development. Nature 444:707–712
3. Price DJ, Kennedy H, Dehay C, Zhou L, Mercier M, Jossin Y, Goffinet AM, Tissir F, Blakey D, Molnar Z (2006) The development of cortical connections. Eur J Neurosci 23: 910–920
4. Sakmann B, Neher E (1984) Patch clamp techniques for studying ionic channels in excitable membranes. Annu Rev Physiol 46:455–472
5. Moody WJ, Bosma MM (2005) Ion channel development, spontaneous activity, and activity-dependent development in nerve and muscle cells. Physiol Rev 85:883–941
6. Moore AR, Filipovic R, Mo Z, Rasband MN, Zecevic N, Antic SD (2009) Electrical excitability of early neurons in the human cerebral cortex during the second trimester of gestation. Cereb Cortex 19:1795–1805
7. Luhmann HJ, Hanganu I, Kilb W (2003) Cellular physiology of the neonatal rat cerebral cortex. Brain Res Bull 60:345–353
8. Luhmann HJ, Reiprich RA, Hanganu I, Kilb W (2000) Cellular physiology of the neonatal rat cerebral cortex: intrinsic membrane properties, sodium and calcium currents. J Neurosci Res 62:574–584
9. Kerkovich DM, Sapp D, Weidenheim K, Brosnan CF, Pfeiffer SE, Yeh HH, Busciglio J (1999) Fetal human cortical neurons grown in culture: morphological differentiation, biochemical correlates and development of electrical activity. Int J Dev Neurosci 17:347–356
10. Khazipov R, Esclapez M, Caillard O, Bernard C, Khalilov I, Tyzio R, Hirsch J, Dzhala V, Berger B, Ben-Ari Y (2001) Early development of neuronal activity in the primate hippocampus in utero. J Neurosci 21:9770–9781
11. Chiu FC, Rozental R, Bassallo C, Lyman WD, Spray DC (1994) Human fetal neurons in culture: intercellular communication and voltage- and ligand-gated responses. J Neurosci Res 38:687–697
12. Johnson MA, Weick JP, Pearce RA, Zhang SC (2007) Functional neural development from human embryonic stem cells: accelerated synaptic activity via astrocyte coculture. J Neurosci 27:3069–3077
13. Picken Bahrey HL, Moody WJ (2003) Early development of voltage-gated ion currents and firing properties in neurons of the mouse cerebral cortex. J Neurophysiol 89:1761–1773
14. Mo Z, Moore AR, Filipovic R, Ogawa Y, Kazuhiro I, Antic SD, Zecevic N (2007) Human cortical neurons originate from radial glia and neuron-restricted progenitors. J Neurosci 27:4132–4145
15. Molnar Z, Metin C, Stoykova A, Tarabykin V, Price DJ, Francis F, Meyer G, Dehay C, Kennedy H (2006) Comparative aspects of cerebral cortical development. Eur J Neurosci 23:921–934
16. Rakic P (1995) A small step for the cell, a giant leap for mankind: a hypothesis of neocortical expansion during evolution. Trends Neurosci 18:383–388
17. Jiang C, Haddad GG (1992) Differential responses of neocortical neurons to glucose and/or O_2 deprivation in the human and rat. J Neurophysiol 68:2165–2173
18. Beck H, Goussakov IV, Lie A, Helmstaedter C, Elger CE (2000) Synaptic plasticity in the human dentate gyrus. J Neurosci 20:7080–7086
19. Broicher T, Speckmann EJ (2012) Living human brain slices: network analysis using voltage sensitive dyes. In Isolated Central Nervous System Circuits (Ed K Ballanyi), Neuromethods Series Vol. 73 (Ed W Walz). Springer Science+Business Media, LLC, New York, NY, pp 285–300
20. Vanhatalo S, Kaila K (2006) Development of neonatal EEG activity: from phenomenology to physiology. Semin Fetal Neonatal Med 11:471–478
21. Kostovic I, Judas M (2006) Prolonged coexistence of transient and permanent circuitry elements in the developing cerebral cortex of fetuses and preterm infants. Dev Med Child Neurol 48:388–393
22. Andre M, Lamblin MD, D'Allest AM, Curzi-Dascalova L, Moussalli-Salefranque F, Nguyen The Tich S, Vecchierini-Blineau MF, Wallois F, Walls-Esquivel E, Plouin P (2010) Electroencephalography in premature and full-term infants. Developmental features and glossary. Neurophysiol Clin 40:59–124
23. Wilson CJ (2008) Up and down states. Scholarpedia 3, p 1410, revision #68992
24. Chauvette S, Volgushev M, Timofeev I (2010) Origin of active states in local neocortical networks during slow sleep oscillation. Cereb Cortex 20:2660–2674
25. Trapp S, Ballanyi K (2012) Autonomic nervous system in vitro: studying tonically active neurons controlling vagal outflow in rodent brainstem slices. In Isolated Central Nervous System Circuits (Ed K Ballanyi), Neuromethods Series Vol. 73 (Ed W Walz). Springer Science+Business Media, LLC, New York, NY, pp 1–59

26. Ruangkittisakul A, Panaitescu B, Secchia L, Bobocea N, Kantor C, Kuribayashi J, Iizuka M, Ballanyi K (2012) Anatomically 'Calibrated' Isolated Respiratory Networks from Newborn Rodents. In Isolated Central Nervous System Circuits (Ed K Ballanyi), Neuromethods Series Vol. 73 (Ed W Walz). Springer Science+Business Media, LLC, New York, NY, pp 61–124
27. Hamill OP, Marty A, Neher E, Sakmann B, Sigworth FJ (1981) Improved patch-clamp techniques for high-resolution current recording from cells and cell-free membrane patches. Pflugers Arch 391:85–100
28. Molleman A (2003) Patch clamping: an introductory guide to patch clamp electrophysiology. Wiley, West Sussex
29. Somjen GG (2002) Ion regulation in the brain: implications for pathophysiology. Neuroscientist 8:254–267
30. Sanchez-Vives MV (2012) Spontaneous rhythmic activity in the adult cerebral cortex in vitro. In Isolated Central Nervous System Circuits (Ed K Ballanyi), Neuromethods Series Vol. 73 (Ed W Walz). Springer Science+Business Media, LLC, New York, NY, pp 263–284
31. Kantor C, Panaitescu B, Kuribayashi J, Ruangkittisakul A, Jovanovic I, Leung V, Lee TF, MacTavish D, Jhamandas JH, Cheung PY, Ballanyi K (2012) Spontaneous neural network oscillations in hippocampus, cortex and locus coeruleus of newborn rat and piglet brain slices. In Isolated Central Nervous System Circuits (Ed K Ballanyi), Neuromethods Series Vol. 73 (Ed W Walz). Springer Science+Business Media, LLC, New York, NY, pp 315–356
32. Kostovic I, Rakic P (1990) Developmental history of the transient subplate zone in the visual and somatosensory cortex of the macaque monkey and human brain. J Comp Neurol 297:441–470
33. Allendoerfer KL, Shatz CJ (1994) The subplate, a transient neocortical structure: its role in the development of connections between thalamus and cortex. Annu Rev Neurosci 17:185–218
34. Kanold PO, Luhmann HJ (2010) The subplate and early cortical circuits. Annu Rev Neurosci 33:23–48
35. Moore AR, Zhou WL, Jakovcevski I, Zecevic N, Antic SD (2011) Spontaneous electrical activity in the human fetal cortex in vitro. J Neurosci 31:2391–2398
36. Frankenhaeuser B, Hodgkin AL (1957) The action of calcium on the electrical properties of squid axons. J Physiol 137:218–244
37. Acker CD, Antic SD (2009) Quantitative assessment of the distributions of membrane conductances involved in action potential backpropagation along basal dendrites. J Neurophysiol 101:1524–1541
38. Zecevic N, Chen Y, Filipovic R (2005) Contributions of cortical subventricular zone to the development of the human cerebral cortex. J Comp Neurol 491:109–122
39. Poolos NP, Jones TD (2004) Patch-clamp recording from neuronal dendrites. Curr Protoc Neurosci Chapter 6: p Unit 6 19
40. Rakic S, Zecevic N (2003) Emerging complexity of layer I in human cerebral cortex. Cereb Cortex 13:1072–1083
41. Neher E, Sakmann B (1992) The patch clamp technique. Sci Am 266:44–51
42. Barbour B (2010) Electronics for electrophysiologists tutorial. http://www.biologie.ens.fr/perso/barbour/
43. Neher E (1992) Correction for liquid junction potentials in patch clamp experiments. Methods Enzymol 207:123–131
44. Burgard EC, Hablitz JJ (1993) Developmental changes in NMDA and non-NMDA receptor-mediated synaptic potentials in rat neocortex. J Neurophysiol 69:230–240
45. Kim HG, Fox K, Connors BW (1995) Properties of excitatory synaptic events in neurons of primary somatosensory cortex of neonatal rats. Cereb Cortex 5:148–157
46. Connors BW, Benardo LS, Prince DA (1983) Coupling between neurons of the developing rat neocortex. J Neurosci 3:773–782
47. Loturco JJ, Blanton MG, Kriegstein AR (1991) Initial expression and endogenous activation of NMDA channels in early neocortical development. J Neurosci 11:792–799
48. Mienville JM, Lange GD, Barker JL (1994) Reciprocal expression of cell-cell coupling and voltage-dependent Na current during embryogenesis of rat telencephalon. Brain Res Dev Brain Res 77:89–95
49. McCormick DA, Prince DA (1987) Post-natal development of electrophysiological properties of rat cerebral cortical pyramidal neurones. J Physiol 393:743–762
50. Beckh S, Noda M, Lubbert H, Numa S (1989) Differential regulation of three sodium channel messenger RNAs in the rat central nervous system during development. EMBO J 8: 3611–3616
51. Goldin AL (2001) Resurgence of sodium channel research. Annu Rev Physiol 63: 871–894
52. Boiko T, Rasband MN, Levinson SR, Caldwell JH, Mandel G, Trimmer JS, Matthews G (2001) Compact myelin dictates the differential targeting of two sodium channel isoforms in the same axon. Neuron 30:91–104

53. Hille B (2001) Ion channels of excitable membranes, 3rd edn. Sinauer, Sunderland
54. Rush AM, Dib-Hajj SD, Waxman SG (2005) Electrophysiological properties of two axonal sodium channels, Nav1.2 and Nav1.6, expressed in mouse spinal sensory neurones. J Physiol 564:803–815
55. Felts PA, Yokoyama S, Dib-Hajj S, Black JA, Waxman SG (1997) Sodium channel alpha-subunit mRNAs I, II, III, NaG, Na6 and hNE (PN1): different expression patterns in developing rat nervous system. Brain Res Mol Brain Res 45:71–82
56. Mechaly I, Scamps F, Chabbert C, Sans A, Valmier J (2005) Molecular diversity of voltage-gated sodium channel alpha subunits expressed in neuronal and non-neuronal excitable cells. Neuroscience 130:389–396
57. Meadows LS, Chen YH, Powell AJ, Clare JJ, Ragsdale DS (2002) Functional modulation of human brain Nav1.3 sodium channels, expressed in mammalian cells, by auxiliary beta 1, beta 2 and beta 3 subunits. Neuroscience 114:745–753
58. Pan F, Beam KG (1999) The absence of resurgent sodium current in mouse spinal neurons. Brain Res 849:162–168
59. Shibata R, Misonou H, Campomanes CR, Anderson AE, Schrader LA, Doliveira LC, Carroll KI, Sweatt JD, Rhodes KJ, Trimmer JS (2003) A fundamental role for KChIPs in determining the molecular properties and trafficking of Kv4.2 potassium channels. J Biol Chem 278:36445–36454
60. Mennerick S, Zorumski CF (2000) Neural activity and survival in the developing nervous system. Mol Neurobiol 22:41–54
61. Zito K, Svoboda K (2002) Activity-dependent synaptogenesis in the adult Mammalian cortex. Neuron 35:1015–1017
62. Zhang ZW (2004) Maturation of layer V pyramidal neurons in the rat prefrontal cortex: intrinsic properties and synaptic function. J Neurophysiol 91:1171–1182
63. Katz LC, Shatz CJ (1996) Synaptic activity and the construction of cortical circuits. Science 274:1133–1138
64. Rakic P, Komuro H (1995) The role of receptor/channel activity in neuronal cell migration. J Neurobiol 26:299–315
65. Arnold SE, Hyman BT, Van Hoesen GW, Damasio AR (1991) Some cytoarchitectural abnormalities of the entorhinal cortex in schizophrenia. Arch Gen Psychiatry 48: 625–632
66. Akbarian S, Kim JJ, Potkin SG, Hetrick WP, Bunney WE Jr, Jones EG (1996) Maldistribution of interstitial neurons in prefrontal white matter of the brains of schizophrenic patients. Arch Gen Psychiatry 53:425–436
67. Kanold PO (2004) Transient microcircuits formed by subplate neurons and their role in functional development of thalamocortical connections. Neuroreport 15:2149–2153
68. Ruangkittisakul A, Secchia L, Bornes TD, Palathinkal DM, Ballanyi K (2007) Dependence on extracellular Ca^{2+}/K^{+} antagonism of inspiratory centre rhythms in slices and en bloc preparations of newborn rat brainstem. J Physiol 584:489–508
69. Agopyan N, Avoli M (1988) Synaptic and non-synaptic mechanisms underlying low calcium bursts in the in vitro hippocampal slice. Exp Brain Res 73:533–540
70. Haas HL, Jefferys JG (1984) Low-calcium field burst discharges of CA_1 pyramidal neurones in rat hippocampal slices. J Physiol 354:185–201
71. Yaari Y, Konnerth A, Heinemann U (1983) Spontaneous epileptiform activity of CA1 hippocampal neurons in low extracellular calcium solutions. Exp Brain Res 51:153–156
72. Rasband MN, Trimmer JS (2001) Developmental clustering of ion channels at and near the node of Ranvier. Dev Biol 236:5–16
73. Berghs S, Aggujaro D, Dirkx R Jr, Maksimova E, Stabach P, Hermel JM, Zhang JP, Philbrick W, Slepnev V, Ort T, Solimena M (2000) betaIV spectrin, a new spectrin localized at axon initial segments and nodes of ranvier in the central and peripheral nervous system. J Cell Biol 151:985–1002
74. Stuart G, Schiller J, Sakmann B (1997) Action potential initiation and propagation in rat neocortical pyramidal neurons. J Physiol 505: 617–632
75. Rasband MN, Peles E, Trimmer JS, Levinson SR, Lux SE, Shrager P (1999) Dependence of nodal sodium channel clustering on paranodal axoglial contact in the developing CNS. J Neurosci 19:7516–7528
76. Inda MC, De Felipe J, Munoz A (2006) Voltage-gated ion channels in the axon initial segment of human cortical pyramidal cells and their relationship with chandelier cells. Proc Natl Acad Sci USA 103:2920–2925
77. Cummins TR, Jiang C, Haddad GG (1993) Human neocortical excitability is decreased during anoxia via sodium channel modulation. J Clin Invest 91:608–615
78. Zhang JH, Gibney GT, Xia Y (2001) Effect of prolonged hypoxia on Na^{+} channel mRNA subtypes in the developing rat cortex. Brain Res Mol Brain Res 91:154–158

79. Xia Y, Haddad GG (1999) Effect of prolonged O_2 deprivation on Na^+ channels: differential regulation in adult versus fetal rat brain. Neuroscience 94:1231–1243
80. Rakic S, Zecevic N (2000) Programmed cell death in the developing human telencephalon. Eur J Neurosci 12:2721–2734
81. Gavrieli Y, Sherman Y, Ben-Sasson SA (1992) Identification of programmed cell death in situ via specific labeling of nuclear DNA fragmentation. J Cell Biol 119:493–501
82. Hodgkin AL, Huxley AF (1952) A quantitative description of membrane current and its application to conduction and excitation in nerve. J Physiol 117:500–544
83. Engel D, Jonas P (2005) Presynaptic action potential amplification by voltage-gated Na^+ channels in hippocampal mossy fiber boutons. Neuron 45:405–417
84. Bikson M, Hahn PJ, Fox JE, Jefferys JG (2003) Depolarization block of neurons during maintenance of electrographic seizures. J Neurophysiol 90:2402–2408
85. Jones RS, Heinemann U (1987) Abolition of the orthodromically evoked IPSP of CA_1 pyramidal cells before the EPSP during washout of calcium from hippocampal slices. Exp Brain Res 65:676–680

Chapter 4

Histological Characterization of Physiologically Determined Fast-Spiking Interneurons in Slices of Primate Dorsolateral Prefrontal Cortex

Kenneth N. Fish, Guillermo Gonzalez-Burgos, Aleksey V. Zaitsev, and David A. Lewis

Abstract

GABA (γ-aminobutyric acid)-ergic interneurons are a structurally and functionally diverse group of cells that, together, regulate the activity of neuronal networks giving rise to the brain oscillations necessary for information processing. The phenotypes of GABAergic neurons can be classified by morphological, electrophysiological, and neurochemical features. However, as further information is gathered about these features and the role(s) each phenotype plays in shaping cortical function, group membership will undoubtedly change. This chapter details a set of methodologies that can be used to characterize GABAergic neurons in the monkey dorsolateral prefrontal cortex (DLPFC). Specifically, it (*1*) provides a brief protocol for the dissection of DLPFC tissue from monkeys, (2) describes electrophysiological measures used to identify GABAergic neurons, (3) details how to fill neurons for morphological analyses, and (4) describes a fluorescence imaging technique for the quantification of fluorescently labeled puncta (putative synaptic components), the colocalization of different labels in the same synaptic structure, and the quantitation of fluorescence intensity in these same structures in brain tissue sections.

Key words: Electrophysiology, GABA, Glutamic acid decarboxylase, Monkey, Morphology, Nonhuman primate, Parvalbumin, Quantitative microscopy, Synapse

1. Introduction

1.1. Background and Aims of Study

GABA (γ-aminobutyric acid)-ergic interneurons are a structurally and functionally diverse group of cells that, together, regulate the activity of neuronal networks giving rise to the brain oscillations necessary for information processing (1, 2). The phenotypes of GABAergic neurons can be classified by morphological,

Klaus Ballanyi (ed.), *Isolated Central Nervous System Circuits*, Neuromethods, vol. 73,
DOI 10.1007/978-1-62703-020-5_4, © Springer Science+Business Media New York 2012

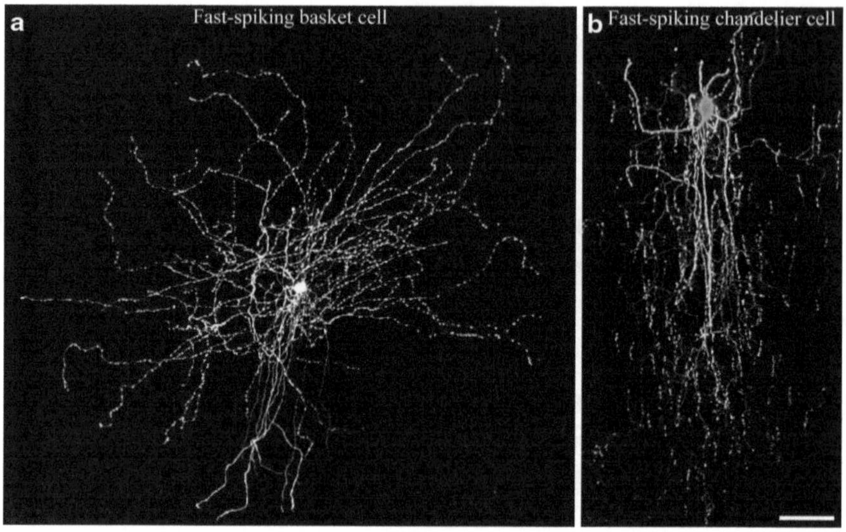

Fig. 1. Morphologically distinct cell types with fast-spiking firing patterns. Confocal reconstructions of neurons exhibiting fast-spiking firing patterns during recording (see Sect. 3.2.2): (**a**) fast-spiking basket cell and (**b**) fast-spiking chandelier cell. Cells were filled with biocytin during intracellular electrophysiological recording and then incubated with streptavidin-Alexa Fluor 633 conjugate (Invitrogen, Carlsbad, CA, dilution 1:500) post fixation for morphological analysis. Bar = 50 μm (Reproduced with permission from Oxford University Press from Zaitsev et al. (4))

electrophysiological, and neurochemical features (3). However, as further information is gathered about these features and the role(s) each phenotype plays in shaping cortical function, group membership will undoubtedly change. This chapter details a set of methodologies that can be used to characterize GABAergic neurons in the monkey dorsolateral prefrontal cortex (DLPFC) (Figs. 1 and 2). Specifically, it (1) provides a brief protocol for the dissection of DLPFC tissue from monkeys, (2) describes electrophysiological measures used to identify GABAergic neurons, (3) details how to fill neurons for morphological analyses, and (4) describes a fluorescence imaging technique for the quantification of fluorescently labeled puncta (putative synaptic components), the colocalization of different labels in the same synaptic structure, and the quantitation of fluorescence intensity in these same structures in brain tissue sections.

1.2. Why Study Monkeys?

Since it is rare that healthy live human brain tissue from patients is available for functional studies, nonhuman model systems are needed to study GABAergic neuron phenotypes. To date, the majority of research characterizing GABAergic neurons has been performed in rodents. However, clear differences exist in cortical GABAergic circuitry between rodents and primates (4–17). For example, in the primate neocortex, the percentage of GABAergic neurons is higher than in rodents (11, 18). In addition, the developmental origin of some interneurons appears to differ between primates and rodents (19, 20), and interneurons with

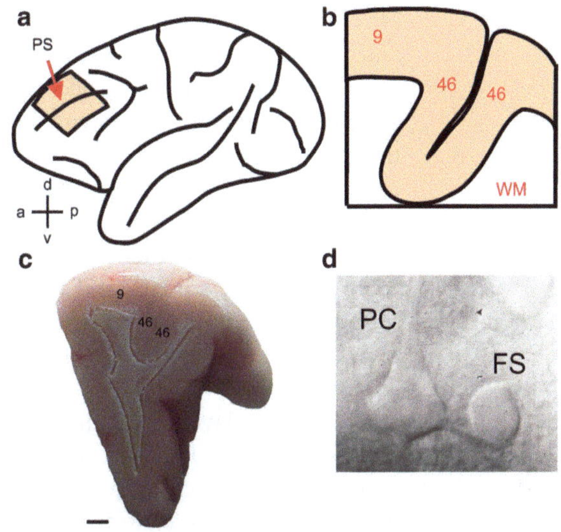

Fig. 2. The monkey principal sulcus. (**a**) Lateral view of the macaque monkey cortex showing the approximate location of the dorsolateral prefrontal cortex (DLPFC) tissue blocks next to the principal sulcus (*PS*) removed during surgery (see Sect. 3.1). (**b**) Schematic view of a typical coronal slice containing portions of DLPFC areas 9 and 46, including the medial and lateral banks of the principal *PS*. (**c**) Low-magnification image of the *PS* from a perfusion-fixed animal (the gray/white matter border was scored for clarity). (**d**) A pyramidal neuron (*PC*) and a fast-spiking (*FS*) interneuron visually identified in layers 2/3 using infrared differential interference contrast video microscopy (see Sect. 2.1.2). Bar in C = 2 mm; *a*, anterior; *p*, posterior; *d*, dorsal; *v*, ventral; and *WM*, white matter.

firing properties unique to primates have been reported (21, 22), just to name a few differences. Thus, in order to interpret human findings, particularly postmortem ones, in the context of the rich body of physiologic, pharmacologic, and anatomic data that exists for rodents, it is necessary to perform complementary studies in nonhuman primates.

1.3. Inhibitory Interneuron Morphology

Inhibitory interneurons vary greatly in their morphology. Dendritic morphology is the most variable feature and cannot reliably be used to classify the different interneuron subpopulations. However, the axonal arborization can reveal the anatomical identity of an interneuron because such cells seem to be particularly specialized to target different domains of the postsynaptic cell membrane, different cortical columns, and different layers within a column (23). Thus, reconstruction of the axonal arbor is an important step for characterizing the identity of cortical neurons in primates. To differentiate morphological groups of interneurons, two readily observed qualitative morphological criteria might be used. These criteria are the distribution of axonal arbors across layers (which reflects the predominant laminar targets of a neuron) and a combination of axonal arborization patterns and terminal branch properties (which may reflect the targeting of different subdomains of cortical neurons)

(4). Also, monkey interneurons have different horizontal (tangential) and vertical (radial) spreads of axons. There are about ten distinct types of interneurons in neocortex, which are well recognized based on axonal properties, and many intermediate varieties. Interneurons that have fast-spiking electrophysiological properties (Sect. 1.4) and contain the Ca^{2+}-binding protein parvalbumin (PV) (Sect. 1.5) comprise two well-defined morphological subtypes: the so-called basket cells and chandelier neurons (Fig. 1). Basket cells specialize in targeting the somata and proximal dendrites of pyramidal neurons and interneurons. The primary axons of these cells give rise to a few long and stout main lateral branches. The latter give rise to linear collaterals at right or oblique angles. The collaterals spread across several layers but almost never reach layer 1. Basket cells include wide (~1,000 μm), medium (~700 μm), and local (~450 μm) arbor morphological varieties, according to their horizontal axonal span (22, 24). The chandelier cells are relatively small cells (axonal arbor width ~250–300 μm). They are recognized based on their characteristic axonal terminals, which form short vertical arrangements of terminals, termed axon cartridges, known to exclusively target the axon initial segment of pyramidal cells (25).

1.4. Fast-Spiking Firing Pattern

The two morphologically distinct subtypes of PV-containing interneurons, specifically basket and chandelier neurons, display distinctive electrophysiological properties. Based on their response to depolarizing DC current injection, PV-positive basket and chandelier cells have been characterized as fast-spiking neurons. These properties include having individual spikes with extremely short duration (duration at half spike amplitude is typically 0.3–0.5 ms) and large after-hyperpolarizing potentials (Fig. 3a). In addition, in response to injection of relatively long depolarizing DC current steps (500–600 ms), fast-spiking neurons can fire at a constant high frequency without showing spike-frequency adaptation (Fig. 3b). Such properties clearly separate the fast-spiking basket and chandelier neurons not only from pyramidal cells but also from other interneuron subtypes. In monkey DLPFC, we recently showed that the fast-spiking firing properties of basket and chandelier neurons are essentially identical (4, 22, 26, 27). However, fast-spiking chandelier cells may have a slightly higher input resistance and consequently lower rheobase (4, 27), meaning they could fire at a higher frequency than fast-spiking basket neurons in response to depolarizing inputs of similar magnitude.

1.5. Molecular Markers of Interneuron Subtypes

Ca^{2+}-binding proteins and several neuropeptides tend to be expressed in different subpopulations of interneurons (28, 29, 65). Although molecular markers do not map perfectly to different interneuron subtypes, their expression is an important general correlate of anatomical and electrophysiological attributes (23, 26). One especially clear correlation is the case of the PV-positive GABA

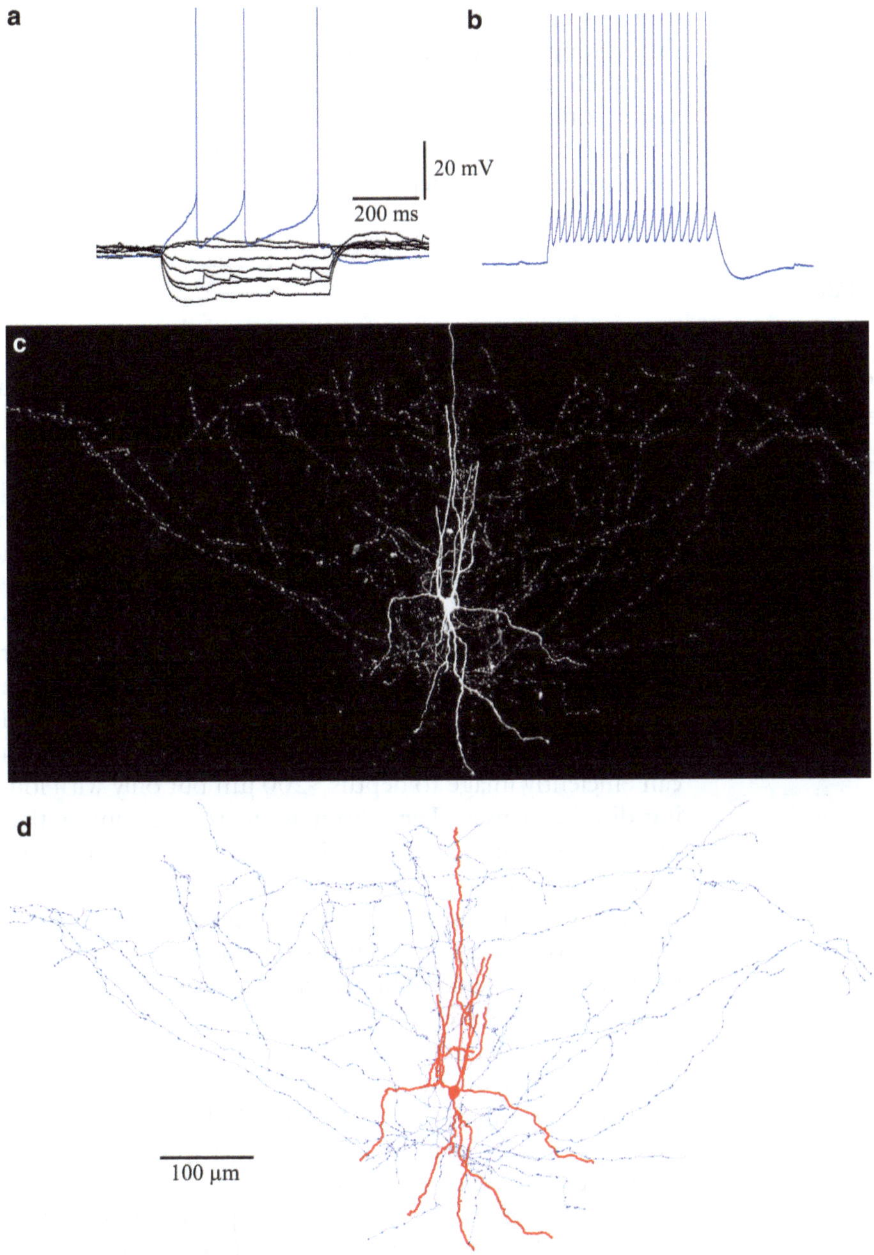

Fig. 3. Fast-spiking basket cell physiology and morphology. Electrophysiological and morphological analyses of the same cell. (**a**) Response of the membrane potential of a fast-spiking interneuron to injection of hyperpolarizing and depolarizing DC current steps. (**b**) Suprathreshold depolarizing DC current steps produce high-frequency firing without spike frequency adaptation in fast-spiking neurons. (**c**) Confocal reconstruction of the same cell recorded in (**a**–**b**). (**d**) Neurolucida reconstruction of **c**. The axon is designated in *blue* and the dendrites and soma in *red* (Reproduced with permission from Oxford University Press from Zaitsev et al. (4)).

neurons since PV expression is found only in fast-spiking cells and fast-spiking properties are found only in PV-containing neurons (both chandelier and basket cells). Interestingly, PV-containing fast-spiking GABA neurons do not express any of the neuropeptides localized to other interneuron subtypes (for instance, somatostatin, vasoactive intestinal peptide, cholecystokinin, or neuropeptide-Y).

2. Materials

2.1. Microscopy Equipment

2.1.1. General Microscope Setup

When applicable, wide-field fluorescence microscopy is an ideal tool for capturing fluorescent images (30, 31). This imaging modality is not normally suited for imaging tissue sections greater than several microns thick because too much out-of-focus blur reaches the detector. But, when the goal is to only reconstruct filled neurons, the signal to background ratio is usually sufficient to image the filled cell by wide-field epifluorescence microscopy. As noted below, if a high-resolution reconstruction is not necessary, the imaging can be performed without resectioning but must be performed using longer-working-distance objectives. When available, multiphoton laser scanning microscopes are excellent for reconstructing filled neurons without resectioning because they can efficiently image to depths >200 μm but only with long-working-distance lenses. The advantage to resectioning is that higher numeric aperture (NA) objectives can be used, providing greater detail. If the goal is to combine dye filling with immunofluorescence to study protein expression within single neurons, there will most likely be too much out-of-focus information even in resectioned tissue for wide-field epifluorescence microscopy. Confocal microscopes are well suited for this scenario. In general, there are two types of confocal microscopes: laser scanning confocal microscopes (LSCMs) and spinning disk confocal microscopes (SDCMs), e.g., the DSU (Olympus, Center Valley, PA, USA; www.olympusamerica.com) or the system by Quorum (Guelph, ON, Canada; www.quorumtechnologies.com). Depending on the configuration, LSCMs also offer greater excitation power above 600 nm that allows for efficient excitation of far-red fluorochromes (e.g., Alexa 647). LSCMs can also improve signal to background due to point illumination (32). LSCMs are an excellent choice for capturing image stacks of filled neurons and counting fluorescently labeled puncta that mark synaptic components because they offer a relatively high axial resolution (~500 nm vs. ~700 nm for the DSU using a 60X 1.42 NA objective). However, with a pinhole size set to 1 Airy unit, which gives the best compromise between depth discrimination and efficiency, ~99% of the photons that enter the objective are rejected. In contrast, ~10x more photons reach the detector in SDCMs (33). Although this increase results from more out-of-focus information reaching the detector, deconvolution algorithms can be used to enhance resolution (Sect. 4.2) (34, 35). In addition, SDCMs use a cooled charge-coupled device (CCD) camera that can have a higher quantum efficiency (>90%) than the photomultiplier tubes (~25–50%) used as detectors on LSCMs (31–33, 36). Furthermore, using the relationship between image signal-to-noise ratio and fluorophore photobleaching as an index of system efficiency, the efficiency of SDCMs consistently exceeds that of LSCMs (32). Finally, as a general rule, 3D multiwavelength image stacks can be collected substantially faster with SDCMs than most LSCMs. The increased speed greatly facilitates the collection of large data sets, which are required for most high-resolution reconstructions of neurons. Thus, SDCMs are exceptional tools for both quantifying numbers of synaptic components and estimating protein levels within synaptic structures in tissue sections.

2.1.2. Microscope Setup for Neuron Visualization and Patch-Clamp Recording

Individual neurons in 350-μm-thick brain slices, used by us, cannot be visualized using standard bright-field microscopy techniques because of the large number of cells closely packed together that act as birefringent-phase objects. Thus, the visualization of a neuron is obscured because of the scattered light produced by cellular structures (e.g., somata) above and below the focal plane. To minimize light scattering and visualize single neurons for recordings in thick brain slices, infrared video microscopy is combined with patch pipette recordings. To more clearly visualize individual neurons, infrared illumination can be combined with optics used for contrast generation, e.g., differential interference contrast (DIC) optics and electronic contrast enhancement (Fig. 2d) (37–41). This combination of techniques was applied for the first time to slices from nonhuman primates in 2000 by Gonzalez-Burgos, Barrionuevo, and Lewis (42).

The majority of microscopy companies offer upright microscopes designed for electrophysiology combined with cell visualization in acutely isolated brain tissue. Microscopes used for electrophysiology studies generally have a third-party fixed stage, and X-Y gross optical positioning is performed using a manual or motorized platform translator that the microscope sits on. Micromanipulators are used to position the pipette used for patching/dye filling. The microscope, fixed stage, and micromanipulators usually sit together on a vibration isolation table/platform. The microscope is usually equipped with one or two low-magnification air objectives (e.g., 2X or 4X); a high-magnification (e.g., 40X or 60X) long-working-distance (~1.8–2.0 mm) water immersion objective designed for infrared illumination, which is used for patching; and port magnification changer so a wide range of magnifications can be achieved without changing the objective. The infrared DIC signal is captured on an infrared-sensitive camera, e.g., a C2400 video camera with video enhancement device (Hamamatsu, Bridgewater, NJ, USA; www.hamamatsu.com).

3. Methods

3.1. Primate Brain Slice Preparation

The focus here is on the tissue from the DLPFC, which is involved in the planning and execution of complex behaviors. DLPFC slices are obtained from young adult (3.5–6 kg; 4–5 years old) male cynomologous monkeys (*Macaca fascicularis*) (27, 42). The monkeys need to be treated according to the guidelines outlined in the "National Institutes of Health Guide for the Care and Use of Laboratory Animals" (or the equivalent guidelines of the country where experimentation is going to be performed). The slices are cut from a small block of high-quality tissue containing a portion of area 9, and both the medial and lateral banks of the principal sulcus (area 46) are obtained from one hemisphere (Fig. 2a–c). Animals recover quickly with no impairments in eating or drinking or overt behavior deficits. Similarly, high-quality tissue (both anatomy and physiology are equal to slices obtained in the initial surgery) can be obtained from a nonhomotropic region of the contralateral hemisphere at a second terminal surgical procedure. A team of 4–5 electrophysiologists records from slices obtained at each experiment so that a substantial amount of data is obtained. In addition, the remainder of the tissue is harvested at the second terminal procedure and is used in molecular and anatomic studies of other brain regions.

The isolated cortical tissue block is placed in ice-cold artificial cerebrospinal fluid (ACSF) slicing solution composed of (in mM) 230 sucrose, 1.9 KCl, 1.2 Na_2HPO_4, 33 $NaHCO_3$, 6 $MgCl_2$, 1 $CaCl_2$, 10 D-glucose, and 2 kynurenic acid; pH 7.3–7.4 gassed with 95% O_2–5% CO_2 ("carbogen"). Slices of 350-μm thickness are cut from each block in the coronal plane using a VT 1000S vibratome (Leica Biosystems, Nussloch, Germany; www.leica-microsystems.com; see Sect. 3.3 for a basic description on how to section) and incubated at room temperature for >2 h in solution consisting of (in mM) 126 NaCl, 2 KCl, 1.2 Na_2HPO_4, 10 D-glucose, 25 $NaHCO_3$, 6.0 $MgCl_2$, and 1.0 $CaCl_2$. For recording, slices are transferred to a submersion-type chamber (see also Chaps. 1-Trapp and Ballanyi (43) and 2-Ruangkittisakul et al. (44)) and superfused with carbogenated ACSF (in mM: 126 NaCl, 2.5 KCl, 1.2 Na_2HPO_4, 25 $NaHCO_3$, 2.0 $CaCl_2$, 1.0 $MgCl_2$, 10 glucose) at 32–33°C. Slices can usually be used for 9–15 h after slicing.

3.2. Electrophysiological and Morphological Identification of PV Neurons

For simplicity, this chapter focuses on fast-spiking PV-expressing GABAergic interneurons in the monkey DLPFC (Sect. 1). However, the methods for visualizing neurons in brain sections and patching can be easily adapted for other cell types and brain regions.

3.2.1. Visualizing Neurons for Whole-Cell Recording

To visualize a neuron, a 350-μm-thick slice is transferred from the submersion "storage" chamber to a glass-bottom recording chamber under the upright microscope. The recording chamber needs to be set up to allow for the slice to be perfused using a peristaltic pump-driven or gravity-based perfusion system (see also Chaps. 1-Trapp and Ballanyi (43) and 2-Ruangkittisakul et al. (44)). In addition to regulating the temperature of the recording chamber, in-line solution heaters can be used to maintain constant temperature. The slice is held in place using a "harp," i.e., a metal U-shaped platinum wire strung with a parallel array of nylon thread (see also Chap. 1-Trapp and Ballanyi (43)). Verify that all infrared-DIC optical elements (polarizer, analyzer, prisms) and the heat filter are correctly positioned in the optical path. Looking through the oculars, focus and adjust the condenser for Kohler illumination by closing the field diaphragm and focusing the condenser lens by adjusting its height until the field diaphragm is in focus. Once in focus and centered, the diaphragm is reopened. Now, the infrared interference filter is placed in position and the light path switched to the infrared-sensitive camera. The camera's offset, gain, and shading corrections are adjusted until an image is obtained that maximizes the dynamic range without saturating pixels and even illumination.

3.2.2. Patch Clamping

Prior to collecting the brain tissue, patch electrodes are made from borosilicate glass capillaries (e.g., 1.5-mm outside diameter and 0.86-mm internal diameter) that are pulled using a multistage process on a micropipette puller (see also Chaps. 1-Trapp and Ballanyi (43), 2-Ruangkittisakul et al. (44), and 3-Moore et al. (45)). For interneurons, the DC resistance of the pipettes in standard recording solution should be ~4–7 MΩ. Just prior to placing the pipette in the micromanipulator, it is filled with an intracellular dye-filling recording solution such as (in mM) 120 K-gluconate, 20 KCl, 10 4-(2-hydroxyethyl)-1-piperazineethanesulfonic acid (HEPES), 2 $MgCl_2$, 0.2 ethylene glycol tetraacetic acid (EGTA), 4 Mg-ATP, 0.3 Tris-ATP, 7 phosphocreatine, and Alexa 568 (100 μM). Instead of using an Alexa dye, another option for visualizing neuronal morphology post fixation is biocytin. Like Alexa dyes, biocytin is highly soluble in aqueous solutions and has a small molecular weight. Although neuron visualization requires a secondary step that utilizes avidin or streptavidin conjugates, the large variety of these reagents provides detection flexibility (e.g., see (46)).

The next step is to identify a neuron for whole-cell recording, but before doing so, a note of the orientation of the slice should be made by drawing a diagram using a 2X or 4X objective, noting any landmarks near the area of interest. This is followed by switching to a higher magnification (e.g., 40X with the magnification changer before the camera set at 1X) to identify a healthy neuron in the slice (Fig. 2d) and increasing the magnification using the magnification changer for approaching the selected neuron with the patch pipette. In general, GABAergic neurons are smaller neurons that have round or oval somata, rather than the large triangular-shaped somas typical of pyramidal neurons (Fig. 2d). In addition, the neuron's soma should be located within a 30-μm region that lies below the first 10–20 μm of the slice. The reason for staying within this region is twofold: (1) if the neuron is too close to the surface, there is a good chance that at least some of the axon/dendrites were damaged when cutting, and (2) the slice will need to be resectioned (see below) for analysis.

As the neuron is approached with the patch electrode, positive (~20 mm Hg) pressure is applied to it (47) for expelling small amounts of solution out of the pipette to keep it clear and to clean the cell surface. Under most conditions, the dye that escapes the pipette prior to patching diffuses rapidly and thus does not hinder postdye-filling imaging. As the pipette tip reaches the neuron, a dimple in the plasma membrane will form and the pressure is quickly released to create a small amount of negative pressure. At this stage, the pipette voltage is changed to approximately −80 mV because negative pipette potentials presumably facilitate seal formation. The resistance of the seal is tested using a voltage step. A seal of at least 1 GΩ is necessary for good-quality recordings, and ideally the seal resistance should exceed 2 GΩ. At the next step, the electrode capacitance transient should be nulled. Using a MultiClamp 700B (Molecular Devices, Sunnyvale, CA, USA; www.moleculardevices.com), it can be done automatically by pressing the "Cp Fast: Auto button."

A typical electrophysiology rig is equipped with a computer-controlled MultiClamp with software (a combined voltage/current clamp and data acquisition system in one compact microcomputer controlled unit). Systems frequently have a convenient feature called "ZAP," which facilitates the rupturing of patches when going whole cell by applying a single +1.3 V pulse for a chosen duration (usually <1 ms). The patch protocol should be run along with increasing suction so that the disrupted membrane is taken up into the pipette. Alternatively, mouth suction can be used, which provides excellent control of pressure. In this case, the strength of suction depends on the pipette size and the particular situation. The pulse suction should be increased gradually from gentle to strong and should be stopped immediately upon observing a sudden appearance of large capacitance transients at the beginning and the end of the test pulse, indicating break-in. Now that entry to the cell cytoplasm has been gained, the holding potential is adjusted until the cell/pipette has 0 current in order to record from the cell at its resting membrane potential (typically more negative than −60 mV). To characterize the intrinsic membrane properties of the neuron, hyper- and depolarizing DC current steps of 500-ms duration are applied in 5–10 pA increments at 0.2 Hz with 2 repetitions (Fig. 3a, b). While determining the spiking frequency of the recorded neuron, the dye will diffuse into the cell soma and subsequently into the dendritic and axonal processes. Because neocortical neurons in slices of the monkey DLPFC seem to lack intrinsic pacemaker properties, in standard recording conditions, they typically do not show spontaneous activity contrary to the behavior of cortical circuits in slices from neonatal and mature rodents and other species, e.g., ferret (see also Chaps. 3-Moore et al. (45), 8-Sanchez-Vives (48), and 11-Kantor et al. (49)). Therefore, to produce neuron activity in the monkey neocortical slices, single neurons are stimulated with injection of current steps (Fig. 3a, b).

In slices from monkey DLPFC, some fast-spiking neurons (either basket or chandelier) display spontaneous excitatory postsynaptic potentials (EPSPs) of very large amplitude (up to 10–15 mV; Fig. 4b, c). In contrast, some fast-spiking neurons as well as pyramidal cells and interneurons of non-fast-spiking subtypes rarely show such kind of large synaptic events. The source and function of such large spontaneous synaptic events are poorly understood, but their presence means that certain fast-spiking neurons readily discharge when provided with some level of continuous subthreshold depolarization (Fig. 4c). Chemical/pharmacological stimulation of network activity in cortical brain slices can be achieved by either modifying the extracellular solution by lowering divalent ion concentrations and increasing [K^+], e.g., from 2.5 to 3.5 mM (50), or applying excitatory receptor agonists such as kainate or carbachol. In certain conditions, such modifications of the brain slice extracellular recording solution produce network activity that mimics patterns observed in vivo such as up-down state transitions or synchronized network oscillations (see also Chaps. 3-Moore et al. (45), 8-Sanchez-Vives (48), 10-Luhmann and Kilb (51), and 11-Kantor et al. (49)).

Fast-spiking cells can be identified by their typical discharge of high-frequency nonaccommodating spikes in response to near-threshold current injection. Once the cell has been confirmed to be fast-spiking and the dye has diffused into the cell for ~5–15 min, the pipette is removed. Longer duration of whole-cell recording leads to the reduction of optical signals from markers (e.g., PV). The critical aspect is the time of dye diffusion within the cell. It is important to wait at least 20–30 min from the break-in to fixation, while the pipette might be withdrawn earlier. The success rate of removing the pipette without destroying the soma is increased with higher seal resistance. To maintain cellular morphology, the pipette is removed in current-clamp mode. The pipette is slowly moved away from the cell to a distance of about 50% of soma size, during which the membrane of the cell forms a tube. The main aim is to make this tube as narrow as possible to prevent damage of the soma. The magnitude of voltage responses to current steps is correlated with this process and should increase. A sudden jump of the

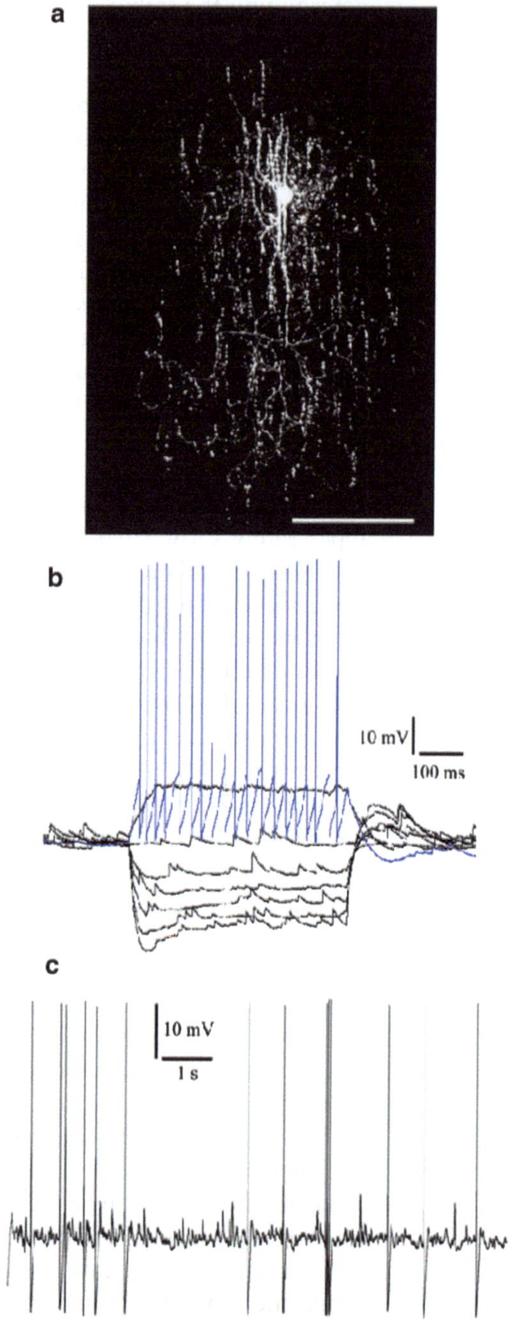

Fig. 4. Spontaneous synaptic activity in fast-spiking neurons. (**a**) Confocal reconstruction of the chandelier cell recorded in **b**. (**b**) Recordings of membrane potential from cell in **a** showing the presence of large spontaneous excitatory postsynaptic potentials. (**c**) Large spontaneous excitatory postsynaptic potentials in a fast-spiking neuron when paired with continuous subthreshold depolarization elicited spiking activity (Reproduced with permission from Oxford University Press from Zaitsev et al. (4)).

electrode potential to ~0 mV indicates that it has been successfully withdrawn. In general, one fast-spiking neuron per slice is identified and filled. If it takes several attempts to identify a fast-spiking neuron, the neuron's location needs to be marked on the slice diagram. Once the pipette is removed, the slice is transferred to a dish and fixed in 4% paraformaldehyde solution. The slice should be fixed for ≥1 day on a rotary shaker (~70 r.p.m.) in the dark at 4°C. After fixation, it should be rinsed for 3×15 min in 0.1 M phosphate-buffered solution (PBS) composed of (in mM) 154 NaCl, 19 KH2PO4, and 81 Na2HPO4 •H_2O. The three exchanges are performed to assure all paraformaldehyde is removed. At this point, the slices can be stored at 4°C in the dark in a solution of 0.1 M PBS+0.05% NaN_3 until they are ready to be processed for immunofluorescence microscopy. For long-term storage, the slices are kept at −20°C or −80°C in storing solution (33% glycerol, 33% ethylene glycol, in 0.1 M PBS).

3.3. Resectioning for Morphological and Molecular Characterization of Interneurons

In general, antibodies, which will be used to detect specific proteins by immunofluorescence microscopy (Sect. 4), do not evenly penetrate the 350-μm-thick slices used in our electrophysiology studies. In addition, the working distances of objectives used for high-magnification, high-resolution image reconstruction are relatively small (~150 μm). Thus, the slices should be resectioned for most antibody-labeling experiments. With that said, when neurons are filled with fluorescent reagents (e.g., Alexa dyes) and if a high-resolution reconstruction of the neuron's morphology is not needed, a fluorescently labeled neuron within a 350-μm section can be imaged with a low-magnification, long-working-distance objective (e.g., 20X 0.75 NA 0.6 mm W.D. objective). For resectioning, the slice is attached to a vibratome chuck bottom down (i.e., the same side that was touching the glass bottom of the recording chamber) using cyanoacrylate glue. Sectioning takes place in PBS (Sect. 3.2.2), which serves to lubricate the sectioning blade, to maintain specimen temperature (if necessary), to preserve the specimen, and to facilitate section retrieval. Once the chuck has been mounted in the vibratome, move the slice close to the blade. The slice is moved vertically (up) into the blade's path, which is along the horizontal axis, and their movements are synchronized such that for each slice advance, the blade will move along the horizontal axis to a predetermined point and then return to the start point. The slice advance is set to 10 μm, and the blade advances such that it will only just touch the slice once it is raised into the blade's path. The vibratome is moved to advance/blade cycle until the slice enters the blade's path. The slice advance is changed to 50 μm and the blade advance position so that it will cut through the entire slice, and the vibratome is activated through 4–5 cycles. The first section will be 60–70 μm thick, while the subsequent sections will be ~50 μm thick. It is useful to place the sections into labeled wells as they are cut so that their position relative to one another is known. In general, 40–50-μm sections can be processed for fluorescence immunocytochemical labeling using procedures that permit labeling throughout the section's z-axis with limited postprocessing shrinkage in this axis (26, 52, 53).

3.4. Antibody Labeling

Markers of axon terminals and dendritic spines appear as immunoreactive (IR) puncta under the light microscope when using immunohistochemistry to visualize them, allowing ready quantification of structure number and density via the application of stereological techniques (54). In addition to quantifying synaptic structure number, IR puncta analysis has been used to garner information about relative protein levels and synaptic structure morphology in multiple species (e.g., see (53, 55–59)). By combining multiple markers in multifluorescence labeling experiments, terminals of neuron subclasses can be identified. For example, in monkey sections stained for PV and both the 65 and 67 kD isoforms of glutamic acid decarboxylase (GAD65 and GAD67, respectively), the PV antibody marks a specific interneuron subpopulation, the GAD markers, which concentrate in synaptic terminals, mark inhibitory terminals, and morphology (PV cartridges vs. noncartridge IR puncta) can be used to differentiate terminals of PV basket cells from those of PV chandelier cells (Fig. 4a).

Aldehyde fixatives react with amines and proteins to generate fluorescent by-products. One way to deal with aldehyde-induced autofluorescence is to reduce –CHO groups to –OH with sodium borohydride ($NaBH_4$). Thus, prior to permeabilizing sections, they are pretreated with fresh 1% $NaBH_4$ in PBS for 30 min at room temperature on a rotary shaker (~70 r.p.m.). Sections are then rinsed 8 × 3 min (multiple exchanges are performed to fully remove the $NaBH_4$) and then blocked/permeabilized with 0.3% Triton X, 10% normal serum (match to the species of the secondary antibody(s)), 1% bovine serum albumin, 0.1% lysine, and 0.1% glycine in PBS (blocking solution) at room temperature on a shaker. After 3 h, the sections are transferred to 10-ml glass scintillation vials, and the blocking solution is replaced with fresh PBS that includes the primary antibody(s). Sections are incubated in primary antibody solution at 4°C for 48 h on a shaker, rinsed 4 × 30 min in PBS, and followed by incubation with secondary antibody(s) (diluted in blocking solution) for 24 h at 4°C on a shaker. Sections are rinsed 4 × 30 min in PBS followed by mounting. Sections should be mounted using #1.5 cover glass to minimize spherical aberration when using high-NA objectives. For mounting, lay a section flat on a slide, remove excess buffer, and cover with mounting solution before the section dries. Since the section will be wet when it is covered with mount, the mount should be water based (e.g., VECTASHIELD HardSet mounting medium, Vector Laboratories, Burlingame, CA; www.vectorlabs.com).

Carefully lay a cover glass over the section making sure not to introduce any air bubbles. Store the sample in the dark at room temperature for ~1 h and then thinly coat the cover glass and slide at their interface with clear nail hardener.

4. Results

4.1. Data Acquisition

4.1.1. Sampling Frequency

For 3D confocal microscopy, it is imperative that the x-, y-, and z-sampling frequencies are sufficient to adequately represent objects of interest in the image stack. The Nyquist sampling theorem explains exactly how great the sampling density (in both x-y and the spacing of z-planes) has to be to record all information from a sample. The theorem is based on the realization that one must sample at a rate greater than twice the maximum frequency in the signal in order to reconstruct the original from the sampled version, otherwise aliasing occurs. The theorem takes into account the numerical aperture of the objective, mounting/immersion media, excitation and emission wavelengths, and the number of excitation photons. There are several online Nyquist rate calculators. In the x-y plane, the sampling rate is determined by pixel size, which should be at least two times smaller than the smallest features that you expect to see in your specimen. As a rule of thumb, for confocal microscopy imaging, sampling distances along the z-axis can be up to 1.7 times the Nyquist ones. Note that oversampling is harmless, unless it results in photobleaching. However, it increases capture and computational time and storage needs.

The resolution of light microscopes is significantly lower in the z-axis than in the x and y dimensions. Resolution, r, can be estimated for x- and y-axes as $r_{x,y} = 0.61\lambda/NA$ (where λ is the wavelength of the emitted light and NA is the numerical aperture of the objective). In contrast, resolution in the z-axis is estimated as $r_z = 2\lambda\eta/NA^2$ (where η is the refractive index of the mounting/immersion media, which are assumed to have the same refractive index, and the refractive index of the cover slip is ≥ that of the mounting media). Thus, for imaging an Alexa488 (emission 519 nm) fluorophore using a 60X oil 1.42 NA objective with immersion oil of refractive index 1.51, x- and y-axis resolution is ~223 nm, while z-axis resolution is ~777 nm. For these parameters, the Nyquist sampling (x, y, z in nm) equals 42, 42, 123. Thus, the spacing between optical z-planes should be ≤209 μm. If imaging was performed on a SDCM using a Hamamatsu ORCA-R2 (6.45 μm x 6.45 μm cell size) and the above parameters, the pixel size would be 0.1075 × 0.1075 μm, which means that information in the x-y plane would be undersampled (sampling is ~2.5 × Nyquist value). Here, a LSCM would have the advantage of being able to zoom in so that each pixel covers an x-y area of 42×42 nm (or smaller), which would be proper Nyquist sampling.

4.1.2. Imaging

In most cases, when making high-resolution reconstructions of dye-filled neurons, several image stacks must be taken in order to make a montage encompassing the entire filled neuron in a brain slice. A motorized x-y stage greatly facilitates making montages. The concentration of dye in filled neurons is usually substantially greater in the soma than in distal regions of axons and dendrites, making it difficult to properly image both regions using the same settings. Therefore, to avoid oversaturating the soma or underexposing the dendrites, the field containing the soma is imaged three times: the first scan uses an exposure time appropriate for the most distal dendrites, while the settings of the second and third passes are appropriate for proximal dendrites and the soma, respectively. The three images for each z-plane are then combined such that the highest nonsaturating pixel value for each point is used in the final image. As for the other fields needed to make a montage of the entire neuron, a single optimum exposure is generally all that is needed.

4.2. Postcapture Data Processing

Prior to neuronal tracing and masking of synaptic proteins to be quantified, it is often desirable to further enhance the resolution of the data via processing with one of several deconvolution algorithms (60). The choice of the appropriate algorithm is at one level empirical, determined by a visual judgment of which algorithm provides the most robust discrimination and 3D representation of the labeled synaptic structures and thus will best serve to estimate structure numbers. However, not all deconvolution approaches are created equal with regard to quantification of fluorescence intensity. Constrained iterative approaches that model the data as point sources, essentially placing scattered photons back at their origin, will allow quantification of fluorescence intensity. One example of an iterative deconvolution algorithm is the Jansson-van Cittert (JVC) algorithm. The JVC deconvolution method is used widely in image reconstruction applications in which a consistent blurring function has degraded an image. JVC uses an iterative approach to minimize the difference between the original distorted image and the convolution of the estimated true signal with the point-spread function (PSF) of the system (61). To ameliorate the effect of noise, it is common to include a "smoothing" step, in which high-frequency noise is filtered out. The JVC method is the basis for the constrained deconvolution methods distributed by several companies, such as Applied Precision (Issaquah, WA, USA; www.api-com), Intelligent Imaging Innovations (Denver, CO, USA; www.intelligent-imaging.com), and Zeiss (Oberkochen, Germany; www.zeiss.com/smt). Other approaches that discard photons (e.g., nearest neighbor algorithms) cannot be used for quantification of fluorescence intensity.

Blind deconvolution should be used for image restoration of SDCM and LSCM data because the PSF, and thus the blur, is not known precisely. A very good software package that has blind deconvolution algorithms tweaked specifically for SDCM and LSCM is AutoQuant. AutoQuant software is directly available through Media Cybernetics (Acton, MA, USA; www.mediacy.com) and is also available as an add-on module in imaging programs such as SlideBook (Intelligent Imaging Innovations).

4.3. Neurolucida Tracing

The purpose of performing neuron reconstruction is to create a realistic, meaningful, and quantifiable model of a selected neuron (Fig. 3d). The method, accuracy, and thoroughness used in performing the reconstruction are dependent on the analytical needs of the researcher. Neurons can be traced in 2D or 3D and on live or captured images manually or using software automation. Described here is how to make high-resolution reconstructions of dye-filled neurons in serial physical sections that are costained for synaptic proteins using Neurolucida, captured 3D image stacks, and computer-assisted manual tracing (see, e.g., Fig. 5b). For those looking for automated approaches, see the methods paper by Rodriguez and colleagues for a demonstration of automated reconstruction of 3D neuronal morphology from LSCM image stacks (62).

Once a montage is made from the image stacks, a reference point is designated and the soma is outlined. Next, place a point where the axon emerges from the soma, and then continue placing points on the axon keeping it in focus as you move away from the soma by changing the z position. Points where the axon branches are designated as bifurcating or trifurcating should be included so that branch complexity can be analyzed. In addition, varicosities (putative terminals) are also specifically marked for posttracing analysis. A new feature in Neurolucida version 9 allows for automated detection of axon thickness so that the tracing line's thickness matches that of the axon. Dendrites are traced in a similar fashion, and for spiny neurons, dendritic spine type (e.g., thin, stubby, mushroom) can be specifically designated. Finally, tissue shrinkage due to tissue processing is most pronounced in the z-axis. If z-axis shrinkage is significantly different between sections, a shrinkage correction factor (which is the reciprocal of the shrinkage that occurs along the axis) is applied. Once a neuron is traced, the Neurolucida Explorer program can be used to measure morphological variables (e.g., somatic area, axonal node total, total surface area of the axon (μm^2), highest order axon segment) of the reconstructions. The morphological characterization can be very extensive. For example, McGarry and colleagues recently used 67 morphological variables of the reconstructions to further divide a subpopulation of somatostatin-positive neocortical interneurons into three groups (63).

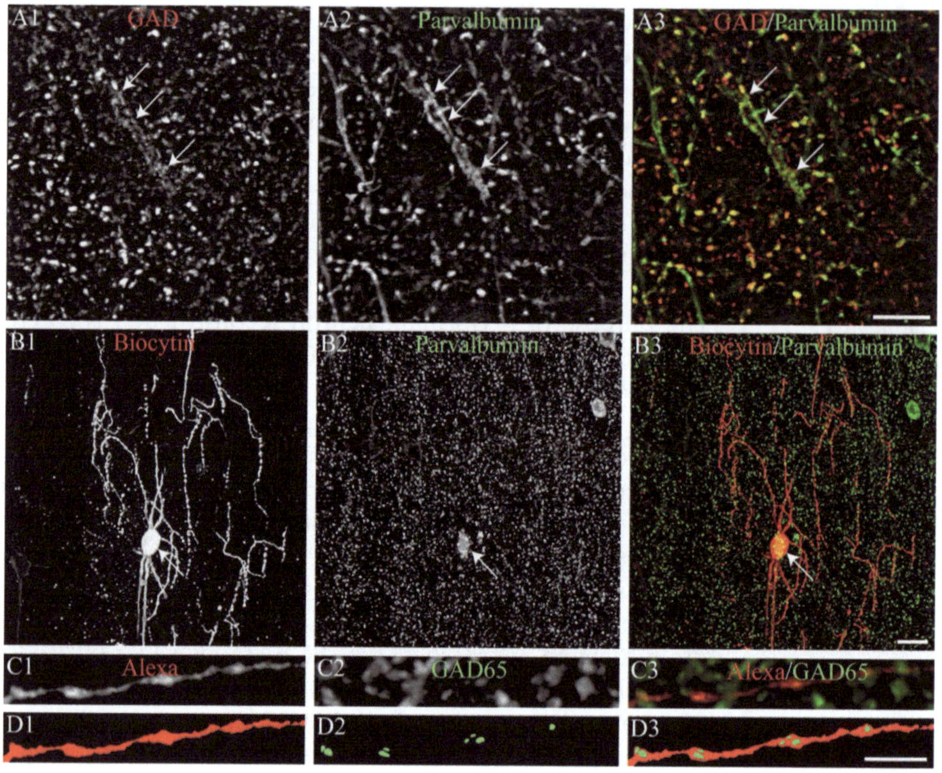

Fig. 5. Molecular characterization of monkey fast-spiking cell. (**a**) Cryostat section (40-μm thin slices) labeled for both glutamic acid decarboxylase (GAD) isoforms (*red*) and parvalbumin (PV) (*green*). Projection image was made from 13 z-planes taken 0.2 μm apart. (**b**) Double immunostaining for biocytin (*red*) and PV (*green*). The cell was filled during recording, during which time it was identified as a fast-spiking neuron. After recording/filling, the slice was fixed and resectioned (see Sect. 3.3). The image is a projection image of all the information within the section and was acquired on a confocal microscope using a 40X 1.3 NA objective. (**c**) Part of a fast-spiking neuron's axon. The neuron was filled with Alexa 568 (*Red*) during recording, fixed, and immunocytochemically labeled for GAD65 (*green*) as described in the text. (**d**) Image masks of the axon section (D1) and IR puncta (D2) colocalized with the axon in **c** pseudocolored corresponding to the original data. Bars: A = 10 μm, B = 20 μm, and D = 5 μm.

4.4. Quantitative Analysis of IR Puncta Colocalized with Filled Neurons

The colocalization frequency of synaptic proteins with putative terminals of dye-filled neurons can be manually determined using 2D and/or 3D image data sets without further processing. However, in order to extract additional information (i.e., terminal shape and fluorescence intensity), it is necessary to perform threshold-based segmentation in which a binary threshold level (either determined visually or by a thresholding algorithm) is set to designate the intensity level of the original grayscale histogram at which binary segregation occurs. Thus, in binary threshold segmentation, pixels (or voxels in 3D images; the smallest resolution unit in the image) are set as being "on" or "off," depending on how they compare to the selected threshold level. A binary image, called a mask, is then used to designate pixels classified as being "on" (Fig. 5d). For each group of adjacent on pixels (i.e., those designating an IR punctum), the corresponding binary image is called an object mask. For data sets in which the signal-to-noise ratio is high, IR puncta have relatively uniform fluorescence intensities, and fluorescence from adjacent puncta is nonoverlapping, a single threshold can be used to make mask objects of individual IR puncta. However, these imaging conditions rarely exist when using fluorescence immunohistochemistry to study synaptic structures or protein levels in postmortem brain sections.

We recently described a novel approach to masking objects in 3D data sets which resulted in improved object masking in thick brain tissue sections (53). Rather than choosing a single set of intensity thresholds to create a binary selection mask, our approach used multiple iterations that systematically varied the intensity thresholds and combined the resultant masks via morphologic selection criteria (see example of masking in Fig. 5c, d). The use of varying intensity thresholds is particularly adapted for the selection of objects without uniform and/or high fluorescence intensities, which results when proteins differ in their abundance within synaptic structures, and can vary substantially over the course of development and as a result of disease states. By combining neuron dye filling and immunofluorescence microscopy with neuronal tracing and the above segmentation methodology, information about how synaptic protein expression differs throughout the axon can be obtained. In addition, these combined techniques can be used to gain knowledge about how the molecular content of individual terminals differs as a result of the postsynaptic target. This information is crucial for understanding how synaptic protein expression corresponds to function.

5. Limitations and Pitfalls

When using fluorescent protein recombinants (e.g., green-fluorescent protein), the relationship between fluorescence intensity and antigen concentration is likely to be linear. In contrast, when using fluorescence immunocytochemistry to label proteins in tissue sections, the absolute relationship of fluorescence intensity to protein abundance can only be determined through conducting rigorous control experiments, and thus linearity should not be presumed. With that said, because CCD camera and photomultiplier tube detection systems are linear across a broad range of fluorescence intensities (53), quantitative fluorescence microscopy has been used in numerous studies to measure relative changes in protein levels (53, 55, 57). In a system where the expression of a protein is only effected in a subpopulation of cells, e.g., GAD_{67} expression is reduced in a subset of PV-positive interneurons in schizophrenia (64), quantitative fluorescence immunohistochemistry will be more informative than techniques (e.g., Western Blot) that treat the tissue as a whole.

Photobleaching refers to the loss of fluorescence that can occur as a result of fluorophores cycling through emission and excitation. As a result, the extent of photobleaching will be proportional to the intensity of the excitation light and the duration of exposure. When it occurs, photobleaching can make it difficult to detect the target protein and, if nonuniform, can bias quantification of fluorescence intensity. Fortunately, the newer generation fluorescence dyes (e.g., the Alexa series dyes) are significantly more resistant to photobleaching than traditional dyes (e.g., FITC), which are destroyed after significantly fewer emission-excitation cycles.

6. Conclusions

Interneurons of the cerebral cortex are a morphologically, electrophysiologically, and molecularly diverse population of cells. This chapter has just touched upon the electrophysiology, immunohistochemistry, and confocal microscopy techniques used to study these cells, specifically in monkeys. These techniques are continuously being exploited to gain as much information as possible about the different interneurons of the cerebral cortex so that classes and subclasses can be identified in order to understand how inhibition shapes cortical function (3).

Acknowledgments

This work was supported by the NIMH (MH051234 and MH084053 to DAL, MH085108 to KNF).

References

1. Somogyi P, Klausberger T (2005) Defined types of cortical interneurone structure space and spike timing in the hippocampus. J Physiol 562:9–26
2. Skaggs WE, McNaughton BL, Permenter M, Archibeque M, Vogt J, Amaral DG, Barnes CA (2007) EEG sharp waves and sparse ensemble unit activity in the macaque hippocampus. J Neurophysiol 98:898–910
3. Ascoli GA, Alonso-Nanclares L, Anderson SA, Barrionuevo G, Benavides-Piccione R, Burkhalter A et al (2008) Petilla terminology: nomenclature of features of GABAergic interneurons of the cerebral cortex. Nat Rev Neurosci 9:557–568
4. Zaitsev AV, Povysheva NV, Gonzalez-Burgos G, Rotaru D, Fish KN, Krimer LS, Lewis DA (2009) Interneuron diversity in layers 2–3 of monkey prefrontal cortex. Cereb Cortex 19:1597–1615
5. Povysheva NV, Zaitsev AV, Rotaru DC, Gonzalez-Burgos G, Lewis DA, Krimer LS (2008) Parvalbumin-positive basket interneurons in monkey and rat prefrontal cortex. J Neurophysiol 100:2348–2360
6. Melchitzky DS, Lewis DA (2008) Dendritic-targeting GABA neurons in monkey prefrontal cortex: comparison of somatostatin- and calretinin-immunoreactive axon terminals. Synapse 62:456–465

7. Bystron I, Blakemore C, Rakic P (2008) Development of the human cerebral cortex: Boulder Committee revisited. Nat Rev Neurosci 9:110–122
8. De Felipe J (2001) Cortical interneurons: from Cajal to 2001. Prog Brain Res 136:215–238
9. De Felipe J, Ballesteros-Yanez I, Inda MC, Munoz A (2006) Double-bouquet cells in the monkey and human cerebral cortex with special reference to areas 17 and 18. Prog Brain Res 154:15–32
10. De Felipe J, Jones EG (1988) A light and electron microscopic study of serotonin-immunoreactive fibers and terminals in the monkey sensory-motor cortex. Exp Brain Res 71:171–182
11. Gabbott PL, Bacon SJ (1996) Local circuit neurons in the medial prefrontal cortex (areas 24a, b, c, 25 and 32) in the monkey: II. Quantitative areal and laminar distributions. J Comp Neurol 364:609–636
12. Gabbott PL, Bacon SJ (1996) Local circuit neurons in the medial prefrontal cortex (areas 24a, b, c, 25 and 32) in the monkey: I. Cell morphology and morphometrics. J Comp Neurol 364:567–608
13. Jones EG (2009) The origins of cortical interneurons: mouse versus monkey and human. Cereb Cortex 19:1953–1956
14. Meyer G (2007) Genetic control of neuronal migrations in human cortical development. Adv Anat Embryol Cell Biol 189:1–111, 1 p preceding
15. Rakic S, Zecevic N (2003) Emerging complexity of layer I in human cerebral cortex. Cereb Cortex 13:1072–1083
16. Yanez IB, Munoz A, Contreras J, Gonzalez J, Rodriguez-Veiga E, De Felipe J (2005) Double bouquet cell in the human cerebral cortex and a comparison with other mammals. J Comp Neurol 486:344–360
17. Cruz DA, Lovallo EM, Stockton S, Rasband M, Lewis DA (2009) Postnatal development of synaptic structure proteins in pyramidal neuron axon initial segments in monkey prefrontal cortex. J Comp Neurol 514:353–367
18. Gabbott PLA, Dickie BGM, Vaid RR, Headlam AJM, Bacon SJ (1997) Local-circuit neurones in the medial prefrontal cortex (areas 25, 32 and 24b) in the rat: morphology and quantitative distribution. J Comp Neurol 377:465–499
19. Letinic K, Zoncu R, Rakic P (2002) Origin of GABAergic neurons in the human neocortex. Nature 417:645–649
20. Molyneaux BJ, Arlotta P, Menezes JR, Macklis JD (2007) Neuronal subtype specification in the cerebral cortex. Nat Rev Neurosci 8:427–437
21. Povysheva NV, Zaitsev AV, Kroner S, Krimer OA, Rotaru DC, Gonzalez-Burgos G, Lewis DA, Krimer LS (2007) Electrophysiological differences between neurogliaform cells from monkey and rat prefrontal cortex. J Neurophysiol 97:1030–1039
22. Krimer LS, Zaitsev AV, Czanner G, Kroner S, Gonzalez-Burgos G, Povysheva NV, Iyengar S, Barrionuevo G, Lewis DA (2005) Cluster analysis-based physiological classification and morphological properties of inhibitory neurons in layers 2–3 of monkey dorsolateral prefrontal cortex. J Neurophysiol 94:3009–3022
23. Markram H, Toledo-Rodriguez M, Wang Y, Gupta A, Silberberg G, Wu C (2004) Interneurons of the neocortical inhibitory system. Nat Rev Neurosci 5:793–807
24. Lund JS, Lewis DA (1993) Local circuit neurons of developing and mature macaque prefrontal cortex: Golgi and immunocytochemical characteristics. J Comp Neurol 328:282–312
25. De Felipe J (1999) Chandelier cells and epilepsy. Brain 122:1807–1822
26. Zaitsev AV, Gonzalez-Burgos G, Povysheva NV, Kroner S, Lewis DA, Krimer LS (2005) Localization of calcium-binding proteins in physiologically and morphologically characterized interneurons of monkey dorsolateral prefrontal cortex. Cereb Cortex 15: 1178–1186
27. Gonzalez-Burgos G, Krimer LS, Povysheva NV, Barrionuevo G, Lewis DA (2005) Functional properties of fast spiking interneurons and their synaptic connections with pyramidal cells in primate dorsolateral prefrontal cortex. J Neurophysiol 93:942–953
28. Kawaguchi Y, Kondo S (2002) Parvalbumin, somatostatin and cholecystokinin as chemical markers for specific GABAergic interneuron types in the rat frontal cortex. J Neurocytol 31: 277–287
29. Cauli B, Audinat E, Lambolez B, Angulo MC, Ropert N, Tsuzuki K, Hestrin S, Rossier J (1997) Molecular and physiological diversity of cortical nonpyramidal cells. J Neurosci 17:3894–3906
30. Swedlow JR, Platani M (2002) Live cell imaging using wide-field microscopy and deconvolution. Cell Struct Funct 27:335–341
31. Murray JM, Appleton PL, Swedlow JR, Waters JC (2007) Evaluating performance in three-dimensional fluorescence microscopy. J Microsc 228:390–405
32. Wang E, Babbey CM, Dunn KW (2005) Performance comparison between the high-speed Yokogawa spinning disc confocal system and single-point scanning confocal systems. J Microsc 218:148–159

33. Sandison DR, Webb WW (1994) Background rejection and signal-to-noise optimization in confocal and alternative fluorescence microscopes. Appl Opt 33:603–615
34. Benveniste M, Schlessinger J, Kam Z (1989) Characterization of internalization and endosome formation of epidermal growth factor in transfected NIH-3 T3 cells by computerized image-intensified three-dimensional fluorescence microscopy. J Cell Biol 109: 2105–2115
35. Hiraoka Y, Agard DA, Sedat JW (1990) Temporal and spatial coordination of chromosome movement, spindle formation, and nuclear envelope breakdown during prometaphase in Drosophila melanogaster embryos. J Cell Biol 111:2815–2828
36. Shaw PJ (2006) Comparison of widefield/deconvolution and confocal microscopy for three-dimensional imaging. In: Pawley JB (ed) Handbook of biological confocal microscopy. Springer, New York, pp 453–467
37. Stuart GJ, Dodt HU, Sakmann B (1993) Patch-clamp recordings from the soma and dendrites of neurons in brain slices using infrared video microscopy. Pflügers Arch 423:511–518
38. Sakmann B, Edwards F, Konnerth A, Takahashi T (1989) Patch clamp techniques used for studying synaptic transmission in slices of mammalian brain. Q J Exp Physiol 74:1107–1118
39. Dodt HU, Eder M, Schierloh A, Zieglgänsberger W (2002) Infrared-guided laser stimulation of neurons in brain slices. Sci STKE 2002(120):pl2
40. Dodt HU, Zieglgänsberger W (1990) Visualizing unstained neurons in living brain slices by infrared DIC-videomicroscopy. Brain Res 537:333–336
41. MacVicar BA, Hochman D (1991) Imaging of synaptically evoked intrinsic optical signals in hippocampal slices. J Neurosci 11:1458–1469
42. Gonzalez-Burgos G, Barrionuevo G, Lewis DA (2000) Horizontal synaptic connections in monkey prefrontal cortex: an in vitro electrophysiological study. Cereb Cortex 10:82–92
43. Trapp S, Ballanyi K (2012) Autonomic nervous system in vitro: studying tonically active neurons controlling Vagal outflow in Rodent brainstem slices. In Isolated Central Nervous System Circuits (Ed K Ballanyi), Neuromethods Series Vol. 73 (Ed W Walz). Springer Science+Business Media, LLC, New York, NY, pp 1–59
44. Ruangkittisakul A, Panaitescu B, Secchia L, Bobocea N, Kantor C, Kuribayashi J, Iizuka M, Ballanyi K (2012) Isolated brainstem respiratory centers from perinatal rodents. In Isolated Central Nervous System Circuits (Ed K Ballanyi), Neuromethods Series Vol. 73 (Ed W Walz). Springer Science+Business Media, LLC, New York, NY, pp 61–124
45. Moore AR, Zhou W, Jakovcevski I, Zecevic N, Antic S (2012) Physiological properties of human fetal cortex in vitro. In Isolated Central Nervous System Circuits (Ed K Ballanyi), Neuromethods Series Vol. 73 (Ed W Walz). Springer Science+Business Media, LLC, New York, NY, pp 125–158
46. Sun XJ, Tolbert LP, Hildebrand JG, Meinertzhagen IA (1998) A rapid method for combined laser scanning confocal microscopic and electron microscopic visualization of biocytin or neurobiotin-labeled neurons. J Histochem Cytochem 46:263–273
47. Stuart G (1998) Patch-pipette recording in brain slices. Curr Protoc Neurosci 6.7:6.7.1–6.7.10
48. Sanchez-Vives MV (2012) Spontaneous rhythmic activity in the adult cerebral cortex in vitro. In Isolated Central Nervous System Circuits (Ed K Ballanyi), Neuromethods Series Vol. 73 (Ed W Walz). Springer Science+Business Media, LLC, New York, NY, pp 263–284
49. Kantor C, Panaitescu B, Kuribayashi J, Ruangkittisakul A, Jovanovic I, Leung V, Lee TF, MacTavish D, Jhamandas J, Cheung PY, Ballanyi K (2012) Early network oscillations in cortex and locus coeruleus of horizontal brain slices from newborn mammals. In Isolated Central Nervous System Circuits (Ed K Ballanyi), Neuromethods Series Vol. 73 (Ed W Walz). Springer Science+Business Media, LLC, New York, NY, pp 315–356
50. Sanchez-Vives MV, McCormick DA (2000) Cellular and network mechanisms of rhythmic recurrent activity in neocortex. Nat Neurosci 3:1027–1034
51. Luhmann HJ, Kilb W (2012) Intact in vitro preparation of the neonatal rodent cerebral cortex-analysis of cellular properties and network activity. In Isolated Central Nervous System Circuits (Ed K Ballanyi), Neuromethods Series Vol. 73 (Ed W Walz). Springer Science+Business Media, LLC, New York, NY, pp 301–314
52. Konopaske GT, Dorph-Petersen KA, Sweet RA, Pierri JN, Zhang ZW, Sampson AR, Lewis DA (2008) Effect of chronic antipsychotic exposure on astrocyte and oligodendrocyte numbers in macaque monkeys. Biol Psychiatry 63:759–765
53. Fish KN, Sweet RA, Deo AJ, Lewis DA (2008) An automated segmentation methodology for quantifying immunoreactive puncta number

and fluorescence intensity in tissue sections. Brain Res 1240:62–72

54. Sweet RA, Fish KN, Lewis DA (2010) Mapping synaptic pathology within cerebral cortical circuits in subjects with schizophrenia. Front Hum Neurosci 4:1–14
55. Sugiyama Y, Kawabata I, Sobue K, Okabe S (2005) Determination of absolute protein numbers in single synapses by a GFP-based calibration technique. Nat Methods 2:677–684
56. Hohensee S, Bleiss W, Duch C (2008) Correlative electron and confocal microscopy assessment of synapse localization in the central nervous system of an insect. J Neurosci Methods 168:64–70
57. Glynn MW, McAllister AK (2006) Immunocytochemistry and quantification of protein colocalization in cultured neurons. Nat Protoc 1:1287–1296
58. Darya K, Ganguly A, Lee D (2009) Quantitative analysis of synaptic boutons in Drosophila primary neuronal cultures. Brain Res 1280: 1–12
59. Bergsman JB, Krueger SR, Fitzsimonds RM (2006) Automated criteria-based selection and analysis of fluorescent synaptic puncta. J Neurosci Methods 152:32–39
60. Wallace W, Schaefer LH, Swedlow JR (2001) A working person's guide to deconvolution in light microscopy. Biotechniques 31:1076–1078
61. Agard DA (1984) Optical sectioning microscopy: cellular architecture in three dimensions. Annu Rev Biophys Bioeng 13: 191–219
62. Rodriguez A, Ehlenberger D, Kelliher K, Einstein M, Henderson SC, Morrison JH, Hof PR, Wearne SL (2003) Automated reconstruction of three-dimensional neuronal morphology from laser scanning microscopy images. Methods 30:94–105
63. McGarry L, Packer A, Fino E, Nikolenko V, Sippy T, Yuste R (2010) Quantitative classification of somatostatin-positive neocortical interneurons identifies three interneuron subtypes. Front Neural Circuits 4:12
64. Lewis DA, Hashimoto T, Volk DW (2005) Cortical inhibitory neurons and schizophrenia. Nat Rev Neurosci 6:312–324
65. Conde ÇF, Lund JS, Jacobowitz DM, Baimbridge KG, Lewis DA (1994) Local circuit neurons immunoreactive for calretinin, calbindin D-28 k, or parvalbumin in monkey prefrontal cortex: distribution and morphology. J Comp Neurol 341:95–116

Chapter 5

Neural Circuits Underlying Circadian Oscillations in Mammals: Clocks in a Dish

Takahiro J. Nakamura, Stephan Michel, Gene D. Block, and Christopher S. Colwell

Abstract

In mammals, the part of the nervous system responsible for most circadian behavior can be localized to a bilaterally paired structure in the hypothalamus known as the suprachiasmatic nucleus (SCN). Understanding this mammalian circadian system will require a detailed multilevel analysis of neural SCN circuits. Many of the techniques and approaches that are used for the analysis of the circuitry driving circadian oscillations in the SCN are similar to those employed in other brain regions. There is, however, one fundamental difference that needs to be taken into consideration, that is, the physiological, cellular, and molecular properties of SCN neurons vary with the time of day. These cells continue to oscillate in the dish, so the daily circadian cycle always needs to be a consideration regardless of the experimental preparation. In this chapter, we will consider the preparations and techniques that we have used to analyze the SCN circuit including acute brain slice, slice culture, and disassociated neurons.

Key words: Biological clock, Bioluminescence, Clock gene, Electrophysiology, Isolated hypothalamus, Suprachiasmatic nucleus

1. Introduction

Almost all organisms, including humans, exhibit daily rhythms in their behavior and physiology. Endogenous cellular networks composed of multiple circadian oscillators generate these rhythms. This network of oscillators provides temporal information to an organism's physiological systems and behavioral outputs. In mammals, the part of the nervous system responsible for most circadian behavior can be localized to a bilaterally paired structure in the hypothalamus known as the suprachiasmatic nucleus (SCN). Many

Klaus Ballanyi (ed.), *Isolated Central Nervous System Circuits*, Neuromethods, vol. 73,
DOI 10.1007/978-1-62703-020-5_5, © Springer Science+Business Media New York 2012

neurons in the SCN are intrinsic oscillators that continue to generate near 24 h rhythms in electrical activity, secretion, and gene expression when isolated from the rest of the organism (1). Neurons within the SCN generate robust, synchronized rhythms in the transcription, translation, and degradation of key "clock genes" in an autoregulatory loop that has an endogenous periodicity of approximately 24 h (2, 3). While other mammalian cell populations can generate rhythms in gene expression, the ability to generate intrinsic circadian rhythms in electrical activity appears to be a unique property of mammalian SCN neurons and one essential to the function of the circadian timing system. Other circadian oscillators, for instance in the heart, have also been identified, and an interplay between the SCN and these peripheral oscillators governs physiological, behavioral, and endocrine rhythms. Despite the existence of other circadian oscillators, the clock within the SCN exhibits a profound effect on temporal organization of behavior and physiology and, thus, represents the "master" oscillator coordinating daily rhythms throughout the mammalian body. Understanding the mammalian circadian system will require detailed understanding of SCN circuits and individual neurons at many levels of analysis.

The SCN is a bilaterally paired nucleus made up of tightly compacted, small-diameter (10–12 μm diameter) neurons just lateral to the third ventricle atop the optic chiasm. Anatomical studies generally support the division of the SCN into at least two subdivisions including a ventral (core) and dorsal (shell) region (4). The ventral neurons are thought to act as an integrator of external input, receiving information from three major pathways: the retinohypothalamic tract (RHT), the geniculohypothalamic tract from the intergeniculate leaflet of the thalamus, and from the raphe nuclei (4). The ventral neurons must integrate the environmental information and relay this information to the rest of the SCN. In contrast, neurons of the dorsal shell appear to generate the most robust circadian oscillations, at least at the level of gene expression (5, 6). The fact that many core projections terminate on shell neurons, further supports the idea that interplay between these two centers is responsible for the output of circadian information from the SCN (7). The outputs of the SCN largely travel to other hypothalamic regions including the subparaventricular zone (8). These hypothalamic relay nuclei send projections throughout the central nervous system and endocrine system, providing multiple pathways by which the SCN can convey temporal information to the brain and body (9).

One key to understanding how neurons in the SCN synchronize to the environment as well as synchronize to each other may lie within the different neuropeptides expressed in the SCN (Fig. 1). For example, the retinal ganglion cells (RGCs) that project to the SCN express the neuropeptide pituitary adenylate

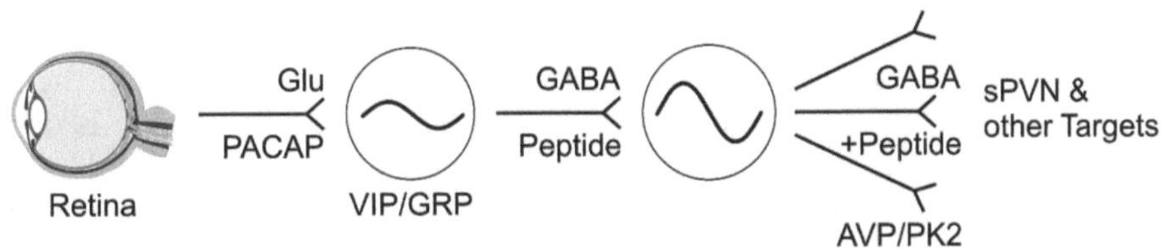

Fig. 1. *Schematic of the light input pathway into the suprachiasmatic nucleus (SCN).* We feel that neuropeptides are critical for communication within this circuit. In the intact system, melanopsin-containing retinal ganglion cells depolarize in response to photic stimulation. The light-induced depolarization and action potential generation result in the release of glutamate and pituitary adenylate cyclase-activating polypeptide (PACAP) onto retinorecipient cells within the SCN. Depending on the species, these cells are concentrated in the ventral region (core) of the SCN and express γ-aminobutyric acid (GABA) as well as the neuropeptides vasoactive intestinal peptide (VIP) or gastrin-releasing peptide (GRP). The neurons in the dorsomedial shell express vasopressin, prokineticin 2 (PK2), as well as GABA. The outputs of the SCN can come from both of these two SCN subdivisions and largely travel to other hypothalamic regions including the subparaventricular zone (sPVN).

cyclase-activating polypeptide (PACAP) as well as the excitatory transmitter glutamate. Many of the retinal recipient neurons within the ventral SCN express the neuropeptide vasoactive intestinal peptide (VIP) as well as the neurotransmitter γ-aminobutyric acid (GABA). The neurons in the dorsomedial shell express vasopressin and GABA. Finally, in the hamster, there is also good evidence for a population of calbindin-positive cells that may function to "gate" the photic response of the SCN population (7). Over the last few years, a number of studies have examined the role of neuropeptides in cellular communication and function of the circadian system. Nevertheless, it is fair to say that we are still in the early stages of understanding the physiological roles played by these peptides in the SCN circuit.

The ability of the SCN neural population to generate neural activity rhythms in isolation from the rest of the organism has been documented in numerous studies using an acute brain slice preparation (10). Moreover, circadian rhythms in electrical discharge have been recorded from cultures made from SCN tissue (11–13). This work has led to the general acceptance of the view that the generation of circadian electrical activity rhythms is not a neural network property in which synaptic feedback information from one neuron to the other is required for the presence of circadian rhythmicity. Instead, these studies are all consistent with the idea that many SCN neurons are stable, self-sustained oscillators that have the intrinsic capacity to generate circadian rhythms. That said, there is evidence that circuit properties contribute to the robustness and precision of SCN circadian oscillations (14). One of the challenges of ongoing work is the development of an understanding of how a molecular feedback loop occurring at the level of transcription and translation interacts with the neuronal plasmalemmal membrane to produce physiological rhythms. It is not known how the molecular feedback loop drives the electrical rhythm in membrane processes, and this is a critical question for the field.

Many of the techniques and approaches that are used for the analysis of the circuit driving circadian oscillations in the SCN are similar to those used in other brain regions. There is one fundamental difference that needs to be taken into consideration, that is, the physiological, cellular, and molecular properties of SCN neurons vary with time of day. These cells continue to oscillate (maintaining phase) in the dish, so the daily cycle needs to be a consideration independent on whether the experiments are conducted in vivo, using in vitro brain slices, or in cultured neurons. For example, circadian oscillations in GABA-mediated synaptic transmission in both presynaptic release and postsynaptic responses have been described (15, 16). Moreover, the neurons show spontaneous circadian oscillations in electrical activity driven, at least in part, by daily rhythms in intrinsic membrane properties in these cells (10, 17). Signaling pathways within these cells oscillate, including dynamic changes of the free cytosolic concentration of Ca^{2+} and cyclic adenosine monophosphate/protein kinase A (cAMP/PKA) activity (18–20). Since circadian oscillations are generated by a transcriptional feedback loop, a wide range of genes are rhythmically regulated including those involved in secretion and synaptic transmission critical for circuits. So, for the circadian system, time of day is a critical variable that always has to be considered. The extent to which these circadian variations influence other brain circuits is not yet known although there is a growing appreciation that molecular oscillations in clock gene expression are widespread within the nervous system (21, 22).

Given that the SCN system comprises a circuit of endogenous oscillators that is synchronized to light, the external lighting conditions for all of these experiments need to be carefully considered. When animals are in a light/dark (LD) cycle, they may exhibit day/night variation in a number of parameters. These rhythms are influenced by the internal circadian system and by the direct effects of light and dark. Consequently, a rhythm recorded under these conditions can be considered a diurnal (i.e., day vs. night), but not necessarily a circadian rhythm, that is, an internally generated rhythm. An important (though little) employed experiment is to reverse the timing of the LD cycle and show that the resulting rhythm follows the change in the light schedule. This confirms that the diurnal rhythm is at least driven by the environmental conditions, but does not necessarily address whether the rhythm has an endogenous origin. In order to isolate the endogenous circadian component driving these differences, it is necessary to hold the organism in constant conditions, that is, constant dark (DD) and constant temperature. In the intact system, the animal's behavior and locomotor activity can be followed with the convention that the onset of activity for a nocturnal organism is defined as circadian time (CT) 12. While activity onset is a convenient and reliable way to follow the endogenous rhythm, other robust circadian outputs could be used including body temperature or heart rate. For generation of brain slices, the LD cycle in which the animal was held prior to being sacrificed is an excellent predictor of the activity phase of the SCN population. Cells in culture for extended periods can be synchronized with temperature cycles or pharmacological manipulations.

For the rest of this chapter, we will consider the techniques used to analyze the SCN circuit including acute brain slice, organotypic slice culture, and disassociated neurons. The material, methods, and approaches in employing each of these techniques including a short description of the advantages and disadvantages will be described. Certainly, other techniques can be usefully applied, but we will focus on the preparations and techniques with which we have working experience.

2. Materials

2.1. Materials for Brain Slices

1. Solutions (concentration in nM): artificial cerebrospinal fluid (ACSF) – 130 NaCl, 26 $NaHCO_3$, 3 KCl, 2 $MgCl_2$, 1.25 NaH_2PO_4, 2 $CaCl_2$, and 10 D-glucose. Modified ACSF (for brain dissection and vibratome slicing): same as ACSF except 5 nM $MgCl_2$ and 1 nM $CaCl_2$. The solutions are prepared the day before the experiment and kept at 4°C and "carbogenated" with carbogen (95% O_2, 5% CO_2) to adjust pH to 7.2–7.4; osmolality is 290–300 mOsm. For long-term recording of slices (>24 h), gentamicin (5 μg/ml) is added to the ACSF.
2. Anesthetic agents and anesthetic chamber.
3. Sterilized dissection tools, guillotine for decapitation, dissecting scissors, forceps, spatulas, and razor blade for trimming.
4. Slice cutting: vibratome (e.g., Microslicer, DTK Model 1500E, Ted Pella, Redding, CA, USA; www.tedpella.com) or Leica VT 1200 S (Wetzlar, Germany; www.leica-microsystems.com) or tissue chopper (e.g., McIlwain Tissue Chopper, e.g., Ted Pella).
5. Electrophysiological recording equipment (patch clamp): vibration isolation table (e.g., Technical Manufacturing Corporation [TMC], Peabody, MA, USA; www.techmfg.com), Faraday cage, manipulator (e.g., Sutter Instrument Company, Novato, CA, USA [www.sutter.com], or Luigs & Neumann, Ratingen, Germany [www.luigs-neumann.com]), slice recording chamber (e.g., Warner Series 20, Hamden, CT, USA; www.warneronline.com), gravity-fed perfusion control (e.g., ALA perfusion systems, Farmingdale, NY, USA; www.alascience.com), gas (CO_2/O_2) control and distribution, vacuum control,

patch-clamp amplifier, for example, the Axon Axopatch 200B (Molecular Devices Sunnyvale, CA, USA; www.moleculardevices.com) or EPC10 (HEKA, Lambrecht, Germany; www.heka.com), and control software (e.g., Axon pClamp or HEKA patchmaster).

6. Electrophysiological recording equipment (long-term multi-unit recording): vibration-damped table, faraday, temperature-controlled perfusion chamber, water bath, perfusion pump, vacuum control, extracellular amplifier (e.g., 5113, Signal Recovery, Oak Ridge, TN, USA; www.signalrecovery.com), and digitizer (e.g., CED 1401 with Spike2 software, CED, Cambridge, England; www.ced.co.uk).
7. Visual patch clamp: upright water immersion microscope equipped with infrared differential interference contrast (IR-DIC; e.g., Zeiss Axioskop FS2plus, Carl Zeiss, Göttingen, Germany; www.zeiss.com or Olympus BX51W, Olympus, Center Valley, PA, USA; www.olympusamerica.com), IR-sensitive charge-coupled device (CCD) camera, for example, C2400 (Hamamatsu, Bridgewater, NJ, USA; www.hamamatsu.com) or VX55 (TILL Photonics, Gräfelfing, Germany; www.till-photonics.com), and video monitor.
8. Ca^{2+} imaging: light source (e.g., monochromator TILL Photonics Polychrome V or Lambda DG-4, Sutter Instrument, Novato, CA, USA; www.sutter.com), cooled CCD camera, for example, Microview model 1,317 × 1,035 pixel (Princeton Instruments, Trenton, NJ, USA; www.princetoninstruments.com), or TILL Imago QE cooled interline CCD (1,376 × 1,040 pixel), and software (e.g., MetaFluor, Molecular Devices).

2.2. Materials for Slice Culture

1. Sterilized dissection tools: guillotine for decapitation, fine forceps, small dissecting scissors, spatulas, scalpels, Petri dish, etc.
2. Anesthetic agents and anesthetic chamber
3. Vibratome (Sect. 2.1.)
4. Dissecting stereomicroscope
5. 35-mm² dish (e.g., Nunc, Thermo Fisher Scientific, Rochester, NY, USA) and membrane inserts (Millicell-CM 30 mm, low height, PICMORG-50, Millipore, Billerica, MA, USA; www.millipore.com)
6. Hank's buffered salt solution (HBSS) supplemented with 0.35 g/l $NaHCO_3$, 10 nM 4-(2-hydroxyethyl)-1-piperazineethanesulfonic acid (HEPES), and antibiotics (50 U/ml penicillin, 50 μg/ml streptomycin)
7. Recording media: serum-free, no $NaHCO_3$, no phenol red, Dulbecco's modified Eagle's medium (D-2902, Sigma-Aldrich,

St. Louis, MO, USA) supplemented with 0.35 g/l $NaHCO_3$, 10 nM HEPES (pH 7.2), B27 Supplement (2%; 17504–010, Invitrogen; Carlsbad, CA, USA; www.invitrogen.com), 0.1 nM luciferin, and antibiotics (25 U/ml penicillin, 25 μg/ml streptomycin)

8. Recording apparatus with incubator

2.3. Materials for Disassociated Cell Culture

1. Sterilized dissection tools: fine forceps, small dissecting scissors, spatulas, scalpels, Petri dish, etc.
2. Anesthetic agents and anesthetic chamber
3. Tissue chopper
4. Dissecting stereomicroscope
5. Glass bottom dish (e.g., Glass-based dish, AGC Techno Glass Co., Ltd., Funabashi, Japan; www.atg.ushop.jp)
6. HBSS supplemented with 0.35 g/l $NaHCO_3$, 10 nM HEPES, and antibiotics (50 U/ml penicillin, 50 μg/ml streptomycin)
7. Culture media: Dulbecco's modified Eagle's medium (D-2902, Sigma-Aldrich; www.sigmaaldrich.com) supplemented with 3.7 g/l NaHCO3, 3.5 g/l D-glucose, B27 Supplement (2%; 17504–010, Invitrogen), 16 mg phenol red, and antibiotics (25 U/ml penicillin, 25 μg/ml streptomycin)
8. Recording media: culture media except phenol red and supplemented with 0.1 nM luciferin
9. Dissociation solution (DS): 5.8 nM $MgCl_2$, 1.6 nM HEPES, 0.2 nM NaOH, 117 nM Na_2SO_4, 3 nM K_2SO_4, 1 mg phenol red
10. Papain DS (pH 7.4): one vial of papain (>100 units, Worthington Biochemical Co., Lakewood, NJ, USA; www.worthingtom-biochem.com) and 0.8 mg L-cysteine in 5 ml DS
11. Stop buffer: heavy inhibitor (pH 7.4) – 0.05 g BSA and 0.05 g trypsin inhibitor in 5 ml DS; light inhibitor (pH 7.4): 0.5 ml heavy inhibitor and 4.5 ml DS
12. Coating solution: 10 μg/ml laminin and 0.1 mg/ml Poly-D-lysine in ddH_2O
13. Inverted microscope with stage top incubator (e.g., Tokai Hit INU-WELS-F1, Fujinomiya, Japan; www.tokaihit.com)
14. Cooled intensified charge-coupled device (CCD) camera system (e.g., Stanford Photonics model XR Mega, Palo Alto, CA, USA; www.stanfordphotonics.com)
15. CO_2 incubator

3. Methods

3.1. Acute SCN Brain Slices

3.1.1. General Features of SCN Slices

The acutely isolated brain slice has the advantages of offering accessibility and control while at the same time preserving many of the synaptic connections of the SCN circuitry. Previous workers have reported that the phase of the rhythms expressed in the SCN brain slice is predicted well by the prior LD cycle. For these reasons, it is an excellent in vitro preparation to characterize synaptic communication in the SCN and to search for diurnal variations. The first concern in the use of acute SCN brain slices is the duration of time for which neural activity can be recorded. For the majority of brain regions, brain slices seem to remain healthy for 6–12 h. Due to this technical limitation, most experiments look for day–night variation by comparing properties of SCN neurons prepared in the day with those prepared during the night. For these experiments, animals are placed in constant conditions and the behavioral rhythm recorded with wheel running activity. The brain slice can then be prepared at different phases of the daily cycle to evaluate circadian regulation. Multiunit activity (MUA) of acute slices has been recorded for longer time spans (24–72 h), and the use of slice culture yields even longer recording times, thus allowing the assessment of period length of the rhythm. The same basic procedure for the preparation of brain slices is used for extracellular or intracellular sharp microelectrode or whole-cell patch-clamp recordings and Ca^{2+} imaging experiments.

3.1.2. Methods for Generating SCN Slices

Animals are kept in light–dark regimes for a time sufficient to synchronize or entrain their circadian rhythm. Entrainment is typically achieved within 2–3 weeks, but it is advisable to validate the phase of the circadian clock of the animal at the time of preparation by monitoring the animal's locomotor activity rhythm. This is essential when using manipulations of the LD cycle, such as shifting the phase in a "jet-lag" experiment or switching to constant darkness. The animals are killed by decapitation at times determined by the LD cycle ("Zeitgeber" time, ZT) or the behavioral rhythms (subjective or circadian time, CT). The effect of the time of preparation itself on the subsequent phase of the rhythm in electrical activity is small (23, 24], but caution should be taken when slices are prepared at night. Procedures need to be performed either in dim far-red light or under IR illumination with the help of an IR viewer.

The brain slice procedures that we use are similar to those previously described (15, 16, 25–28). Successful recording requires fast removal of the brain. The longer the time between decapitation and submersion of the brain into the ice-cold slice solution, the poorer the quality of the neurons in terms of resting membrane potential and other properties. The goal is to complete this part of the procedure in 1–2 min. Similarly, the success rate in >24-h recordings of acute brain slices is improved by minimizing the total time it takes to transfer the slices to the recording chamber (4–6 min). It is also critical to cut the optic nerves very carefully in an early stage of dissection to prevent any strain on the optic chiasm during preparation. Specifically, brains are dissected and placed for ~1 min in ice-cold carbogenated ACSF. After trimming the brain to a block of tissue containing the hypothalamus, a vibratome is used for making coronal slices with a thickness of 250–350 μm for patch-clamp and Ca^{2+} imaging experiments under visual control (see also Chaps. 1-Trapp and Ballanyi (29), 3-Moore et al. (30), and 8-Sanchez-Vives (31)). Successful sectioning of SCN slices using the vibratome requires a firm attachment of the brain to the cutting block. This is a common problem for beginners. It should be avoided that the brain moves when the blade hits the tissue. Cutting should be done with the optic chiasm facing the blade. The right combination of forward speed of the blade, vibration frequency, and, if applicable, amplitude of oscillation are critical for cutting smoothly through the tissue. Brains from older mice, or especially rats, require slower forward speed and higher-frequency oscillation of the blade. The cyanoacrylate ("superglue") used to fix the brain to the cutting chamber of the vibratome can form filaments when ACSF is poured into the chamber. This can impede cutting when these filaments are located between the blade and the brain tissue. This can be avoided by dripping a few drops of ACSF onto the brain before filling the chamber.

For >24-h MUA recordings of electrical activity from SCN neurons, a tissue chopper is preferred over a vibratome for producing 400–500-μm-thick slices to minimize the time of preparation. Brain slices containing the SCN are then placed for at least 1 h before recording in our standard ACSF. Patch-clamp and Ca^{2+} imaging experiments can be performed at room temperature (22–25°C), but slices need to be warmed up to 36°C for 30 min right after cutting to activate endopeptidases. For MUA recordings, slices are immediately transferred to a recording chamber that is continuously superfused (1.5 ml/min) with warm (35.5°C) carbogenated ACSF with added antibiotic (gentamicin, 40 mg/ml). SCN slices are mechanically stabilized with either a metal fork or a nylon grid, preventing movement but allowing access for recording electrodes (see also Chap. 1-Trapp and Ballanyi (29)).

While most acute slices from other brain regions will be used for a short period of time, the analysis of circadian rhythms requires longer recording times (at least 32 h). The cultured slice would be more appropriate for longer recording times, but the acute slice offers many advantages when interested in photic entrainment and resetting mechanisms. Also, the culture condition can impair electrode or dye access to cells. To optimize the time acute slices stay healthy in the recording dish, conditions must be as stable as possible. Some of the problems that can occur and jeopardize

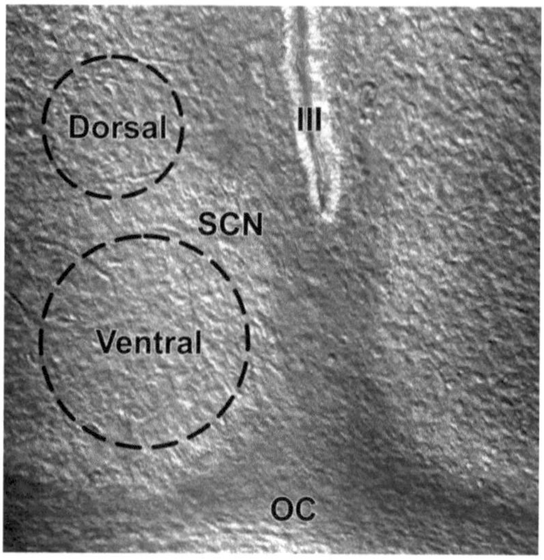

Fig. 2. *Infrared differential interference contrast (IR-DIC) video image of mouse SCN.* We have labeled the regions that we take to be dorsal or ventral in our experiments. The third ventricle (III) and optic chiasm (OC) are used as references. Most of our electrophysiological recording in the brain slice are made using this imaging technology to localize the recording site.

long-term recordings are fluctuating bath fluid level, condensation due to temperature differences and fluctuations in ambient temperature, mechanically instable slice tissue or electrodes, salt forming at the aspiration needle, and fluctuations in the vacuum.

We have made use of a number of techniques to analyze the acute brain slice preparation. In the next section, we will briefly describe the techniques that we have found most useful for the analysis of the circadian circuit. Some of these techniques can also be usefully applied to the slice culture and disassociated SCN neurons that we will describe later.

3.2. Optical Approaches

3.2.1. IR-DIC Video Microscopy

Video microscopy allows visualization of live, unstained cells in brain slice preparations (32, 33). The brain slice rests in a perfusion chamber and is illuminated with IR light. Images are taken with DIC optics and contrast-enhanced video microscopy. Figure 2 shows a view of SCN neurons in a brain slice as seen with IR-DIC video microscopy. The optic chiasm, third ventricle, and SCN are clearly visible. At higher magnification, it is possible to distinguish the borders and major processes of cells from the surface to about 50 – 100 μm deep into a brain slice. This imaging technology is capable of targeting specific cells for electrophysiological analysis. The experimenter can clearly identify cells of the SCN and even distinguish different cell populations within the SCN. In addition, IR-DIC microscopy has improved the frequency of successful recordings and allows the careful visual positioning of iontophoretic and stimulating electrodes. These are significant technological advantages for this type of study.

3.2.2. Ca^{2+} Imaging

Ca^{2+} imaging techniques allow dynamic measurements of the free cytosolic Ca^{2+} concentration inside neurons and have been particularly important in understanding the relationship between membrane events and transcriptional regulation within the SCN. To carry out Ca^{2+} imaging in the brain slice, we use a cooled CCD camera that is added to the Olympus fixed stage microscope (18). Two different methods are used to load the cells with the dye. In the first approach, slices are incubated in 10 μM of the membrane-permeant acetoxymethyl (AM) form of the Ca^{2+}-sensitive fluorescent dye fura-2 (fura-2-AM; Invitrogen) at 37°C for 10 min. In a recent modification of this technique (34), the SCN slice is exposed briefly (1 min) to stock solution of fura-2-AM (1 nM), and subsequently loading is continued for 1 h at room temperature using the diluted fura-2-AM in ACSF (10 μM). We used this modified loading protocols successfully in older rats (6 months) and old mice (2 years). The alternative approach uses membrane-impermeant fura-2 (1 nM pentapotassium salt) which is loaded into cells via the whole-cell-recording patch pipette. The fluorescence of fura-2 is excited alternatively at wavelengths of 357 nM and 380 nM by means of a high-speed (1.2-ms switching capability) wavelength switcher (Lambda DG-4). With the optics in many objectives, there is a big difference in the transmission of light at 340 and 380 nM. This makes it difficult to bring fluorescence measurements from both 340 nM and 380 nM into the dynamic range of the camera. Accordingly, the dye can be excited with 357-nM light instead of 340 nM in these experiments. Image analysis software allows the selection of "regions of interest" within the field from which measurements of dynamic changes of fura-2 fluorescence intensity are taken. In order to minimize bleaching of the dye, the intensity of excitation light and sampling frequency should be kept as low as possible (see also Chaps. 2-Ruangkittisakul et al. (35) and 11-Kantor et al. (36)).

Studies with fura-2 are technically easy, but have potential problems. Firstly, it is not possible to confidently resolve from which cell types (neurons vs. glia) the data are collected. Secondly, in practice, it appears to be more difficult to load cells in slices from older animals, and, like with all AM dyes applied via the bath, only cells located at the surface of the slice take up the dye. Loading dye via the patch pipette (Fig. 3) can solve these problems. However, this approach is much more laborious, and it may be best to use this technique only when necessary to confirm a result obtained with the membrane-permeant form of this dye. Although fura-2 is the standard dye we use for ratiometric Ca^{2+} measurements, there are some good arguments for employing other dyes. We have made some use of the visible light Ca^{2+} indicator Oregon Green BAPTA (Invitrogen). We obtain good loading of SCN slices with this dye, and one should see less phototoxicity. The use of visible rather than UV excitation should produce less autofluorescence. Since the cells

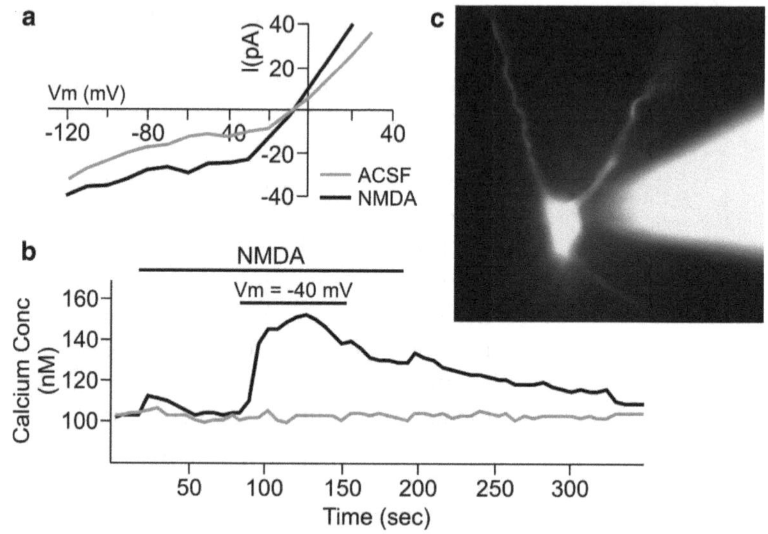

Fig. 3. *SCN neuron loaded with the fluorescent Ca^{2+} dye fura-2 through patch pipette allowing simultaneous measurements of cytosolic Ca^{2+} and membrane current.* The agonist of one subgroup of ionotropic glutamate receptors N-Methyl-D-aspartate (NMDA, 25 μM) was bath applied. (**a**) *Top panel* shows the current/voltage (IV) curve before (*gray*) and upon (*black*) NMDA treatment. (**b**) Panel shows estimated Ca^{2+} concentrations. Because of the voltage dependence of NMDA receptor-mediated currents, the Ca^{2+} response is greatly increased by moving the cell's membrane potential to −40 mV. (**c**) Panel shows example of SCN neuron filled with fura-2.

in the SCN are known to undergo daily rhythms in metabolism, the possibility of daily rhythms in autofluorescence must be considered although we have found that autofluorescence is responsible for less than 5% of the signal. Finally, there are good arguments for the use of 2-photon ("multiphoton") laser scanning microscopy. This approach offers several important advantages compared to conventional fluorescent microscopy with a CCD camera. All forms of imaging in thick brain tissue are limited by light scatter, and longer-wavelength light scatters less than shorter-wavelength light. For this reason, the ability of 2-photon microscopy to make use of long-wavelength light offers significant advantages for the resolution of structures deep in a brain slice. Furthermore, in 2-photon microscopy, the indicator dye is excited in a nonlinear manner. The requirement for two near-coincident photons to achieve excitation of the dye means that only focused light reaches the required intensities and that scattered light does not excite the dye. This nonlinear excitation results in a significant reduction in photodamage and bleaching. This reduction in the damage caused by the excitation light is likely to be a critical advantage for experiments that involve taking measurements of Ca^{2+} for extended periods of time (see also Chaps. 2-Ruangkittisakul et al. (35) and 11-Kantor et al. (36)).

The free concentration of cytosolic Ca^{2+} can be calculated from the ratio (R) of fluorescence intensities at 357 and 380 nM, using the following equation: $[Ca^{2+}] = K_d \times Sf \times (R - R_{min})/(R_{max} - R)$ (37). The value for K_d is set at 135 nM while values for R_{min} and R_{max} are determined with both in vitro and in vivo calibration methods. Initially, the in vitro method is used to make estimate values. With this method, rectangle glass capillaries are filled with high Ca^{2+} (fura-2 + 10 nM Ca^{2+}), low Ca^{2+} (fura-2 + 10 nM EGTA), and a control saline without fura-2. The fluorescence (F) at 380-nM excitation of the low Ca^{2+} solution is imaged, and the gain of the camera adjusted to maximize the signal. These camera settings will then remain fixed and background subtracted measurements made with 380- and 357-nM excitation of the three solutions: R_{min} = F357 nM in low Ca^{2+}/F380 in low Ca^{2+}, R_{max} = F357 in high Ca^{2+}/F380 in high Ca^{2+}, and Sf = F380 in low Ca^{2+}/F380 in high Ca^{2+}. For the more accurate "in vivo" calibration methods (38), cells are loaded via the patch pipette using solutions inside the electrode similar to the normal internal solution but containing either no Ca^{2+} (20 nM EGTA) or 10 nM Ca^{2+} for R_{min} and R_{max}, respectively. To obtain estimates of the effective K_d, three different EGTA/Ca^{2+} solutions are used with calculated free Ca^{2+} of 111 nM (10 nM EGTA/3.5 nM Ca^{2+}), 207 nM (10 nM EGTA/5 nM Ca^{2+}), and 483 nM (10 nM EGTA/7 nM Ca^{2+}). By calibrating in the slice preparation with solutions of ionic strength similar to those used for measurements, some of the uncertainties associated with calibration of Ca^{2+} indicators can be avoided.

3.3. Electrophysiological Approaches

3.3.1. Long-Term Multiunit and Single-Unit Extracellular Recordings

Extracellular electrical activity of SCN neurons in freshly prepared brain slice can be measured by extracellular glass, metal, or suction electrodes. Glass microelectrodes have been used to record the neuronal activity of single SCN neurons for short times (typically <20 min). The accumulative data of repeated measurements represent the activity of the population of SCN neurons (39–43). This method is rather labor intensive, but can be useful to determine neural activity at peak or trough of the daily rhythm. The use of stationary metal electrodes (Fig. 4) allows the automated continuous recording of MUA for >24 h (26, 44–46). Electrodes are fabricated from platinum/iridium wire (ø 50–75 μm), which are mechanically stabilized within glass capillaries, electrically isolated by resin, and polished at the tip. Recently, the use of suction electrodes has been described showing stable long-term recordings of SCN neuronal activity (up to 92 h) and a good yield of distinguishable single units in the MUA signal (45). MUA recordings with stationary electrodes require extremely stable conditions for slice, perfusion, and electrode placement.

Slices are placed on a nylon mesh fixed on a metal platform grid made from titanium or platinum. The platform is mounted in a diamond-shaped chamber optimized for laminar flow of the bath medium. The bath level is adjusted to 0.1 mm above the slice surface and held stable throughout the experiment. A peristaltic pump

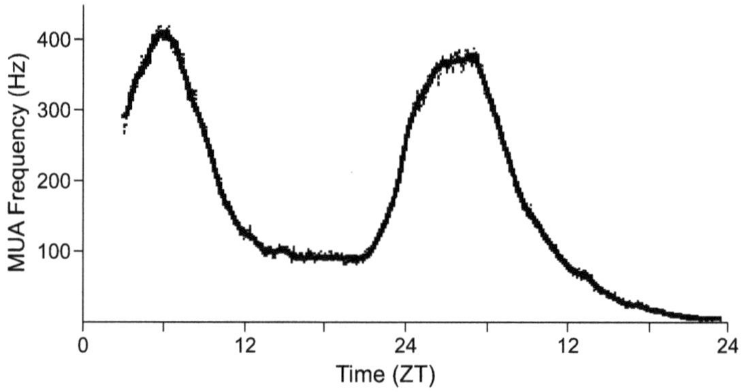

Fig. 4. *Rhythms in multiunit activity (MUA) recorded in mouse SCN.* Acute SCN brain slices exhibit a typical circadian rhythm in spontaneous electrical activity with peak activity during the mid of the projected light phase (mid-subjective day) and low activity during the projected dark phase (mid-subjective night).

provides even inflow of medium into the chamber, and a steel cannula 18-gauge needle placed perpendicular to the surface and connected to a regulated vacuum aspirates the medium at the other end of the flow-through chamber. The chamber is mounted on a vibration-damped table inside a Faraday cage. Metal or suction electrodes as well as the slice hold-down are placed with the help of micromanipulators (e.g., MM33, Märzhäuser, Wetzlar, Germany; www.marzhauser.com).

Neuronal signals are amplified and bandwidth filtered (e.g., See Methods) (46). Action potentials with signal-to-noise ratio of 2:1 (noise <5 μV from baseline) are selected by "spike triggers" and counted electronically every 10 s. The positions of the electrodes and spike trigger settings should not be changed during the experiment. Data analysis of MUA includes smoothing of raw data to determine the phase of the peak and trough of the electrical activity rhythm. Peak width of the MUA waveforms can be determined between the upward and downward flank at the half-maximum frequency, representing the duration of the electrical circadian output signal generated by the SCN.

For isolation of single units from MUA data, the complete waveform of action potentials above a given threshold is digitized and recorded (Spike-2, CED). Offline analysis of spike parameters (amplitude, latency) is used to validate distinct spike waveforms in a cluster analysis (46). Putative single units identified in distinct clusters are verified displaying an event-free refractory period (0–50 ms) in the interspike interval histogram. Similar to MUA data, single unit firing rate can be binned and plotted over time. The timing and width of the single-unit activity pattern is calculated on smoothed data.

3.3.2. Sharp Microelectrode Recording

Sharp microelectrode techniques were used in the initial analysis of the physiology of SCN neurons by Kim and Dudek (47), but have not been used much recently. We occasionally make use of methods that are similar to those previously used (48). Intracellular recordings of membrane potential or currents are obtained from SCN neurons using glass microelectrodes filled with 3 M K acetate. Electrode resistances typically vary from 70 to 100 MΩ. The signals are amplified by an Axoclamp 2A amplifier in bridge or current clamp mode.

After a neuron is impaled, a baseline recording is obtained to ensure that the cell's membrane properties are stable. This baseline membrane potential is maintained throughout the course of the experiment by manually adjusting the DC current. Hyperpolarizing DC current pulses (0.3–0.4 nA, 50-ms duration, 0.5 pulse/s repetition rate) were applied throughout all experiments to monitor conductance changes. The cell's membrane properties are monitored every 10 min by determining its current–voltage relationship by injection of depolarizing and hyperpolarizing pulses (100-ms duration); the cell's input resistance is determined from the linear portion of the current–voltage plots. Membrane excitability is also measured as the response of the neuron to depolarizing current pulses (0.1-nA steps, 500-ms duration). Electrophysiological information on the effects of experimental treatments on action potential and after hyperpolarization duration and amplitude is obtained by analyzing the first action potential evoked by the current pulses. PClamp software (Molecular Devices, Sunnyvale, CA, USA) is used for data analysis.

3.3.3. Whole-Cell Patch-Clamp Recording

The methods for patch-clamp recording are similar to those described previously (15, 16, 25–28). Slices are placed in a custom-designed chamber attached to the fixed stage upright microscope. The slice is held down with thin nylon threads glued to a platinum wire and submerged in continuously flowing, carbogenated ACSF (25°C) at 2 ml/min. Recording whole-cell currents in voltage clamp with patch electrodes from visually identified cells in 350-μm-thick slices is a standard procedure in our research group. This method provides a number of important advantages over "blind patch" and "thin slice" techniques. Individual neurons can be visualized up to 150 μm below the surface of the slice using IR-DIC optics. Thus, the recording pipette can be aimed at a particular cell. In the SCN, this means that specific cell populations can be targeted for electrophysiological analysis. Visualization also permits more accurate placement of iontophoretic pipettes and stimulating electrodes. The thicker slice also makes tissue handling easier than with 120–150-μm-thin slices. Furthermore, the synaptic physiology can be studied because much of the local connectivity remains intact in these thicker slices. Subsequent staining with biocytin permits more precise cell identification of the dendritic processes.

A disadvantage of this preparation is the necessity for using tissue from younger animals. Cells are best visualized in slices from animals younger than postnatal day (P) 30. Although cells can be visualized up to and beyond P60 (in mice), in the older tissue it is more difficult to clearly distinguish cells especially deeper into the tissue. Whole-cell recordings also tend to be of better quality in tissue from animals younger than P30.

Electrodes are pulled on a multistage puller (See Methods). DC patch electrode resistances in the bath are typically 3–6 MΩ when filled with standard "intracellular" solution (see below). Whole-cell recordings are obtained with an Axon Instruments 200B amplifier and monitored online with pCLAMP (Molecular Devices See Methods). To minimize changes in offset and/or liquid junction potentials with changing ionic conditions, the ground path uses an ACSF agar bridge to a Ag/AgCl ground well. Cells are approached with slight positive electrode pressure (20–30 mm H_2O), and offset potentials are corrected. The pipette is lowered to the vicinity of the membrane keeping the positive pressure. After forming a high-resistance seal (2–10 GΩ) by applying slight negative pressure, a second (fast) pulse of negative pressure is used to break the membrane. While entering the whole-cell mode, a repetitive test pulse of 10 mV is delivered to the cell in a passive potential range (−60 to −70 mV). Once the whole-cell recording configuration is established, whole-cell capacitance is estimated, integrating current transients produced by voltage steps according to standard methods (25). Whole-cell capacitance and electrode resistance are neutralized and compensated (50–80%) using the pulse. Data acquisition is then initiated. The series- and input-resistances are monitored throughout the experiment by checking the response to small pulses in a passive membrane potential range. Data should be rejected if series resistance is >20 MΩ or if either value changes significantly (>20%) during the course of the experiment.

The extracellular solutions used will vary according to the goals of the experiment. The standard solution is described above. When necessary, various combinations of ionic solutions plus specific blocking agents are used to maximize the presence of a particular ion current under study and minimize the presence of others. Solution exchanges within the slice are achieved by a rapid gravity-fed delivery system. In our system, the effects of bath-applied drugs begin within 15 s and are typically complete by 1–2 min. The composition of the internal solutions used to isolate different intrinsic (if necessary), ligand-gated, or synaptic currents will also vary according to the particular conditions and goals of the experiment. The standard solution to block K^+ currents in the patch pipette contains (in nM) 125 Cs-methanesulfonate, 9 EGTA, 8 HEPES, 8 MgATP, 4 NaCl, 3 KCl, and 1 $MgCl_2$. The pH of this solution is adjusted to 7.25–7.3 and the osmolality is between

280 and 290 mOsm. In some experiments, in which activity of voltage-sensitive Ca^{2+} channels is studied, GTP (1 nM), phosphocreatine (10 nM), and leupeptin (0.1 nM) were added. With this internal solution, stable Ca^{2+} currents can be recorded for up to 30 min. EGTA is omitted from the experiments using the patch electrode to load Ca^{2+} indicators.

A common concern in voltage-clamp electrophysiological experiments is the adequacy of the space clamp in cells in slices, that is, the ability to maintain voltage control of the membrane at sites distant to the recording electrode. For most studies using this technique, the ability to maintain voltage control can be improved by the blockade of most voltage-gated currents and the use of cells from younger tissue which do not have large dendritic trees. Maintenance of reasonably good space clamp is difficult in older cells with larger amplitude membrane currents. For all cells, smooth transitions in the current–voltage relationships can be used as indicators of good voltage control. In addition, any cell which shows a series resistance >20 MΩ or tail currents which do not decay rapidly with a single exponential should not be used. It is important to be aware of these problems and carefully monitor the adequacy of space-clamp conditions. Another technical concern is that while the whole-cell patch-clamp technique offers improved voltage control, it also dialyzes the inside of the cell. This can result in the loss of membrane currents that are, for example, highly dependent upon phosphorylation. One solution to this problem is, for comparison, to run some experiments with sharp-electrode intracellular recording techniques and without ion channel blockers. An alternative possibility is to use the "perforated patch" technique (26).

3.3.4. Evoked EPSCs and Stimulation Techniques

Electrical stimulation can be used to induce local excitatory postsynaptic currents (EPSCs) that are mediated by both NMDA- and AMPA/KA-type ionotropic glutamate receptors (Fig. 5) (25, 49). Local stimulation is applied with bipolar electrodes constructed from twisted Teflon-coated silver wires (0.2 mm diameter exposed at the tip, tip separation 0.2–0.5 mm). The electrode is placed 0.5–1.0 mm from the recording pipette. Constant current square pulses (50–1,000-μA intensity, 10–100-μs duration, 1 pulse/4–5 s) are used to induce short latency, graded amplitude EPSCs. In some cases, the stimulation frequency can be increased to up to 100 Hz. This should induce temporally summated long-duration EPSCs which will likely have a larger NMDA-receptor component. The threshold for EPSC onset is determined, and a series of intensities from just above threshold for the EPSC to maximum amplitude is used. Changes in EPSC characteristics are determined prior to and, at selected intervals, after drug administration. Local stimulation activates cut input and output fibers as well as local neurons. The EPSC is contaminated by activation of $GABA_A$ receptors when the stimulus is applied in the SCN. To reduce the contribution of

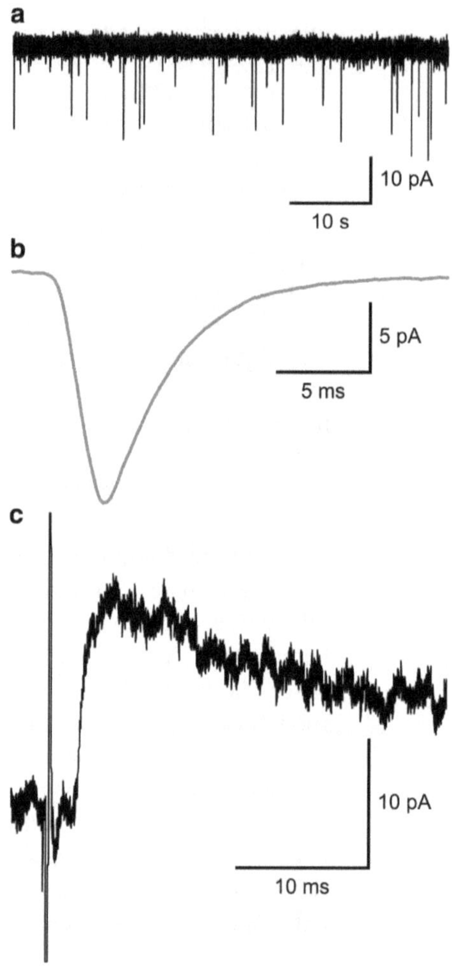

Fig. 5. *Examples of EPSCs recorded in SCN neurons in an acute brain slice.* (**a**) *Top panel* shows an example of spontaneous excitatory postsynaptic currents (EPSCs) recorded from a ventral SCN neuron during the night. Experiments were carried out in presence of the voltage-dependent Na^+ channel blocker tetrodotoxin (TTX) and the $GABA_A$ receptor blocker bicuculline. (**b**) *Middle panel* shows example of average sEPSC waveform. (**c**) *Bottom panel* shows an example of the response evoked by stimulation of the optic nerve. The stimulus artifact is followed with a delay by the synaptic response. The optic nerve is stimulated with the use of a custom-made suction electrode; constant current pulses (biphasic square wave) of 0.1–1.0-mA intensity and 0.5–2.0-ms duration are applied in 30-s intervals. The current trace is an average of five individual traces and is mostly mediated by AMPA-type glutamate receptors.

$GABA_A$ receptors, the blocker bicuculline can be added to the ACSF. To prevent action potentials, QX-314 (1 nM) can be used in the recording pipette. Specific antagonists of ionotropic glutamate receptors, for example, AP5 or CNQX, can be bath applied to isolate EPSCs mediated mainly by activation of either AMPA/KA or NMDA receptors, respectively. The optic nerve is stimulated with the use of a custom-made suction electrode; constant current pulses (biphasic square wave) of 0.1–1.0-mA intensity and 0.5–2.0-ms duration are applied in 30-s intervals.

3.4. Cell Identification

3.4.1. Reconstruction of Cellular Morphology

The morphology of cells that have been electrophysiologically analyzed can be examined using intracellular marking with biocytin. The methods for tissue processing of biocytin-filled cells are well established (50) and have been used to study the SCN (41, 51). After electrophysiological analysis, 350-μm-thick slices containing biocytin-filled cells are fixed by overnight immersion in paraformaldehyde (4%) and picric acid (0.2%) in phosphate-buffered saline (PBS). The slices are then washed with Tris-buffered saline for 1 h and processed histochemically for biocytin staining. The purpose of staining recorded neurons is to identify the type of cell. Staining can be performed in whole mounts.

Immunohistochemical processing can be combined with biocytin histochemistry. After the fixed whole mount is processed for biocytin, the tissue is cut frozen into 50-μm sections. Sections are incubated with the primary antibody (diluted with 3% normal goat serum plus 0.1% Triton X-100 in PBS) for 48 h at 4°C. The tissue is then incubated for 1 h in biotinylated secondary antibody at room temperature, followed by 1 h incubation in an avidin–biotin–peroxidase complex at room temperature (ABC method, Vector labs, Burlingame, CA). Sections are incubated in diaminobenzidine for 5–20 min. All previous steps are followed by rinses in PBS (3 × 5 min each). The tissue is mounted on slides, dehydrated, and cleared in alcohols and xylenes.

3.4.2. Molecular Characterization of Cells

Electrophysiological and molecular techniques can be combined in order to examine gene expression in SCN neurons from which currents have been recorded (Fig. 6). The first step in analyzing gene expression in single cells involves harvesting the cytoplasmic mRNA from SCN cells before the application of the amplification protocol. In order to do this, a few modifications to the normal patch-clamp procedure are required. Great care is needed to create an RNase-free working environment. It is best to use gloves and make exclusive use of RNase-free reagents. After carrying out electrophysiological analysis, the cytoplasm is collected into the patch pipette by applying negative pressure under visual control. After collection of cytoplasm, the tip of the electrode is broken into a tube for reverse transcription as described below. Because the PCR uses DNA as template, the mRNA must be transcribed into its complementary DNA sequence, the cDNA, using the enzyme reverse transcriptase (RT). This enzyme needs dNTPs and a primer that binds to the mRNA to start the elongation. The primer can be an oligo-dT primer or a random hexamer primer. Because mRNAs have a poly-A tail, using the complementary oligo-dT primer leads to a more specific and efficient reverse transcription of mRNAs. Sometimes the mRNA has a complicated secondary structure, and

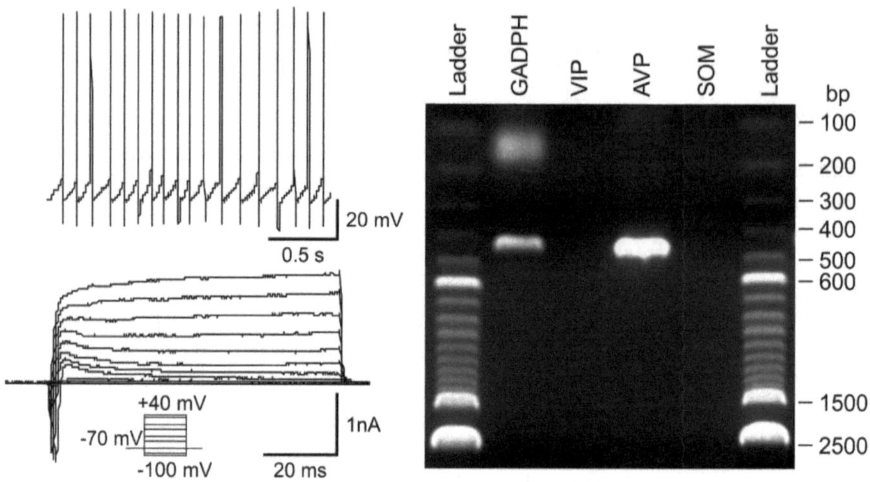

Fig. 6. *Patch-clamp and single-cell real-time polymerase chain reaction (RT-PCR): Physiological and gene expression from a single cell in the dorsal SCN region from an acute brain slice. Top left*, current clamp measurement of membrane potential showing the spontaneous firing rate of SCN neuron. Most SCN neurons show spontaneous activity during the day. *Bottom left*, voltage-clamp measurement of inward and outward currents. *Right*, after harvesting the cytoplasm from this neuron, RT-PCR technique reveals the mRNA expression for the housekeeping gene of glyceraldehyde 3-phosphate dehydrogenase (GAPDH) and vasopressin (AVP), but not VIP or somatostatin (SOM). Data collected by T. Bullmann.

the RT cannot read through this sequence. Then the cDNAs contain only short sequences complementary to the 5′ parts of the mRNAs. In this case, the use of random hexamer primers is recommended. These primers bind statistically on all RNA sequences. Using the random hexamer primers the cDNAs contain sequences complementary to the mRNA sequence 3′ from a complicated secondary structure but also from rRNAs. Because mRNAs and rRNAs are competitors, the use of random hexamer primers is less efficient than oligo-dT primers. Both oligo-dT and hexamer primers can be successfully used to detect mRNA coding for VIP, VP, CalB, and the housekeeping gene glyceraldehyde 3-phosphate dehydrogenase (GAPDH). The oligo-dT primers and random primers were purchased from Invitrogen based on published sequences. RT was performed using the Sensiscript RT Kit from Qiagen (Valencia, CA, USA).

After RT, the subsequent PCR with specific primers and the HotStar Taq Master Mix Kit (Qiagen) was performed as multiplex PRC with the RT product. Firstly, 20 cycles (94°C, 15 min, activation of polymerase; 94°C, 35 s, splitting of DNA strands; 60°C, 30 s, annealing; 72°C, 120 s, elongation) of PCR were performed followed by a final elongation period of 7 min at 72°C. The product of the first amplification is purified using the QIAquick PCR Purification Kit (Qiagen) and eluted. The purified first PCR product is then used as a template in the second round of PRC amplification. In this second round, primers for VIP, VP, CalB, and GAPDH were individually amplified with the same PCR program but with 35 cycles. Thus, at the end of the second round, each cell

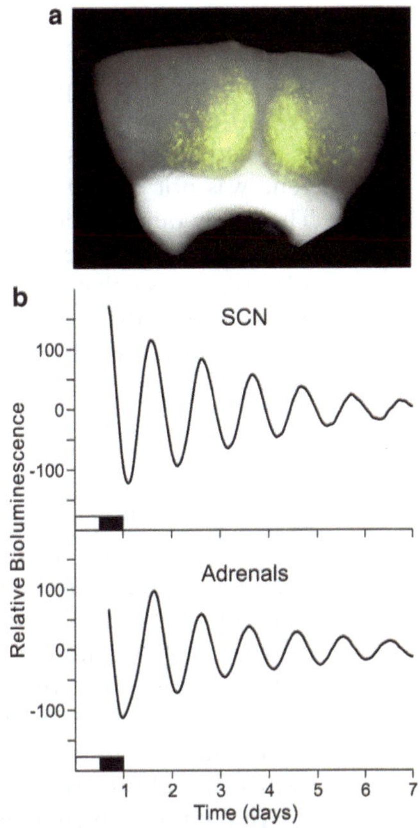

Fig. 7. *Circadian rhythm of PER2::LUC activity from explanted SCN culture.* (**a**) Overlay image showing Per2-expressing cells (*yellow*) within the SCN depicted in the bright-field image of the organotypic slice. (**b**) Representative record of bioluminescence showing circadian profile of PER2::LUC expression rhythm in a cultured SCN explant from mice. Data are also shown from adrenal explants for comparison. Bioluminescence counts are plotted relative to the light/dark cycle prior to sacrifice, with light onset = CT 0.

produced four PCR products. Each amplification product was then run on a 1.2% agarose gel in parallel with molecular weight markers and stained with ethidium bromide.

The very high sensitivity of the RT-PCR technique used makes it prone to contamination artifacts. It is best to run a range of controls for both the harvesting of RNA and the RT-PCR reaction. One standard control is to collect the sample run through the protocol without the reverse transcriptase. Another control is to demonstrate that there is no gene expression from SCN cells in which a giga-ohm seal is formed between the pipette and the membrane without rupturing the membrane. This negative control indicates that, under your experimental conditions, you do not get contamination from the bath or by passing through cells in the slice on the way to patching a targeted cell. In one study, of the 24 cells sampled in the central SCN region, we were able to detect GAPDH in all neurons, VP in seven neurons, and VIP in four neurons. All of the negative controls failed to show any PRC product. Because PCR allows single copies to be detected, it was possible, using appropriate controls, to show a true negative within a cell (i.e., if a single copy can be detected in a given gene-specific assay and negative results are still obtained, it can be stated that this gene was not expressed within the cell at the time of sampling). Therefore, our data demonstrate that in the central SCN region of the mouse brain slice, we can detect the mRNA of both VP and VIP. We expect that if we focused on the ventral population the proportion of VIP + neurons will increase. All techniques designed to detect mRNA suffer from the potential drawback that mRNA expression does not necessarily mean protein synthesis.

3.5. Organotypic SCN Slice Cultures

3.5.1. General Features of SCN Slice Cultures

Organotypic slice cultures are widely used for neuroscience research and have proven useful for the analysis of the circadian circuit (see also Chap. 14-Biggs et al. (52)). The culturing system has been often used to record rhythms of long-term neural activity (e.g., on a multielectrode array dish), neuropeptide and hormone release, and bioluminescence activity. Originally, the "roller tube culture" method of SCN culture was developed in order to measure neuropeptide release from the SCN (53, 54). However, the constant rotation is difficult to combine with electrophysiological or imaging techniques. In addition, this technique requires young tissue (<P10) so that the cultured slice can flatten and attach onto a cover slip.

In contrast, "membrane culture" method is quite simple and easily applied to electrophysiology study and optical imaging studies and can be used with tissue from older animals (55). Therefore, the membrane culture is the method of choice for most circuit analysis. We will describe the membrane method for optical recording of adult mouse and rat SCN culture. Most of our experiments (Fig. 7) have been done using PERIOD2::LUCIFERASE knockin mice (PER2::LUC mice) generated in Dr. Takahashi's lab (56). A variety of other mouse lines are available to record circadian rhythms of clock genes (*Per1-luc* rats and mice, *Bmal1-luc* mice, etc.). Furthermore, the physiological and imaging approaches described under the acute brain slice could also be applied to the slice culture preparation. The advantages of this technique are that the slices live for a number of cycles required for circadian analysis. The circuit appears to be intact. One of the main disadvantages of this technique is the difficulty in applying drug treatments. The circadian oscillations in these cells can be reset by media changes and careful controls must be carried out to ensure that the treatment and not just the solution change are responsible.

3.5.2. Methods for Generating SCN Slice Cultures

The culture procedures that we use are similar to those previously described (57–59). Anesthetized PER2::LUC mice are sacrificed by decapitation. The brain is removed and placed in chilled HBSS. The brain is trimmed to put on a vibratome and sliced in the coronal plane at a thickness of 300–400 μm. The bilateral SCN and a minimum of surrounding tissue are isolated from the slice using scalpels with the aid of a dissecting microscope. The SCN slice is placed at the liquid interface on membranes in 35-mm^2 dishes containing 1.0-ml recording media. The dish is sealed by a round cover glass using vacuum grease. The dish is placed into a recording apparatus inside a light-tight 36.5°C chamber with light/temperature/humidity control.

We are using LumiCycle (Actimetrics, Wilmette, IL, USA) optical recording apparatus which has 32 channels and self-contained head-on photon counting module (Hamamatsu, Bridgewater, NJ, USA). More recording apparatuses are commercially available: *LM-2400* (Hamamatsu), *LM-300* (Hamamatsu), *Kronos* (Atto Co. Tokyo, Japan), and *POLARstar* (BMG LABTECH, Ortenberg, Germany). The bioluminescence signal can be recorded every 10 min for at least 7 days. The resulting data are normalized by subtraction of the 24-h running average from the raw data and then smoothed with a 2-h running average (Origin Lab 7.0; OriginLab Inc., Northampton, MA). The circadian period is evaluated to take each peak point from smoothed data. The amplitude of each oscillation is determined by the difference between the lowest and highest points in each cycle. The same preparation can be used with a cooled intensified CCD (ICCD) camera to record PER2 expression rhythms of individual SCN cells within an intact network.

3.6. Bioluminescence Imaging in Dispersal Cell Culture for SCN Neurons

3.6.1. General Features of Dispersal Cell Culture for SCN Neurons

Dispersal cell cultures are a useful tool for analysis of SCN neurons (11–13). Since single neurons can generate circadian oscillations, the cell culture preparation can be used to examine the ionic mechanisms underlying circadian rhythms as well as signal transduction mechanisms by which the membrane and molecular timekeeping mechanisms interact. While this preparation gives the experimenter a high degree of control including the ability to transfect the cells with various genetic constructs, any results from this analysis has to be tempered by the understanding that the cells may have been altered by the culturing processes and findings have to be validated in more intact preparations. Single-cell bioluminescence recording from the SCN dispersal cells of PER2::LUC mice (Fig. 8) can also be used. The use of a cooled ICCD camera (Hamamatsu) allows high-resolution luminescence images.

3.6.2. Methods for Generating Dispersal Cell Culture for SCN Neurons

A glass bottom dish is coated with the coating solution (laminin and poly-D-lysine) at 37°C for 1–24 h. The dish is washed with ddH_2O for three times and then air-dried under UV in the clean bench. P4–6 PER2::LUC mice are anesthetized and decapitated. The brain tissue is removed and placed in chilled HBSS. The brain is trimmed to put on a tissue chopper and sliced in the coronal plane at a thickness of 200–250 μm. The bilateral SCN and a minimum of surrounding tissue are isolated from the slices using fine

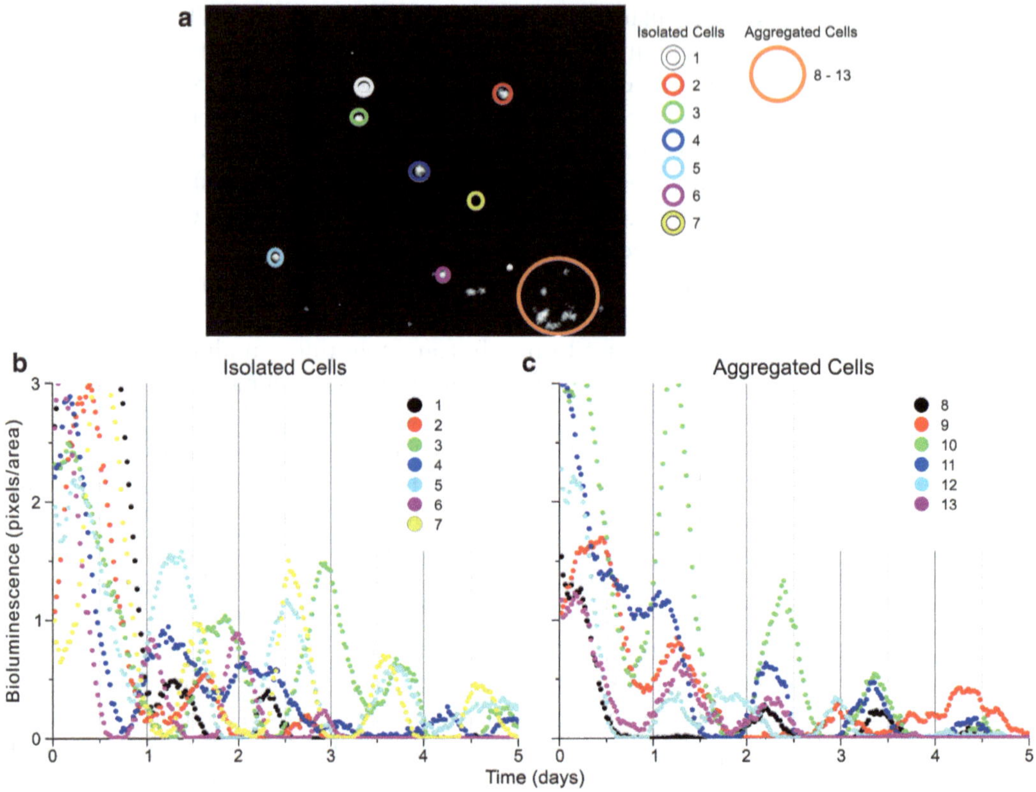

Fig. 8. *Bioluminescence recording of PER2::LUC rhythm in dispersed SCN cells*. (**a**) Bioluminescence image of PER2::LUC in dispersal SCN cells cultured on a special electron beamed patterned cover slip which can isolate individual cells. (**b**, **c**) Isolated (**b**) or aggregated cells (**c**) are picked and the rhythms are plotted. The aggregated cells showed a greater degree of phase synchrony than the isolated cells.

forceps with the aid of a dissecting microscope. The SCN are placed into a 0.5-ml tube with 100-μl papain DS and incubated at 35°C for 30 min. The SCN are washed with prewarmed (35–37°C) heavy inhibitor, light inhibitor, and culture media in order and then placed into a 0.5-ml tube with 20-μl culture media. The cells are pipetted with media several times to disperse the cells. Each medium containing the SCN is put onto the center of the slip and incubated at 37°C for 30 min until cells are stuck down on the cover slip. Culture media are added by appropriate volume and placed in CO_2 incubator (36.5°C, 5% CO_2, 95% O_2). The half volume of medium is changed three times a week.

3.6.3. Bioluminescence Recording with Cooled ICCD Camera System

The cell culture procedures that we use are similar to those previously described (60, 61). We sometimes add 10% NCS to help the cells to grow. We usually collect four SCN at a time and pool them into the same tube. The medium is replaced with recording media before recording. The culture dish is placed onto the microscope stage top incubator (36.5°C, 5% CO_2, 95% O_2) housed inside a light-tight box. Images are acquired with the cooled ICCD camera at 15 frames/s, filtered in real time to eliminate single-image noise events (e.g., cosmic rays), and stored as 5-min summed images. For analysis, a 1-h moving average is then applied, images are converted to 8 bit, and one image per 30 min is used for analysis of single-cell data. For further analysis, several software packages are available to analyze rhythmic expression patterns.

4. Summary and Future Directions

As neuroscientists, an essential goal of our research is to understand how the nervous system works at all levels of organization. We now understand a large number of basic principles underlying the properties of neurons and their synaptic interconnections. We can accurately describe neural interactions, and in many cases, we can use imaging techniques to know what brain regions are active in particular behaviors. Still, there is a giant gap in our understanding of how behaviors emerge from neurons and the circuits they form. Filling this gap requires an understanding of how neurons are organized in circuits and networks generate the complex interactions of electrical and chemical signals that ultimately generate rhythmic behaviors.

The overall goal of our research is to understand the circuit-level properties of neurons that generate circadian rhythms. Circadian rhythms offer a special opportunity for a study of neural circuits. Although individual neurons can generate the basic periodicity, other properties emerge from interactions within local circuits (14). To study this circuit, we are attempting to develop a set of in vitro circadian models of increasing cellular complexity, from single cells to geometrically defined assemblies of oscillator neurons that make up the intact system. We hope to study how the canonical properties of circadian clocks emerge through multioscillator interactions. Much of the resources and efforts in the field of circadian rhythms have been devoted to understanding the molecular genetic mechanisms that underlie these oscillations. We feel that the field of chronobiology would benefit from expanding its focus on genetic mechanisms to include more work on how molecular rhythms drive rhythms in electrical activity as well as how the SCN circuit functions as an ensemble. In order to affect this expansion, we need to bring together the insights and talents of neuroscientists from a number of backgrounds including anatomy, physiology, and modeling. New insights emerging from outside of the field seem likely. The partnership between more physiologically oriented scientists with those with quantitative/model expertise is likely to pay dividends, and indeed, this has started to occur (62, 63). Hopefully, our experience with the SCN preparations, as described in this chapter, will help further work in this important area.

References

1. Hastings MH, Herzog ED (2004) Clock genes, oscillators, and cellular networks in the suprachiasmatic nuclei. J Biol Rhythms 19:400–413
2. Maywood ES, O'Neill JS, Reddy AB, Chesham JE, Prosser HM, Kyriacou CP, Godinho SI, Nolan PM, Hastings MH (2007) Genetic and molecular analysis of the central and peripheral circadian clockwork of mice. Cold Spring Harb Symp Quant Biol 72:85–94
3. Takahashi JS, Hong HK, Ko CH, McDearmon EL (2008) The genetics of mammalian circadian order and disorder: implications for physiology and disease. Nat Rev Genet 9:764–775
4. Morin LP, Shivers KY, Blanchard JH, Muscat L (2006) Complex organization of mouse and rat suprachiasmatic nucleus. Neuroscience 137:1285–1297
5. Yan L, Okamura H (2002) Gradients in the circadian expression of Per1 and Per2 genes in the rat suprachiasmatic nucleus. Eur J Neurosci 15:1153–1162
6. Hamada T, Antle MC, Silver R (2004) Temporal and spatial expression patterns of canonical clock genes and clock-controlled genes in the suprachiasmatic nucleus. Eur J Neurosci 19:1741–1748
7. Antle MC, Silver R (2005) Orchestrating time: arrangements of the brain circadian clock. Trends Neurosci 28:145–151
8. Abrahamson EE, Moore RY (2001) Suprachiasmatic nucleus in the mouse: retinal innervation, intrinsic organization and efferent projections. Brain Res 916:172–191
9. Kalsbeek A, Palm IF, La Fleur SE, Scheer FAJL, Perreau-Lenz S, Ruiter M, Kreier F, Cailotto C, Buijs RM (2006) SCN outputs and the hypothalamic balance of life. J Biol Rhythms 21:458–469
10. Schaap J, Pennartz CM, Meijer JH (2003) Electrophysiology of the circadian pacemaker in mammals. Chronobiol Int 20:171–188
11. Welsh DK, Logothetis DE, Meister M, Reppert SM (1995) Individual neurons dissociated from rat suprachiasmatic nucleus express independently phased circadian firing rhythms. Neuron 14:697–706
12. Herzog ED, Geusz ME, Khalsa SB, Straume M, Block GD (1997) Circadian rhythms in mouse suprachiasmatic nucleus explants on multimicroelectrode plates. Brain Res 757:285–290
13. Honma S, Shirakawa T, Katsuno Y, Namihira M, Honma K (1998) Circadian periods of single suprachiasmatic neurons in rats. Neurosci Lett 250:157–160
14. Welsh DK, Takahashi JS, Kay SA (2010) Suprachiasmatic nucleus: cell autonomy and network properties. Annu Rev Physiol 72:551–577
15. Itri J, Colwell CS (2003) Regulation of inhibitory synaptic transmission by vasoactive intestinal peptide (VIP) in the mouse suprachiasmatic nucleus. J Neurophysiol 90:1589–1597
16. Choi HJ, Lee CJ, Schroeder A, Kim YS, Jung SH, Kim JS, Kim DY, Sohn EJ, Han HC, Hong SK, Colwell CS, Kim YI (2008) Excitatory actions of GABA in the suprachiasmatic nucleus. J Neurosci 28:5450–5459
17. Kuhlman SJ, McMahon DG (2006) Encoding the ins and outs of circadian pacemaking. J Biol Rhythms 21:470–481
18. Colwell CS (2000) Circadian modulation of calcium levels in cells in the suprachiasmatic nucleus. Eur J Neurosci 12:571–576
19. Ikeda M, Sugiyama T, Wallace CS, Gompf HS, Yoshioka T, Miyawaki A, Allen CN (2003) Circadian dynamics of cytosolic and nuclear Ca^{2+} in single suprachiasmatic nucleus neurons. Neuron 38:253–263
20. O'Neill JS, Maywood ES, Chesham JE, Takahashi JS, Hastings MH (2008) cAMP-dependent signaling as a core component of the mammalian circadian pacemaker. Science 320:949–953
21. Lamont EW, Robinson B, Stewart J, Amir S (2005) The central and basolateral nuclei of the amygdala exhibit opposite diurnal rhythms of expression of the clock protein Period2. Proc Natl Acad Sci USA 102:4180–4184
22. Wang LM, Dragich JM, Kudo T, Odom IH, Welsh DK, O'Dell TJ, Colwell CS (2009) Expression of the circadian clock gene Period2 in the hippocampus: possible implications for synaptic plasticity and learned behaviour. ASN Neuro 1(3)
23. Gillette MU (1986) The suprachiasmatic nuclei: circadian phase-shifts induced at the time of hypothalamic slice preparation are preserved in vitro. Brain Res 79:176–181
24. vanderLeest HT, Vansteensel MJ, Duindam H, Michel S, Meijer JH (2009) Phase of the electrical activity rhythm in the SCN in vitro not influenced by preparation time. Chronobiol Int 26:1075–1089
25. Colwell CS (2001) NMDA-evoked calcium transients and currents in the suprachiasmatic nucleus: gating by the circadian system. Eur J Neurosci 13:1420–1428
26. Itri JN, Michel S, Vansteensel MJ, Meijer JH, Colwell CS (2005) Fast delayed rectifier potassium current required for circadian neural activity. Nat Neurosci 8:650–656

27. Itri JN, Vosko AM, Schroeder A, Dragich JM, Michel S, Colwell CS (2010) Circadian regulation of A-type potassium currents in the suprachiasmatic nucleus. J Neurophysiol 103: 632–640
28. Michel S, Itri J, Han JH, Gniotczynski K, Colwell CS (2006) Regulation of glutamatergic signalling by PACAP in the mammalian suprachiasmatic nucleus. BMC Neurosci 7:15
29. Trapp S, Ballanyi K (2012) Autonomic nervous system in vitro: studying tonically active neurons controlling vagal outflow in rodent brainstem slices. In Isolated Central Nervous System Circuits (Ed K Ballanyi), Neuromethods Series Vol. 73 (Ed W Walz). Springer Science+Business Media, LLC, New York, NY, pp 1–59
30. Moore AR, Zhou WL, Jakovcevski I, Zecevic N, Antic SD (2012) Properties of acutely isolated fetal (Human) brain tissues. In Isolated Central Nervous System Circuits (Ed K Ballanyi), Neuromethods Series Vol. 73 (Ed W Walz). Springer Science+Business Media, LLC, New York, NY, pp 125–158
31. Sanchez-Vives MV (2012) Spontaneous rhythmic activity in the adult cerebral cortex in vitro. In Isolated Central Nervous System Circuits (Ed K Ballanyi), Neuromethods Series Vol. 73 (Ed W Walz). Springer Science+Business Media, LLC, New York, NY, pp 263–284
32. MacVicar BA (1984) Infrared video microscopy to visualize neurons in the in vitro brain slice preparation. J Neurosci Methods 12:133–139
33. Dodt HU, Zieglgänsberger W (1994) Infrared videomicroscopy: a new look at neuronal structure and function. Trends Neurosci 17(11):453–458
34. Irwin RP, Allen CN (2007) Calcium response to retinohypothalamic tract synaptic transmission in suprachiasmatic nucleus neurons. J Neurosci 27:11748–11757
35. Ruangkittisakul A, Panaitescu B, Secchia L, Bobocea N, Kantor C, Kuribayashi J, Iizuka M, Ballanyi K (2012) Isolated brainstem respiratory centers from perinatal rodents. In Isolated Central Nervous System Circuits (Ed K Ballanyi), Neuromethods Series Vol. 73 (Ed W Walz). Springer Science+Business Media, LLC, New York, NY, pp 61–124
36. Kantor C, Panaitescu B, Kuribayashi J, Ruangkittisakul A, Jovanovic I, Leung V, Lee TF, MacTavish D, Jhamandas JH, Cheung PY, Ballanyi K (2012) Spontaneous neural network oscillations in hippocampus, cortex and *locus coeruleus* of newborn rat and piglet brain slices. In Isolated Central Nervous System Circuits (Ed K Ballanyi), Neuromethods Series Vol. 73 (Ed W Walz). Springer Science+Business Media, LLC, New York, NY, pp 315–356
37. Grynkiewicz G, Poenie M, Tsien RY (1985) A new generation of Ca^{2+} indicators with greatly improved fluorescence properties. J Biol Chem 260:3440–3450
38. Neher E (1995) The use of fura-2 for estimating Ca buffers and Ca fluxes. Neuropharmacology 34:1423–1442
39. Groos G, Hendriks J (1982) Circadian rhythms in electrical discharge of rat suprachiasmatic neurones recorded in vitro. Neurosci Lett 34:283–288
40. Green DJ, Gillette R (1982) Circadian rhythm of firing rate recorded from single cells in the rat suprachiasmatic brain slice. Brain Res 245:198–200
41. Jiang ZG, Yang Y, Liu ZP, Allen CN (1997) Membrane properties and synaptic inputs of suprachiasmatic nucleus neurons in rat brain slices. J Physiol 499:141–159
42. Gribkoff VK, Pieschl RL, Wisialowski TA, van den Pol AN, Yocca FD (1998) Phase shifting of circadian rhythms and depression of neuronal activity in the rat suprachiasmatic nucleus by neuropeptide Y: mediation by different receptor subtypes. J Neurosci 18: 3014–3022
43. Bouskila Y, Dudek FE (1993) Neuronal synchronization without calcium-dependent synaptic transmission in the hypothalamus. Proc Natl Acad Sci USA 90:3207–3210
44. Albus H, Bonnefont X, Chaves I, Yasui A, Doczy J, van der Horst GT, Meijer JH (2002) Cryptochrome- deficient mice lack circadian electrical activity in the suprachiasmatic nuclei. Curr Biol 12:1130–1133
45. Brown TM, Banks JR, Piggins HD (2006) A novel suction electrode recording technique for monitoring circadian rhythms in single and multiunit discharge from brain slices. J Neurosci Methods 156:173–181
46. VanderLeest HT, Houben T, Michel S, Deboer T, Albus H, Vansteensel MJ, Block GD, Meijer JH (2007) Seasonal encoding by the circadian pacemaker of the SCN. Curr Biol 17:468–473
47. Kim YI, Dudek FE (1991) Intracellular electrophysiological study of suprachiasmatic nucleus neurons in rodents: excitatory synaptic mechanisms. J Physiol 444:269–287
48. Colwell CS, Levine MS (1995) Excitatory synaptic transmission in neostriatal neurons: regulation by cyclic AMP-dependent mechanisms. J Neurosci 15:1704–1713
49. Kim YI, Choi H-J, Colwell CS (2006) BDNF regulation of NMDA receptor-mediated

synaptic currents in suprachiasmatic nucleus neurons. J Neurosci Res 84:1512–1520
50. Kita H, Armstrong W (1991) A biotin-containing compound N-(2-aminoethyl) biotinamide for intracellular labeling and neuronal tracing studies: comparison with biocytin. Neurosci Methods 37:141–150
51. Colwell CS (2000) Rhythmic coupling among cells in the suprachiasmatic nucleus. J Neurobiol 43:379–388
52. Biggs JE, Lu VB, Kim H, Lai A, Todd KG, Ballanyi K, Colmers WF, Smith PA (2012) Defined medium organotypic cultures of spinal cord put 'Pain in a Dish'. In Isolated Central Nervous System Circuits (Ed K Ballanyi), Neuromethods Series Vol. 73 (Ed W Walz). Springer Science+Business Media, LLC, New York, NY, pp 405–435
53. Tominaga K, Okamur H, Inouye ST (1994) Organotypic slice culture of the suprachiasmatic nucleus. Neurosci Biobehav Rev 18:601–604
54. Shinohara K, Honma S, Katsuno Y, Abe H, Honma K (1995) Two distinct oscillators in the rat suprachiasmatic nucleus in vitro. Proc Natl Acad Sci USA 92:7396–7400
55. Yamazaki S, Straume M, Tei H, Sakaki Y, Menaker M, Block GD (2002) Effects of aging on central and peripheral mammalian clocks. Proc Natl Acad Sci USA 99: 10801–10806
56. Yoo SH, Yamazaki S, Lowrey PL, Shimomura K, Ko CH, Buhr ED, Siepka SM, Hong HK, Oh WJ, Yoo OJ, Menaker M, Takahashi JS (2004) PERIOD2:LUCIFERASE real-time reporting of circadian dynamics reveals persistent circadian oscillations in mouse peripheral tissues. Proc Natl Acad Sci USA 101:5339–5346
57. Yamazaki S, Takahashi JS (2005) Real-time luminescence reporting of circadian gene expression in mammals. Methods Enzymol 393:288–301
58. Nakamura TJ, Sellix MT, Menaker M, Block GD (2008) Estrogen directly modulates circadian rhythms of PER2 expression in the uterus. Am J Physiol Endocrinol Metab 295:E1025–E1031
59. Nakamura TJ, Sellix MT, Kudo T, Nakao N, Yoshimura T, Ebihara S, Colwell CS, Block GD (2010) Influence of the estrous cycle on clock gene expression in reproductive tissues: effects of fluctuating ovarian steroid hormone levels. Steroids 75:203–212
60. Aujard F, Herzog ED, Block GD (2001) Circadian rhythms in firing rate of individual suprachiasmatic nucleus neurons from adult and middle-aged mice. Neuroscience 106:255–261
61. Kwak Y, Lundkvist GB, Brask J, Davidson A, Menaker M, Kristensson K, Block GD (2008) Interferon-gamma alters electrical activity and clock gene expression in suprachiasmatic nucleus neurons. J Biol Rhythms 23:150–159
62. Mirsky HP, Liu AC, Welsh DK, Kay SA, Doyle FJ 3rd (2009) A model of the cell-autonomous mammalian circadian clock. Proc Natl Acad Sci USA 106:11107–11112
63. Vasalou C, Herzog ED, Henson MA (2009) Small-world network models of intercellular coupling predict enhanced synchronization in the suprachiasmatic nucleus. J Biol Rhythms 24:243–254

Chapter 6

Osmoregulatory Circuits in Slices and En Bloc Preparations of Rodent Hypothalamus

Tevye J. Stachniak, Jessica R. Sudbury, Eric Trudel, Katrina Y. Choe, Sorana Ciura, and Charles W. Bourque

Abstract

Osmoregulatory neural networks in the mammalian hypothalamus are responsible for keeping plasma osmolality near a constant set point. This process involves primary osmosensory neurons located in the *organum vasculosum lamina terminalis* (OVLT) and effector neurons located elsewhere in the hypothalamus and in other parts of the brain. Hypothalamic effector neurons include vasopressin (VP)- and oxytocin (OT)-releasing magnocellular neurosecretory cells (MNCs) in the supraoptic (SON) and paraventricular nuclei. Osmotically induced changes in action potential discharge by MNCs cause proportional changes in VP and OT release into the blood to modulate water and sodium excretion at the kidney. This osmotic regulation of firing rate in MNCs plays a critical role in the maintenance of body fluid balance. In this chapter, we describe spontaneously active in vitro preparations of rodent hypothalamus that have been developed for the specific purpose of studying how osmosensory neurons in the OVLT and other hypothalamic structures regulate MNCs in the SON.

Key words: Explant, Hypothalamus, Magnocellular, Osmoreceptor, Osmosensory, OVLT, Oxytocin, Suprachiasmatic nucleus, Supraoptic nucleus, Vasopressin

1. Introduction

1.1. The Hypothalamus: The "Brain of the Brain"

The hypothalamus is a phylogenetically old part of the vertebrate brain. It is often referred to as the "brain of the brain," because it has a significant impact on functions in many other parts of the central nervous system (CNS) and because it is responsible for vital homeostatic functions that emerged early during evolution (1). For example, in mammals, the hypothalamus affects mood, directs reproductive behavior, and coordinates the responses to stress. It is also responsible for the control of circadian rhythms (2, 3) and

Klaus Ballanyi (ed.), *Isolated Central Nervous System Circuits*, Neuromethods, vol. 73,
DOI 10.1007/978-1-62703-020-5_6, © Springer Science+Business Media New York 2012

primordial emotions, such as thirst and hunger (4, 5). Moreover, through its direct role in the control of the pituitary and the autonomic nervous system, the hypothalamus regulates a wide array of physiological parameters, including blood pressure (6, 7), body temperature (8, 9), as well as fluid and electrolyte balance (10). Accordingly, a constellation of diseases and symptoms can be caused by tumors, vascular stroke, or physical trauma affecting this part of the brain (11). The hypothalamus is an area of intense research, and the development of reduced preparations of rodent hypothalamus that can be maintained alive in vitro has played a key role in advancing our understanding of the pathophysiology of the hypothalamus. The objective of this chapter is to provide a synthesis of the methodology used in our laboratory for studies on hypothalamic circuits involved in the central nervous control of body fluid homeostasis.

1.2. The Hypothalamus: A Compact and Complex Structure

As implied by its name, the hypothalamus lies beneath the thalamus, on either side of the third ventricle, and occupies the ventral-most part of the brain, directly above the pituitary gland. In rats and mice, the rostral pole of the hypothalamus is defined by the position of the optic chiasma, the *lamina terminalis,* and the anterior commissure, while its caudal end is defined by the mamillary complex. The hypothalamus houses several cytoarchitecturally well-defined nuclei that contain neurons involved in the control of specific homeostatic functions. The hypothalamus also contains less well-defined areas that also comprise important but often more diffusely organized groups of neurons. In either case, these nuclei (or areas) commonly include intermingled subsets of functionally and chemically distinct neurons that project both within and outside the hypothalamus (1). In addition to "conventional" neurons and glial cells, the hypothalamus contains two distinct classes of neuroendocrine cells, specifically the tuberoinfundibular neurosecretory cells (TNCs) and the magnocellular neurosecretory cells (MNCs) (12). The TNCs, located in various hypothalamic areas, for example, the preoptic area, parts of the paraventricular nucleus (PVN), or the arcuate nucleus, project axons mainly toward the external zone of the median eminence where they form neurohemal contacts with vessels of the hypothalamo-hypophysial portal system (1). The axon terminals of TNCs release hypophysiotropic hormones (e.g., corticotropin-, gonadotropin-, and thyrotropin-releasing hormones) into this system to regulate hormone synthesis and release from endocrine cells in the anterior lobe of the pituitary. The MNCs, located mainly in parts of the PVN and in the supraoptic nucleus (SON) (13), project axons through the internal zone of the median eminence (14) and into the posterior lobe of the pituitary (i.e., the neurohypophysis), where their nerve endings release either oxytocin (OT) or vasopressin (VP, antidiuretic hormone) into the general circulation (15).

Although the hypothalamus is a small and anatomically compact structure, its organization is highly complex. As noted above, many hypothalamic nuclei or areas contain neurons involved in more than one function. Even single neuron types can be implicated in multiple homeostatic roles (16). It is therefore quite common to observe a substantial overlap in the wiring diagrams that characterize anatomical networks involved in the control of distinct hypothalamic functions.

1.3. Advantages and Disadvantages of In Vitro Hypothalamic Preparations

The relatively small size of the hypothalamus (about 6 × 6 × 3 mm in rat) represents a potential advantage for in vitro studies because the proximal location of various nuclei increases the likelihood that a combination of structures of interest can be maintained in a single preparation (e.g., in a single 400-μm-thick acute brain slice). A significant disadvantage, however, is that 300–500-μm consecutive sections of hypothalamus (in any plane) are almost always anatomically distinct (17), which significantly reduces the number of "equivalent" slices that can be obtained from the brain of a single animal. This is similar to slices containing the active medullary inspiratory center (Chap. 2-Ruangkittisakul et al. (18)), *locus coeruleus* (Chap. 11-Kantor et al. (19)), or circadian oscillators (Chap. 5-Nakamura et al. (20)). Contrary, various slices of basically similar neural network content can be obtained from diverse CNS regions such as the hippocampal formation, cerebellum, and neocortex as outlined elsewhere in this issue (see Chaps. 1-Trapp and Ballanyi (21), 3-Moore et al. (22), 4-Fish et al. (23), 7-McKay et al. (24), 8-Sanchez-Vives (25), 9-Broicher and Speckmann (26), and 10-Luhmann and Kilb (27)). In some cases, this problem can be lessened through the use of 100–200-μm thin slices and/or by separating left and right sides of the slices into "hemisections." Finally, the heterogeneity of neuronal subtypes found in various nuclei is a difficult challenge that must be resolved for functional analyses of hypothalamic networks. Unlike in layered structures, such as neocortex or cerebellum, where different types of neurons can have recognizable shapes and positions, it is often difficult to tell the difference between local interneurons, projection neurons, and neurosecretory neurons by visual inspection alone in hypothalamic slice preparations (28). Because the functional and molecular phenotype of the cell being recorded is often unknown at the time of recording, sophisticated protocols are commonly implemented to determine the identity of the neuron after the recording is completed. For example, a neuron being studied by sharp microelectrode or whole-cell recording of membrane potential (changes), or underlying ion currents, can be labeled by a dye included in the recording patch pipette, and its phenotype can later be defined by immunocytochemical staining (29). Alternately, a neuron to be recorded could be preselected by the live detection of a fluorescent molecule such as green fluorescent protein (GFP) whose expression is being driven by a cell-specific gene promoter (30–32).

1.4. In Vitro Analysis of MNCs: Targeting the Supraoptic Nucleus

In this review, we will outline procedures that have been developed in our laboratory for the specific analysis of hypothalamic neuron networks involved in the regulation of VP and OT release. VP and OT are peptide hormones synthesized in the somata of distinct subsets of MNCs (33, 34). In mammals, bolus release of OT into the circulation induces uterine contractions at birth and triggers episodic milk ejection during lactation (15). In some species like rats, OT also operates as a natriuretic hormone (35, 36) and therefore contributes to osmoregulation (10). OT and VP are released by the axon terminals of MNCs in the neurohypophysis in response to Ca^{2+} influx induced by the arrival of action potentials initiated in the somata (37). Hormone release into the circulation is dictated entirely by electrical neuronal activity (38). A refined understanding of the central mechanisms controlling VP and OT secretion can therefore be gained from an analysis of the factors regulating action potential firing in VP and OT MNCs. Anatomical studies have shown that VP and OT MNCs can both be found (1) in the SON, (2) in the magnocellular subdivision of the PVN, and (3) in smaller accessory magnocellular cell groups (e.g., *nucleus circularis*) (13). However, studies aimed specifically at MNCs are often focused on the SON because this structure is easy to identify compared to accessory cell groups and because unlike the PVN it lacks other types of neurons. Indeed, in addition to MNCs, the PVN comprises interneurons, TNCs, and other types of projection neurons (29, 39).

1.5. Cutting Plane Affects Connectivity in Hypothalamic Slices

Recordings from putative MNCs in hypothalamic slices were first reported in 1978 (40, 41). These initial studies were conducted on coronal slices prepared according to methods adapted from Yamamoto and McIlwain (42, 43). Subsequent studies performed on SON MNCs in coronal slices suggested the absence of spontaneous network activity (Fig. 1). Specifically, addition of the voltage-activated Na^+ channel blocker tetrodotoxin to the recording solution (Sect. 2.1) has little impact on the basal frequency of spontaneous excitatory postsynaptic currents (EPSCs) (44). In other words, spontaneous synaptic events observed in SON neurons in coronal slices consist exclusively of miniature (quantal) EPSCs. However, studies by Hatton and coworkers have shown that SON MNCs in slices of rat hypothalamus cut in the horizontal plane (with the brain affixed to the mounting stage dorsal cortex side down) retain intact afferent connections originating from various structures including basal forebrain (45), olfactory system (46, 47), and the tuberomamillary nucleus (TMN) (48, 49). Neurons in the TMN of such slices were also demonstrated to receive functional afferent connections from the diagonal band of Broca (DBB), the lateral preoptic area (LPOA), and the lateral hypothalamic area (50). Studies, performed by Tasker and colleagues, have also revealed the existence of functional afferent connectivity to

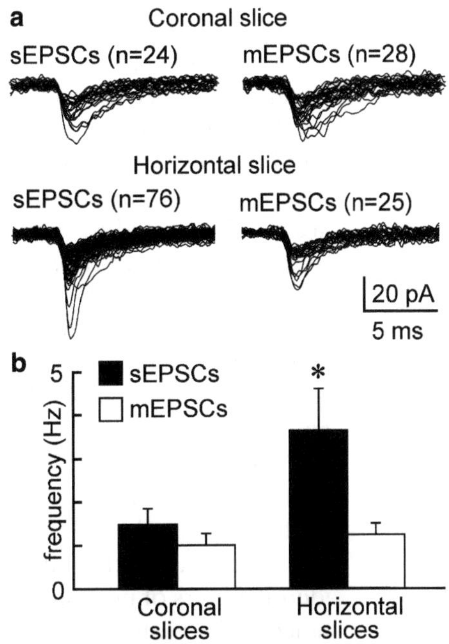

Fig. 1. *Spontaneous synaptic events in rat hypothalamic slices.* (**a**) Each panel superimposes all excitatory postsynaptic currents (EPSCs) detected during 15-s segments of whole-cell patch-clamp recordings obtained from an MNC (magnocellular neurosecretory cell) in coronal (*upper row*) or angled horizontal slices of adult rat hypothalamus (*lower row*). *Panels on the left* show spontaneous EPSCs (sEPSCs) recorded under control conditions, whereas panels on the right show miniature EPSCs (mEPSCs) recorded from the same cells in the presence of the voltage-activated Na^+ channel blocker tetrodotoxin (0.5 μM). Numbers in brackets indicate the total count of events recorded in each panel. (**b**) Bar graphs summarize mean (± s.e.m.) frequencies recorded under all conditions (Data from ref (44)). Note that sEPSC frequency is significantly ($*P < 0.05$, two-way repeated measures ANOVA) higher than mEPSC frequency in horizontal slices, indicating spontaneous network activity in this, but not the other model (Adapted with permission from (44)).

TNCs and MNCs in the PVN in horizontal slices of rat hypothalamus (51). Thus in contrast to coronal slices, in which little functional afferent connectivity to the SON is retained, horizontal slices provide a greater potential for the analysis of hypothalamic circuitry involved in the regulation of SON MNC functions.

1.6. In Vitro Analysis of Hypothalamic Networks Regulating VP MNCs

Our studies have largely focused on VP MNCs in the SON, and we shall thus review techniques pertaining to the analysis of these particular neurons. In mammals, VP is released under hyperosmotic conditions to promote water reabsorption from the kidney in order to maintain systemic osmolality near the osmotic set point (10). VP is also released during hypovolemia to defend against decreases in blood pressure (52), during hyperthermia to prevent the osmotic impact of fluid loss due to sweating (25, 53), and during sleep to minimize the impact of evaporative water loss while the possibility of water intake is curtailed (54). Our laboratory has designed various rodent preparations that are suitable for the analysis of how homeostatic perturbations modulate the activity of VP MNCs at a cellular and synaptic level (17, 55–58). Specifically, we are interested in how neurons in the suprachiasmatic nucleus (SCN) may mediate the circadian regulation of VP release (59) and how neurons in the preoptic area may contribute to the thermal (60) and hemodynamic control of VP neurons (61, 62). We are also interested in how osmosensory neurons in the *organum vasculosum lamina terminalis* (OVLT) contribute to the osmotic and thermal modulation of VP MNCs (44, 63). Moreover, our lab has examined some of the properties of TNCs and projection neurons in the arcuate nucleus (56), and ongoing work is exploring the role of glial cells in the local control of MNCs in the OVLT and SON.

Most of these studies have made use of two in vitro models: acute hypothalamic slices (Sects. 2.3, 2.4, 2.5, 2.6, and 2.7) and acute superfused hypothalamic explants (Sect. 2.8). In each case, we have found that maintaining intact connectivity between identified sets of neurons is a prerequisite for the presence of spontaneous electrical activity patterns that closely reflect those observed in vivo (15). In the following sections, we will describe in detail the methodology and materials used in the production of these two models and provide examples of the retained electrical properties and connectivity.

2. Materials and Methods

2.1. Composition of Solutions

The composition of the solutions used in the acute hypothalamic slices and explants is noted below. These include saline solutions used specifically for slicing procedures, artificial cerebrospinal fluid (ACSF) of varying composition used to superfuse the tissues during the experiments, and intracellular solutions used to fill the micropipettes used for whole-cell patch-clamp recording (see also Chaps. 1-Trapp and Ballanyi (21), 2-Ruanglittisakul et al. (18), and 3-Moore et al. (22)). In all cases, the final osmolality is adjusted by adding appropriate amounts of mannitol. Rat ACSF (295 ± 2 mosmol/kg) comprises (in mM) 120 NaCl, 26 $NaHCO_3$, 10 D-glucose, 3 KCl, 1.23 $NaH_2PO_4{\cdot}H_2O$, 1.48 $MgCl_2{\cdot}6H_2O$, and 1–2.4 $CaCl_2$. Mouse-slicing saline (345 ± 5 mosmol/kg) comprises (in mM) 87 NaCl, 25 $NaHCO_3$, 25 D-glucose, 75 sucrose, 2.5 KCl, 1.25 $NaH_2PO_4{\cdot}H_2O$, and 7 $MgCl_2{\cdot}6H_2O$. Mouse ACSF (312 ± 2 mosmol/kg) comprises (in mM) 128 NaCl, 26 $NaHCO_3$, 10 D-glucose, 3 KCl, 1.25 $NaH_2PO_4{\cdot}H_2O$, 1.25 $MgCl_2{\cdot}6H_2O$, and 1–2.4 $CaCl_2$. Rat intracellular patch pipette solution (285 ± 2 mosmol/kg, pH 7.3) comprises (in mM) 110 K^+ gluconate, 10 HEPES, 1 $MgCl_2{\cdot}6H_2O$, 1 EGTA, 0.1 $CaCl_2$, 10 KCl, 4 $Na_2{\cdot}ATP$, and 1 Na·GTP. Mouse internal patch pipette solution

(290 ± 2 mosmol/kg; pH 7.3) comprises (in mM) 130 K^+ gluconate, 10 HEPES, 1 $MgCl_2 \cdot 6H_2O$, and 1 EGTA.

2.2. Removal of the Brain

In all cases described below, 21–31 days old mice or 30–300 days old rats are first deeply anesthetized with halothane and then sacrificed by decapitation with a rodent guillotine (model #550012; Harvard Apparatus, Saint-Laurent, QC, Canada; www.harvardbioscience.ca). The head is immediately placed in 0–4°C cold species-specific ACSF (rat) or slicing solution (mouse) for 10–30 s to reduce temperature (Sect. 2.1). An extensive craniotomy is then performed to expose as much of the dorsal and lateral aspects of the brain from the caudal cervical area all the way to the olfactory bulbs. A transverse cut is then made just behind the olfactory bulbs using small scissors, and the rostral pole of the brain is then gently lifted just enough to visualize and cut the optic nerves before any tension might be caused by further lifting. Once the optic nerves are cut, the rostral pole of the brain is lifted slightly more to expose and cut the internal carotid arteries. The brain is then completely removed from the cranial vault by cutting the cranial nerves. With practice, this procedure takes <90 s to complete. Stretching of the optic nerves and tracts is to be avoided, as this damages the OVLT and SON and thus greatly reduces the number and quality of recordings that can be obtained.

2.3. Angled Horizontal Slices Preserve Osmoregulatory Circuits in Rodents

Casual inspection of a rodent brain atlas quickly reveals that the OVLT and the SCN do not lie in the same stereotaxic horizontal plane as the SON. In order to develop an acute slice preparation that could be used for analysis of the osmotic and circadian regulation of SON MNCs, we systematically explored the impact of cutting “horizontal” slices of rat brain at various angles (44). As illustrated in Fig. 2, cutting the rat brain at an angle near 38° relative to the dorsal surface of the cortex was found to provide optimal conservation of connectivity between the OVLT, SCN, and SON. Despite the almost inevitable variability in the exact position of the brain once glued onto the mounting stage, we found that equivalent and reliable results can be obtained from rat brain with cutting angles of 35–45°. Slices obtained using this procedure also retain much of the ventral preoptic area and several other hypothalamic nuclei. Analogous studies subsequently explored the cutting angles required to obtain similar slices from mouse brain. Because of the smaller dimension of the mouse brain (compared to rat), it is generally easier to retain more structures within a single slice of mouse hypothalamus, and thus, one would presume that the cutting angle would be less critical. Although this is true, our findings revealed that a cutting angle of 32° provided optimal access to the structures of interest (OVLT, SCN, SON) in slices prepared from mouse brain. The precise procedures used in each case are outlined in the next sections.

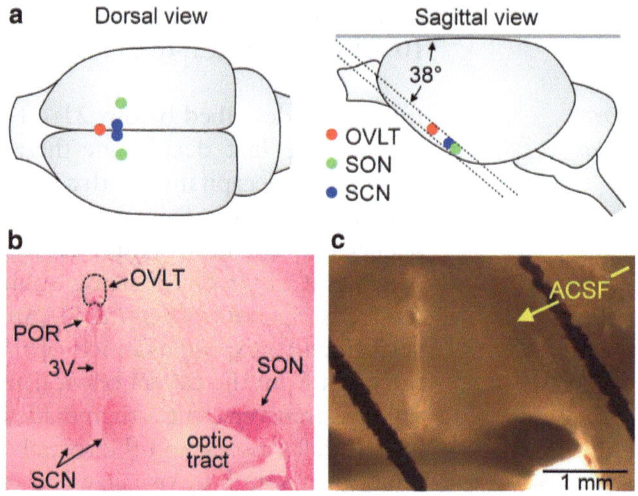

Fig. 2. *Anatomy of the angled horizontal slice preparation.* (**a**) Sketches illustrate the relative positions of the OVLT (*organum vasculosum lamina terminalis*), SCN (suprachiasmatic nuclei), and SON (supraoptic nuclei) in a dorsal (*left*) and side (*right*) view of the adult rat brain. *Dotted lines* illustrate the plane of sections used to make a horizontal slice angled ~38° relative to the dorsal surface of the cortex. (**b**) Micrograph of a histological section taken through the same plane that has been stained with *neutral red* (44). Landmarks including the third ventricle (3 V) and preoptic recess (POR) are shown. (**c**) Photograph of an equivalent transilluminated live preparation. The two diagonal dark bands are nylon threads of a platinum harp used to stabilize the slice in the recording chamber. The direction of artificial cerebrospinal fluid (ACSF) flow is indicated by an *arrow* (Panel **B** is reproduced with permission from (44)).

2.4. Angled Slices of Rat Hypothalamus

Following extraction (Sect. 2.2), the brain is immersed in ice-cold rat ACSF gassed with a mixture of 95% O_2–5% CO_2 (carbogen). The ventral brain region is trimmed with a razor blade into a block suitable for sectioning by making coronal cuts 3 mm rostral to the optic chiasma, and just rostral to the pons, and by making sagittal cuts to remove the temporal lobes. The resulting block of tissue is glued by the dorsal surface to an angled (35°) mounting block so that the rostral pole slants upward. The assembly is then placed on the stage of a VT1200 vibratome (Leica Biosystems, Nussloch, Germany; www.leica-microsystems.com) with the ventral surface facing the razor blade and submerged in ice-cold carbogenated ACSF. A fresh and clean razor blade is preinstalled in the slicer and calibrated for minimal z-axis motion using the Vibrocheck device (Leica Biosystems). The first cut (speed setting 0.14 mm/s, vibration amplitude 1.5 mm) is made just caudal to the optic chiasma such that portions of the optic tracts visible in the discarded tissue form a small V shape. A single 400-μm slice is then cut and transferred dorsal side up to a custom-made glass-bottomed recording chamber. The submerged slice is immobilized using a small "harp," that is, a C-shaped platinum wire fitted with thin nylon threads (see also Chap. 1-Trapp and Ballanyi (21)), and continuously superfused with carbogenated ACSF (32°C) at a rate of 1–3 ml/min. The slice is generally left alone for >45 min prior to recording, to allow any cellular debris to wash clear of the recording field. Although each brain yields only one slice that maintains the nuclei of interest, the gross architecture of the slice has a stereotypic appearance that allows easy identification of several structures of interest (Fig. 2).

2.5. Angled Slices of Mouse Hypothalamus

Following extraction, the brain is placed in a dish containing ice-cold carbogenated mouse-slicing solution (Sect. 2.1) (64). After a few seconds, the brain is quickly "rinsed" in a second dish containing the same solution. The brain is then placed on filter paper that has been soaked with slicing solution, and a razor blade is used to make coronal cuts to remove the rostral pole of the brain (to retain ~2–3 mm of optic nerves) and the caudal part of the brain from a position corresponding to where the optic tracts enter the temporal lobes. The dorsal side of the resulting block of brain is then blotted on dry tissue paper and glued dorsal side down with optic nerves pointing upward on a prechilled mounting stage angled at 32° with respect to the horizontal plane. Similar to generating rat slices, the stage is then placed in the bath compartment of the VT1200 vibrating slicer (Sect. 2.4) containing ice-cold and carbogenated mouse-slicing solution. A first cut is made to remove the ventral aspect of the tissue by aligning the blade so that it makes contact with the optic tracts at a point just caudal to the optic chiasma. The blade is advanced slowly (speed setting 0.08 mm/s, vibration amplitude 1.5 mm). A single 300-μm-thick section is then cut at the same speed and transferred into a 250-ml glass beaker containing 32°C warm, carbogenated mouse-slicing solution. The slice is placed onto a custom-made floating net that keeps it ~5 mm beneath the surface of the solution and protects it from the mechanical disturbance caused by continuous gassing. After 30–45 min of incubation, the slice is transferred to a glass-bottomed recording chamber where it is superfused at 2.5 ml/min with warm (32°C) carbogenated mouse ACSF. Recordings generally begin after 30 min of superfusion following procedures equivalent to those outlined above for rats.

2.6. Patch-Clamp Recording from Hypothalamic Neurons in Slices

Glass capillaries (1.2 mm outer ø) are pulled on a horizontal micropipette puller (P-87, Sutter Instrument Company, Novato, CA, USA; www.sutter.com) and filled with the appropriate internal pipette solution. To assist with postfixation immunocytochemical identification of recorded cells (see below), neurobiotin (0.1%w/v; Vector Laboratories, Burlingame, CA, USA; www.vectorlabs.com) may be added to the internal solution, adjusting osmolality accordingly. Pipette DC resistances in ACSF are typically 4–6 MΩ for mouse and 2–4 MΩ for rat. Electrical stimulation of the OVLT or SCN is accomplished by passing current pulses (20–60 μA, 0.5–1 ms) through a bipolar stimulating electrode made from Teflon-coated platinum/iridium wires (50 μm outer ø; A-M Systems, Sequim, WA, USA; www.a-msystems.com), inserted into the structure of interest. Stimuli are delivered via an isolated DS2 stimulation device (Digitimer Ltd., Welwyn Garden City, England; www.digitimer.com), triggered using a digital interface (Digidata 1440; Molecular Devices, Sunnyvale, CA, USA; www.moleculardevices.com), and controlled by a personal computer running Clampex 10 software (Molecular Devices).

Cells targeted for recording are visualized on a black and white video monitor using an upright microscope (BX51WI, Olympus, Center Valley, PA, USA; www.olympusamerica.com) coupled to a charge-coupled device (CCD) digital video camera (Coolsnap cf2, Röper Scientific, Ottobrunn, Germany; www.roperscientific.de). Electrodes are visually guided to each cell using a motorized micromanipulator (Model MX-7600; Siskiyou Corporation, Grants Pass, OR, USA; www.siskiyou.com). VP and OT neurons are indistinguishable by visual inspection. In Wistar rats, ~60% of MNCs in the SON are VP-containing (33, 34). This proportion rises to 70% in Long Evans rats (65) and 85% in C57/BL6 mice (17). The distinctive (but slightly overlapping) distributions of VP and OT MNCs in the SON allow one to selectively target VP-rich or OT-rich regions of the nucleus (Fig. 3). However, definitive identification of the cells requires either direct observation of a gene-specific reporter in the live preparation (Fig. 3) or the use of postrecording immunocytochemistry.

In all cases, positive pipette pressure (10–20 mmHg) is maintained inside the recording patch pipette while manipulating it toward the cell in order to clear local debris and avoid contaminating the tip. A slight sweeping motion can be used to ensure a clean access to the targeted part of the cell membrane before making physical contact. Once contact is made, positive pressure is then removed, and slight negative pressure is applied as required to establish a high-resistance electrical seal (typically 3–20 GΩ). After breaking this seal to establish the whole-cell-recording configuration, the pipette series resistance is typically 5–15 MΩ. Increases in series resistance are a common problem in most patch-clamp experiments using slices (see also Chap. 3-Moore et al. (22)). We found that the best approach to minimize this problem is to use the pipettes with the lowest resistance possible. Typically, this means using pipettes that are just small enough to avoid aspirating the entire cell in the process of seal formation. Although it is possible to reorient the electrode to correct the impact of drift if necessary, this is almost never a "cure" for increases in series resistance.

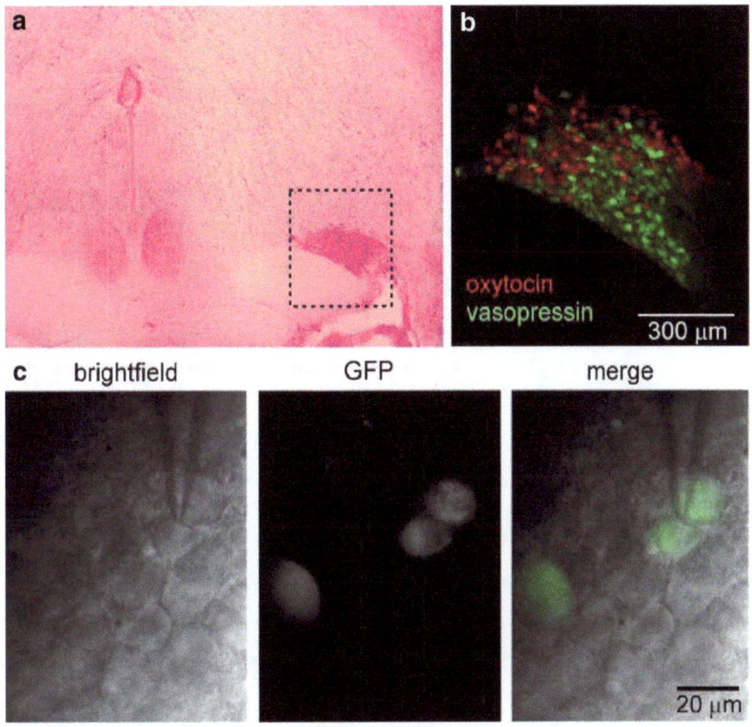

Fig. 3. *Identification of vasopressin (VP) and oxytocin (OT) neurons in angled horizontal slices of adult rat hypothalamus.* (**a**) Image shown in Fig. 2b is repeated to highlight the SON (*boxed region*). (**b**) A fluorescence image of the corresponding area where increased brightness reveals immunostained VP and OT neurons. OT- and VP-rich parts of the SON are identified by a *dotted line*. (**c**) Panels show a higher magnification view of the SON with a recording pipette entering the field of view from the *top*. The *left image* shows a bright-field view, whereas the *middle image* shows enhanced green fluorescent protein (GFP) expressed in VP neurons. The *right panel* superimposes both images (merge) to show that the cell targeted for recording is VP-positive.

2.7. Functional Connectivity in Angled Horizontal Slices

Our electrophysiological analyses have revealed that extensive functional connectivity between the SCN, SON, and OVLT is preserved in angled horizontal slices (17, 44, 59). Specifically, antidromic stimulation experiments in this slice type prepared from rat brain have confirmed that neuronal somata in the OVLT project axons to both the SCN and the SON. The same approach confirmed that axons from SCN neurons project to the OVLT and SON and that axons from SON neurons project through the median eminence in transit to the neurohypophysis (44). Electrical stimulation of the OVLT or SCN in slices from either rats or mice was found to evoke monosynaptic excitatory and inhibitory responses in neurons recorded in the SON and SCN (17, 44, 59). More importantly, functional activity-dependent circuitry is retained in this preparation since tetrodotoxin (Sect. 1.5) reduces the frequency of spontaneous EPSCs (sEPSCs) in MNCs (Fig. 1). Moreover, as

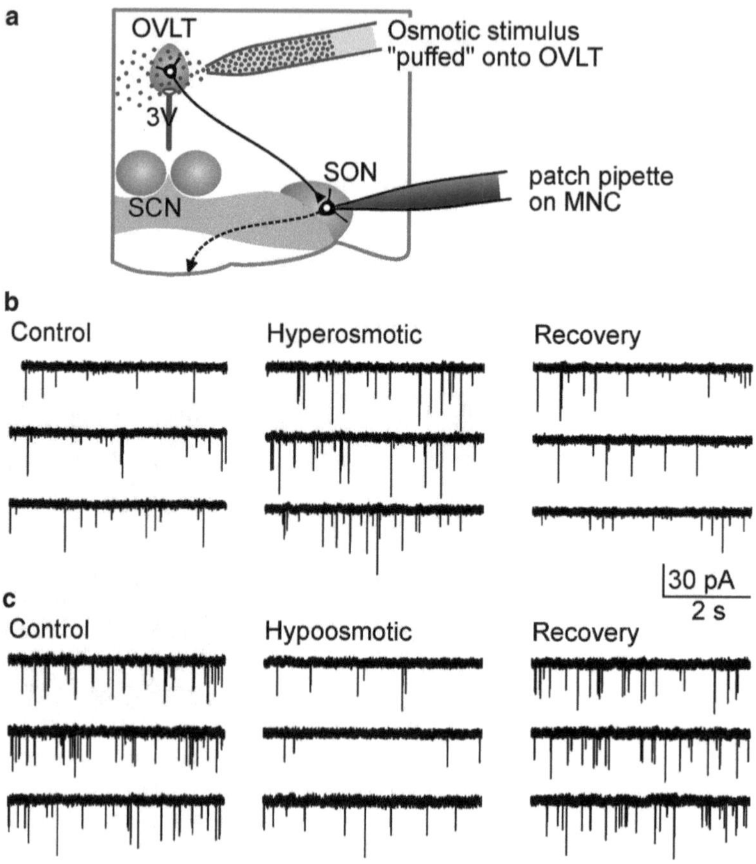

Fig. 4. *Osmoregulatory network activity in horizontal slices of rat hypothalamus*. (**a**) shows a sketch of the experimental configuration. Structures illustrated include the OVLT, SON, SCN, and 3 V (see Fig. 2). (**b**) Each panel shows three consecutive samples of sEPSC activity recorded from a SON neuron before, during, and after the application of a hyperosmotic stimulus to the OVLT. (**c**) Experiment on a different SON neuron (same layout) showing the effect of applying a hypoosmotic stimulus to the OVLT. Note that hyperosmolality increases sEPSC frequency and hypoosmolality reduces sEPSC frequency. Calibration applies to b and c.

illustrated in Fig. 4, local application of a hyperosmotic solution to the OVLT to excite osmosensory neurons increases the frequency of sEPSCs in SON MNCs. Conversely, reducing the basal firing rate of OVLT neurons, for example, with local applications of γ-aminobutyric acid or hypoosmotic solutions, reduces the frequency of sEPSCs in MNCs (44).

These findings indicate that osmosensory neurons in the OVLT remain functionally connected to MNCs in the SON in angled horizontal slice preparation. A recent study from our laboratory has used angled horizontal slices to demonstrate that SCN-mediated changes in OVLT-MNC interactions contribute to the circadian regulation of VP-releasing neurons (59). So much of the local circuitry controlling osmoregulatory circuits is functionally preserved in these slices.

2.8. Superfused Hypothalamic Explants

Superfused acute hypothalamic explants consist of an ~2-mm-thick slab of the ventral medial aspect of the adult rat and mouse brain. Superficial structures in this preparation are kept alive via delivery of warmed carbogenated ACSF. The main advantages of this preparation are that it is relatively unperturbed and cells located within 250 μm of the surface can display normal electroresponsive properties for 12–24 h. Because of its exceptional mechanical stability, this preparation is ideally suited for extracellular single unit recording and for sharp microelectrode intracellular recording. Because no slicing is performed in the immediate neighborhood of the nuclei of interest, neurons in explants are not subjected to the partial loss of their dendritic tree, and they retain a reasonably normal anatomical relationship with the ventral *glia limitans*. The main disadvantages of the preparation are that cells lying deeper than about 250 μm below the surface are not viable. The preparation is therefore best suited for the analysis of neurons and network connections lying against the ventral surface of the brain. Lastly, because of its thickness, patch-clamp recording (if necessary) must be performed using the "blind" approach (66) (see also Chap. 2-Ruangkittisakul et al. (18)).

The preparation of hypothalamic explants is relatively simple and requires no special instrument (Fig. 5a). The brain is first extracted as described for slices (Sect. 2.2) and then glued dorsal cortex facing down onto the bottom of an inverted 35-mm plastic petri dish using cyanoacrylate adhesive (Fig. 5b). A few drops of cold carbogenated ACSF are applied to the ventral surface to keep the tissue moist. A razor blade is then used to manually procure a 1.5–2.5-mm-thick slab of the ventral hypothalamus, taking care to avoid any folding of the tissue (Fig. 5c–e). The explant is then placed in a recording chamber consisting of a hollowed Plexiglass cylinder (14 mm inner ø) filled with Sylgard elastomer (Fig. 5f). The explant is pinned to this base using fine insect pins (model 26001–35; Fine Science Tools Inc., North Vancouver, BC; www.finescience.com) (Fig. 5g–h). Subsequently, a tube is positioned to deliver carbogenated ACSF (0.8–1.5 ml/min) at 32°C over the preparation. It normally takes <90 s to perform these first steps. Next, the surface of the brain is prepared for recording. This part of the procedure involves using fine forceps to carefully remove any overlying membranes, in particular the arachnoid mater, and small blood vessels from the area of interest. This requires a clear view of the transilluminated preparation at 10–30X using a quality binocular microscope (e.g., model SMZ645, Nikon Canada Ltd., Mississauga, ON, Canada; www.nikoninstruments.com) as membranes must be removed without touching any part of the brain surface. Figure 5g–h shows top views of the explant with incident illumination before and after it has been pinned, and after membranes overlying the supraoptic, preoptic, and frontal lobe regions have been removed. The removal of membranes and blood vessels

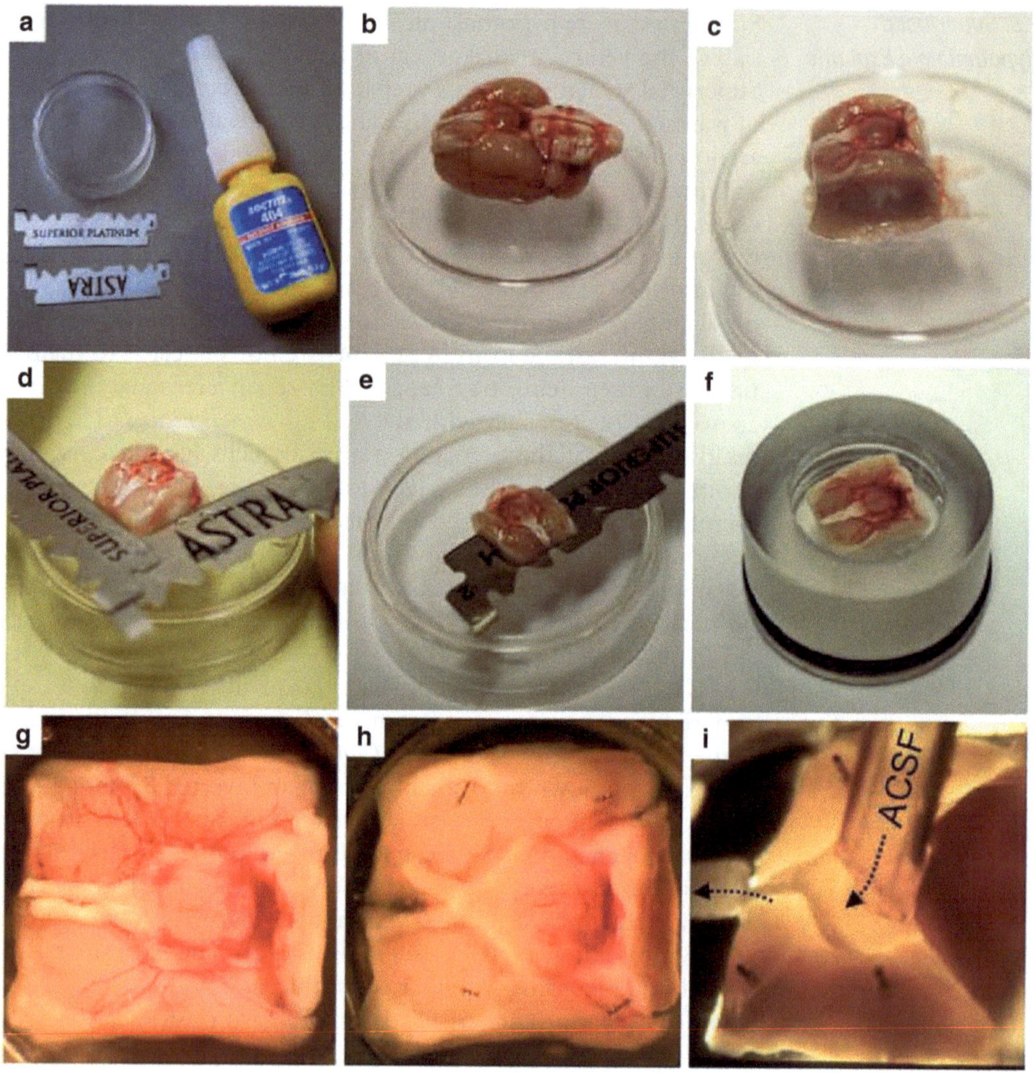

Fig. 5. *Steps involved in the preparation of rat hypothalamic explants* (See text for details).

is extended as far as possible laterally and at positions lying downstream in the desired path of drainage. The time required for this step can vary from one to several minutes, and the total length of time taken is not a negative factor for the ultimate quality of the preparation as long as fresh carbogenated ACSF is being delivered to the relevant area and as the surface of the brain is not disturbed by any of the tools. Once this "cleaning" step is completed, the delivery tube is moved to its final position, usually located ~1 mm upstream of the target recording area. The tube is placed so that its bottom lip makes gentle contact with the tissue. Once in position, this tube should not be moved until the end of the experiment. It is important to make sure that fluid does not collect or stagnate around and over the preparation. In fact, ACSF delivered to the target area should flow away as fast as possible, and the removal of waste fluid can be facilitated by placing a small cotton wick downstream of the explant to guide the waste ACSF toward the exit port of the chamber. The preparation is then allowed to equilibrate for 1–3 h before recording begins. As for dissection, optimal viewing conditions for electrode placement are achieved with transillumination via a fiber-optic guide placed immediately beneath the recording chamber (Fig. 5i).

Because of their extended viability, useful recordings can be obtained from hypothalamic explants for periods of time that greatly exceed those normally obtained from acute slice preparations (typically <8 h). Although spontaneous electrical activity can be recorded as soon as 60 min after the preparation is established, we have found that an equilibration time of 2–3 h (or more) provides optimal performance. Superfused hypothalamic explants allowed to equilibrate for at least 3 h display outstanding mechanical stability and allow extracellular and intracellular recordings of high quality to be obtained for several hours. Moreover, as the preserved tissue remains relatively intact, hypothalamic networks comprising neuronal somata and axons lying along the ventral surface are functionally maintained, and corresponding physiological responses can be detected for extended periods of time. As illustrated in Fig. 6, local applications of osmotic stimuli to OVLT neurons cause changes in the firing rate of MNCs in the SON. Note that in these experiments the explants are fixed onto a slanted base in a way that prevents the spatial spread of solutions applied specifically to the OVLT.

3. Notes and Potential Pitfalls

1. The solutions used for mouse and rat hypothalamic preparations are adjusted to different osmolalities (Sect. 2.1), and we wish to highlight this point. The osmotic set point for rats lies near 292–298 mosmol/kg (depending on strain), whereas the osmotic set point for C57/BL6 mice lies near 312 mosmol/kg (10). We have noted significant improvements in the viability and function of SON MNCs and OVLT neurons in mouse slices prepared and maintained using solutions adjusted to 312 mosmol/kg. Whether this is important only for hypothalamic osmoregulatory circuits or other brain areas remains to be determined.
2. We make patch-clamp recordings from rat and mouse preparations using pipettes filled with solutions that are 10–22 mosmol/kg lower than the corresponding ACSF (Sect. 2.1). Because of our interest in the mechanisms that couple osmotically induced changes in cell volume to ion channel

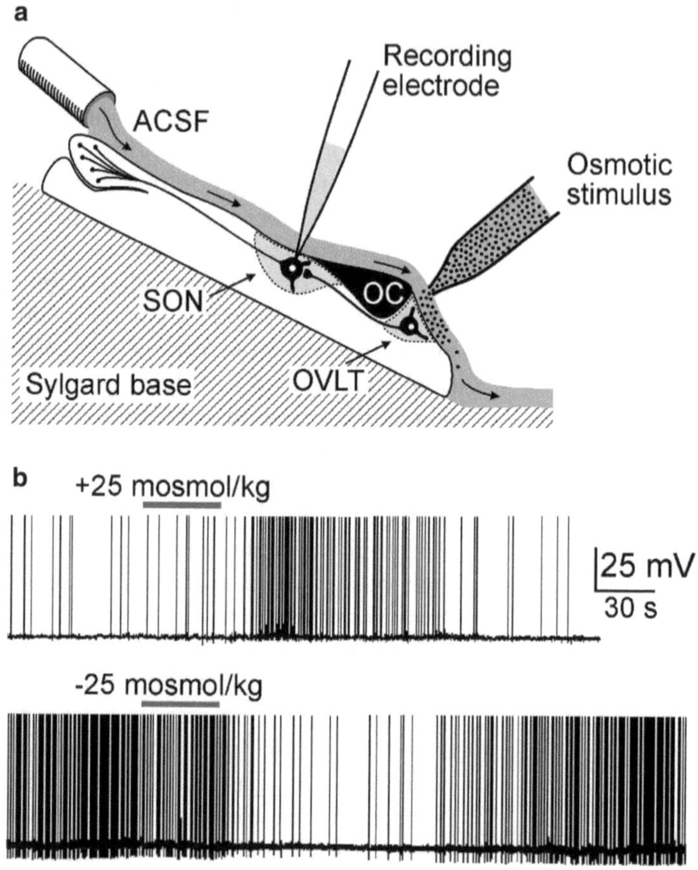

Fig. 6. *Osmoregulatory network activity in superfused rat hypothalamic explants.* (**a**) Schematic diagram illustrating a typical experimental configuration as viewed in the sagittal plane. An explant prepared as shown in Fig. 5 is fixed to the slanted Sylgard base of the recording chamber, and the superfusion tube is placed at a high position upstream of the recording site. Here, a SON neuron is impaled for membrane potential recording with a sharp microelectrode, while a small pipette delivering osmotic stimuli is positioned over the OVLT. Panels in (**b**) illustrate changes in intracellular action potential firing in response to hypertonic (*upper panel*) and hypotonic stimuli (*lower panel*). These responses are mediated entirely through changes in excitatory synaptic drive (changes in sEPSC frequency) as the solutions applied to the OVLT do not contact any part of the SON neurons.

modulation, for example (67), we closely monitor changes in cell volume using imaging methods during many of our experiments. As reported by others (68), we have noticed that the use of internal and external solutions of equal osmolality is commonly associated with a progressive swelling of the cell being recorded after the whole-cell configuration is established. This swelling is very likely due to the small positive hydrostatic pressure caused by the height of the fluid column inside the pipette. In many types of neurons, this may be without consequence, but this is not the case in osmosensitive neurons! We have found empirically that using mildly hypoosmotic internal solutions, which favors a net efflux of water from the cell to the extracellular compartment, can abrogate the emergence of progressive and unprovoked changes in cell volume (see also Chap. 3-Moore et al. (22)). Although this effect is more pronounced during recordings from freshly isolated cells, where total cell volume is smaller than in situ due to the absence of a significant dendritic compartment, we continue to use mildly hypoosmotic solutions when recording from neurons in slices.

3. Extracting the brain using the procedure described above (Sect. 2.2) normally causes the stalk of the pituitary gland to be severed at the level of the median eminence, and thus, the distal axonal processes of MNCs are lost. This does not appear to be of concern for most applications since there are no afferents known to originate from within the pituitary and since the spontaneous electrical activity patterns of MNCs in slices and explants (17, 40, 41, 57, 58) are very similar to those observed in vivo (15). However, if desired, it is possible to keep the full extent of the MNC axons (and pituitary gland) intact during the preparation of superfused hypothalamic explants (49). To do so simply requires that care be taken to release the pituitary from the *sella turcica* by cutting the *dura mater* that overlies the gland (i.e., the *diaphragma sellae*). This step should be performed just after severing the internal carotid arteries in the procedure described above (Sect. 2.2), and care should be taken to avoid excessive stretching of the pituitary stalk.

4. Electrophysiological recordings from superfused hypothalamic explants have confirmed that high quality recordings can be obtained from cells up to a depth of 200–250 μm from the surface of the tissue. Intracellular recordings from MNCs in the SON in this range display intrinsic membrane properties that are equivalent to those observed during intracellular recordings in vivo (69). Neurons impaled at greater depths generally lack spontaneous action potential firing and have a tendency to depolarize when repetitive firing is induced by current injection. Although the presence of decaying underlying tissue does not appear to compromise the natural activity and properties of surface neurons, it is unlikely that cells lying at depths exceeding 250 μm can contribute in a meaningful way to the network activity observed in this preparation.

5. Our extracellular solutions use Ca^{2+} concentrations ranging from 1 to 2.4 mM (Sect. 2.1). Ca^{2+} concentrations between 1.0 and 1.2 mM are systematically used by our lab in studies involving extracellular recordings of single unit activity and in experiments performed on acutely isolated neurons (not discussed in this chapter). As observed by Pittman and colleagues (70), we have found that using lower (and more physiological) superfusate Ca^{2+} concentrations yields more natural levels of

spontaneous electrical activity. This could be due to a number of reasons including (1) surface charge screening effect causing an increase in action potential threshold, (2) amplified activation of Ca^{2+}-dependent K^+ conductances, and (3) blockade of resting pacemaker currents mediated by nonselective cation channels (71). Higher levels of Ca^{2+} (2–2.4 mM) are generally used in studies where an enhancement of specific Ca^{2+}-dependent processes or where reduced overall excitability becomes an advantage (e.g., studies targeting intrinsic mechanisms). This issue is discussed in more detail elsewhere in this book (Chaps. 2-Ruangkittisakul et al. (18), 3-Moore et al. (22), and 8-Sanchez-Vives (72)).

4. Conclusions

Acute preparations of adult rat and mouse hypothalamus provide an excellent opportunity to investigate many aspects of the cellular and synaptic basis implicated in the central control of body fluid homeostasis under near-physiological conditions. Hypothalamic explants and slices both allow functionally intact networks of identified neurons to be studied in vitro, with a great deal of experimental control.

Acknowledgments

Work in our lab is supported by operating grants MOP-9939 and MOP-82818 from the Canadian Institutes of Health Research and by a James McGill Professor Chair to CWB. JRS and SC were recipients of Canada Graduate Scholarships. TJS was recipient of a CIHR doctoral Award and ET received a Doctoral Award from the Heart and Stroke Foundation of Canada. The Research Institute of the McGill University Health Centre is supported by the Fonds de la Recherche en Santé du Québec.

References

1. Morgane PJ, Panksepp J (1979) Handbook of the hypothalamus, vol 1, Anatomy of the hypothalamus. Marcel Dekker, New York, p 726
2. Okamura H (2007) Suprachiasmatic nucleus clock time in the mammalian circadian system. Cold Spring Harb Symp Quant Biol 72:551–556
3. Reppert SM, Weaver DR (2002) Coordination of circadian timing in mammals. Nature 418:935–941
4. Denton DA, McKinley MJ, Weisinger RS (1996) Hypothalamic integration of bodyfluid regulation. Proc Natl Acad Sci USA 93:7397–7404
5. Dietrich MO, Horvath TL (2009) Feeding signals and brain circuitry. Eur J Neurosci 30:1688–1696
6. Stocker SD, Keith KJ, Toney GM (2004) Acute inhibition of the hypothalamic paraventricular nucleus decreases renal sympathetic nerve activity and arterial blood pressure in

water-deprived rats. Am J Physiol Regul Integr Comp Physiol 286:R719–R725

7. Toney GM, Chen QH, Cato MJ, Stocker SD (2003) Central osmotic regulation of sympathetic nerve activity. Acta Physiol Scand 177:43–55
8. Boulant JA (2000) Role of the preoptic-anterior hypothalamus in thermoregulation and fever. Clin Infect Dis 31:S157–S161
9. Boulant JA (1981) Hypothalamic mechanisms in thermoregulation. Fed Proc 40:2843–2850
10. Bourque CW (2008) Central mechanisms of osmosensation and systemic osmoregulation. Nat Rev Neurosci 9:519–531
11. Edwards OM, Clark JD (1986) Post-traumatic hypopituitarism. Six cases and a review of the literature. Medicine (Baltimore) 65:281–290
12. Pittman QJ, Blume HW, Renaud LP (1981) Connections of the hypothalamic paraventricular nucleus with the neurohypophysis, median eminence, amygdala, lateral septum and midbrain periaqueductal gray: an electrophysiological study in the rat. Brain Res 215:15–28
13. Sherlock DA, Field PM, Raisman G (1975) Retrograde transport of horseradish peroxidase in the magnocellular neurosecretory system of the rat. Brain Res 88:403–414
14. Alonso G, Assenmacher I (1981) Radioautographic studies on the neurohypophysial projections of the supraoptic and paraventricular nuclei in the rat. Cell Tissue Res 219:525–534
15. Poulain DA, Wakerley JB (1982) Electrophysiology of hypothalamic magnocellular neurones secreting oxytocin and vasopressin. Neuroscience 7:773–808
16. Cunningham ET Jr, Sawchenko PE (1991) Reflex control of magnocellular vasopressin and oxytocin secretion. Trends Neurosci 14:406–411
17. Sharif-Naeini R, Ciura S, Stachniak TJ, Trudel E, Bourque CW (2008) Neurophysiology of supraoptic neurons in C57/BL mice studied in three acute in vitro preparations. Prog Brain Res 170:229–242
18. Ruangkittisakul A, Panaitescu B, Secchia L, Bobocea N, Kantor C, Kuribayashi J, Iizuka M, Ballanyi K (2012) Isolated brainstem respiratory centers from perinatal rodents. In Isolated Central Nervous System Circuits (Ed K Ballanyi), Neuromethods Series Vol. 73 (Ed W Walz). Springer Science+Business Media, LLC, New York, NY, pp 61–124
19. Kantor C, Panaitescu B, Kuribayashi J, Ruangkittisakul A, Jovanovic I, Leung V, Lee TF, MacTavish D, Jhamandas JH, Cheung PY, Ballanyi K (2012) Spontaneous neural network oscillations in hippocampus, cortex and *locus coeruleus* of newborn rat and piglet brain slices. In Isolated Central Nervous System Circuits (Ed K Ballanyi), Neuromethods Series Vol. 73 (Ed W Walz). Springer Science+Business Media, LLC, New York, NY, pp 315–356
20. Nakamura TJ, Michel S, Block GD, Colwell CS (2012) Neural circuits underlying circadian oscillations in mammals: clocks in a dish. In Isolated Central Nervous System Circuits (Ed K Ballanyi), Neuromethods Series Vol. 73 (Ed W Walz). Springer Science+Business Media, LLC, New York, NY, pp 183–210
21. Trapp S, Ballanyi K (2012) Autonomic nervous system *in vitro*: studying tonically active neurons controlling vagal outflow in rodent brainstem slices. In Isolated Central Nervous System Circuits (Ed K Ballanyi), Neuromethods Series Vol. 73 (Ed W Walz). Springer Science+Business Media, LLC, New York, NY, pp 1–59
22. Moore AR, Zhou WL, Jakovcevski I, Zecevic N, Antic SD (2012) Physiological properties of human fetal cortex *in vitro*. In Isolated Central Nervous System Circuits (Ed K Ballanyi), Neuromethods Series Vol. 73 (Ed W Walz). Springer Science+Business Media, LLC, New York, NY, pp 125–158
23. Fish KN, Gonzales-Burgos G, Zaitsev AV, Lewis DA (2012) Histological characterization of physiologically determined fast spiking interneurons in slices of the primate dorsolateral prefrontal cortex. In Isolated Central Nervous Circuits (Ed K Ballanyi), Neuromethods Series Vol. 73 (Ed W Walz). Springer Science+Business Media, LLC, New York, NY, pp 159–181
24. McKay BE, Tadayonnejad R, Anderson DM, Engbers JDT, Fernandez FR, Iftinca M, Turner RW (2012) Establishing in vivo like activity in rat cerebellar cells maintained in vitro. In Isolated Central Nervous System Circuits (Ed K Ballanyi), Neuromethods Series Vol. 73 (Ed W Walz). Springer Science+Business Media, LLC, New York, NY, pp 233–262
25. Sharif-Naeini R, Ciura S, Bourque CW (2008) TRPV1 gene required for thermosensory transduction and anticipatory secretion from vasopressin neurons during hyperthermia. Neuron 58:179–185
26. Broicher T, Speckmann EJ (2012) Living human brain slices: network analysis using voltage sensitive dyes. In Isolated Central Nervous System Circuits (Ed K Ballanyi), Neuromethods Series Vol. 73 (Ed W Walz). Springer Science+Business Media, LLC, New York, NY, pp 285–300
27. Luhmann HJ, Kilb W (2012) Intact in vitro preparation of the neonatal rodent cerebral cortex-analysis of cellular properties and network

activity. In Isolated Central Nervous System Circuits (Ed K Ballanyi), Neuromethods Series Vol. 73 (Ed W Walz). Springer Science+Business Media, LLC, New York, NY, pp 301–314

28. Stern JE (2001) Electrophysiological and morphological properties of pre-autonomic neurones in the rat hypothalamic paraventricular nucleus. J Physiol 537:161–177
29. Hoffman NW, Tasker JG, Dudek FE (1991) Immunohistochemical differentiation of electrophysiologically defined neuronal populations in the region of the rat hypothalamic paraventricular nucleus. J Comp Neurol 307:405–416
30. Spergel DJ, Krüth U, Shimshek DR, Sprengel R, Seeburg PH (2001) Using reporter genes to label selected neuronal populations in transgenic mice for gene promoter, anatomical, and physiological studies. Prog Neurobiol 63:673–686
31. Ueta Y, Fujihara H, Serino R, Dayanithi G, Ozawa H, Matsuda K, Kawata M, Yamada J, Ueno S, Fukuda A, Murphy D (2005) Transgenic expression of enhanced green fluorescent protein enables direct visualization for physiological studies of vasopressin neurons and isolated nerve terminals of the rat. Endocrinology 146:406–413
32. van den Pol AN, Yao Y, Fu LY, Foo K, Huang H, Coppari R, Lowell BB, Broberger C (2009) Neuromedin B and gastrin-releasing peptide excite arcuate nucleus neuropeptide Y neurons in a novel transgenic mouse expressing strong Renilla green fluorescent protein in NPY neurons. J Neurosci 29:4622–4639
33. Vandesande F, Dierickx K (1975) Identification of the vasopressin producing and of the oxytocin producing neurons in the hypothalamic magnocellular neurosecretory system of the rat. Cell Tissue Res 164:153–162
34. Swaab DF, Nijveldt F, Pool CW (1975) Distribution of oxytocin and vasopressin in the rat supraoptic and paraventricular nucleus. J Endocrinol 67:461–462
35. Huang W, Lee SL, Sjoquist M (1995) Natriuretic role of endogenous oxytocin in male rats infused with hypertonic NaCl. Am J Physiol 268:R634–R640
36. Verbalis JG, Mangione MP, Stricker EM (1991) Oxytocin produces natriuresis in rats at physiological plasma concentrations. Endocrinology 128:1317–1322
37. Bicknell RJ (1988) Optimizing release from peptide hormone secretory nerve terminals. J Exp Biol 139:51–65
38. Dreifuss JJ et al (1971) Action potentials and release of neurohypophysial hormones in vitro. J Physiol 215(3):805–817
39. Swanson LW, Sawchenko PE (1983) Hypothalamic integration: organization of the paraventricular and supraoptic nuclei. Annu Rev Neurosci 6:269–324
40. Haller EW, Brimble MJ, Wakerley JB (1978) Phasic discharge in supraoptic neurones recorded from hypothalamic slices. Exp Brain Res 33:131–134
41. Hatton GI, Armstrong WE, Gregory WA (1978) Spontaneous and osmotically stimulated activity in slices of rat hypothalamus. Brain Res Bull 3:497–508
42. Yamamoto C, McIlwain H (1966) Electrical activities in thin sections from the mammalian brain maintained in chemically-defined media in vitro. J Neurochem 13:1333–1343
43. Yamamoto C, McIlwain H (1966) Potentials evoked in vitro in preparations from the mammalian brain. Nature 210:1055–1056
44. Trudel E, Bourque CW (2003) A rat brain slice preserving synaptic connections between neurons of the suprachiasmatic nucleus, organum vasculosum lamina terminalis and supraoptic nucleus. J Neurosci Methods 128:67–77
45. Hatton GI, Yang QZ (2002) Synaptic potentials mediated by alpha 7 nicotinic acetylcholine receptors in supraoptic nucleus. J Neurosci 22:29–37
46. Hatton GI, Yang QZ (1989) Supraoptic nucleus afferents from the main olfactory bulb–II. Intracellularly recorded responses to lateral olfactory tract stimulation in rat brain slices. Neuroscience 31:289–297
47. Yang QZ, Smithson KG, Hatton GI (1995) NMDA and non-NMDA receptors on rat supraoptic nucleus neurons activated monosynaptically by olfactory afferents. Brain Res 680:207–216
48. Weiss ML, Yang QZ, Hatton GI (1989) Magnocellular tuberomammillary nucleus input to the supraoptic nucleus in the rat: anatomical and in vitro electrophysiological investigations. Neuroscience 31:299–311
49. Yang QZ, Hatton GI (1994) Histamine mediates fast synaptic inhibition of rat supraoptic oxytocin neurons via chloride conductance activation. Neuroscience 61:955–964
50. Yang QZ, Hatton GI (1997) Electrophysiology of excitatory and inhibitory afferents to rat histaminergic tuberomammillary nucleus neurons from hypothalamic and forebrain sites. Brain Res 773:162–172
51. Boudaba C, Schrader LA, Tasker JG (1997) Physiological evidence for local excitatory synaptic circuits in the rat hypothalamus. J Neurophysiol 77:3396–3400

52. Dunn FL, Brennan TJ, Nelson AE, Robertson GL (1973) The role of blood osmolality and volume in regulating vasopressin secretion in the rat. J Clin Invest 52:3212–3219
53. Forsling ML, Ingram DL, Stanier MW (1976) Effects of various ambient temperatures and of heating and cooling the hypothalamus and cervical spinal cord on antidiuretic hormone secretion and urinary osmolality in pigs. J Physiol 257:673–686
54. Forsling ML (2000) Diurnal rhythms in neurohypophysial function. Exp Physiol 85: 179S–186S
55. Stachniak TJ, Bourque CW (2006) Visually guided whole cell patch clamp of mouse supraoptic nucleus neurons in cultured and acute conditions. Am J Physiol Regul Integr Comp Physiol 291:R68–R76
56. MacMillan SJ, Bourque CW (1993) Intracellular recordings from neurons of the arcuate nucleus in superfused explants of rat hypothalamus. Neuroendocrinology 57:159–166
57. Bourque CW, Renaud LP (1983) A perfused in vitro preparation of hypothalamus for electrophysiological studies on neurosecretory neurons. J Neurosci Methods 7:203–214
58. Bourque CW, Renaud LP (1983) In vitro neurophysiology of identified rat hypothalamic 'neuroendocrine' neurons. Neuroendocrinology 36:161–164
59. Trudel E, Bourque CW (2010) Central clock excites vasopressin neurons by waking osmosensory afferents during late sleep. Nat Neurosci 13:467–474
60. Nakashima T, Hori T, Kiyohara T, Shibata M (1985) Osmosensitivity of preoptic thermosensitive neurons in hypothalamic slices in vitro. Pflüger's Arch 405:112–117
61. Aradachi H, Honda K, Negoro H, Kubota T (1996) Median preoptic neurones projecting to the supraoptic nucleus are sensitive to haemodynamic changes as well as to rise in plasma osmolality in rats. J Neuroendocrinol 8: 35–43
62. Stocker SD, Toney GM (2005) Median preoptic neurones projecting to the hypothalamic paraventricular nucleus respond to osmotic, circulating Ang II and baroreceptor input in the rat. J Physiol 568:599–615
63. Richard D, Bourque CW (1995) Synaptic control of rat supraoptic neurones during osmotic stimulation of the organum vasculosum lamina terminalis in vitro. J Physiol 489:567–577
64. Bischofberger J, Engel D, Li L, Geiger JR, Jonas P (2006) Patch-clamp recording from mossy fiber terminals in hippocampal slices. Nat Protoc 1:2075–2081
65. Rhodes CH, Morrell JI, Pfaff DW (1981) Immunohistochemical analysis of magnocellular elements in rat hypothalamus: distribution and numbers of cells containing neurophysin, oxytocin, and vasopressin. J Comp Neurol 198:45–64
66. Kolaj M, Yang CR, Renaud LP (2000) Presynaptic $GABA_B$ receptors modulate organum vasculosum lamina terminalis-evoked postsynaptic currents in rat hypothalamic supraoptic neurons. Neuroscience 98:129–133
67. Zhang Z, Kindrat AN, Sharif-Naeini R, Bourque CW (2007) Actin filaments mediate mechanical gating during osmosensory transduction in rat supraoptic nucleus neurons. J Neurosci 27:4008–4013
68. Heidelberger R, Zhou ZY, Matthews G (2002) Multiple components of membrane retrieval in synaptic terminals revealed by changes in hydrostatic pressure. J Neurophysiol 88:2509–2517
69. Bourque CW, Renaud LP (1991) Membrane properties of rat magnocellular neuroendocrine cells in vivo. Brain Res 540:349–352
70. Pittman QJ, Hatton JD, Bloom FE (1981) Spontaneous activity in perfused hypothalamic slices: dependence on calcium content of perfusate. Exp Brain Res 42:49–52
71. Zhang Z, Bourque CW (2006) Calcium permeability and flux through osmosensory transduction channels of isolated rat supraoptic nucleus neurons. Eur J Neurosci 23: 1491–1500
72. Sanchez-Vives MV (2012) Spontaneous rhythmic activity in the adult cerebral cortex *in vitro*. In Isolated Central Nervous System Circuits (Ed K Ballanyi), Neuromethods Series Vol. 73 (Ed W Walz). Springer Science+Business Media, LLC, New York, NY, pp 263–284

Chapter 7

Establishing In Vivo-Like Activity in Rat Cerebellar Cells Maintained In Vitro

Bruce E. McKay, Reza Tadayonnejad, Dustin M. Anderson, Jordan D.T. Engbers, Fernando R. Fernandez, Mircea Iftinca, and Ray W. Turner

Abstract

Purkinje cells of the cerebellar cortex and neurons in the deep cerebellar nuclei (DCN) were among the first central neurons to be studied extensively through the use of in vitro preparations. Yet, the degree to which the patterns of action potential ("spike") output recorded in these cells in vitro match those recorded in vivo has been a matter of debate and uncertainty. We have identified relatively simple steps that can be applied to the preparation of cerebellar tissue slices or to recording conditions that increase the probability of recording spike output in vitro that more closely matches that found in the live animal. Of particular importance are considerations of the rapid development of Purkinje cell structure and spike output properties during the period used for patch-clamp recordings, reintroducing key synaptic inputs to the in vitro preparation and establishing appropriate reversal potentials for ion species. We also report that storing water to be used for preparing medium in polyethylene carboys introduces a contaminant that decreases the amplitude of T-type Ca^{2+} currents and the ability to generate rebound burst discharge in DCN cells. Controlling for these factors restores spontaneous tonic firing, increases the gain to parallel fiber input, uncovers bistable rhythmic behavior in Purkinje cells, and increases the probability for generating rebound discharge in DCN cells.

Key words: Brain slices, Cerebellum, Deep cerebellar nuclear neurons, Patch-clamp, Postnatal development, Purkinje cells, Reversal potentials, Spontaneous spike output, Temperature, Water storage

1. Introduction

1.1. Cerebellar Neurons Are Studied in a Variety of Preparations

One major aim of using in vitro preparations is to understand the properties of individual cells that shape the spontaneous activity of the neural circuit in which they function. The cerebellum was among the first cortical structures to receive the attention of electrophysiologists given the early interest in motor systems. As such,

Klaus Ballanyi (ed.), *Isolated Central Nervous System Circuits*, Neuromethods, vol. 73,
DOI 10.1007/978-1-62703-020-5_7, © Springer Science+Business Media New York 2012

it has been somewhat of a proving ground for comparing recordings that are obtained in the intact circuitry of anesthetized or awake animals in vivo to a wide range of in vitro techniques that span from dissociated cells to organotypic cultures through in vitro slices and even intact brainstem–cerebellar preparations (1–5). The cerebellum was also among the first central structures to see intracellular, intradendritic, and whole-cell patch-clamp recordings, Ca^{2+} imaging in vitro, and recently 2-photon imaging in vivo (6–11). As for many other areas, our understanding of the activity of cerebellar neurons in vivo has also progressed from one of interpreting all unit discharge as being synaptically driven, i.e., that parallel fiber input elicits all "simple" action potentials ("simple spikes") in Purkinje cells, to recognizing that intrinsic membrane properties allow many neurons to spontaneously discharge spikes. For instance, we now know that Purkinje cell simple (Na^{+}-dependent) spikes occur spontaneously both in vitro and in vivo in the absence of synaptic inputs, with physiological extracellular K^{+} levels and without chemical excitatory stimulation (1–3, 12–15). Establishing the conditions necessary to recreate physiological (in vivo) activity patterns in vitro has proven to be a learning process as new techniques and preparations have been applied.

1.2. Discrepancies in Spike Output Properties Between Preparations

The relationship between the spontaneous activity of cerebellar cells recorded in vitro to those found in vivo can often be unclear and at times so different as to divide opinions on what constitutes a physiologically relevant output. This has proven to be particularly true for Purkinje cells, in which a slow oscillatory trimodal activity recorded in vitro (10, 16, 17) does not match the relatively tonic firing pattern reported in vivo (14, 18) (for a description of trimodal output, see Results (Sect. 4.1) and Fig. 6). Interpreting some of these results will be complicated by the use of anesthetics in vivo (19–21), while the use of patch-clamp recordings for in vitro analyses often requires the use of early postnatal animals, increasing the potential for Purkinje cell activity recorded in vitro to differ from the adult animals used for in vivo recordings. A similar discrepancy has arisen for the activity of "deep cerebellar nuclei" (DCN) cells, where the potential to generate a rebound spike burst following inhibitory synaptic input (22, 23) or for T-type voltage-activated Ca^{2+} channels to contribute to rebound responses is still under investigation (24–27). Through our work on in vitro tissue slices of rat cerebellar Purkinje cells and their downstream targets in the DCN, we have identified several key considerations that allow one to record activity in vitro that is much more in line with that reported in vivo for these output cells. This chapter reviews several facets of maintaining cerebellar slices in vitro and applying patch-clamp recording techniques that optimize the ability to record activity more representative of that found in vivo.

1.3. Overview of Cerebellar Circuitry

Our work has centered on cerebellar cortical Purkinje neurons and large diameter (presumed excitatory output) neurons contained in the DCN to understand the contributions of intrinsic properties and synaptic inputs to the firing properties of these neurons. To provide perspective on the connectivity and central importance of these cell types, we include here a brief overview of cerebellar circuitry (for a more detailed description, see (28)). Inputs to the cerebellum ascending from the brainstem arrive as two principal types: climbing fibers that originate in the inferior olive and mossy fibers that originate from numerous structures that include the spinocerebellar, vestibular, reticular, and pontine nuclei. Climbing fibers synapse onto the somata and proximal dendrites of upward of ten Purkinje neurons within the cerebellar cortex. Notably, individual adult Purkinje cells receive input from only a single climbing fiber, but this input is one of the most powerful excitatory synaptic inputs within the central nervous system. Mossy fibers send collateral branches to cells of the DCN (the fastigial (medial), globose/emboliform (interpositus), and dentate (lateral) nuclei) and ascend upward to synapse onto granule cells within the granular layer of cerebellar cortex. The axons of granule cells ascend into the molecular layer (superficial to the Purkinje cell layer) to bifurcate and produce parallel fibers oriented parallel to the surface of the cortex but perpendicular to the orientation of Purkinje neuron dendrites. This architecture enables individual parallel fibers to contact hundreds of Purkinje neurons, with each Purkinje neuron receiving inputs from 50,000 to 100,000 parallel fibers. Also contained within the cerebellar cortex are two inhibitory interneurons: basket cells, whose axon terminals form plexuses known as "pinceaux" around the axon hillock of Purkinje neurons, and stellate cells, which form inhibitory contacts on Purkinje cell dendrites. The third type of inhibitory cell in cerebellar cortex is Golgi cells located in the granular layer, with a major projection of inhibitory synapses to synaptic glomeruli. The sole output of the cerebellar cortex is the Purkinje cell axon. The axon descends through the white matter and uses the neurotransmitter γ-aminobutyric acid (GABA) to form inhibitory synapses onto DCN cells (spinocerebellum, cerebrocerebellum) or onto neurons of the lateral vestibular nuclei (vestibulocerebellum). The DCN cells are embedded within the cerebellar white matter and provide the major output from cerebellum (29–31). Interestingly, DCN cells must convert a massive inhibitory projection from multiple Purkinje cells to an excitatory output to other motor nuclei. This translation is accomplished through the expression of distinct intrinsic biophysical membrane properties that provide the capacity to generate a rebound depolarization and increased spike firing following an inhibitory stimulus (32–34). The activity patterns of Purkinje cells are taken to be critical to theories of cerebellar function, as a change in Purkinje cell firing frequency, whether it is a transient elevation in spike frequency or a pause in firing, will impact how DCN cells respond to cortical input (7, 35–38).

2. Materials

2.1. Animals

Sprague–Dawley rats were purchased from Charles River Laboratories (Charles River, Quebec, Canada) as rat dams with litters at an age of postnatal day (P) 10, or timed-pregnant to arrive during late gestation. Rats were maintained in standard colony conditions within the Animal Resources Centre. All procedures were completed in accordance with the guidelines established by the Canadian Council on Animal Care.

2.2. Solutions

Chemicals were obtained from Sigma (St. Louis, MO; www.sigmaaldrich.com) unless otherwise stated. The composition of artificial cerebrospinal fluid (aCSF) was (in mM) 125 NaCl, 3.25 KCl, 1.5 $CaCl_2$, 1.5 $MgCl_2$, 25 $NaHCO_3$, and 25 d-glucose, pH adjusted to 7.4 upon gassing with carbogen (95% O_2, 5% CO_2 (see Sect. 3.2.2 for rationale)). On-cell (cell-attached) recordings were obtained using patch electrodes containing either a 4-(2-hydroxyethyl)-1-piperazine-ethanesulfonic acid (HEPES) pH-buffered medium similar to aCSF (in mM) (150 NaCl, 3.25 KCl, 1.5 $CaCl_2$, 1.5 $MgCl_2$, 10 HEPES, and 25 D-glucose, pH 7.4) or the K-gluconate-based internal solution below. The patch-recording electrolyte for whole-cell current-clamp membrane potential (V_m) recordings consisted of (in mM) 130 K-gluconate, 0.1 ethylene glycol tetraacetic acid (EGTA), 10 HEPES, 7 NaCl, 0.3 $MgCl_2$ and pH = 7.3 upon adjustment with 1 N KOH. 5 mM di-tris-creatine phosphate, 2 mM Tris-ATP, and 0.5 mM Na-GTP were added daily to the recording electrolyte from frozen stock solutions. Patch electrode solution for voltage-clamp recordings of T-type Ca^{2+} channel current consisted of (mM) 100 CsCl, 10 KCl, 10 EGTA, 7 NaCl, 0.3 $MgCl_2$, and 10 HEPES, with tetrodotoxin (TTX, 200 nM), CsCl (1–2 mM), and tetraethylammonium (TEA, 5 mM) in the superfusate (no junction potential subtracted in voltage-clamp recordings). Neurobiotin (0.1%; Vector Laboratories, Burlingame, CA, USA; www.vectorlabs.com) was added to the patch pipette solution in some experiments to enable visualization of cell morphology during subsequent histological processing.

2.3. Equipment

Brain slices were prepared using a vibratome (Ted Pella, Redding, CA, USA; www.tedpella.com) or a VT1000S (Leica Biosystems, Nussloch, Germany; www.leica-microsystems.com). Visualization of brain slices was performed with a Axioskop FS-2 microscope (Zeiss, Oberkochen, Germany; www.zeiss.com/smt) equipped with infrared-differential interference contrast (IR-DIC) optics and a Newvicon camera (Dage MTI, Michigan City, IN, USA; www.dagemti.com) linked to a video monitor (Hitachi VM9012U). Recordings were made with a variety of amplifiers available from Axon Instruments (Molecular Devices Inc., Sunnyvale, CA, USA; www.moleculardevices.com) including Multiclamp 700 series amplifiers, Axoclamp 2 series amplifiers, and Axopatch 200 series amplifiers. pCLAMP software (versions 8 through 10; Molecular Devices) was used for data collection, and some data analysis, with select analyses, was completed in MATLAB (Mathworks, Natick, MA, USA; www.mathworks.com). Pipettes were pulled from thick-walled (fiber-filled) borosilicate glass (1.5 mm O.D.; A-M Systems, Sequim, WA, USA; www.a-msystems.com) with a Sutter P-87 puller (Sutter Instrument Company, Novato, CA, USA; www.sutter.com). Electrodes had DC resistances in aCSF of 4–8 MΩ for somatic recordings and 8–12 MΩ for dendritic recordings. A bipolar concentric stimulating electrode (Frederick Haer, Bowdoin, ME, USA; www.fh-co.com) was used to evoke synaptic responses using stimuli delivered with an isolated stimulator (Model DS2, Digitimer Ltd, Welwyn Garden City, England; www.digitimer.com). In some experiments, the stimulator was controlled by a Master-8 (A.M.P.I., Jerusalem, Israel; www.ampi.co.il). Confocal imaging was done using a FV300 BX50 confocal microscope with FluoView software (Olympus, Center Valley, PA, USA; www.olympusamerica.com).

3. Methods

3.1. Generation of Cerebellar Slices and Recordings from Cerebellar Neurons

Detailed descriptions of our methods can be found in the previous chapters (23, 26, 39–41) but are summarized here.

3.1.1. Preparation of Tissue Slices

Rats were anesthetized with a single subcutaneous injection of sodium pentobarbital (65 mg/kg; MTC Pharmaceuticals, Joliet, IL, USA; www.mtcresearch.com). After establishing that a sufficient depth of anesthesia was reached, with no overt responses to tail or foot pinch, the rat was decapitated, the posterior skull removed, and the cerebellum dissected out and immediately bathed in ice-cold aCSF (Sect. 2.2). For Purkinje cell recordings, the cerebellum was blocked in the sagittal plane by removing one cerebellar hemisphere with a scalpel cut and then mounted with cyanoacrylate glue on an ice-cold vibratome cutting tray (Sect. 2.3). Parasagittal tissue slices of 300-μm thickness were obtained from the cerebellar vermis in carbogenated ice-cold aCSF. Parasagittal tissue sections from the cerebellar vermis were used exclusively as slices in this plane are more likely to maintain the flat dendritic trees of Purkinje cells. For DCN cell recordings, either parasagittal or coronal sections (each of 200-μm thickness) were prepared. We have not found it necessary during slice preparation to use aCSF containing high sucrose or glutamate receptor antagonists to obtain high-quality slices (compare Chaps. 1-Trapp and Ballanyi (42), 3-Moore et al. (43), and 8-Sanchez-Vives (44)). Following cutting, slices were incubated at 35°C for 30–45 min and subsequently maintained at room temperature (~22°C) for no more than 4–5 h.

3.1.2. Electrophysiology

Selection of Cells

Slices were transferred to a heated (35°C) recording chamber on the stage of the Zeiss Axioskop FS-2 microscope (Sect. 2.3) and kept in place with a nylon-strung platinum harp (see also Chap. 1-Trapp and Ballanyi (42)). The tissue was continuously perfused with aCSF at ~2–3 ml/min as a submerged preparation. Cells were visualized with IR-DIC optics and a camera linked to a video monitor (Sect. 2.3). Juvenile Purkinje cells from P0 to ~ P9 rats were identified by their comparatively larger somata relative to neighboring cell types and their intermediate position between the internal and external granule cell layers. Adult Purkinje cells were identified by their characteristic morphology and position within the cerebellar cortex. To ensure a diverse sample of Purkinje cells, recordings were not restricted to specific lobules or regions within lobules. Unless otherwise noted, all DCN cell recordings were obtained from the interpositus nucleus. The spontaneous behavior of Purkinje cells and DCN neurons was examined in the absence of holding current. The occurrence of rebound bursts in DCN cells was defined as including all spikes following an inhibitory stimulus that exceeded a rate of more than two standard deviations beyond the mean tonic firing rate for 1 s preceding the stimulus (23). Peak rebound frequencies were determined for the first five spikes immediately following a hyperpolarization and plotted as the increase above the preceding tonic firing level.

Whole-Cell Recordings

Whole-cell current-clamp data were acquired at a sampling rate of 25 kHz and low-pass-filtered at 10 kHz, whereas voltage-clamp data were acquired at 10 kHz and low-pass-filtered at 2–5 kHz. Resting V_m of Purkinje cells was maintained at −70 mV with hyperpolarizing current for most experiments. Recordings were accepted for use only for initial electrical "giga-seal" formations of 1 GΩ and were immediately rejected, if series resistance after establishing the whole-cell configuration was >15 MΩ. After whole-cell "break-in," cells were rejected, if the average V_m during spontaneous activity was more depolarized than −45 mV. Series resistance was compensated with the bridge balance circuitry of the intracellular amplifier (Sect. 2.3). Outside-out patches were rejected, if they did not have a resistance of >1 GΩ as determined by a small voltage step during patch formation, or if current levels were unstable during the first 30 s of recording. Outside-out patch recordings were rejected post hoc, if they had a resistance of <1 GΩ (as determined by a small voltage step) at the end of the experiment, or if rundown of current was evident.

3.1.3. Extracellular Synaptic Stimulation

Activation of Parallel Fiber and Climbing Fiber Synapses on Purkinje Cells

Parallel fiber and climbing fiber responses were pharmacologically isolated with a mixture of the antagonist of NMDA-type ionotropic glutamate receptors (2R)-amino-5-phosphonovaleric acid (AP5, 25 μM), the $GABA_A$ receptor antagonist picrotoxin (50 μM), and the $GABA_B$ receptor antagonist CGP 55845 (1 μM). Parallel fiber responses were evoked with the stimulating electrode (Sect. 2.3), positioned in the mid-molecular layer ~60 μm from the Purkinje cell soma, and were identified by their graded response to stimulus intensity (45). Climbing fiber responses were evoked with the stimulating electrode positioned in the granular layer ~40 μm from the Purkinje cell soma and were identified by their all-or-none response to a threshold amount of stimulation (45).

Activation of Purkinje Cell Synapses on DCN Neurons

$GABA_A$ergic inhibitory synaptic inputs to DCN cells were isolated pharmacologically with the ionotropic glutamate receptor antagonist DNQX (10 μM), AP5 (25 μM) and CGP 55845 (1 μM). Inhibitory postsynaptic potentials (IPSPs) were evoked by activating Purkinje cell afferents to DCN cells by stimulating just dorsal to the recording site and outside of the interpositus nucleus itself. Synaptic responses reversed polarity between −70 and −75 mV, which approximated the calculated equilibrium potential for Cl^-ions (E_{Cl}) of −76 mV. For a few cells, stimulation resulted in antidromic activation of spike discharge. In these cases, the electrode was repositioned such that only IPSPs were evoked, and if this was not successful, the cell was discarded. The intensity of stimulation was adjusted such that a single stimulus was sufficient to block the generation of an action potential in these spontaneously discharging neurons.

3.1.4. Purkinje Cell Morphology and Imaging

To obtain morphological data on recorded Purkinje cells, whole-cell recordings were maintained for at least 10 min to ensure a complete wash-in of Neurobiotin via diffusion from the patch electrode solution (Sect. 2.2). Following recording, slices were transferred to 4% paraformaldehyde and fixed for several days at 4°C. Slices were "washed" in 0.1 M phosphate buffer solution (PBS) for several hours; then placed in a solution of PBS, Triton X-100 (0.1%), and streptavidin-Cy3 (1:1,500); and kept in the dark. Slices were gently agitated for several hours to ensure thorough exposure to streptavidin-Cy3 and then left overnight at 4°C. Slices were washed in PBS, mounted on poly-d-lysine gel-coated slides, coverslipped with antifade medium (90% glycerol/PBS/0.1% p-phenylenediamine; pH 10), sealed with nail polish, and stored at −20°C. Purkinje cells were imaged on a confocal microscope (Sect. 2.3) with laser intensity set high enough to resolve individual spines. Images were acquired with a step size (z-axis) of 0.5 μm at a resolution of 1,024 × 1,024 and Kalman filtered.

3.1.5. Dynamic Clamp

"Dynamic clamp" is a recording procedure that allows current associated with specific ion channels to be injected according to the relative position of V_m with respect to the reversal potential for that current. Dynamic clamp was performed using the Linux-based Real-Time eXperimental Interface (RTXI), which is a result of the merging of three different open-source dynamic clamp systems (46–50). Alternatives to RTXI include QuB, which is Windows-based and specialized for simulating kinetic ion channel models (51), whereas another implementation called "G-Clamp" uses Real-Time LabVIEW (52). Each of these dynamic clamp systems are in development, and the readers are encouraged to explore the documentation for each to determine which suits their needs best.

3.1.6. Analysis

Mean values are reported as mean ± S.E.M.

3.2. In Vitro Recording Conditions

One factor that makes it difficult to compare Purkinje and DCN cell activity between studies has been the successful application of an unusually wide range of in vitro techniques that include dissociated cells, organotypic slice cultures, acute slice preparations, and intact brainstem–cerebellar preparations (5, 9, 10, 12, 53–56). Each of these in vitro models draws upon different methodologies that must be considered when comparing activity. The most obvious of these would be the reduction in morphology for acutely dissociated cells (1) but also the potential for reverberatory activity in relatively intact circuits in the intact brainstem–cerebellar preparation (5). An increasingly recognized variable is the potential for at least subtle species-specific differences, with cerebellar recordings in vitro performed in guinea pigs, rats, or mice (1, 4, 5, 24, 26, 54, 55, 57, 58), while in vivo recordings are typically derived from cats, guinea pigs, rats, or mice (7, 18, 20, 34, 37, 59–61). As our experience is entirely in the area of acute slices from rat cerebellum, we will focus attention on this preparation.

We have found several factors that influence cell activity when nervous tissue is maintained as a slice preparation, resulting in spike output patterns that do not accurately reflect activity in vivo. For DCN cells, many of these effects can be found in relation to the activity of T-type Ca^{2+} channels or regulation of Ca^{2+}-dependent (K_{Ca}-type) K^+ channels by Ca^{2+} influx. T-type channels belong to the Cav3 family of low voltage-activated (LVA) Ca^{2+} channels, with three isoforms (Cav3.1, Cav3.2, Cav3.3) that are differentially expressed in Purkinje and DCN neurons (26, 62–64). These channels are important to the output of both cell types, with T-type channels helping govern subthreshold responses in Purkinje cells (64–66) and increases in the frequency of spike discharge during rebound bursts of spike output following periods of membrane hyperpolarization in DCN cells (27, 56, 57).

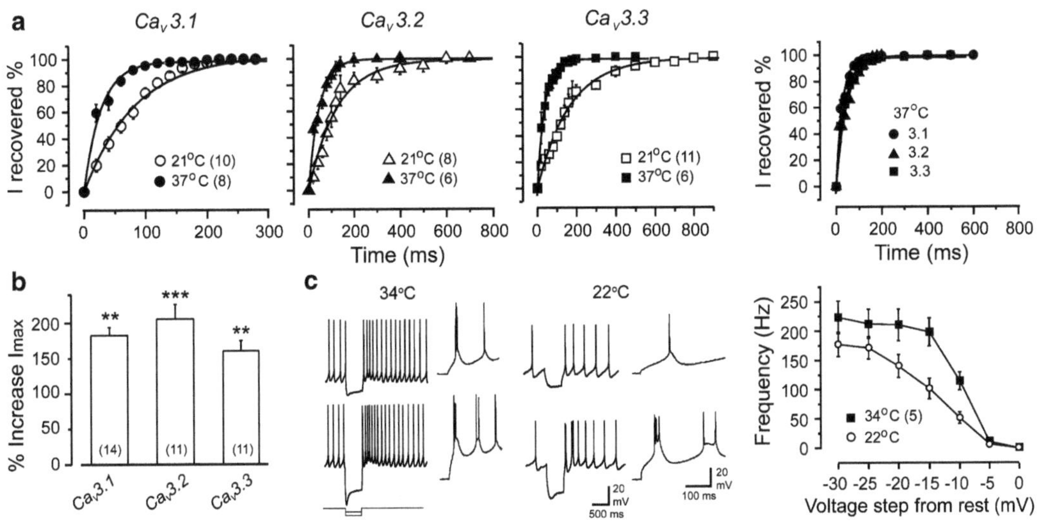

Fig. 1. Effects of recording conditions on expressed whole-cell-recorded Cav3 Ca^{2+} currents and spike output in "deep cerebellar nuclei" (DCN) cells in vitro. (**a**) A differential rate of recovery from inactivation between Cav3 channel isoforms present at room temperature (21°C) is absent at physiological temperatures in tsA-201 human embryonic kidney (HEK) cells (37°C records are superimposed at *right*). Recovery was tested using a 2-s conditioning pulse to −30 mV followed by a variable recovery period at −120 mV and subsequent test pulse to −30 mV. (**b**) The maximum Cav3 current evoked in tsA-201 cells increases by up to 210% of control at 37°C compared to 21°C. (**c**) Representative recordings of rebound bursts in a DCN "transient burst" cell following current-evoked hyperpolarizations to −75 mV and −90 mV (500-ms pulse duration). Both tonic and rebound frequencies are higher at 34°C compared to 22°C, with significant shifts in spike amplitude and burst duration at 22°C. *Insets* in (**c**) show traces of recordings immediately following the membrane hyperpolarization on an expanded time scale (Data in (**a, b**) were modified and reproduced with permission from (67)).

3.2.1. Temperature

A fundamental element that is often overlooked is the temperature, at which cells are recorded in brain slices. Voltage-clamp recordings are improved by working at room temperature, but this can mask intrinsic activity present under current-clamp conditions. For instance, studies conducted on Cav3 Ca^{2+} channels at room temperature in heterologous expression systems had identified differences in the kinetics between three Cav3 channel isoforms (67–69) that are often used to interpret their relative contributions to neuronal activity. For instance, the rate of recovery from inactivation differs between the three Cav3 isoforms when expressed in tsA-201 human embryonic kidney (HEK) cells, with Cav3.1 current exhibiting a much faster rate of recovery than Cav3.2 or Cav3.3 (Fig. 1a) (68). This is important, in that T-type Ca^{2+} channels are nearly fully inactivated near resting potential. The rate of recovery from inactivation is thus an important factor in determining the amount of T-type Ca^{2+} current available to generate a rebound depolarization in DCN cells following an inhibitory stimulus.

By comparing the properties of Cav3 currents recorded at room temperature and physiological temperature, we found that differences in the rate of recovery from inactivation were entirely absent at 37°C, with all isoforms exhibiting a fast and equivalent rate of recovery (Fig. 1a) (67). We also found that the maximal Cav3 current in the HEK cells was increased by up to 210% when temperature was raised from room to physiological temperature (Fig. 1b) as previously reported in neuronal recordings (70, 71). This increase in current would be predicted to affect the spontaneous subthreshold activity of Purkinje cells (65) and rebound burst intensity in DCN cells (27). In fact, lowering the temperature from 37°C to 21°C decreased the magnitude of rebound bursts in DCN cells over a range of prestep hyperpolarizations (Fig. 1c). The maximal rebound spike frequency attained immediately following a current-evoked hyperpolarization to –75 mV (equivalent to E_{Cl}) in "transient burst" DCN cells was reduced in half from 199 ± 23.8 Hz to 102 ± 17.9 Hz ($n = 5$) by shifting from 37°C to 21°C (Fig. 1c). The rate of tonic firing present at a resting V_m of –60 mV was also reduced from 13 ± 2.7 Hz to 6 ± 1.2 Hz ($n = 5$). Similarly, a slow "trimodal" oscillation often detected in rat Purkinje cells in vitro at physiological temperatures (10, 17) was reversibly blocked when temperature was shifted between physiological and room temperature (17) (B.E. McKay, unpublished observations). Maintaining a temperature between 33°C and 35°C is thus recommended for recording activity that reflects the output capability of a given cerebellar cell in vitro.

3.2.2. Reversal Potentials

A second important consideration is the composition of the electrolyte when using patch electrodes for whole-cell recordings. In this patch-recording configuration, there is a substantial, if not complete, ionic exchange between the cell (at least close to the soma) and electrode within minutes of forming low resistance access to its cytosol. This is important in that the concentration gradient for ions between the inside and outside of a cell is established entirely by the patch electrode electrolyte and aCSF, thereby defining the reversal potentials (E_{rev}) for the major ions Na^+, K^+, Ca^{2+}, and Cl^-. This is key to DCN cell recordings where the magnitude of $GABA_A$ receptor-mediated IPSPs depends on the value of E_{Cl}, which will define the extent to which inhibition can promote recovery from Cav3 channel inactivation that occurs near resting V_m.

To adjust the recording electrolyte, one must balance ionic concentrations with other electrode contents and still maintain a correct osmolarity and pH at physiological temperatures. After carefully considering all factors, we converged on a set of solutions that provide an acceptable range of E_{rev} for all ionic species. The composition of aCSF was (in mM) 125 NaCl, 3.25 KCl, 1.5 $CaCl_2$, 1.5 $MgCl_2$, 25 $NaHCO_3$, and 25 d-glucose, which adjusts to a pH of 7.4 upon gassing with carbogen (95% O_2, 5% CO_2) (Sect. 2.2). The patch-recording electrolyte for whole-cell current-clamp V_m recordings consisted of (in mM) 130 K-gluconate, 0.1 EGTA, 10

HEPES, 7 NaCl, 0.3 $MgCl_2$, and pH = 7.3 upon adjustment with 1 N KOH. 5 mM di-tris-creatine phosphate, 2 mM Tris-ATP, and 0.5 mM Na-GTP were added daily to the patch electrode electrolyte from frozen stock solutions (Sect. 2.2). With these solutions, the calculated E_{rev} at 35°C were E_{Na}: +55 mV, E_K: −97 mV, and E_{Cl}: −76 mV, with a theoretical junction potential of ~10 mV subtracted from all current-clamp recordings. We find that with these solutions the tonic firing rate and rebound frequency increases in cerebellar cells in vitro are equivalent whether recording in on-cell mode (when cell contents are undisturbed) or in whole-cell configuration (23).

3.2.3. Cytosolic Ca^{2+} Buffering

Dialysis of cells by the patch-recording electrolyte will also redefine the Ca^{2+} buffering capacity of a cell. The influence of including different concentrations of Ca^{2+} buffers (EGTA) in the patch electrolyte (Sect. 2.2) can be readily seen in recordings of cerebellar cells. We found that increasing the concentration of EGTA from 0.1–1.0 mM to 5–10 mM in whole-cell recordings from Purkinje cells increased repetitive Ca^{2+}-Na^+ spike bursts (40), a result that reflects a loss of the repolarizing influence of spike-activated afterhyperpolarizations (AHPs). We also found that using internal EGTA above 0.1 mM can affect the ability for Ca^{2+} influx to activate AHPs in DCN cells, causing cells to shift to a slow oscillatory swing of V_m and high-frequency discharge (M. L. Molineux, unpublished observations). Similar results were reported upon internal perfusion of another frequently used type of Ca^{2+} buffer, 1,2-bis(o-aminophenoxy)ethane-N,N,N′,N′-tetraacetic acid (BAPTA) (55), block of either N-type Ca^{2+} channels (72) or SK-type K^+ channels (55, 73), revealing a coupling between N-type Ca^{2+} influx and SK channels to generate a large portion of the AHP. The early convention of using 5 mM EGTA in patch-recording electrolyte then helps account for slow oscillatory swings of V_m reported in some of the first whole-cell patch recordings in DCN cells (74).

3.2.4. Water Storage

We have encountered another issue in preparing aCSF for tissue slices that proves to be critical to slice health and specifically to T-type Ca^{2+} currents and the responses to which these currents contribute. All of the water we use to prepare medium is generated by feeding distilled water through a Barnstead Nanopure Infinity unit (Series 8965) that is subsequently stored in a large volume Corning Pyrex aspirator bottle (i.e., Corning 1,220–5). With this arrangement, slices are consistently healthy and provide highly reproducible results. By comparison, we have on two separate occasions experienced difficulties with slices maintained in aCSF prepared from water that was stored for any period of time in a translucent rigid polypropylene carboy (i.e., Nalgene Rectangular Carboy, LDPE-4 or LPE polyethylene). This holds true for recordings from cerebellar Purkinje cells or DCN cells (Fig. 2) and for

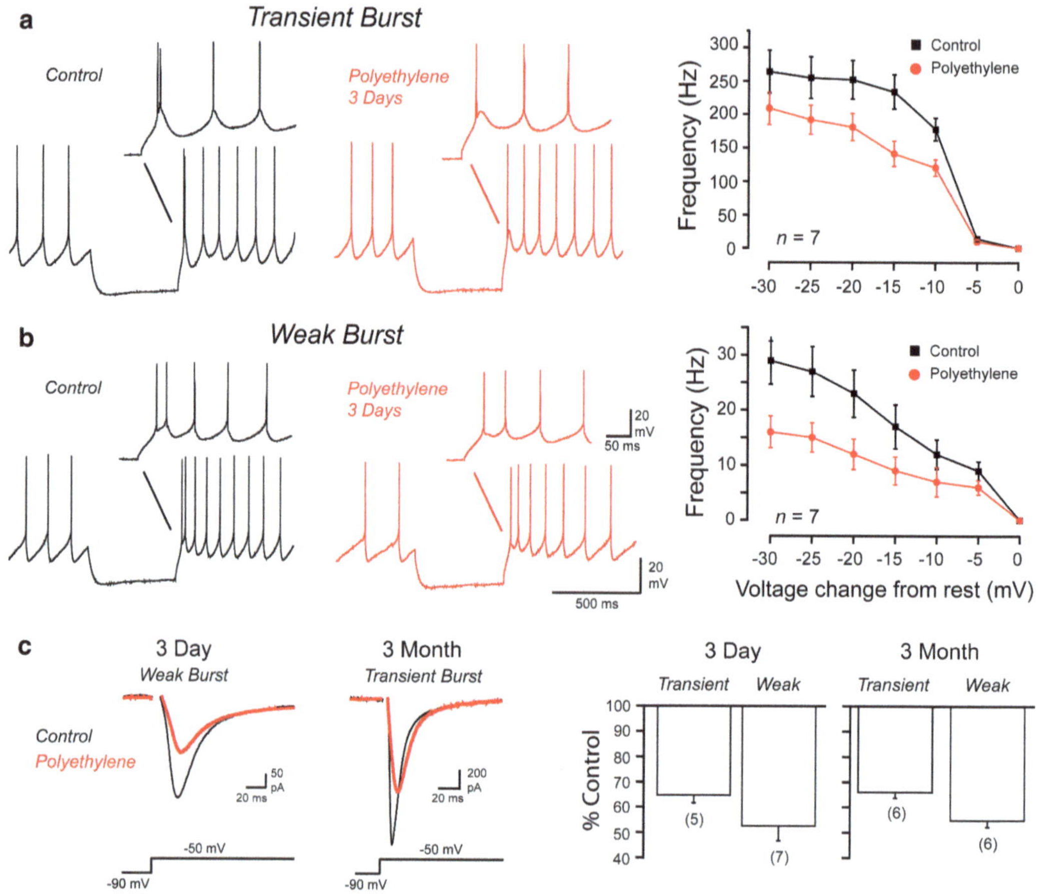

Fig. 2. Effects of storage of distilled water for slice superfusate on rebound frequencies and T-type Ca^{2+} currents in DCN cells. (**a**, **b**) Representative recordings and frequency plots for "transient burst" (**a**) and "weak burst" (**b**) cells in response to current-evoked hyperpolarizations (500-ms pulse duration) from a resting potential of −60 mV. The frequency of rebound bursts is reduced in both phenotypes by switching from artificial cerebrospinal fluid (aCSF) prepared from water stored in a Pyrex container to aCSF prepared from water stored for 3 days in a polyethylene carboy (translucent LDPE-4). Recordings of rebound responses were evoked from a prestep hyperpolarization to −80 mV. *Insets* show expanded views of rebound increases in frequency immediately following the hyperpolarization. On the *right* are plots of the mean values of rebound frequency above baseline firing rates following a range of current-evoked membrane hyperpolarizations for the two sources of water (frequency of discharge: control vs. polyethylene significant at $p < 0.05$ for most membrane voltages). (**c**) Voltage-clamp recordings of T-type Ca^{2+} currents evoked by voltage steps from −90 mV to −50 mV for DCN "weak burst" and "transient burst" cells. Note that perfusing aCSF containing water stored in a polyethylene carboy reduces T-type current peak amplitude by a similar amount for either burst phenotype whether stored for 3 days or for 3 months. The capacitance artifacts in LVA current recordings of **C** were digitally removed.

hippocampal CA_1 pyramidal cells (R. W. Turner, unpublished observations). We have not tracked down the exact nature of the problem, but it may reflect the release of the endocrine disruptor Bisphenol A or its breakdown products from plastic containers. Indeed, these compounds have been reported to affect N- and L-type Ca^{2+} channels (75, 76), Ca^{2+}-dependent CREB activation (77), and both K_{Ca} and ATP-sensitive K^+ (K_{ATP}) channels (78).

Here, we simply provide some tests to illustrate the magnitude of problems that can arise from the source of water used to prepare the superfusate.

Examples of rebound bursts recorded in a DCN 'Transient Burst' or 'Weak Burst' cell in aCSF prepared from water stored in a Pyrex container are shown in Fig. 2a, b. Conducting experiments in this medium consistently generated a rebound increase in spike frequency of ~250 Hz above baseline firing rates for transient burst cells (Fig. 2a) and ~30 Hz above baseline for weak burst cells (Fig. 2b) immediately following hyperpolarization through current injection to a V_m of −80 mV for 500 ms. But after 10 min exposure to water that had been stored in a polyethylene carboy for 3 days, the rebound burst frequency decreased (Fig. 2a, b), with no significant effect on input resistance. When tested from a hyperpolarized level −15 mV below resting V_m (i.e., near E_{Cl} = −76 mV), the rebound frequency of transient burst cells was reduced by ~35% (Fig. 2a) and in weak burst cells by ~50% (Fig. 2b). Attempts to reverse these effects by subsequent return to aCSF prepared from water stored in Pyrex container failed to allow recovery of rebound bursts ($n = 3$; not shown). Storage of water in a polyethylene carboy for 3 months produced the same relative results, with no change in the magnitude of decrease in rebound frequencies ($n = 11$; data not shown). By comparison, cell health (as defined by the ability to obtain long-lasting stable whole-cell recordings) and rebound burst capacity were very consistent for either burst phenotype if water was stored in a Pyrex container for 1 day or after months of storage.

The effects of polyethylene-stored water appear to reflect a change in the intrinsic excitability of DCN cells, as all recordings were conducted in the presence of AP5 (25 μM), DNQX (10 μM), and picrotoxin (50 μM) (Sect. 3.1.3). Given that previous work has shown that T-type Ca^{2+} currents are evoked in DCN cells subsequent to imposed membrane hyperpolarizations (26, 27, 56, 79), we extended these tests to Cav3 currents isolated under voltage-clamp. T-type Ca^{2+} currents were evoked using preconditioning voltage steps to −90 mV for 500 ms and then a step to −50 mV (Fig. 2c). We found that in aCSF prepared from water stored in a Pyrex container, T-type currents fell into two groups of 154 ± 44 pA ($n = 13$) and 1,064 ± 161 pA ($n = 11$), as reported for recordings from weak burst or transient burst cells, respectively (27). However, perfusion of aCSF prepared from water stored in a polyethylene carboy for either 3 days or 3 months led to ~35–50% reduction in T-type current within 10 min (Fig. 2c). As found for rebound frequency increases, the degree of reduction of Cav3 current by polyethylene-stored water was again greater for "weak burst" cells (Fig. 2b, c).

These results indicate the presence of a contaminant released by polyethylene storage containers (at least of the LPDE-4 and

LPE classes) that affects the availability of Cav3 currents. The effects appear to be relevant to both Cav3.1 and Cav3.3 Ca^{2+} channel isoforms, as these have been shown to associate with "transient burst" and "weak burst" phenotypes in DCN cells, respectively (26). Nevertheless, the effects of the contaminant are not restricted to T-type currents, as we first detected effects by polyethylene storage containers on the site for initiation of Na^{+} spike discharge in CA_1 hippocampal pyramidal cells (R. W. Turner, unpublished observations). The time of exposure necessary to produce these effects can be very short, as exposing freshly distilled water to transport in a polyethylene container for as little as 10 min was sufficient to significantly alter the physiology of hippocampal slices. We have also encountered significant effects of this water on the general health and ability to patch-clamp cerebellar Purkinje cells, although this was not systematically documented (J. D. T. Engbers, unpublished observations). The source/storage of water used to prepare superfusate for slice experiments can thus affect ion channels that contribute to spike output patterns, potentially affecting the reproducibility of results between laboratories.

4. Results

4.1. Reconciling In Vivo and In Vitro Activity of Purkinje and DCN Neurons

Purkinje cells in vivo are often reported to generate "simple spikes" (Sect. 1) at a constant rate of ~40 Hz that are punctuated by climbing fiber-evoked complex spikes at a rate of ~1 Hz (14, 80). In contrast, Purkinje cells recorded in acute cerebellar slices in vitro tend to generate simple (Na^{+}) spikes at substantially higher frequencies and spontaneously generate a "trimodal" pattern of output (10, 17). During trimodal output, Purkinje cells have three characteristic and sequential states that repeat in a cyclic manner: a period of Na^{+} spike output, a period of Ca^{2+}-Na^{+} burst output (e.g., Na^{+} spikes grouped into high-frequency bursts via an underlying dendritic Ca^{2+} spike), and finally a silent period. The occurrence of spontaneous Ca^{2+}-Na^{+} bursts in Purkinje cells in acute brain slices was described in the first in vitro recordings (9, 10) and has been widely studied over the past three decades as one of the normal patterns of spike output. However, the dissimilarity in in vivo versus in vitro Purkinje cell action potential discharge patterns has been taken by some investigators as evidence that Purkinje cells in acute brain slices are "unhealthy." We found that there are two very significant differences between the typical Purkinje cell that is recorded in vivo and one recorded in vitro. These differences are the developmental stage of the animal, in which Purkinje cells are studied, and the state of activity of climbing fiber synaptic input (40, 41). We discuss below the importance of these two factors to obtaining "in vivo-like" Purkinje cell patterns of action potential discharge in acute brain slices.

4.2. Young or Old Slices?

Purkinje cell recordings in vivo are typically conducted in adult animals, whereas in stark contrast, Purkinje cell recordings in vitro are typically obtained in brain slices prepared from animals that are at most only a few weeks old (typically <3 weeks). The choice of animal in both cases is based on the goal of cell identification. For in vivo electrophysiological recordings in adult animals, recording and stimulating electrodes are guided to specific types of neurons based on their location as predicted from a stereotaxic atlas relevant to mature animals. For in vitro electrophysiology, recording and stimulating electrodes are guided visually to the cell of interest, typically via video imaging using IR-DIC optics, with subsequent electrophysiological confirmation of neuronal identity (see also Chaps. 1-Trapp and Ballanyi (42), 3-Moore et al. (43), and 8-Sanchez-Vives (44)).

Because visual identification of neurons is central to targeting neurons of interest and to obtaining a high-quality seal for patch-clamp recording, factors that impair visualization of neurons must be minimized. A major factor here is the presence of glial cells and the development of perineuronal nets in the extracellular space that obscure the light path, which in rat cerebellum becomes increasingly prominent beyond ~ P17 (81). Purkinje cell bodies and dendrites are thus best visualized when tissue slices are obtained from rats younger than P17, even though it is possible to record from either region from much older animals (41). However, below P17 the Purkinje cell is undergoing a massive growth of its dendrites and a remodeling of expressed ion channels, which both contribute to a remarkable change in electrophysiological properties over the period of time when acute tissue slices are typically prepared.

4.3. Postnatal Development of Purkinje Cell Morphology and Electrophysiology

Cerebellar Purkinje cells are remarkable for the extent of their postnatal development (82, 83). At the time of parturition for rats and mice, Purkinje cells are small with only short multipolar perisomatic dendrites. They are poorly innervated and generate mundane spike output patterns. Yet after only a few postnatal weeks, Purkinje cells have grown immensely, attaining the elaborate "flat dendritic tree" structure renowned in these cells (Fig. 3). In the adult, Purkinje cells are contacted by an overwhelming number of synaptic inputs and generate a wide array of spike output patterns. Recognizing the earliest time point at which Purkinje cells attain the structure, intrinsic properties, and synaptic physiology representative of "mature" cells is key to balancing the benefits of studying spike output in tissue slices that are young enough to enable adequate visualization of cells. One method for determining such time points is described below.

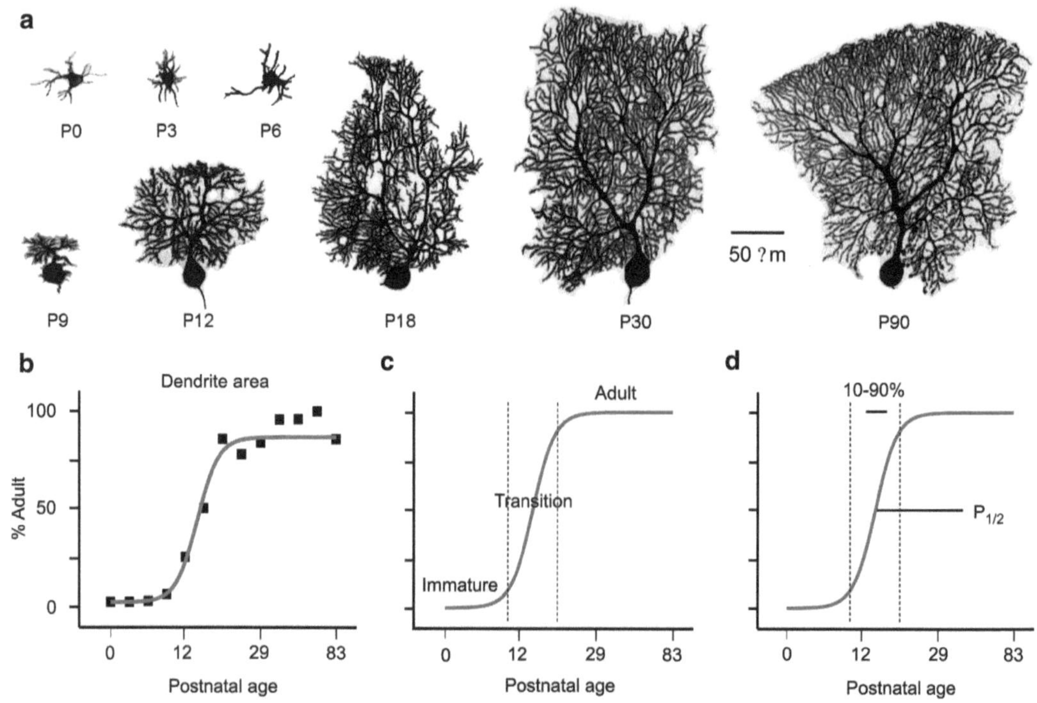

Fig. 3. Development of Purkinje cell morphology. (**a**) Two-dimensional projections of stacks of confocal images of rat Purkinje cells from postnatal day (P) 0 to P90 filled with Neurobiotin and detected with streptavidin-Cy3. (**b**) Time course of development for mean cross-sectional surface area of Purkinje cell dendrites normalized to adult values (age increments are P0, 3, 6, 9, 12, 15, 18, 23, 29, 35, 41, 53, 83) to illustrate the accuracy of a sigmoidal curve fit to the anatomical data. (**c**, **d**) Plot of percentage of adult-like character versus postnatal age for a fictive characteristic to define Purkinje cell developmental windows and fit parameters. (**c**) Postnatal age windows corresponding to immature, transition, and adult stages separated by *dashed lines*. (**d**) The sigmoidal curve fit is valuable for defining the postnatal age range during which a characteristic matures from 10% to 90% adult-like character and the postnatal age at which a characteristic reaches 50% adult-like character ($P_{1/2}$) (Panels (**a**) and (**b**) are adapted with permission from (41)).

4.3.1. Purkinje Cell Structural Development

We completed a study of the development of the rat Purkinje cell which revealed the extent of postnatal changes for a variety of Purkinje cell morphological and electrophysiological characteristics (41). To fully illustrate the developmental sequence and identify boundary points consistent with adult Purkinje cells, we applied simple curve fits to our data set. To illustrate, consider the growth of the dendritic tree presented in Fig. 3. Purkinje cells begin their postnatal development with a perisomatic dendritic configuration, with multiple dendrites extending <50 μm from the soma (P0 to ~ P9) (Fig. 3a). By P9, some Purkinje cells have completely retracted their perisomatic dendrites and developed a single primary stem dendrite (Fig. 3a). The principal dendrite increases markedly in size over the next ~1 week, rapidly reaching its adult configuration (Fig. 3a). When the amount of adult-like character of dendritic structure (e.g., percentage of terminal dendritic area) is plotted against postnatal age, it is apparent that Purkinje cell development can be divided into 3 principal stages: an immature stage, a transition stage, and an adult stage (Fig. 3b, c).

4.3.2. Purkinje Cell Electrophysiological Development

The exponential growth of the Purkinje cell dendrite is extremely important to deciding the appropriate age of rats from which slices should be prepared. As Fig. 3 clearly illustrates, an exponential growth of the dendrite begins at ~ P12 and persists until ~ P18, after which there is an approach to a plateau in further growth. This information must be considered together with the fact that development of perineuronal nets and glial investment in the cerebellum occurs at ~ P17 (81). Thus, to optimize the visualization of neurons, many investigators choose to study Purkinje cells prior to ~ P17. The extent to which this choice affects slice recordings is emphasized by the finding that changes in Purkinje cell morphology were paralleled by a shift in the intrinsic membrane properties as well as patterns of evoked and spontaneous spike output (41). Throughout the stable immature stage (≤P9), intracellular depolarizing current pulses evoked a transient low-threshold burst discharge of action potentials followed by low-frequency action potentials (Fig. 4a). However, during the transition stage (P15), Purkinje cells progressed through an underdeveloped form of repetitive high-threshold bursts (Fig. 4a). The pattern of output ultimately matured to a stable adult pattern (P60) that consisted of a series of spikes that gradually inactivated and that lead into a period of repetitive high-threshold Ca^{2+}-Na^{+} burst output (repetitive somatic Na^{+} spikes terminated by a dendritic Ca^{2+} spike) (Fig. 4a).

The Na^{+} spike and Ca^{2+}-Na^{+} burst characteristics within these spike phenotypes were quantified and revealed a similar 3-stage developmental sequence (Fig. 4) (41). We found that the Na^{+} spike waveform began as a broad and low amplitude event with no discernible fast AHP and matured to a large amplitude spike with an extremely narrow half-width and robust fast AHP by ~ P18 (Fig. 4b–d) (41). The maturation of Ca^{2+}-Na^{+} burst output characteristics was equally impressive, first appearing in select P12 cells coincident with the presence of a significant dendritic structure and decrease in input resistance (Figs. 4c, d and 5a–c) (41). With increasing age, the duration of Ca^{2+}-Na^{+} bursts increased, and the Na^{+} spikes within bursts were generated at higher frequencies and in greater numbers (Figs. 4c–e and 5c) (41). In a similar manner, the number of cells that exhibited spontaneous trimodal activity at physiological temperatures increased exponentially from ~ P12 through P18 and thus at a rate that paralleled the development of dendritic structure and changes in spike output (Fig. 4f).

Given the important relationship between dendritic surface area, cell input resistance, distribution of ion channels, and the synapses associated with an ever expanding dendrite, this time period is associated with marked variability in spike properties in general and overall an "immature" phenotype of spike output characteristics. Although such rapid change in cell structure and spike output properties presents an interesting window of opportunity to address specific neurobiological questions, it is not suitable for obtaining

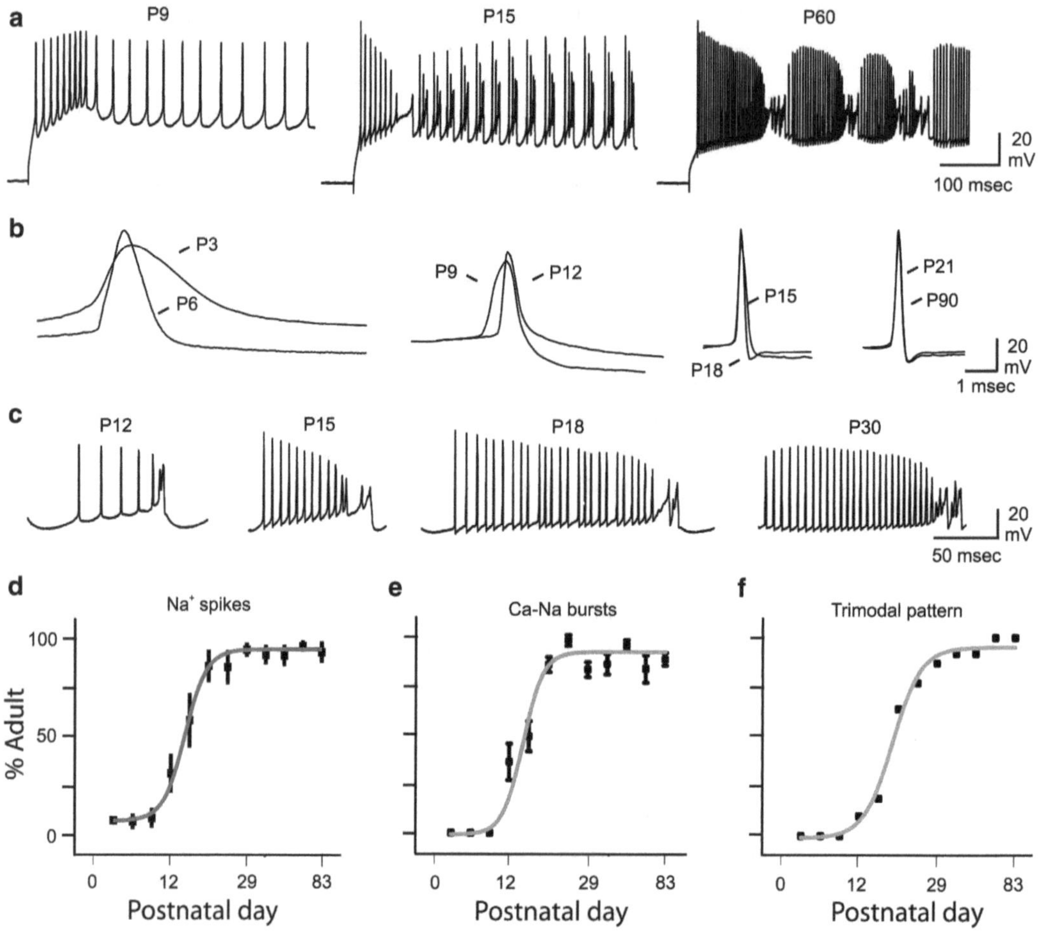

Fig. 4. Development of Purkinje cell electrophysiological characteristics. (**a**) Representative examples of Purkinje cell spike output in response to square wave current injections during the immature stage (P9), transition stage (P15), and adult stage (P60). (**b**) Representative single Na^+ spikes from different ages superimposed on the same time and voltage scales. (**c**) Representative single Ca^{2+}-Na^+ bursts from different ages evoked during square wave current pulses. (**d–f**) Time courses of development for Na^+ spike characteristics. (**d**) Ca^{2+}-Na^+ burst characteristics (**e**) and the incidence of expression of a trimodal pattern of spontaneous output (**f**) (age increments as in Fig. 4b). The plot in (**d**) is a composite analysis drawn from data for the threshold for spike discharge, spike frequency at threshold, rate of repolarization, and AHP area. The plot in (**e**) is a composite analysis drawn from data for burst duration, Na^+ spike frequency within bursts, and the number of spikes per burst (Adapted with permission from (41)).

"adult-like" or "in vivo-like" spike output from Purkinje cells. The differences in spike output patterns of immature and mature Purkinje cells in fact help reconcile important differences between Purkinje cells studied in a variety of acute and chronic preparations. For instance, Purkinje cells that are extracted from late-embryonic or early postnatal animals and then maintained for long periods in culture have a low-threshold bursting pattern of spike output similar to that seen in Purkinje cells in acute slices prepared from < P9 animals (41, 66). In contrast, the high-threshold bursting pattern is noted only in Purkinje cells measured in acute slices extracted

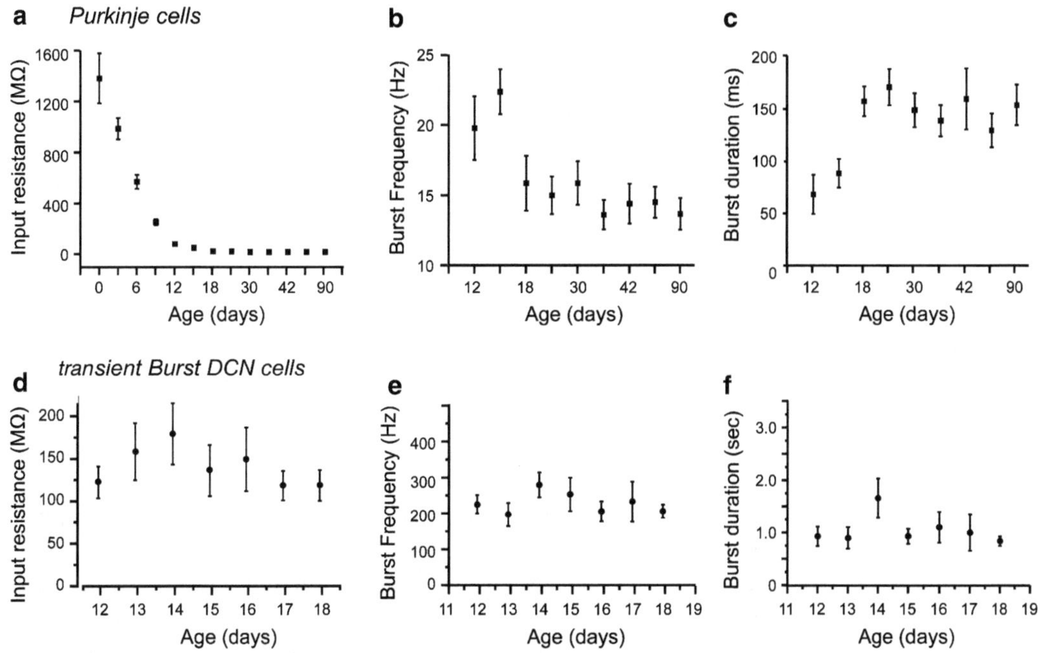

Fig. 5. Development of intrinsic membrane and burst output properties in Purkinje and DCN cells. (**a**) Purkinje cell input resistance decreases dramatically between P0 to P15 to reach adult levels by P18. (**b**, **c**) The frequency of Na^+ spikes (**b**) and the duration (**c**) of Ca^{2+}-Na^+ bursts in Purkinje cells change to adult levels over the time frame of P12-P25. (**d–f**) Transient burst DCN cells do not show an identifiable trend in input resistance or spike output over the time window of P12–P18. (**d**) Input resistance showed no identifiable trend over time, and neither the frequency (**e**) nor duration (**f**) of rebound bursts changed through the developmental window typically used for patch-clamp recordings. Rebound bursts in (**e, f**) were evoked from a resting potential of −60 mV with a current pulse injection to −80 mV (500 ms pulse duration). Plots in (**d–f**) indicate the maximum frequency and total duration of rebound firing that statistically exceed the mean tonic firing rate preceding the stimulus. Cells per data point for **a**, n=7–31; **b**, n=7–15; **c**, n=12–23; **d**, n=7–22; and **e,f**, n=3–13 (Data in (**a–c**) adapted with permission from (41)).

from rats or mice that are at least a few weeks old (17, 41). As indicated above, the trimodal pattern becomes increasingly apparent with age at physiological temperatures (Fig. 4f), while Purkinje cells will convert from a trimodal pattern to tonic firing by simply shifting to room temperature, further increasing the potential for different results between laboratories.

4.4. Maturation of DCN Cells

In contrast to the extensive postnatal development of Purkinje cell structure and characteristics of action potential output, DCN cells appear to exhibit adult-like electrophysiological properties at least by the end of the second postnatal week (84, 85). For instance, Purkinje cells exhibit a substantial reduction in input resistance up to P18, a parameter that likely reflects the parallel increase in membrane surface area associated with growth of the dendritic tree (Figs. 3b and 5a). By comparison, the mean input resistance in DCN cells over this critical window shows no apparent or significant pattern of change (Fig. 5d). A similar result emerges when one compares high-threshold burst output in these two cerebellar cell types. For Purkinje cells, there was a developmental progression toward a lower frequency of Ca^{2+}-Na^{+} burst output in combination with a longer duration of individual spike bursts (Fig. 5b, c). In comparison, the frequency and duration of spike output during rebound burst discharge that incorporates Ca^{2+} channel activation (as evoked following hyperpolarizing current pulses) (24, 86) did not vary across the postnatal developmental time period examined (Fig. 5e,f).

4.5. Control of Purkinje Cell Discharge by Climbing Fiber Inputs In Vitro

Recognizing the remarkable intrinsic structural and electrophysiological changes that occur in Purkinje cells addresses one part of the discrepancy between the characteristics of action potential output that can be measured in vivo versus in vitro. Synaptic input is the second major consideration, and there are two parts to this argument. The first part is to consider that synaptic inputs to Purkinje cells also undergo substantial postnatal development. The second part is to appreciate that in acute brain slices the climbing fiber, which is the most significant synaptic input to the Purkinje cell (from the perspective of injected current), is electrically silent in the brain slice unless artificially activated. When climbing fibers first begin "wiring up" onto Purkinje cells, there are in fact multiple innervations – upward of three or four climbing fibers synapse onto each Purkinje cell. When activated, these fibers trigger their typical all-or-none postsynaptic response, with quantized amplitudes equal to the number of individual climbing fibers contacting the cell (87, 88). Supernumerary climbing fibers are eventually lost, leaving adult-like innervation of one climbing fiber per Purkinje cell by ~ P14. Once mature, a single activated climbing fiber results in a several nanoampere current injection into the Purkinje cell, resulting in a powerful depolarization and complex spike response. The conditions that result in a climbing fiber discharge in vivo are debated, but it has been shown that climbing fibers discharge in vivo at ~1 Hz (80). Importantly, we found that repetitive activation of the climbing fiber synapse in electrophysiologically "mature" Purkinje cells (e.g., about 4 weeks old) at ~1 Hz in vitro restored in vivo-like patterns of action potential discharge (Fig. 6) (40).

4.6. Normalizing Purkinje Cell Output

Reintroducing climbing fiber input blocked trimodal output within ~10 stimuli and reduced Na^{+} spike frequency to a new stable and lower baseline of tonic firing (Fig. 6b–h, i). In essence, reintroducing climbing fiber discharge to the Purkinje cell normalized its output to one resembling that expected in in vivo recordings. Instead of the slow ramping up in spike frequency noted in Fig. 6h (TE, TL, and B), spike frequency was now stable at a lower frequency than it otherwise was in the spontaneous trimodal pattern (Fig. 6h, labeled CF). In support of the physiological relevance of these results, it was shown that blocking climbing fiber discharge in rats in vivo via inactivation of the inferior olive uncovered an oscillatory "in vitro-like" trimodal behavior of Purkinje cells (18). Moreover, the reduction of Na^{+} spike firing by climbing fiber input unraveled

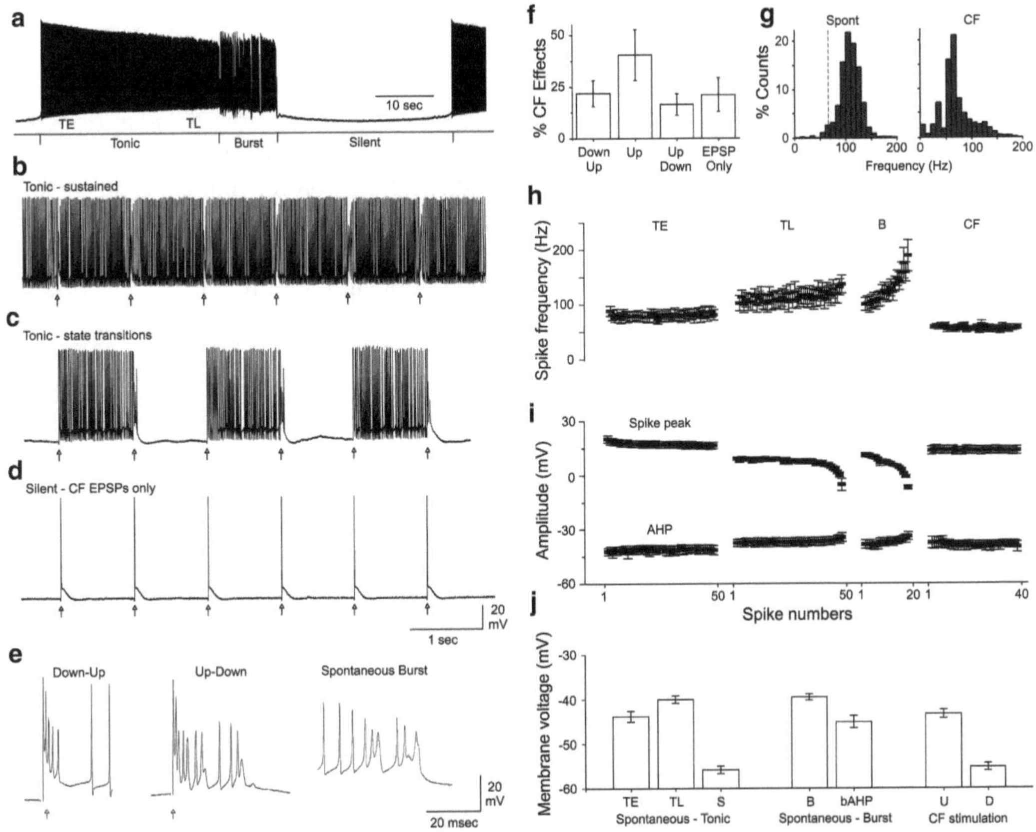

Fig. 6. Climbing fiber discharge determines pattern of spontaneous output, restoring in vivo-like patterns of Purkinje cell activity in cerebellar slices. (**a**) The trimodal pattern consists of a Na^+ spike phase (*tonic*), a Ca^{2+}-Na^+ burst phase (*burst*), and a quiescent phase (*silent* (S)). The tonic phase is subdivided into tonic-early (TE) and tonic-late (TL). (**b–d**) Electrically evoked climbing fiber discharge at 1 Hz (*arrows*) abolishes the burst phase of trimodal output. During the tonic phase, climbing fiber discharge results in either sustained sodium spike output (**b**) or periodic transitions between "up" and "down" states (**c**). During the silent phase, climbing fiber discharge activates complex spike EPSPs only (**d**). (**e**) Expanded views of spike output during a climbing fiber-evoked downstate to upstate transition (*left panel*) and an upstate to downstate transition (*middle panel*). The end of a spontaneous Ca^{2+}-Na^+ burst is shown for comparison (*right panel*). (**f**) Plots of the incidence of the four responses to climbing fiber discharge. (**g**) Na^+ spike frequency histograms for spontaneous output (*left panel*) and during climbing fiber discharge at 1 Hz (*right panel*) for a representative Purkinje cell. The *dashed line* in the spontaneous histogram corresponds to the peak frequency count of the climbing fiber histogram. (**h, i**) Average spike frequency (**h**) and peak Na^+ spike voltage (**i**; *upper traces*) and AHP troughs (**i**; *lower traces*) during trimodal activity and after initiating CF discharges. (**j**) Average membrane voltages during TE, TL, and S phases of trimodal output, spontaneous Ca^{2+}-Na^+ bursts (B) and burst AHPs (bAHPs), and during CF-evoked "up" (U) and "down" (D) states. (**a–e**) and (**g**) are from the same Purkinje cell. Purkinje cells per data point for (**f, h–j**), $n = 10$ (Adapted from (40)).

the ability for strong depolarizing inputs to toggle a Purkinje cell between "up" and "down" states (40, 89, 90), as noted by some investigators in vivo (19, 37).

Importantly, we found that all of these effects could be reproduced by selectively activating the postsynaptic membrane through repetitive intracellular injection of current designed to simulate the climbing fiber excitatory postsynaptic current (EPSC) (40). These tests by us demonstrated that the influence of climbing fiber inputs on Purkinje cell firing can be achieved in the absence of any synaptic inhibitory influence (40). These data provide further evidence that the patterns of spike output measured from Purkinje cells in vitro are not necessarily due to health considerations but rather due to the absence of climbing fiber input that proves to exert a rapid and substantial net inhibitory influence on intrinsically generated trimodal activity.

4.7. Parallel Fiber Gain

An important additional effect of climbing fiber input was that the resulting lower frequency of Na^+ spike output was associated with a markedly improved gain of parallel fiber inputs in vitro (Fig. 7) (40, 89), as had been reported in vivo (91–93). Thus, a lower Na^+ spike frequency during low-frequency climbing fiber activation allowed small parallel fiber synaptic inputs to evoke a substantially greater relative increase in firing frequency than when applied to a

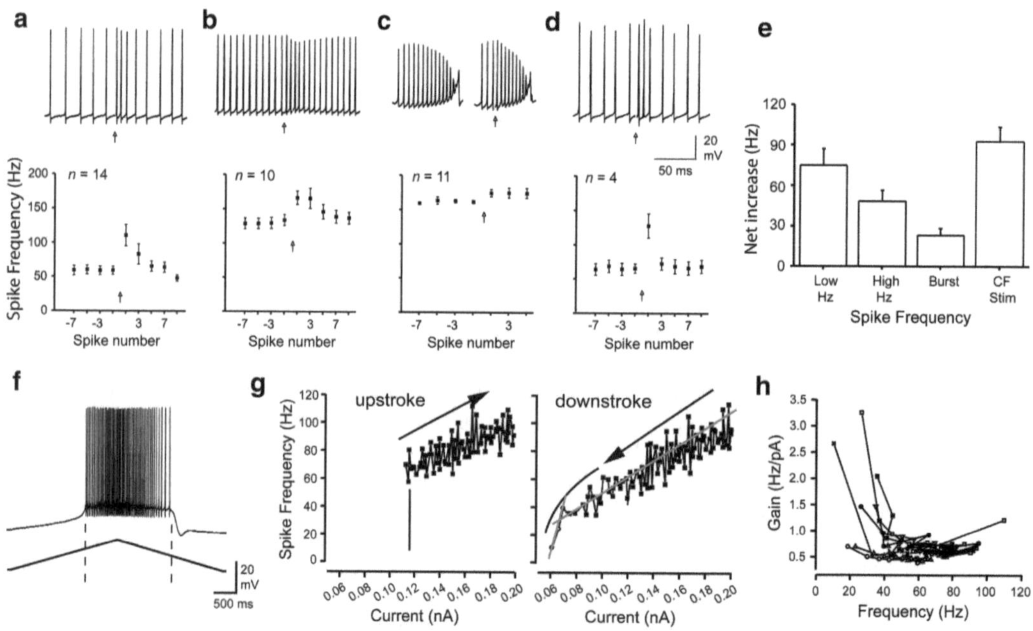

Fig. 7. Enhancement of Purkinje cell gain and readout of parallel fiber inputs at low spike frequencies. (**a**–**d**) Representative recordings and plots of spike frequency when parallel fiber stimuli (*arrows*) are delivered during low frequency (**a**: ~60 Hz), high frequency (**b**: ~130 Hz), burst (**c**: ~160 Hz before and after parallel fiber stimulation), and during climbing fiber-normalized spike output (**d**). Spikes preceding and following parallel fiber stimuli are assigned numbers <0 and >0, respectively. (**e**) Bar plots of the mean increase in parallel fiber-evoked output during different frequencies of spontaneous firing (as for data in **a**–**e**). (**f**–**h**) The increase in parallel fiber gain is due to positive feedback following a spike, consistent with a saddle homoclinic bifurcation during the transition from firing to rest. (**f**) A triangular ramp depolarizing current injection from −75 mV (4 s, 100 pA/s) reveals hysteresis of spike firing, with prolonged firing to lower frequencies on the falling phase of the ramp (compare *dashed lines*). (**g**) Plots of spike frequency in a representative recording illustrating a continuous F-I relationship only on the downstroke of the ramp, with a high initial threshold for spike discharge on the upstroke (*vertical line*) and lower spike frequencies during the downstroke associated with a region of high gain (compare slope of *gray lines*). *Black lines* illustrate the changes in firing frequency as current injection increases (*left panel*) or decreases (*right panel*). (**h**) Plots of instantaneous gain as a function of firing frequency for seven different Purkinje cells determined using the ramp protocol in (**f**), revealing the sharp increase in gain at lower frequencies of firing (Adapted from (40) and (89)).

cell exhibiting higher-frequency Na^+ spike discharge or trimodal activity (Fig. 7a–e). Once again, these effects could be entirely reproduced by exclusively activating the postsynaptic membrane through current injections that only simulated parallel fiber or climbing fiber EPSCs (40). An analysis of the dynamics of Purkinje cell firing that could account for this activity revealed that Purkinje cells undergo a saddle-node bifurcation during the transition between rest and firing states and a saddle homoclinic bifurcation from firing to rest (89). This finding is supported by a hysteresis of spike firing in Purkinje cells, a property revealed during ramp depolarizing current injections as prolonged low-frequency firing only during the downstroke of the ramp (10, 89) (Fig. 7f, g). As noted above, Purkinje cells exhibit membrane bistability after climbing fibers normalize Na^+ spike firing to a lower stable level, evident as bidirectional climbing fiber-induced transitions between rest and firing states (Fig. 6c). This bistable region proves to be associated with an extremely high gain in instantaneous spike output, predicted to be due to positive feedback (i.e., a depolarizing afterpotential) following an action potential (Fig. 7g, h). Thus, the actions of climbing fiber activation on bistability and parallel fiber gain could be traced to intrinsic membrane properties that can be defined using models of the dynamics of spike firing (89).

4.8. Purkinje Cell Properties Versus Behavioral Development

To fully anchor the importance of preparing acute brain slices from rats that will yield "adult-like" spike output characteristics, it has merit to compare the time windows for developmental aspects identified above with those of cerebellar-dependent behaviors – the ultimate readout of "adult-like" cerebellar function. The adult cerebellum underlies the control of posture and balance, the coordination of limb positions, and the learning of some conditioned responses. Although many parts of the brain are expected to contribute to even the simplest behaviors, several behaviors that are significantly dependent on cerebellar function have been identified by means of focal X-irradiation to the cerebellum. These include postural control (e.g., as measured via pelvis elevation during the quadruped stance), balance control (e.g., noted during rearing without forelimb support), coordination of limb position (e.g., observed during the mid-air righting response and by measuring the number of falls from a narrow plank due to foot misplacements), and conditioned eyeblink learning (82, 94). The development of cerebellar-dependent behaviors proves to follow the 3-stage sequence identified above for cellular metrics of cerebellar maturation. The progression of behavioral maturation through the 3 stages is also right-shifted by several days relative to the maturation of cellular parameters, ostensibly suggesting that a critical amount of maturation in cerebellar structure and function must occur prior to the generation of mature cerebellar-dependent behaviors. Of the behaviors noted above, conditioned eyeblink learning appears to define the endpoint of behavioral maturation of the cerebellum. Specifically, it has been shown in rats that at P17 the eyeblink response cannot be conditioned, at P20 the acquisition of conditioning occurs at an intermediate rate, and at P24 the rate of acquisition appears mature (95). The maturation of the cerebellar-dependent conditioned eyeblink response toward the end of the fourth postnatal week in rats also corresponds nicely to the maturation identified above for cellular parameters, underscoring the importance of preparing cerebellar tissue slices from "behaviorally adult-like rats" near the fourth postnatal week.

In summary, the acute cerebellar slice preparation can yield Purkinje cells with in vivo-like firing properties when due consideration is given to postnatal development and the importance of regular activation of the climbing fiber synapse. When cerebellar slices are prepared from rats at least ~4 weeks old and regular (~1 Hz) climbing fiber stimulation is provided to the cell, in vivo-like patterns of Purkinje cell action potential discharge can be obtained in vitro. In particular, the ability to block the trimodal pattern and restore tonic firing using only intracellular injection of simulated climbing fiber EPSCs provides a simple and efficient means to produce in vivo-like patterns of spike output in Purkinje cells.

4.9. Background Conductances In Vivo

A final consideration in analyzing activity in acute (cerebellar) slice preparations is the growing recognition that one must also take into account the fact that neurons in the awake and behaving animal are constantly bombarded by synaptic inputs (96, 97). The associated increase in membrane conductance can cause the neurons to enter a high conductance state with a lower membrane time constant (97, 98), bringing about a change in the spatiotemporal dynamics of neurons that can affect how inputs are processed and output is generated (96, 99–101). Above, we discussed how injection of current waveforms in vitro can mimic the effects of single synaptic inputs on spike output. However, this method cannot replicate the effects of conductance changes associated with a large number of inputs, as would be seen in vivo. A relatively new method, called dynamic clamp (Sect. 3.1.5), allows for the addition of artificial conductances to a neuron in vitro (102).

Unlike current-clamp protocols, where current is set to a constant level, dynamic clamp uses a feedback loop to adjust current levels as a function of membrane voltage, simulating a membrane conductance. To perform accurately, this computation must be made in real time requiring either specialized hardware or a computer with a real-time operating system and software. A well-developed software alternative is RTXI, which is a result of the merging of three different open-source dynamic clamp systems (Sect. 3.1.5) (46–50). Dynamic clamp can be used to examine how in vivo-like inputs affect information processing in neurons (99–101, 103, 104). In the cerebellum, several groups have used dynamic clamp to examine the origin of the complex spike in Purkinje cells (36), the contribution of SK-type K^+ channels to spike output (73), and the effect of synaptic input on Purkinje cell (103) and DCN cell output (104). For example, to determine the different effects of NMDA and AMPA inputs on DCN cell output, Gauck and Jaeger (104) used dynamic clamp to simulate different sets of synaptic conductances (Fig. 8a–c). When these conductances were added to the DCN cell, different combinations of inputs produced different spike outputs (Fig. 8d). With dynamic clamp, conductances being

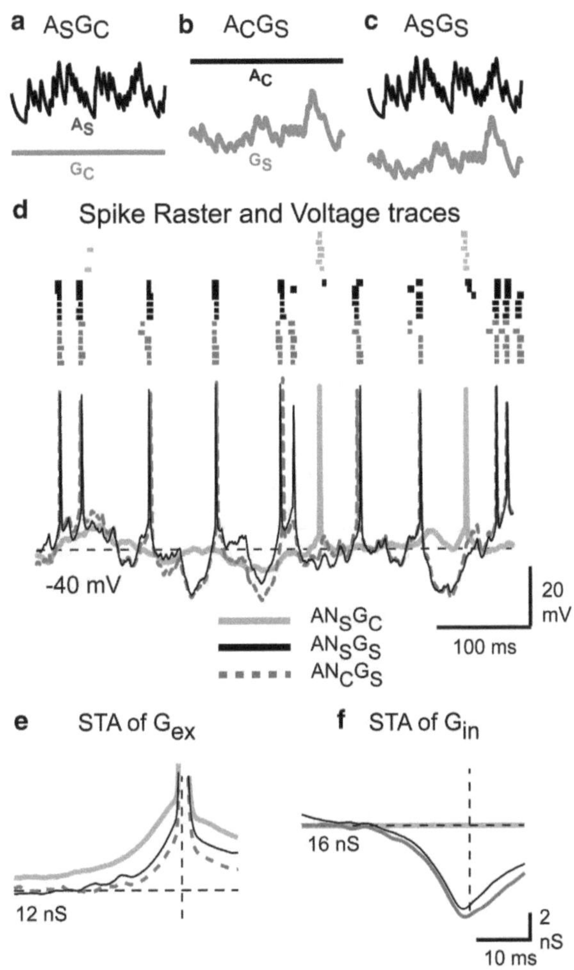

Fig. 8. Use of dynamic clamp to examine the contribution of different types of synaptic input to spike output. (**a–c**) Different sets of simulated synaptic conductances were used as stimuli to DCN cells. In each case, either constant or amplitude-modulated excitatory (A_C or A_S, respectively) or inhibitory (G_C or G_S) conductances were used in different combinations. (**d**) When presented to the same neuron, the different stimuli produced spike trains. If a constant inhibitory conductance was present, there was a large reduction in spike output. However, both amplitude-modulated and constant excitatory conductances resulted in similar outputs when presented in combination with amplitude-modulated inhibitory conductance, suggesting that spike output is being controlled by the inhibitory conductances. (**e, f**) Spike-triggered averages (STA) of the synaptic conductances confirm that spike output is a result of a decrease in inhibition, rather than increases in excitation (Adapted with permission from (104)).

added to the neuron can be analyzed, so spike-triggered averages of the synaptic inputs were performed to determine how features of the synaptic inputs govern neuronal output (Fig. 8e, f). It was observed that in cases where inhibitory conductance was not constant, spikes resulted from a decrease in inhibitory conductance, rather than an increase in excitation. This allowed the authors to make inferences about how DCN neurons could process inhibitory inputs from Purkinje cells in vivo.

5. Conclusions

This review exemplifies with studies on two types of cerebellar cells that it is possible for neuronal spike responses measured in vitro to approach that expected for in vivo recording conditions. However, it is necessary to consider factors relevant to normal cell activity, including temperature and the ionic reversal potentials established by the patch-recording electrolyte. For cerebellar recordings, specific attention must be paid to the remarkable postnatal development of Purkinje cells over the time frame often used for patch recordings and in restoring climbing fiber input to block an intrinsic trimodal pattern of activity that will otherwise emerge in vitro. Once these steps are taken, in vitro recordings can provide extensive insight into aspects of membrane excitability (i.e., Purkinje cell bistability, climbing fiber-induced toggling, or parallel fiber gain) that can be more difficult to record or quantify in the higher conductance state of cells within an active network.

Acknowledgments

We gratefully acknowledge M. Kruskic and L. Chen for expert technical assistance, G. W. Zamponi for collaborative contributions to the work reviewed here, and D. Jaeger for sharing data for representation. Supported by grants from the Canadian Institute of Health Research (CIHR) to RWT and GWZ and studentship or postdoctoral support through the Alberta Heritage Foundation for Medical Research (AHFMR) (FRF, RT, BEM, JDTE, MI), CIHR (BEM, JDTE), NSERC (BEM), an Achievers in Medical Sciences award (DMA), T. Chen Fong awards (DMA, JDTE), and an Izaak Walton Killam award (DMA). RWT is an AHFMR Scientist.

References

1. Raman IM, Bean BP (1999) Ionic currents underlying spontaneous action potentials in isolated cerebellar Purkinje neurons. J Neurosci 19:1663–1674
2. Thach WT (1970) Discharge of cerebellar neurons related to two maintained postures and two prompt movements. I. Nuclear cell output. J Neurophysiol 33:527–536
3. Thach WT (1970) Discharge of cerebellar neurons related to two maintained postures and two prompt movements. II. Purkinje cell output and input. J Neurophysiol 33: 537–547
4. Mouginot D, Gähwiler BH (1995) Characterization of synaptic connections between cortex and deep nuclei of the rat cerebellum in vitro. Neuroscience 64:699–712
5. Llinas R, Mühlethaler M (1988) Electrophysiology of guinea-pig cerebellar nuclear cells in the in vitro brain stem-cerebellar preparation. J Physiol 404:241–258
6. Edwards FA, Konnerth A, Sakmann B, Takahashi T (1989) A thin slice preparation for patch clamp recordings from neurones of the mammalian central nervous system. Pflüger's Arch 414:600–612
7. Schultz SR, Kitamura K, Post-Uiterweer A, Krupic J, Häusser M (2009) Spatial pattern coding of sensory information by climbing fiber-evoked calcium signals in networks of neighboring cerebellar Purkinje cells. J Neurosci 29:8005–8015
8. Sugimori M, Llinas RR (1990) Real-time imaging of calcium influx in mammalian cerebellar Purkinje cells in vitro. Proc Natl Acad Sci USA 87:5084–5088
9. Llinas R, Sugimori M (1980) Electrophysiological properties of in vitro Purkinje cell dendrites in mammalian cerebellar slices. J Physiol 305:197–213
10. Llinas R, Sugimori M (1980) Electrophysiological properties of in vitro Purkinje cell somata in mammalian cerebellar slices. J Physiol 305:171–195
11. Stuart G, Häusser M (1994) Initiation and spread of sodium action potentials in cerebellar Purkinje cells. Neuron 13:703–712
12. Raman IM, Gustafson AE, Padgett D (2000) Ionic currents and spontaneous firing in neurons isolated from the cerebellar nuclei. J Neurosci 20:9004–9016
13. Llinas R, Yarom Y (1981) Properties and distribution of ionic conductances generating electroresponsiveness of mammalian inferior olivary neurones in vitro. J Physiol 315: 569–584
14. Bell CC, Grimm RJ (1969) Discharge properties of Purkinje cells recorded on single and double microelectrodes. J Neurophysiol 32:1044–1055
15. Häusser M, Clark BA (1997) Tonic synaptic inhibition modulates neuronal output pattern and spatiotemporal synaptic integration. Neuron 19:665–678
16. McKay BE, Turner RW (2004) Kv3 K^+ channels repolarize calcium spikes to shape burst output in rat Purkinje cells. Eur J Neurosci 20:729–739
17. Womack M, Khodakhah K (2002) Active contribution of dendrites to the tonic and trimodal patterns of activity in cerebellar Purkinje neurons. J Neurosci 22: 10603–10612
18. Cerminara NL, Rawson JA (2004) Evidence that climbing fibers control an intrinsic spike generator in cerebellar Purkinje cells. J Neurosci 24:4510–4517
19. Lowenstein Y, Mahon S, Chadderton P, Kitamura K, Sompolinsky H, Yarom Y, Hausser M (2006) Purkinje cells in awake behaving animals operate in stable upstate membrane potential. Nat Neurosci 9:461–463
20. Bengtsson F, Jorntell H (2007) Ketamine and xylazine depress sensory-evoked parallel fiber and climbing fiber responses. J Neurophysiol 98:1697–1705
21. Schonewille M, Khosrovani S, Winkelman BH, Hoebeek FE, De Jeu MT, Larsen IM, Van der Burg J, Schmolesky MT, Frens MA, De Zeeuw CI (2006) Purkinje cells in awake behaving animals operate at the upstate membrane potential. Nat Neurosci 9:459–461, author reply 461
22. Alvina K, Walter JT, Kohn A, Ellis-Davies G, Khodakhah K (2008) Questioning the role of rebound firing in the cerebellum. Nat Neurosci 11:1256–1258
23. Tadayonnejad R, Mehaffey WH, Anderson D, Turner RW (2009) Reliability of triggering postinhibitory rebound bursts in deep cerebellar neurons. Channels 3:1–7
24. Zheng N, Raman IM (2009) Ca currents activated by spontaneous firing and synaptic disinhibition in neurons of the cerebellar nuclei. J Neurosci 29:9826–9838
25. Zheng N, Raman IM (2009) Synaptic inhibition, excitation, and plasticity in neurons of the cerebellar nuclei. Cerebellum 9:56–66
26. Molineux ML, McRory JE, McKay BE, Hamid J, Mehaffey WH, Rehak R, Snutch TP, Zamponi GW, Turner RW (2006) Specific T-type calcium channel isoforms are associated

with distinct burst phenotypes in deep cerebellar nuclear neurons. Proc Natl Acad Sci USA 103:5555–5560
27. Molineux ML, Mehaffey WH, Tadayonnejad R, Anderson D, Tennent AF, Turner RW (2008) Ionic factors governing rebound burst phenotype in rat deep cerebellar neurons. J Neurophysiol 100:2684–2701
28. Ito M (1984) The cerebellum and neural control. Raven, New York
29. Garwicz M, Ekerot CF, Jorntell H (1998) Organizational principles of cerebellar neuronal circuitry. News Physiol Sci 13:26–32
30. Chan-Palay V (1977) Cerebellar dentate nucleus: organization, cytology and transmitters. Springer, Berlin
31. De Zeeuw CI, Berrebi AS (1995) Postsynaptic targets of Purkinje cell terminals in the cerebellar and vestibular nuclei of the rat. Eur J Neurosci 7:2322–2333
32. Kitai ST, McCrea RA, Preston RJ, Bishop GA (1977) Electrophysiological and horseradish peroxidase studies of precerebellar afferents to the nucleus interpositus anterior. I. Climbing fiber system. Brain Res 122:197–214
33. McCrea RA, Bishop GA, Kitai ST (1978) Morphological and electrophysiological characteristics of projection neurons in the nucleus interpositus of the cat cerebellum. J Comp Neurol 181:397–419
34. Rowland NC, Jaeger D (2008) Responses to tactile stimulation in deep cerebellar nucleus neurons result from recurrent activation in multiple pathways. J Neurophysiol 99: 704–717
35. De Schutter E, Steuber V (2009) Patterns and pauses in Purkinje cell simple spike trains: experiments, modeling and theory. Neuroscience 162:816–826
36. Davie JT, Clark BA, Hausser M (2008) The origin of the complex spike in cerebellar Purkinje cells. J Neurosci 28:7599–7609
37. Loewenstein Y, Mahon S, Chadderton P, Kitamura K, Sompolinsky H, Yarom Y, Hausser M (2005) Bistability of cerebellar Purkinje cells modulated by sensory stimulation. Nat Neurosci 8:202–211
38. Khaliq ZM, Raman IM (2005) Axonal propagation of simple and complex spikes in cerebellar Purkinje neurons. J Neurosci 25:454–463
39. McKay BE, Molineux ML, Mehaffey WH, Turner RW (2005) Kv1 K^+ channels control Purkinje cell output to facilitate postsynaptic rebound discharge in deep cerebellar neurons. J Neurosci 25:1481–1492
40. McKay BE, Engbers JD, Mehaffey WH, Gordon GR, Molineux ML, Bains JS, Turner RW (2007) Climbing fiber discharge regulates cerebellar functions by controlling the intrinsic characteristics of purkinje cell output. J Neurophysiol 97:2590–2604
41. McKay BE, Turner RW (2005) Physiological and morphological development of the rat cerebellar Purkinje cell. J Physiol 567:829–850
42. Trapp S, Ballanyi K (2012) Autonomic nervous system in vitro: studying tonically active neurons controlling vagal outflow in rodent brainstem slices. In Isolated Central Nervous System Circuits (Ed K Ballanyi), Neuromethods Series Vol. 73 (Ed W Walz). Springer Science+Business Media, LLC, New York, NY, pp 1–59
43. Moore AR, Zhou WL, Jakovcevski I, Zecevic N, Antic SD (2012) Physiological properties of human fetal cortex in vitro. In Isolated Central Nervous System Circuits (Ed K Ballanyi), Neuromethods Series Vol. 73 (Ed W Walz). Springer Science+Business Media, LLC, New York, NY, pp 125–158
44. Sanchez-Vives MV (2012) Spontaneous rhythmic activity in the adult cerebral cortex in vitro. In Isolated Central Nervous System Circuits (Ed K Ballanyi), Neuromethods Series Vol. 73 (Ed W Walz). Springer Science+Business Media, LLC, New York, NY, pp 263–284
45. Konnerth A, Llano I, Armstrong CM (1990) Synaptic currents in cerebellar Purkinje cells. Proc Natl Acad Sci USA 87:2662–2665
46. Christini DJ, Stein KM, Markowitz SM, Lerman BB (1999) Practical real-time computing system for biomedical experiment interface. Ann Biomed Eng 27:180–186
47. Dorval AD, Christini DJ, White JA (2001) Real-time linux dynamic clamp: a fast and flexible way to construct virtual ion channels in living cells. Ann Biomed Eng 29:897–907
48. Butera RJ Jr, Wilson CG, Delnegro CA, Smith JC (2001) A methodology for achieving high-speed rates for artificial conductance injection in electrically excitable biological cells. IEEE Trans Biomed Eng 48: 1460–1470
49. Raikov I, Preyer A, Butera RJ (2004) MRCI: a flexible real-time dynamic clamp system for electrophysiology experiments. J Neurosci Methods 132:109–123
50. Bettencourt JC, Lillis KP, Stupin LR, White JA (2008) Effects of imperfect dynamic clamp: computational and experimental results. J Neurosci Methods 169:282–289
51. Milescu LS, Yamanishi T, Ptak K, Mogri MZ, Smith JC (2008) Real-time kinetic modeling of voltage-gated ion channels using dynamic clamp. Biophys J 95:66–87
52. Kullmann PH, Wheeler DW, Beacom J, Horn JP (2004) Implementation of a fast 16-Bit

dynamic clamp using LabVIEW-RT. J Neurophysiol 91:542–554
53. Mouginot D, Bossu JL, Gahwiler BH (1997) Low-threshold Ca^{2+} currents in dendritic recordings from Purkinje cells in rat cerebellar slice cultures. J Neurosci 17:160–170
54. Jahnsen H (1986) Electrophysiological characteristics of neurones in the guinea-pig deep cerebellar nuclei in vitro. J Physiol 372:129–147
55. Aizenman CD, Linden DJ (1999) Regulation of the rebound depolarization and spontaneous firing patterns of deep nuclear neurons in slices of rat cerebellum. J Neurophysiol 82:1697–1709
56. Muri R, Knöpfel T (1994) Activity induced elevations of intracellular calcium concentration in neurons of the deep cerebellar nuclei. J Neurophysiol 71:420–428
57. Hurlock EC, Bose M, Pierce G, Joho RH (2009) Rescue of motor coordination by Purkinje cell-targeted restoration of Kv3.3 channels in Kcnc3-null mice requires Kcnc1. J Neurosci 29:15735–15744
58. Uusisaari M, Obata K, Knöpfel T (2007) Morphological and electrophysiological properties of GABAergic and non-GABAergic cells in the deep cerebellar nuclei. J Neurophysiol 97:901–911
59. Jorntell H, Ekerot CF (2002) Reciprocal bidirectional plasticity of parallel fiber receptive fields in cerebellar Purkinje cells and their afferent interneurons. Neuron 34:797–806
60. Chen FP, Evinger C (2006) Cerebellar modulation of trigeminal reflex blinks: interpositus neurons. J Neurosci 26:10569–10576
61. Shin SL, Hoebeek FE, Schonewille M, De Zeeuw CI, Aertsen A, De Schutter E (2007) Regular patterns in cerebellar purkinje cell simple spike trains. PLoS One 2:e485
62. Talley EM, Cribbs LL, Lee JH, Daud A, Perez-Reyes E, Bayliss DA (1999) Differential distribution of three members of a gene family encoding low voltage-activated (T-type) calcium channels. J Neurosci 19:1895–1911
63. McKay BE, McRory JE, Molineux ML, Hamid J, Snutch TP, Zamponi GW, Turner RW (2006) Ca(V)3 T-type calcium channel isoforms differentially distribute to somatic and dendritic compartments in rat central neurons. Eur J Neurosci 24:2581–2594
64. Hildebrand ME, Isope P, Miyazaki T, Nakaya T, Garcia E, Feltz A, Schneider T, Hescheler J, Kano M, Sakimura K, Watanabe M, Dieudonne S, Snutch TP (2009) Functional coupling between mGluR1 and Cav3.1 T-type calcium channels contributes to parallel fiber-induced fast calcium signaling within Purkinje cell dendritic spines. J Neurosci 29:9668–9682
65. Swensen AM, Bean BP (2003) Ionic mechanisms of burst firing in dissociated Purkinje neurons. J Neurosci 23:9650–9663
66. Cavelier P, Bossu JL (2003) Dendritic low-threshold Ca^{2+} channels in rat cerebellar Purkinje cells: possible physiological implications. Cerebellum 2:196–205
67. Iftinca M, McKay BE, Snutch TP, McRory JE, Turner RW, Zamponi GW (2006) Temperature dependence of T-type calcium channel gating. Neuroscience 142:1031–1042
68. Klockner U, Lee JH, Cribbs LL, Daud A, Hescheler J, Pereverzev A, Perez-Reyes E, Schneider T (1999) Comparison of the Ca^{2+} currents induced by expression of three cloned alpha1 subunits, alpha1G, alpha1H and alpha1I, of low-voltage-activated T-type Ca^{2+} channels. Eur J Neurosci 11:4171–4178
69. Perez-Reyes E (2003) Molecular physiology of low-voltage-activated t-type calcium channels. Physiol Rev 83:117–161
70. Coulter DA, Huguenard JR, Prince DA (1989) Calcium currents in rat thalamocortical relay neurones: kinetic properties of the transient, low-threshold current. J Physiol 414:587–604
71. Nobile M, Carbone E, Lux HD, Zucker H (1990) Temperature sensitivity of Ca currents in chick sensory neurones. Pflüger's Arch 415:658–663
72. Alvina K, Khodakhah K (2008) Selective regulation of spontaneous activity of neurons of the deep cerebellar nuclei by N-type calcium channels in juvenile rats. J Physiol 586:2523–2538
73. Feng SS, Jaeger D (2008) The role of SK calcium-dependent potassium currents in regulating the activity of deep cerebellar nucleus neurons: a dynamic clamp study. Cerebellum 7:542–546
74. Czubayko U, Sultan F, Thier P, Schwarz C (2001) Two types of neurons in the rat cerebellar nuclei as distinguished by membrane potentials and intracellular fillings. J Neurophysiol 85:2017–2029
75. Takashi Y, Toyoko H, Mayuko O, Akira A, Yoshihiko F (2003) Non-genomic modulation of dopamine release by bisphenol-A in PC12 cells. J Neurochem 87:1499–1508
76. Wozniak AL, Bulayeva NN, Watson CS (2005) Xenoestrogens at picomolar to nanomolar concentrations trigger membrane estrogen receptor-alpha-mediated Ca^{2+} fluxes and prolactin release in GH3/B6 pituitary tumor cells. Environ Health Perspect 113:431–439
77. Quesada I, Fuentes E, Viso-Leon MC, Soria B, Ripoll C, Nadal A (2002) Low doses of the

endocrine disruptor bisphenol-A and the native hormone 17beta-estradiol rapidly activate transcription factor CREB. FASEB J 16:1671–1673

78. Ito Y, Sato S, Son M, Kondo M, Kume H, Takagi K, Yamaki K (2002) Bisphenol A inhibits Cl^- secretion by inhibition of basolateral K^+ conductance in human airway epithelial cells. J Pharmacol Exp Ther 302:80–87
79. Gauck V, Thomann M, Jaeger D, Borst A (2001) Spatial distribution of low- and high-voltage-activated calcium currents in neurons of the deep cerebellar nuclei. J Neurosci 21:RC158
80. Armstrong DM, Rawson JA (1979) Activity patterns of cerebellar cortical neurones and climbing fibre afferents in the awake cat. J Physiol 289:425–448
81. Morales E, Fernandez FR, Sinclair S, Molineux ML, Mehaffey WH, Turner RW (2004) Releasing the peri-neuronal net to patch-clamp neurons in adult CNS. Pflüger's Arch 448:248–258
82. Altman J, Bayer SA (1997) Development of the cerebellar system. CRC Press, New York
83. Sillitoe RV, Joyner AL (2007) Morphology, molecular codes, and circuitry produce the three-dimensional complexity of the cerebellum. Annu Rev Cell Dev Biol 23:549–577
84. Gardette R, Debono M, Dupont JL, Crepel F (1985) Electrophysiological studies on the postnatal development of intracerebellar nuclei neurons in rat cerebellar slices maintained in vitro. II. Membrane conductances. Brain Res 352:97–106
85. Gardette R, Debono M, Dupont JL, Crepel F (1985) Electrophysiological studies on the postnatal development of intracerebellar nuclei neurons in rat cerebellar slices maintained in vitro. I. Postsynaptic potentials. Brain Res 351:47–55
86. Molineux ML, Fernandez FR, Mehaffey WH, Turner RW (2005) A-type and T-type currents interact to produce a novel spike latency-voltage relationship in cerebellar stellate cells. J Neurosci 25:10863–10873
87. Crepel F (1971) Maturation of climbing fiber responses in the rat. Brain Res 35:272–276
88. Scelfo B, Strata P (2005) Correlation between multiple climbing fibre regression and parallel fibre response development in the postnatal mouse cerebellum. Eur J Neurosci 21:971–978
89. Fernandez FR, Engbers JDT, Turner RW (2007) Firing dynamics of cerebellar Purkinje cells. J Neurophysiol 98:278–294
90. Williams SR, Christensen SR, Stuart GJ, Häusser M (2002) Membrane potential bistability is controlled by the hyperpolarization-activated current I_H in rat cerebellar Purkinje neurons in vitro. J Physiol 539:469–483
91. Ebner TJ, Bloedel JR (1981) Role of climbing fiber afferent input in determining responsiveness of Purkinje cells to mossy fiber inputs. J Neurophysiol 45:962–971
92. Ebner TJ, Yu QX, Bloedel JR (1983) Increase in Purkinje cell gain associated with naturally activated climbing fiber input. J Neurophysiol 50:205–219
93. Latham A, Paul DH (1971) Spontaneous activity of cerebellar Purkinje cells and their responses to impulses in climbing fibres. J Physiol 213:135–156
94. Christian KM, Thompson RF (2003) Neural substrates of eyeblink conditioning: acquisition and retention. Learn Mem 10:427–455
95. Stanton ME, Fox GD, Carter CS (1998) Ontogeny of the conditioned eyeblink response in rats: acquisition or expression? Neuropharmacology 37:623–632
96. Softky WR, Koch C (1993) The highly irregular firing of cortical cells is inconsistent with temporal integration of random EPSPs. J Neurosci 13:334–350
97. Destexhe A, Rudolph M, Pare D (2003) The high-conductance state of neocortical neurons in vivo. Nat Rev Neurosci 4:739–751
98. van Vreeswijk C, Sompolinsky H (1996) Chaos in neuronal networks with balanced excitatory and inhibitory activity. Science 274:1724–1726
99. Fernandez FR, White JA (2008) Artificial synaptic conductances reduce subthreshold oscillations and periodic firing in stellate cells of the entorhinal cortex. J Neurosci 28:3790–3803
100. Fernandez FR, White JA (2009) Reduction of spike afterdepolarization by increased leak conductance alters interspike interval variability. J Neurosci 29:973–986
101. Wolfart J, Debay D, Le Masson G, Destexhe A, Bal T (2005) Synaptic background activity controls spike transfer from thalamus to cortex. Nat Neurosci 8:1760–1767
102. Sharp AA, O'Neil MB, Abbott LF, Marder E (1993) Dynamic clamp: computer-generated conductances in real neurons. J Neurophysiol 69:992–995
103. Jaeger D, Bower JM (1999) Synaptic control of spiking in cerebellar Purkinje cells: dynamic current clamp based on model conductances. J Neurosci 19:6090–6101
104. Gauck V, Jaeger D (2003) The contribution of NMDA and AMPA conductances to the control of spiking in neurons of the deep cerebellar nuclei. J Neurosci 23:8109–8118

Chapter 8

Spontaneous Rhythmic Activity in the Adult Cerebral Cortex In Vitro

Maria V. Sanchez-Vives

Abstract

The cerebral cortex in vivo generates different patterns of rhythmic activities with frequency rates ranging from below 1 Hz to fast frequencies well above 10 Hz. Some of these activities occur in the absence of external input and are a consequence of recurrent connectivity within the cortical network. The cerebral cortex in vitro maintains its recurrent connectivity, at least partly, and thus is able to generate some spontaneous rhythmic patterns as far as there is a certain level of intrinsic excitability in the network. An artificial cerebrospinal fluid (ACSF) with an ionic composition that mimics that in situ (Sanchez-Vives and McCormick, Nature Neuroscience. 3:1027, 2000) provides sufficient excitatory drive for the cortical network in vitro to generate not only slow (<1 Hz) rhythmic activity similar to the one that occurs during slow-wave sleep but also fast rhythms (10–80 Hz). Here, the methods and techniques used to prepare and study active cortical slices from adult animals are described, as well as the importance of methodological variables like temperature and oxygenation.

Key words: Brain slices, Cortical rhythms, Ferret, Oscillatory activity, Oxygen, Spontaneous activity, Temperature modulation, Visual cortex

1. Introduction

"*Investigation of the mammalian brain would be greatly furthered if portions from it could be studied by electrophysiological methods while they were completely isolated from the body, as the peripheral nerves and spinal ganglia can be studied.*" This text was written in 1957 by the pioneer in the use of cortical brain slices for electrophysiological measurements, Henry McIlwain, in the first paper describing an in vitro preparation of cortical slices where electrophysiology could be performed (1). In that preparation, neurons could successfully maintain their membrane potential. Ten years later, Yamamoto and McIlwain showed that also synaptic transmission was preserved

Klaus Ballanyi (ed.), *Isolated Central Nervous System Circuits*, Neuromethods, vol. 73,
DOI 10.1007/978-1-62703-020-5_8,

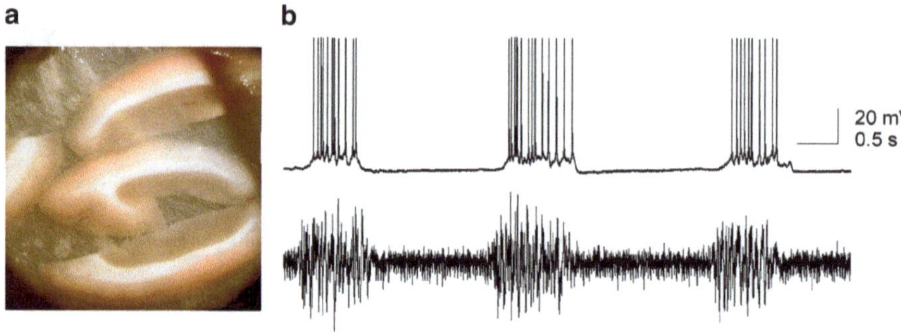

Fig. 1. *Spontaneous rhythmic activity in the cerebral cortex* in vitro. (**a**) Ferret cortical brain slices in an "interface" chamber. (**b**) Top trace, intracellularly recorded slow oscillations in the slice. Bottom trace, synchronous multiunit activity (MUA, high-pass filtered) illustrating the neuronal firing of the local circuit.

in vitro (2). Nowadays, more than half a century after the first study, acutely isolated brain slices still offer some of the same aforementioned advantages over the studies in vivo, advantages that are mostly based on the accessibility of the cortical network for its study with different techniques. In the case of brain slices from cerebral cortex, all cortical layers are accessible if the tissue is cut perpendicular to the pia mater, which is the meningeal layer closest to the brain tissue. Despite the fact that brain slice techniques were described in the late 1950s, the first electrophysiological characterization of cortical neuronal types in vitro appeared in 1982 (3). Since then, new recording techniques have been applied to brain slices, with turning points being the application of patch-clamp recordings to slices (4), the use of imaging techniques like voltage-sensitive dyes, or on calcium fluorescence for the analysis of population firing, or the use of multielectrode arrays. Another important advantage regarding cortical slices is not related to the methods of recording but to the activity generated by the preparation. In the study by Sanchez-Vives and McCormick (5), a preparation of a cortical slice that generated rhythmic slow activity almost identical to the one occurring in the cerebral cortex during slow-wave sleep (6) was described (Fig. 1). Nowadays, the use of "active" cortical slices has become quite extended.

2. Materials

Listed below are some of the materials used in my laboratory for studying cortical brain slices:

- *Microtomes.* Thermo Scientific, model Microm HM 650 V http://www.thermoscientific.com/
- Vibratome Mod: series 1000 http://www.vibratome.com/

- *Blades.* VALET Auto Strop Blades from World Precision Instruments http://www.wpiinc.com/
- *Interface Chamber.* Tissue slice recording chamber (No. 21000–03) con TR-200 temperature controller (No. 21052–00) from Fine Science Tools http://www.finescience.de/
- Some of its characteristics can be found at http://www.muromachi.com/fst/ResEqpGrp/21000-03.htm
- *Electrode Puller.* Sutter Instruments model P-97 http://www.sutter.com/
- *Glass Electrodes for Intracellular Recordings.* Here are included two possible types:
  - Harvard Apparatus GC100FS-10. 1.0 mm outer diameter, 0.5 mm inner diameter
  - http://www.harvardapparatus.com/webapp/wcs/stores/servlet/catalog_11051_10001_-1_HAI
  - World Precision Instruments model 1B100F-4

3. Methods

3.1. Obtaining and Maintaining Cortical Brain Slices from Adult Animals

Here, some aspects of generating spontaneously active cortical brain slices are discussed. In order to obtain such slices, the animals should be deeply anesthetized (e.g., intraperitoneal sodium pentobarbital 40 mg/kg for ferrets) and decapitated with either an animal guillotine or scissors (in the case of mice). The dura mater should be carefully opened to facilitate extraction of the brain. This is particularly important in larger animals such as ferrets, where the dura is quite thick and dura strips could damage the brain during the process of slicing. The brain should then be removed fast and cautiously and be placed in ice-cold cutting solution (kept in the fridge at 4°C before using). We gas our different ACSFs with carbogen (95% O_2 and 5% CO_2). Various laboratories use iced solution that has been kept in the freezer for 30 min at −70°C and has been smoothed in a grinder. The last feature may be more adequate for <1-month-old animals than for adults. In adult animals, we have not found differences in the network activity reported here between both procedures. The ionic content of the solution used to place the brain and next cut the slices follows procedures described by Aghajanian and Rasmussen (7). In this solution, NaCl has been replaced by sucrose, a technique developed to prevent passive Cl^- entry and subsequent cell swelling and lysis. The ionic composition of this artificial cerebrospinal fluid (ACSF) is (in mM) 252 sucrose, 2.5 KCl, 2 $MgSO_4$, 1.25 Na_2HPO_4, 2 $CaCl_2$, 26 $NaHCO_3$, and 10 dextrose, gassed with carbogen to reach a steady-state pH of 7.4. Some authors find good results by perfusing the animal through the heart with ice-cold sucrose-based solution prior to removal of the brain (8) (see also Chap. 1-Trapp and Ballanyi (9)). Other authors propose to replace NaCl by glycerol instead (10).

The trimmed piece of brain to be sliced is glued (Loctite Super-Glue3, gel) to the tray of the vibratome. We have used different vibratomes with similar results for the purposes that we describe here (Vibratome TPI Series 1000 Semi-Automatic Tissue Sectioning System or Microm HM 650 V). When cutting 400-μm-thick slices of cerebral cortex with the vibratome, the brain tissue block should be mechanically stable, the speed of the blade advancement should be slow (~500 μm–1 mm/s), and the horizontal amplitude of the movement should be wide, even the maximal allowed by the machine. A fast blade advancement can ruin the slices. One possible risk is that the tissue would be smashed instead of cut, which has fatal consequences. Another risk is to detach the brain piece from the tray. If slices are too large or if they contain unwanted, non-cortical structures, they can be trimmed down with a scalpel blade.

After preparation, slices are placed in an interface-style recording chamber (e.g., Fine Sciences Tools, Foster City, CA) and bathed for 30 min in carbogenated ACSF containing (in mM) 63 NaCl, 126 sucrose; 2.5 KCl, 2 $MgSO_4$, 1.25 Na_2HPO_4, 2 $CaCl_2$, 26 $NaHCO_3$, and 10 dextrose; pH 7.4. Bath temperature is initially maintained at 34–35°C. The importance of temperature on cortical activity is discussed in detail below. After 30 min, the solution is changed to carbogenated ACSF containing (in mM) 126 NaCl; 2.5 KCl, 2 $MgSO_4$, 1.25 Na_2HPO4, 2 $CaCl_2$, 26 $NaHCO_3$, and 10 dextrose; pH 7.4. After 2 h of recovery, the ACSF is replaced by "in vivo-like" ACSF containing 3.5 KCl, 1 $MgSO_4$, and 1.2 $CaCl_2$, the rest of the components being the same as above. Under these conditions, cortical slices should start oscillating spontaneously generating "Up and Down states" (Sects. 1 and 3.1.) within 20–40 min (Fig. 1) although the probability will vary depending on the species used (see below).

In order for spontaneous rhythmic activity to occur, it is desirable that the cortical network is preserved as much as possible. Still, thinner slices of 350-μm thickness can generate rhythmic activity. Regarding the length and width of the slice, there is no evidence that there is an optimal size for a cortical slice to generate slow oscillations in their LFP (local field potential). Given that mechanical stability is a critical factor during recordings, not using too large slices (>10 mm) is favored. Even when slices of 1–2-mm width can still oscillate, it is not desirable to prepare them at this size, since there is higher chance of mechanical damage to be inflicted during the trimming. Importantly, to increase the probability for the tissue to generate spontaneous rhythms, the animals should be >1 month old.

3.2. Electrophysiological Techniques

3.2.1. Multiunit Recording

Multiunit recording from a slice just requires an electrode, for example, a tungsten electrode (FHC, 2–4 MΩ) or a pulled glass pipette filled up with a solution of 0.9% NaCl. The electrode should be connected to a headstage of a high-impedance preamplifier which is placed close to the preparation. In a later stage, the signal should be amplified (around ×1,000) and high-pass filtered (above 200 Hz) for multiunit recordings. If one wants to record the LFP, it is better to leave the signal as unfiltered as possible, both in the low- and high-frequency ranges. The signal can always be digitally filtered later, if needed. In case of persistent 50–60 Hz noise from electrical interference (50 Hz/60 Hz depending on location) that cannot be eliminated, a "Hum Bug" (Digitimer Ltd., http://www.digitimer.com) can be helpful. It can be used both with intra- and extracellular recordings.

3.2.2. Sharp Microelectrode Intracellular Recording

Interface recording chambers are not ideal for patch (whole-cell or otherwise) recordings, given that the ACSF level is maintained very low. The brain slice is embedded in ACSF from the filter paper below, and the surface of the slice is maintained humid thanks to the steam-saturated environment. For that reason, if intracellular recordings are needed, one may want to consider sharp electrode recordings. Sharp pipettes are generated in pullers such as the one from Sutter Instruments (P-97 micropipette puller). Electrodes can be beveled (Sutter Instruments beveler) in order to decrease their impedance while hardly affecting the size of their tip. Beveling them to a final resistance around 50–75 MΩ can allow a good access to the neurons. Micropipettes can be filled with 1.5–2.0 M potassium acetate, which has an advantage over KCl, since it does not affect chloride reversal potential and thus fast synaptic inhibition involving γ-aminobutyric acid (GABA) receptors. How to impale and record neurons is not the subject of this chapter, but it can be found in other sources (e.g., (11)).

3.2.3. Data Analysis

For intra- and extracellular, single or multiple channel recording, we use a CED (Cambridge Electronic Design http://www.ced.co.uk/) acquisition unit (Micro or Power 1401 depending on the number of channels to be acquired). We use as well their software, Spike 2, during the acquisition as well as for online and offline data analysis, including built-in analysis routines or customly written scripts. From Spike 2, we also export to MATLAB (http://www.mathworks.com/) for further data analysis.

4. Results

4.1. Spontaneous Rhythmic Activity in Cortical Slices

The slow rhythmic activity in cortical slices from adult mammals comprises periods of depolarization and action potential discharge ("Up states") of a duration that in average is around 500 ms,

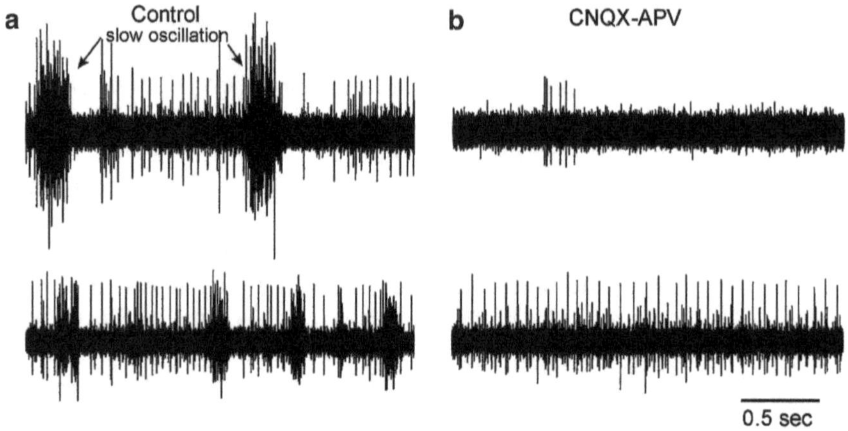

Fig. 2. *Spontaneous firing in a ferret cortical slice.* (**a**) Extracellular MUA recording from layer 5 in ferret prefrontal cortex in "in vivo-like" artificial cerebrospinal fluid (ACSF) as in (5). There are two Up states pointed to by arrows. Note that some units fire during Down states. (**b**) Activity in the same location after synaptic excitation was blocked with 20 μM CNQX and 50 μM APV. Note that in spite of the synaptic blockade, some spontaneous neuronal firing remains. Lower traces are recordings from a different site (Reproduced with permission from (16)).

interspersed with periods of silence ("Down states") (Figs. 1, 2, and 3) resulting in an oscillatory frequency around 0.3 Hz. Underpinning the spike firing during the Up states, there are bursts of synaptic potentials that are clearly visible when the neuron is maintained hyperpolarized by means of injecting hyperpolarizing DC current (Fig. 3e and 5 top). The absolute values of duration of Up/Down states and frequency of the oscillation can vary depending on the cortical area and the functional state of the network, varying, for example, with bath temperature (12) and with the excitatory/inhibitory balance, such that lesser inhibition results in shorter Up states (13).

Simultaneous recordings across different layers by means of arrays of tungsten electrodes revealed that infragranular layers, and in particular layer 5, were leading the start of each Up state that then reaches an almost simultaneous peak of firing along a vertical column (5). A simple setup with just two electrodes is illustrated on Fig. 4a. The initiation of Up states in layer 5 has been a consistent finding, both in vitro and in vivo (14, 15).

In the cortical model proposed in (16), a population of 1,024 excitatory and 256 inhibitory neurons with recurrent connections between them generated slow rhythmic discharge that resembled in many regards that observed experimentally. In this model, there are neurons that are more excitable than others due to their ionic conductances. These neurons are spontaneously active during silent periods. Occasionally, a sufficient number of closely adjacent spontaneously active neurons fire together, triggering a cascade of recurrent excitation that locally brings the network into the firing regime of the oscillation (Up state). Our experimental observations

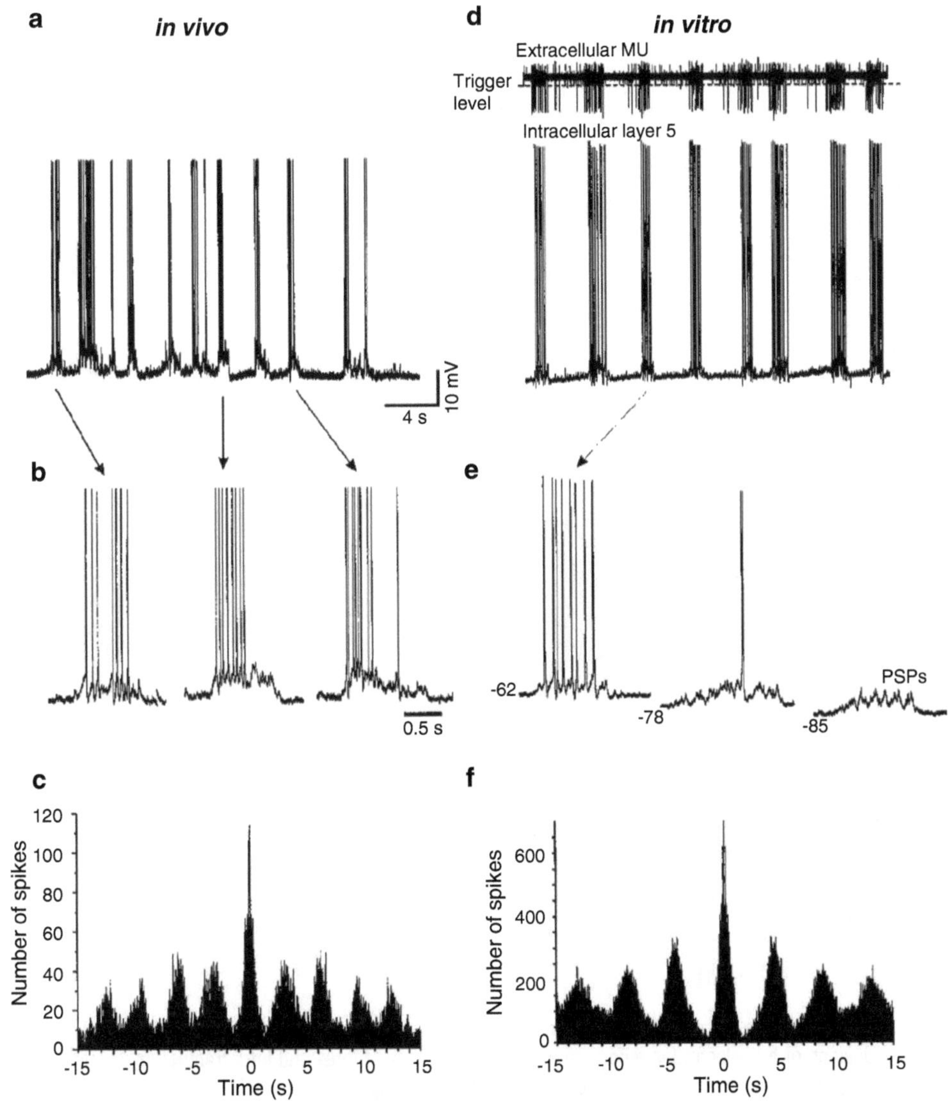

Fig. 3. *Generation of the slow oscillation in vivo and* in vitro. (**a**) Intracellular recording in primary visual cortex of a halothane-anesthetized cat reveals rhythmic sequence of depolarized and hyperpolarized membrane potentials. (**b**) Expansion at faster time scale of three of the depolarizing events or Up states for clarity. (**c**) Autocorrelogram of intracellular recording reveals marked periodicity of about one cycle per 3 s. (**d**) Simultaneous intracellular and extracellular recordings of the slow oscillation in ferret visual cortical slices. The intracellular recording is from a layer 5 intrinsically bursting neuron. (**e**) Depolarizing events or Up states at three different membrane potentials. (**f**) Autocorrelogram of intracellular recording in D shows marked periodicity of approximately once per 4 s (Reproduced with permission from (5)).

suggest that neurons in layer 5 could play such role, given that some spontaneous activity can be recorded in this layer during Down states (Fig. 2b) and they lead the initiation of Up states. Once the Up state is reached, it is maintained by a balanced excitatory/inhibitory recurrent loop. Such activity in a reverberatory local circuit could self-perpetuate; however, we know that Up

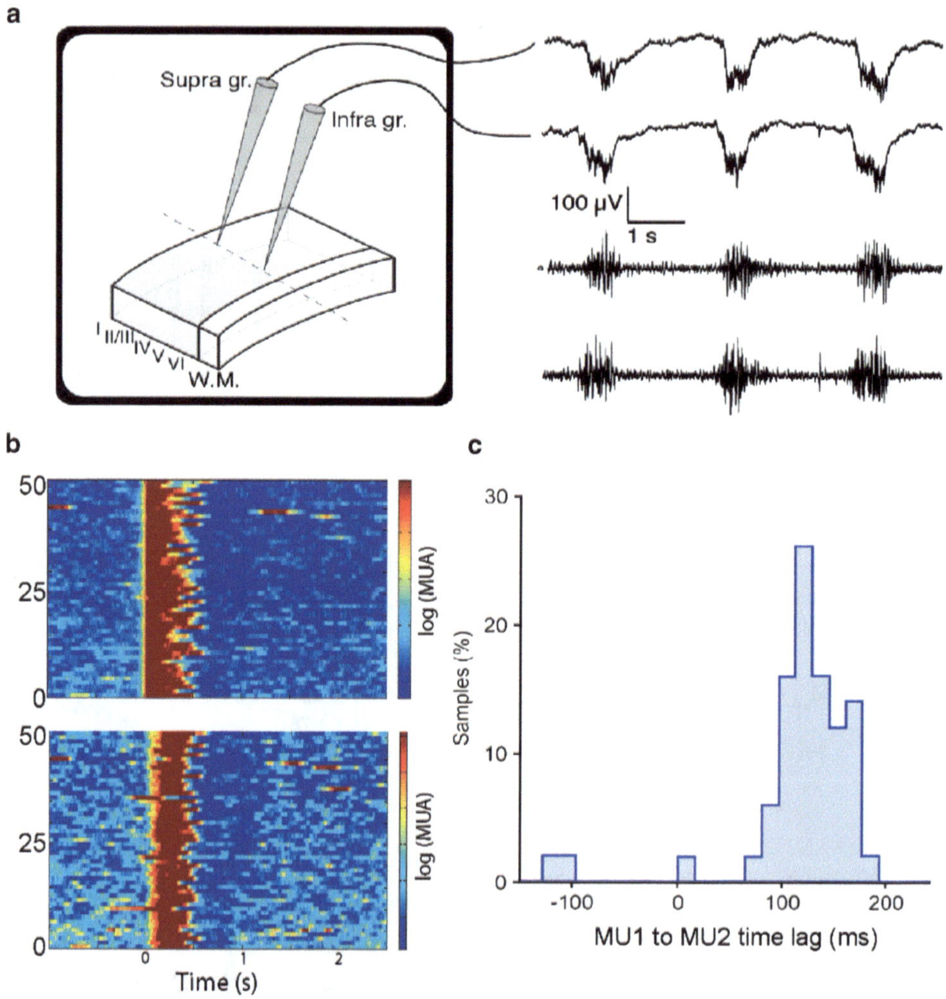

Fig. 4. *Simultaneous multiple recordings for detection of rhythmic wave propagation.* (**a**) Vertical propagation: schematics illustrating the position of two electrodes recording simultaneously from supra- and infragranular layers (*left*). Local field potential recordings (*top*) and corresponding MUA recordings, following high-pass filtering (*bottom*), from supra- and infragranular layers. (**b**) Horizontal propagation: recordings obtained by two electrodes located in layer 5 and located 1 mm away from each other. The same wave can be recorded from both locations and be used to calculate the speed of propagation of the wave. Raster plots (50 Up states) display in each row the color coded time courses of log(MUA) which is a measure of the relative firing rates between Up and Down states centered around the detected Up-state onsets. For details around the log (MUA) calculations, see (13). (**c**) Histogram of time lags between both electrodes for each detected Up state in (**b**) (All recordings were obtained from ferret cortical slices).

states have a certain duration (*ca.* 0.2–1.5 s in vitro) before the system switches to a Down state.

What mechanisms can bring the network to a halt during the Up state? One of the constraints of such mechanisms is that they should be highly synchronous across the network, more than the Up-state initiation (17). As said above, different network mechanisms shape the emergent activity, and several of them have been proposed to be the key to the end of the Up states. Some of these mechanisms include arrival of excitation (18, 19), synaptic depression (20), thalamic disfacilitation (21), activation of K^+ currents (5, 16), or extracellular K^+ dynamics (22). The time course of the slow afterhyperpolarization observed in intracellular recordings during Down states (see Fig. 7c in (5)) suggests that slow K^+ currents have a relevant participation in the termination of Up states and maintenance of Down states. Different mechanisms involving K^+ currents have been proposed, including ATP-dependent K^+ current (23), $GABA_B$ receptor-mediated responses (24), as well as Ca^{2+}- and Na^+-dependent K^+ currents (5), hypothesis that was tested in a cortical model (16). Slow K^+-mediated afterhyperpolarizations can also be modulated by neurotransmitters that control the transition from sleep to awake (25, 26). Slow K^+ currents, and in particular Na^+-dependent K^+ current (27), could be a feasible candidate to underlie Down states.

Slow oscillatory waves propagate along the cerebral cortex network in the anesthetized cat at a speed of around 100 mm/s (6). This propagation has been characterized in humans during slow-wave sleep, and it most often occurs from front to back (28). The same study (28) reports that oscillations can as well be generated at any point in the cortex, and it can propagate in any direction. Some of the properties of the slow oscillations as a traveling wave in vivo are preserved in neocortical slices (5) as well as in three-layered olfactory cortex (29), although in the last one, propagation speed is one order of magnitude faster (114 mm/s) than in neocortex (~10 mm/s in ferret neocortical slices). The speed of propagation of the wave reflects the horizontal propagation of the locally generated Up states along the cerebral cortex, by horizontal meaning along cortical layers and perpendicular to the cortical columns. The speed of wave propagation increases linearly with the progressive removal of inhibition and nonlinearly when the activity transforms into epileptiform by inhibition removal (30). The use of at least two electrodes allows to measure the traveling of the propagating wave and to calculate the time lags between the initiations of each Up state recorded from both electrodes (Fig. 4b, c). This method presents some difficulties, since the wave does not always travel in the same direction, and the Up states may be eventually generated in between the two electrodes and travel on both directions, thus causing confounds in the analysis. Some details on the analysis of propagating waves can be found in (30). A similar experimental approach can be used to study vertical propagation of activity (in the same column and across cortical layers) as in Fig. 4a (5).

Interestingly, (31) found that gamma oscillations occurred during the activated phases of the slow oscillations in anesthetized cats. This was confirmed in the visual cortex of anesthetized ferrets (32). Cortical slices did not seem to be able to generate spontaneous beta/gamma rhythms similar to those in vivo (32), suggesting that such oscillatory activity might be dependent on specific neuromodulators for its generation in the local cortical circuit. However, the analysis of the activity generated in the visual cortex ferret slices during slow oscillatory activity by applying multitaper spectral methods revealed that the local cortical network contains the necessary elements to generate high-frequency rhythms (16). Still, neuromodulators potently modulate beta/gamma-range activity, and in particular, cholinergic agents such as charbachol induce a robust upregulation of high frequencies occurring during Up states.

In some species, cortical slices of a lateral extension of few square millimeters and 400-μm thickness seem to contain enough cortical network to generate slow (<1 Hz) oscillations similar to those occurring during slow-wave sleep, while others do not. This leads to the following question: What is necessary for an isolated cortical network to generate slow oscillations? Clearly, there are numerous factors that may contribute to shape the emergent activity: ionic currents from all excitatory and inhibitory neurons involved, patterns of synaptic connectivity, properties of synaptic connections, excitatory/inhibitory balance, synaptic plasticity, etc. However, there are a few elements that seem to be critical for the generation of rhythmic activity in a cortical slice. As a result of our experimental and modeling studies, we propose three critical factors: (1) a network with recurrent connections, (2) a certain level of excitability, and (3) a mechanism to end the Up states such as the activation of slow K^+ currents.

4.2. Species Dependence of Slow Oscillations

As described above, spontaneous slow rhythmic activity in brain slices was first obtained from ferret visual cortex (5). Oscillatory activity was detected not only in visual cortex but also in other cortical areas such as the prefrontal cortex (33, 34). Spontaneous activity was also reported in deafferented cortical slabs from cat cerebral cortex (35). Ongoing activity recorded from the slabs was different though from the one found in cortical slices. An interesting difference was that activity in cortical slabs was reported to be dependent on the size of the slab, and small slabs (10 × 6 mm) very seldom presented spontaneous Up states or depolarizing events (2–6 min^{-1}). Only occasionally, more regular and rhythmic activity similar to the one reported in vitro was found (35). On the contrary, in the slices maintained in vitro, even very small pieces of cortex (<2 × 5 mm, 400 μm thick) were able to yield slow spontaneous activity of similar frequencies to the ones reported in vivo (~0.3 Hz). The frequency of oscillations in the cortical slices was independent of the slice size.

A relevant aspect to discuss here is the generation of oscillations in cortical slices from different animal species (36). We have found differences across species when recording from cortical slices under the same experimental conditions as the ones reported in (5, 37). These conditions include identical ACSF composition (124 mM NaCl, 3.5 mM KCl, 1 mM $MgSO_4$, 1.25 mM Na_2HPO_4, 1–1.2 mM $CaCl_2$, 26 mM $NaHCO_3$, and 10 mM dextrose, carbogenated for a pH of 7.4) and same interface chamber from Fine Science Tools (Sect. 2). The only difference was the animal species from which the slices were obtained. Initially, we evaluated the existence of organized, rhythmic activity by recording multiunit activity with one or more tungsten electrodes. This activity was recorded from the surface of the slice (Fig. 1) and allowed the testing in different areas of the slice without interfering with the emergent activity. Comparisons across species are often difficult to make between different laboratories, since the procedures of obtaining slices may be different, and so are the recording chambers, the electrode type, or the exact ionic composition of the ACSF.

Slices obtained from different areas of the *ferret* neocortex and also from piriform cortex (29) (part of the paleocortex) generate rhythmic activity in the majority of cases. This is true, even if the neocortical slices are obtained following different orientations (sagittal or coronal). This rhythmic activity, which is not always identical, shares the properties of recruiting a large neuronal population in every Up state and of propagating for several millimeters along the slice. Steriade et al. (6) reported for the cortex of the anesthetized cat in vivo that 88% of the neurons that were recorded in vivo participated in the slow rhythmic activity. Likewise, we have seldom found any intracellularly recorded neuron in ferret neocortical slices that does not participate in this rhythmic network activity, even if sparsely, and this is true for both excitatory and inhibitory neurons (with all the biases that the intracellular recording techniques may imply).

In three occasions, we had the opportunity to record from *cat* cortical slices, and they did also generate prominent rhythmic activity, by that meaning a high similarity to the one generated in vivo in the anesthetized animal. Since the size of the sample is small, we cannot provide average values for the Up-state parameters. However, as illustrated in Fig. 5, the duration of Up states was around 0.5 s and the frequency below 1 Hz. The reliability of cortical slices to generate spontaneous slow oscillations decreased when they were obtained from the cortex of other species such as rats, guinea pigs, or mice. Even when cortical slices from rat, guinea pig, or mouse do eventually generate slow rhythmic activity, these oscillations are not as robust and regular as the ones emerging from ferret cortical slices under the conditions that we described. It is important to remind here that ferrets are not rodents but carnivorous mammals belonging to the family Mustelidae (wisels).

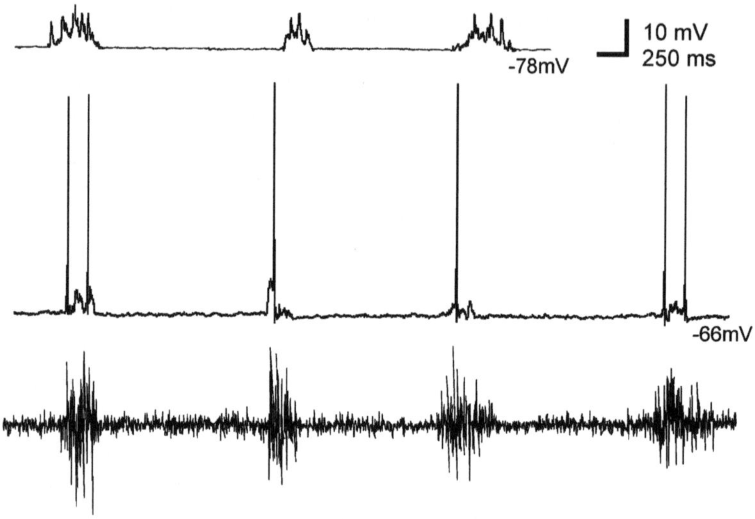

Fig. 5. *Slow oscillations in slices from cat cerebral cortex. Top*, subthreshold recording of a series of three Up states. Hyperpolarizing DC current has been injected to maintain membrane potential at −78 mV. *Middle*, from the same neuron, four suprathreshold Up states. No DC current was injected here, the membrane potential being now of −66 mV. Note how these intracellularly recorded Up states are correlated with simultaneously recorded MUA (*bottom*), revealing that population activity increases during Up states and almost disappears during Down states (Reproduced with permission from Research Signpost (36)).

There are few publications in the literature, where spontaneous rhythmic activity in *guinea pig* cerebral cortex is reported, and those are mostly concerned with entorhinal cortex, in that case in an isolated brain preparation (38). However, to our knowledge, no slow rhythmic activity generated in guinea pig neocortical slices has been reported. During the last years, we have recorded from a total of 200 slices obtained from the cortex of adult guinea pigs (visual cortex in 23 cases and prefrontal cortex in 177 cases). Out of these slices, 55–70% yielded slow oscillatory activity similar to the one reported for the ferret neocortex, with a frequency of Up-state occurrence between 0.1 and 0.5 Hz and a duration of 0.25–1.2 s. In some cases, this spontaneous activity showed a prominent rhythmicity and robustness, like the case illustrated in Fig. 6; however, this was not always the case. The lower percentage of slices generating rhythmic, synchronized activity make the guinea pig a less reliable species for these studies. As described in the ferret, presumably inhibitory neurons such as fast-spiking neurons (39, 40) also participate during Up states in synchrony with the rest of the population, which is illustrated in Fig. 6c for the guinea pig prefrontal cortex.

In rodents (rat and mouse), the recording of slow oscillations from adult neocortical slices in the conditions that we describe in this chapter (interface chamber recordings) was also more difficult to obtain than in ferret or cat. Spontaneous slow oscillations have

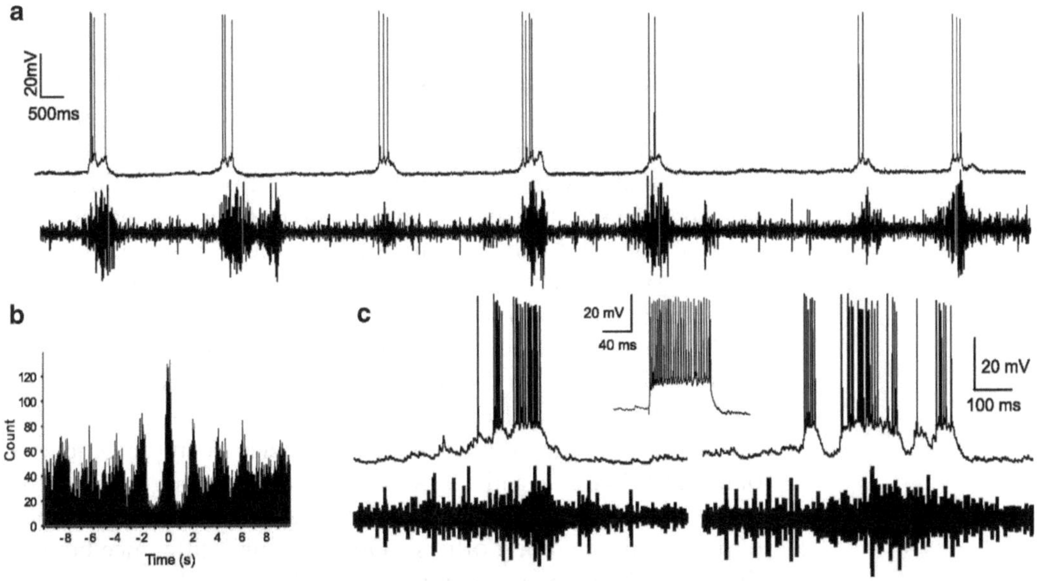

Fig. 6. *Slow oscillations in guinea pig prefrontal cortex slices.* (**a**) Intracellular recording (*top*) and MUA recording (*bottom*) showing rhythmic activity at <1 Hz frequency. (**b**) Autocorrelogram of MUA in (**a**), illustrating a 0.5 Hz rhythmicity. (**c**) *Top traces*, two Up states recorded from a fast-spiking neuron (see response to depolarizing pulse in central panel, illustrating its high frequency of discharge (320 Hz) typical of FS neurons). Note that this inhibitory neuron also discharges in phase with MUA (*bottom trace*) (Reproduced with permission from Research Signpost (36)).

also been reported in *rat* neocortex in vitro. The oscillations are well known to occur in vivo, for example, in the prefrontal area (41), barrel cortex (42, 43), or piriform cortex (44), and they have also been described as a proper sleep rhythm in the rat hippocampus (45). This spontaneous activity in the rat cortical network cannot be confused with the epileptiform activity induced by lowering the Mg^{2+} concentration in the ACSF (46). Reports on slow (<1 Hz), coordinated activity in the rat neocortical network as such are more unusual, although more extreme ionic changes in the ACSF (6.25 mM KCl and 0.5 mM $MgSO_4$) generate activation of the cortical microcircuits (47), giving rise to synchronized activity in different groups of neurons, although the extent of recruited population or propagation of the activity have not been reported. It can be concluded that ACSF requires further modifications probably because higher excitability is necessary in the network in order to generate activation of at least neuronal clusters.

In the *mouse*, higher concentrations of K^+ (5 mM instead of 3 mM) and therefore a still higher excitability have to be induced in order for the slow oscillations to occur with a frequency of 0.1 Hz (48), a frequency lower than the one occurring in vivo (49). The in vivo cortex from the mouse anesthetized with ketamine generates slow oscillatory activity. In vitro, slow rhythmic activity was eventually recorded using the same conditions as the ones described above (Sect. 3), but the activity was neither as robust nor propagated as in the ferret. How can we explain these functional differences between species? Even when for the sake of simplification we tend to assume that there is high similarity between the cerebral cortex of different species, studies of comparative anatomy have revealed that indeed profound differences do exist. These differences include basic features that are determinant of the emergent activity such as the density of neurons, patterns of connectivity, density of synaptic connections, cell types, proportion of inhibitory/excitatory cells, etc. (50, 51). It is difficult to determine which are the critical factors in an isolated cortical network that result in the successful or unsuccessful generation of a particular pattern of emergent activity. This difficulty resides in the fact that a large number of factors participate in the generation of cortical rhythmic activity as well as in determining its characteristics (duration and frequency of the Up states, amplitude, or speed of propagation, among others). One structural difference between the cerebral cortex of different species that may play a critical role in shaping the emerging activity at least in cortical slices is the existence of "patchy" horizontal connectivity, which is missing in rodents (50, 51), while it exists in cats (52), ferrets (53), and primates (54). A network with long horizontal connections and clustering or cliquishness of synaptic contacts approaches what has been called a "small-world" network, a pattern or connectivity that promotes synchronized activity in distally located neurons (55). The existence of clusters of synaptic contacts may be crucial in order to reach a critical number of excited neurons that through reverberation in the local circuit results in the generation of an Up state that then propagates along the network (16). Given that slow oscillatory activity in the cortex relies importantly on the compactness of local cortical connectivity (56), the absence of long patchy connections in the rodents may result in our experimental observations that cortical slices from rodents display Up states that appeared to be more focal and less successful in traveling long distances, therefore lacking synchrony along the cortical slices. The dense interconnectivity between cortical areas and through the thalamocortical loop may compensate for the lack of patchy connectivity when in vivo, since slow oscillations in the in vivo rodent cortex are quite robust and reliable (49).

5. Critical Factors for Spontaneous Activity in Adult Cortex Slices

5.1. Ionic ACSF Composition

Recurrent connectivity in the local circuit is necessary for an isolated piece of cortical network with no active inputs to generate self-sustained activity (Sect. 1) (5, 16, 33, 35, 57). This recurrence exists in the cerebral cortex, and neuronal loops can sustain reverberatory activity (58). Still, for activity to start in the absence of external inputs, there should be a critical level of excitability in the network that allows for spontaneous firing of individual neurons which can eventually lead to the initiation of oscillations (59). The modification of the ionic concentrations of the ACSF mimicking the cerebrospinal fluid provides such level of excitability. We proved this point by demonstrating that in the modified ACSF and upon blockade of fast excitatory transmission with the ionotropic glutamate receptor blockers 6-cyano-7-nitroquinoxalene-2,3-dione (CNQX, 20 μM) and DL-2-amino-5-phosphonovaleric acid (dl-APV, 50 μM), there is spontaneous, non-organized firing of individual neurons in the cortical network (Fig. 2). However, this firing does not give rise to rhythmic activity where there is a temporal organization of the neuronal population firing, a phenomenon that requires an active excitatory connectivity.

The fact that the ACSF does not normally contain neuromodulators and that the neurotransmitter systems are not functioning in the slices means that the levels of neurotransmitters in vitro are probably negligible in comparison with those in vivo. On the other side, the ionic composition of the cerebrospinal fluid in which the brain in situ is bathed and that of the classically used ACSFs to maintain slices in vitro (in mM: 124 NaCl, 2.5 KCl, 2 $MgSO_4$, 1.25 Na_2HPO_4, 2 $CaCl_2$, 26 $NaHCO_3$, and 10 dextrose) is not exactly the same (60, 61). A critical difference concerns the Ca^{2+} concentration of the ACSF. The extracellular Ca^{2+} concentration measured in brain interstitial fluid is ~1.2 mM (62), whereas often in vitro studies are using ACSF with 2 mM or more Ca^{2+} (see Chaps. 2-Ruangkittisakul et al. (63) and 11-Kantor et al. (64)). This fact was already reported in (65), in which a Ca^{2+} concentrations of 1.2 mM was found to be critical to be able to record "chattering cells" (66) or fast rhythmic bursting cells in the cortex in vitro. However, the ACSF used in (65) containing (in mM) 124 NaCl, 2.5 KCl, 2 $MgSO_4$, 1.25 NaH_2PO_4, 1.2 $CaCl_2$, 26 $NaHCO_3$, 10 dextrose, even when suitable for recording from specific cortical neurons (chattering cells), did not result in the generation of any rhythmic spontaneous activity in the slice. In order to come up with an ACSF that mimicked the actual ionic *milieu* in which neurons are in the brain, different compositions were tried, as well as "cocktails" with different neurotransmitters (adenosine, isoproterenol, charbachol, etc.). Eventually, it was observed that adult ferret cortical slices placed in an interface chamber with just an ionically modified ACSF (in mM: 124 NaCl, 3.5 KCl, 1 $MgSO_4$, 1.25 NaH_2PO4, 1.2 $CaCl_2$, 26 $NaHCO_3$, 10 dextrose) and with no neurotransmitters added generated a rhythmic spontaneous activity with a frequency of around 0.3 Hz. It was observed that when the concentration of $CaCl_2$ in the solution was decreased from 1 mM to 0.5 mM, the frequency of the oscillations decreased to 0.1 Hz. The oscillatory activity at 0.3 Hz was very similar to the one originally described by Steriade's group in 1993 (6). Simultaneous multiunit recordings were performed in the slices in the vicinity of the intracellularly recorded neuron, revealing that the periods of activity or Up states were the result of the simultaneous activation of a large group of neurons in a relatively synchronized way (Figs. 2b and 3) and that in between Up states, there was hardly any activity in the in vitro cortical network. This finding reinforced the theory proposed by Steriade et al. (6) as a result of their in vivo experiments, that the origin of the slow (<1 Hz) oscillations was in the cortical network, and this cortical activity later recruits the thalamocortical loop.

5.2. Temperature

Temperature is a crucial parameter in homeostasis in the general physiology of the organism, and thus, of the nervous system. For that reason, its effects on different parameters of neural activity have been measured since the early days of electrophysiology. At the cellular level, changes in temperature affect membrane potential, input resistance, action potential amplitude, duration, and propagation and ionic currents (67–70). Synaptic transmission is also affected by changes in temperature, both in invertebrates and mammals (67, 71–73). Having a large impact on the properties of individual neurons and on synaptic transmission, it is evident that temperature must also have a large effect on network activity. Lowering temperature decreases synchronization in the hippocampal network (74, 75), while higher temperatures increase gamma synchrony in the hippocampus (76). Furthermore, metabolic processes are temperature-dependent (77), affecting numerous processes that determine activity. The sensitivity of synaptic transmission to temperature is such that changes in brain temperature associated to physiological conditions (warming up of arterial blood with muscle exercise, lowering of temperature during swimming) are enough to modify the amplitude of synaptic potentials in the hippocampus (78, 79).

In spite of all this evidence, the importance of the temperature in the slice recording chamber during in vitro recordings is often overlooked (see Chaps. 2-Ruangkittisakul et al. (63) and 11-Kantor et al. (64)). It should be taken into account that any particular phenomenon described at a certain temperature (e.g., room temperature) may not be the same at all temperatures. However, it is often assumed that any studied process under a particular condition is very similar or even the same in an in vivo functioning brain. But in the brain in situ, apart from other differences (presence of neurotransmitters, functioning inputs, etc.), the temperature will be 37°C. Varying the temperature in the bath between 26°C and 41°C results in a strong modulation of the cortical spontaneous emergent activity (12). Slow oscillations that comprise Up and

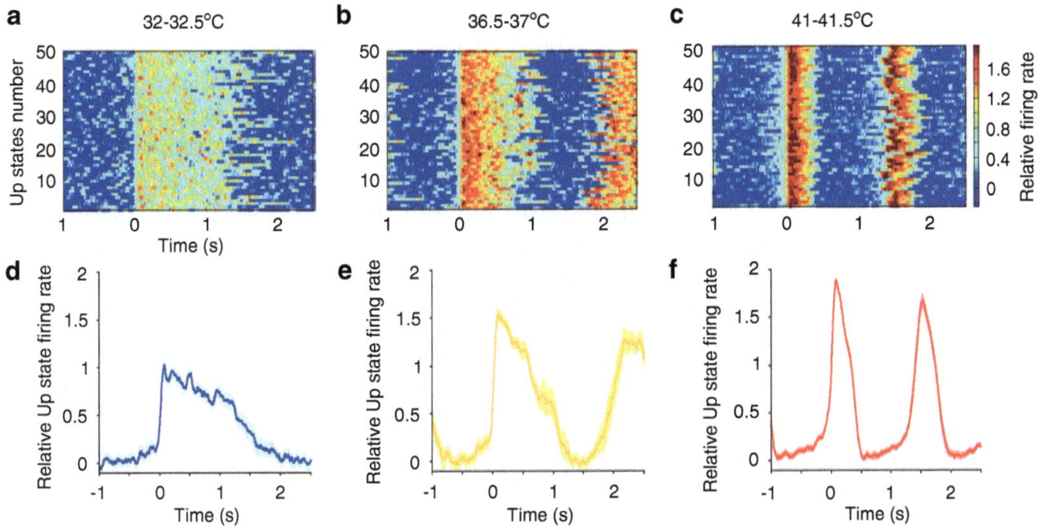

Fig. 7. *Effects of temperature on oscillatory activity in the cortical network.* (**a**, **b**, **c**) Raster plots display in each row the time courses of "relative firing rates" (coded by brightness) centered around the detected Up-state onsets in the specific temperature interval. (**d**, **e**, **f**) Average relative firing rate across detected Up states and centered around Down-to-Up state transitions, as in raster plots of panel (**b**). Shaded strips are the related s.e.m. In all panels, the *left* (**a**), *central* (**b**), and *right* (**c**) columns correspond to different temperatures (32°C, 37°C, and 41°C, respectively) (Reproduced with permission from (12)).

Down states are modulated by temperature. Up states become shorter for warmer temperatures and longer with cooling (Fig. 7). Firing rate during Up states is as well robustly modulated by temperature, increasing with higher temperatures. Given that the sparse firing rate during Down states hardly varied with temperature, what is observed is a progressive merging of Up and Down states for temperatures around 30°C. Below 30°C and down to 26°C, the firing progressively shifted from rhythmic to continuous firing. The slope of the Down-to-Up transitions, which reflects the speed of recruitment of the local network, is progressively steeper for higher temperatures. Altogether, for higher temperatures, activity during Up states becomes more synchronized; Up states showing a steep Down-to-Up transition and high firing rates are concentrated in short Up states, although epileptiform discharges do not appear (12). There is though a tendency to spreading depression (a wave of large neuronal discharge that propagates slowly, in the order of mm/min, and that is followed by silencing of the network) while at >40°C. The relation between increased synchronization and epilepsy is such that cooling has been used to treat epilepsy (80, 81). Regarding speed of propagation of the wave of activity, there is a moderate increase in speed of propagation with temperature. This slight increase could be unexpected since axonal conduction is known to increase with temperature (5%/C°) (69, 82). However, propagation of activity in the cortical network depends also on other factors like excitatory/inhibitory balance, reverberation in the network, and excitability of the network (16, 29, 83). The larger excitability of neurons in the lower range of temperatures (68) and their closeness to threshold probably could compensate factors like the slower axonal conduction at cooler temperatures. The balance of opposite factors results in a moderate increase in the propagation speed with temperature.

Interestingly, fast rhythms (10–80 Hz) were particularly sensitive to temperature. Broadband high-frequency fluctuations in the local field potential are maximal for recordings at physiological temperatures (36–38°C) (12). This suggests that physiological mechanisms are concertedly tuned to generate large-amplitude synaptic activity fluctuations in the network during normal physiological function. Overall, we find that maintaining cortical slices at physiological temperature is critical in order for the generated activity to be analogous to the one in vivo.

5.3. Tissue Oxygen Levels

As mentioned above, the experiments presented here were carried out in an interface-style recording chamber, where oxygen reaches the slice via diffusion from the 95% content of the carbogenated ACSF and mainly by diffusion from the humidified and rich in oxygen (85–90%) air to which the upper slice surface is exposed. For that purpose, the warm water compartment below the recording

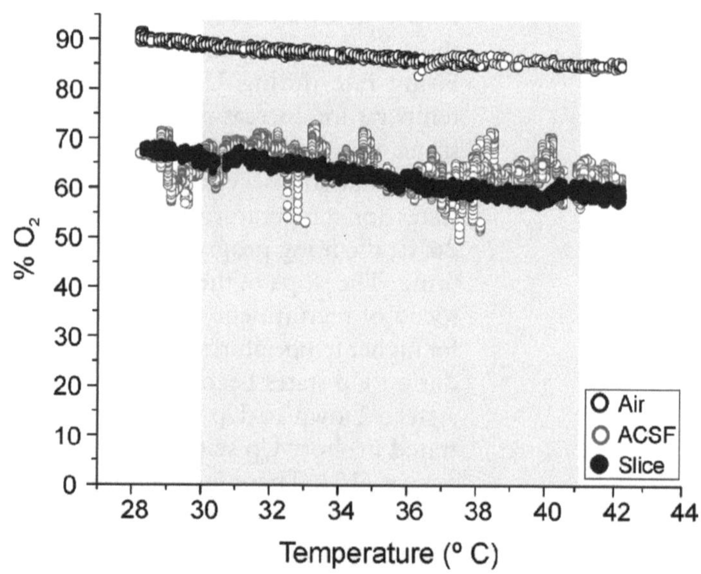

Fig. 8. *Tissue oxygen levels at different temperatures.* Cortical slices were recorded from an interface-style chamber and oxygen levels were measured with an optode (tip diameter 50 μm; Microx TX3, PreSens, Germany). The percentages of oxygen in the air surrounding the slices (*white circles*), in the ACSF (*gray circles*), and within layer 5 of an active visual cortex slice (*black circles*) are represented for a ramp of temperatures between 28°C and 42°C. The complete ramp took 90 min. Each symbol corresponds to 1 s. The larger variability in the measurements taken in ACSF is probably due to the flow dynamics (Reproduced with permission from (12)).

chamber is also carbogenated. The local tissue oxygen level is a critical factor for the generation and propagation of network activity (84), and we checked that the effects of increasing temperature on the slice activity were not secondary to oxygen deprivation (12). Oxygen was measured with an optode (tip diameter 50 μm; Microx TX3, PreSens, Germany) during ramps of temperature between 28°C and 42°C (Fig. 8). This microsensor is designed and was calibrated to be used in liquid, tissue (response time 1 s), and air (response time 0.5 s). Oxygen in the air around the slice was of 90% at 28°C and dropped to 85% at 42°C. Oxygen levels measured in the ACSF in which the slice was bathed were of 66% at 28°C and dropped to 60% when at 42°C (see Chap. 2-Ruangkittisakul et al. (63)). Oxygen level was also measured in layer 5 of visual cortex slices (n = 3) at approximately 100-μm depth during a temperature ramp (28–42°C). In 2 out of these 3 slices, concurrent spontaneous activity was recorded (supplemental data Fig. 6 in (12)). While at 30°C, the oxygen level in an active slice was of 65.5–66.5% (n = 3) and decayed progressively while increasing the temperature, reaching 59–61% while at 42°C (12). An advantage of interface-style recording chambers with respect to submerged-style chamber is that the oxygen levels in the interface-style chamber while at 42°C was the same as the one recorded in an optimized submerged-style chamber while at 32°C (84).

Acknowledgments

Based on work supported by the Ministerio de Ciencia e Innovación de España (BFU2011-27094). MVSV's position is supported by ICREA (Institut Catala de Recerca i Estudis Avançats) at IDIBAPS (Institut d'Investigacions Biomèdiques August Pi i Sunyer). I would like to thank all my collaborators, in particular, R. Reig, M. Ruiz Mejías, V. F. Descalzo, M. Mattia, and M. Winograd.

References

1. Li CL, Mc Ilwain H (1957) Maintenance of resting membrane potentials in slices of mammalian cerebral cortex and other tissues in vitro. J Physiol 139:178–190
2. Yamamoto C, McIlwain H (1966) Potentials evoked in vitro in preparations from the mammalian brain. Nature 210:1055–1056
3. Connors BW, Gutnick MJ, Prince DA (1982) Electrophysiological properties of neocortical neurons in vitro. J Neurophysiol 48:1302–1320
4. Edwards FA, Konnerth A, Sakmann B, Takahashi T (1989) A thin slice preparation for patch clamp recordings from neurons of the mammalian central nervous system. Pflüger's Arch 414:600–612
5. Sanchez-Vives MV, McCormick DA (2000) Cellular and network mechanisms of rhythmic recurrent activity in neocortex. Nat Neurosci 3:1027–1034
6. Steriade M, Nunez A, Amzica F (1993) A novel slow (<1 Hz) oscillation of neocortical neurons in vivo: depolarizing and hyperpolarizing components. J Neurosci 13:3252–3265
7. Aghajanian GK, Rasmussen K (1989) Intracellular studies in the facial nucleus illustrating a simple new method for obtaining viable

motoneurons in adult rat brain slices. Synapse 3:331–338

8. Cunningham MO, Davies CH, Bühl EH, Kopell N, Whittington MA (2003) Gamma oscillations induced by kainate receptor activation in the entorhinal cortex in vitro. J Neurosci 23:9761–9769
9. Trapp S, Ballanyi K (2012) Autonomic nervous system in vitro: studying tonically active neurons controlling vagal outflow in rodent brainstem slices. In Isolated Central Nervous System Circuits (Ed K Ballanyi), Neuromethods Series Vol. 73 (Ed W Walz). Springer Science+Business Media, LLC, New York, NY, pp 1–59
10. Ye JH, Zhang J, Xiao C, Kong JQ (2006) Patch-clamp studies in the CNS illustrate a simple new method for obtaining viable neurons in rat brain slices: glycerol replacement of NaCl protects CNS neurons. J Neurosci Methods 158:251–259
11. Finkel A et al (1993) The Axon guide for electrophysiology and biophysics. Axon Instruments Inc, Foster City
12. Reig R, Mattia M, Compte A, Belmonte C, Sanchez-Vives MV (2010) Temperature modulation of slow and fast cortical rhythms. J Neurophysiol 103:1253–1261
13. Sanchez-Vives MV, Mattia M, Compte A, Perez-Zabalza M, Winograd M, Descalzo VF, Reig R (2010) Inhibitory modulation of cortical up states. J Neurophysiol 103: 1253–1261
14. Chauvette S, Volgushev M, Timofeev I (2010) Origin of active states in local neocortical networks during slow sleep oscillation. Cereb Cortex 20:2660–2674
15. Sakata S, Harris KD (2009) Laminar structure of spontaneous and sensory-evoked population activity in auditory cortex. Neuron 64: 404–418
16. Compte A, Sanchez-Vives MV, McCormick DA, Wang XY (2003) Cellular and network mechanisms of slow oscillatory activity (<1 Hz) and wave propagations in a cortical network model. J Neurophysiol 89:2707–2725
17. Volgushev M, Chauvette S, Mukovski M, Timofeev I (2006) Precise long-range synchronization of activity and silence in neocortical neurons during slow-wave oscillations. J Neurosci 26:5665–5672
18. Shu Y, Hasenstaub A, Badoual M, Bal T, McCormick DA (2003) Barrages of synaptic activity control the gain and sensitivity of cortical neurons. J Neurosci 23:10388–10401
19. Haider B, Duque A, Hasenstaub AR, McCormick DA (2006) Neocortical network activity in vivo is generated through a dynamic balance of excitation and inhibition. J Neurosci 26:4535–4545
20. Bazhenov M, Timofeev I, Steriade M, Sejnowski TJ (2002) Model of thalamocortical slow-wave sleep oscillations and transitions to activated States. J Neurosci 22:8691–8704
21. Contreras D, Timofeev I, Steriade M (1996) Mechanisms of long-lasting hyperpolarizations underlying slow sleep oscillations in cat corticothalamic networks. J Physiol 494: 251–264
22. Frohlich F, Bazhenov M, Timofeev I, Steriade M, Sejnowski TJ (2006) Slow state transitions of sustained neural oscillations by activity-dependent modulation of intrinsic excitability. J Neurosci 26:6153–6162
23. Cunningham MO, Pervouchine DD, Racca C, Kopell NJ, Davies CH, Jones RS, Traub RD, Whittington MA (2006) Neuronal metabolism governs cortical network response state. Proc Natl Acad Sci USA 103:5597–5601
24. Mann EO, Kohl MM, Paulsen O (2009) Distinct roles of $GABA_A$ and $GABA_B$ receptors in balancing and terminating persistent cortical activity. J Neurosci 29:7513–7518
25. Brumberg, J. C., Sanchez-Vives, M. V. & Mccormick, D. A. (2000). Waking up the sleeping slice. In Society for Neuroscience Abstarcts, vol. 26, pp. 1966
26. Schwindt PC, Spain WJ, Foehring RC, Chubb MC, Crill WE (1988) Slow conductances in neurons from cat sensorimotor cortex in vitro and their role in slow excitability changes. J Neurophysiol 59:450–467
27. Bhattacharjee A, Kaczmarek LK (2005) For K^+ channels, Na^+ is the new Ca^{2+}. Trends Neurosci 28:422–428
28. Massimini M, Huber R, Ferrarelli F, Hill S, Tononi G (2004) The sleep slow oscillation as a traveling wave. J Neurosci 24:6862–6870
29. Sanchez-Vives MV, Descalzo VF, Reig R, Figueroa NA, Compte A, Gallego R (2008) Rhythmic spontaneous activity in the piriform cortex. Cereb Cortex 18:1179–1192
30. Sanchez-Vives MV, Mattia M, Compte A, Perez-Zabalza M, Winograd M, Descalzo VF, Reig R (2011) Inhibitory modulation of cortical up states. J Neurophysiol 104: 1314–1324
31. Steriade M, Contreras D, Amzica F, Timofeev I (1996) Synchronization of fast (30–40 Hz) spontaneous oscillations in intrathalamic and thalamocortical networks. J Neurosci 16: 2788–2808
32. Hasenstaub A, Shu Y, Haider B, Kraushaar U, Duque A, McCormick DA (2005) Inhibitory postsynaptic potentials carry synchronized

frequency information in active cortical networks. Neuron 47:423–435
33. McCormick DA, Shu Y, Hasenstaub A, Sanchez-Vives M, Badoual M, Bal T (2003) Persistent cortical activity: mechanisms of generation and effects on neuronal excitability. Cereb Cortex 13:1219–1231
34. Sanchez-Vives MV, Reig R, Descalzo VF et al. (2002) In: FENS (Federation European Neurosciences Societies). Eur J Neurosci, Paris
35. Timofeev I, Grenier F, Bazhenov M, Sejnowski TJ, Steriade M (2000) Origin of slow cortical oscillations in deafferented cortical slabs. Cereb Cortex 10:1185–1199
36. Sanchez-Vives MV, Reig R, Winograd M et al. (2007) In: Timofeev I (ed) Mechanisms of spontaneous active states in the neocortex, (Research Signpost), Kerala, India, pp 23–44
37. Reig R, Gallego R, Nowak LG, Sanchez-Vives MV (2006) Impact of cortical network activity on short-term synaptic depression. Cereb Cortex 16:688–695
38. Dickson CT, Biella G, de Curtis M (2003) Slow periodic events and their transition to gamma oscillations in the entorhinal cortex of the isolated guinea pig brain. J Neurophysiol 90:39–46
39. McCormick DA, Connors BW, Lighthall JW, Prince DA (1985) Comparative electrophysiology of pyramidal and sparsely spiny stellate neurons of the neocortex. J Neurophysiol 54: 782–806
40. Nowak LG, Azouz R, Sanchez-Vives MV, Gray CM, McCormick DA (2003) Electrophysiological classes of cat primary visual cortical neurons in vivo as revealed by quantitative analyses. J Neurophysiol 89:1541–1566
41. Gonzalez-Burgos G, Barrionuevo G, Lewis DA (2000) Horizontal synaptic connections in monkey prefrontal cortex: an in vitro electrophysiological study. Cereb Cortex 10:82–92
42. Reig R, Sanchez-Vives MV (2007) Synaptic transmission and plasticity in an active cortical network. PLoS One 2:e670
43. Petersen CC, Hahn TT, Mehta M, Grinvald A, Sakmann B (2003) Interaction of sensory responses with spontaneous depolarization in layer 2/3 barrel cortex. Proc Natl Acad Sci USA 100:13638–13643
44. Fontanini A, Spano P, Bower JM (2003) Ketamine-xylazine-induced slow (<1.5 Hz) oscillations in the rat piriform (olfactory) cortex are functionally correlated with respiration. J Neurosci 23:7993–8001
45. Wolansky T, Clement EA, Peters SR, Palczak MA, Dickson CT (2006) Hippocampal slow oscillation: a novel EEG state and its coordination with ongoing neocortical activity. J Neurosci 26:6213–6329
46. Kawaguchi Y (2001) Distinct firing patterns of neuronal subtypes in cortical synchronized activities. J Neurosci 21:7261–7272
47. Silberberg G, Wu C, Markram H (2004) Synaptic dynamics control the timing of neuronal excitation in the activated neocortical microcircuit. J Physiol 556:19–27
48. van Drongelen W, Koch H, Marcuccilli C, Pena F, Ramirez JM (2003) Synchrony levels during evoked seizure-like bursts in mouse neocortical slices. J Neurophysiol 90:1571–1580
49. Ruiz-Mejias M et al. (2011) Slow and fast rhythms generated in the cerebral cortex of the anesthetized mouse. J Neurophysiol 106(6):2910–2921
50. DeFelipe J (2005) In: Casanova MF (ed) Neocortical modularity and the cell minicolumn. Nova, New York, pp 57–91
51. DeFelipe J, Alonso-Nanclares L, Arellano JI (2002) Microstructure of the neocortex: comparative aspects. J Neurocytol 31:299–316
52. Gilbert CD, Wiesel TN (1983) Clustered intrinsic connections in cat visual cortex. J Neurosci 3:1116–1133
53. Rockland KS (1985) Anatomical organization of primary Visual Cortex (Area 17) in the Ferret. J Comp Neurol 241:225–236
54. Lund JS, Yoshioka T, Levitt JB (1993) Comparison of intrinsic connectivity in different areas of Macaque monkey cerebral cortex. Cereb Cortex 3:148–162
55. Watts DJ, Strogatz SH (1998) Collective dynamics of 'small-world' networks. Nature 393:440–442
56. Sanchez-Vives MV, Compte A (2005) Structural statistical properties of the connectivity could underlie the difference in activity propagation velocities in visual and olfactory cortices, In: Lecture notes in computer science Springer Lecture Notes in Computer Science, 2005, Volume 3561/2005, 133–142, DOI: 10.1007/11499220_14 Springer Verlag Berlin Heidelberg133
57. Cossart R, Aronov D, Yuste R (2003) Attractor dynamics of network UP states in the neocortex. Nature 423:283–288
58. de Lorente Nó R (1949) In: Fulton JF (ed) Physiology of the nervous system. Oxford University Press, New York, pp 228–330
59. Latham PE, Richmond BJ, Nelson PG, Nirenberg S (2000) Intrinsic dynamics in neuronal networks. I. Theory. J Neurophysiol 83:808–827
60. Yamaguchi T (1986) Cerebral extracellular potassium concentration change and cerebral

impedance change in short-term ischemia in gerbil. Bull Tokyo Med Dent Univ 33:1–8
61. Zhang ET, Hansen AJ, Wieloch T, Lauritzen M (1990) Influence of MK-801 on brain extracellular calcium and potassium activities in severe hypoglycemia. J Cereb Blood Flow Metab 10:136–139
62. Hansen AJ (1985) Effect of anoxia on ion distribution in the brain. Physiol Rev 65:101–148
63. Ruangkittisakul A, Panaitescu B, Secchia L, Bobocea N, Kantor C, Kuribayashi J, Iizuka M, Ballanyi K (2012) Isolated brainstem respiratory centers from perinatal rodents. In Isolated Central Nervous System Circuits (Ed K Ballanyi), Neuromethods Series Vol. 73 (Ed W Walz). Springer Science+Business Media, LLC, New York, NY, pp 61–124
64. Kantor C, Panaitescu B, Kuribayashi J, Ruangkittisakul A, Jovanovic I, Leung V, Lee TF, MacTavish D, Jhamandas JH, Cheung PY, Ballanyi K (2012) Spontaneous neural network oscillations in hippocampus, cortex, and *locus coeruleus* of newborn rat and piglet brain slices. In Isolated Central Nervous System Circuits (Ed K Ballanyi), Neuromethods Series Vol. 73 (Ed W Walz). Springer Science+Business Media, LLC, New York, NY, pp 315–356
65. Brumberg JC, Nowak LG, McCormick DA (2000) Ionic mechanisms underlying repetitive high-frequency burst firing in supragranular cortical neurons. J Neurosci 20:4829–4843
66. Gray CM, McCormick DA (1996) Chattering cells: superficial pyramidal neurons contributing to the generation of synchronous oscillations in the visual cortex. Science 274:109–113
67. Thompson SM, Masukawa LM, Prince DA (1985) Temperature dependence of intrinsic membrane properties and synaptic potentials in hippocampal CA1 neurons in vitro. J Neurosci 5:817–824
68. Volgushev M, Vidyasagar TR, Chistiakova M, Yousef T, Eysel UT (2000) Membrane properties and spike generation in rat visual cortical cells during reversible cooling. J Physiol 522:59–76
69. Hodgkin AL, Katz B (1949) The effect of temperature on the electrical activity of the giant axon of the squid. J Physiol 109:240–249
70. Borst JG, Sakmann B (1998) Calcium current during a single action potential in a large presynaptic terminal of the rat brainstem. J Physiol 506:143–157
71. Asztely F, Erdemli G, Kullmann DM (1997) Extrasynaptic glutamate spillover in the hippocampus: dependence on temperature and the role of active glutamate uptake. Neuron 18:281–293
72. Volgushev M, Vidyasagar TR, Chistiakova M, Eyzel UT (2000) Synaptic transmission in the neocortex during reversible cooling. Neuroscience 98:9–22
73. Fujii S, Sasaki H, Ito K, Kaneko K, Kato H (2002) Temperature dependence of synaptic responses in guinea pig hippocampal CA_1 neurons in vitro. Cell Mol Neurobiol 22: 379–391
74. Javedan SP, Fisher RS, Eder HG, Smith K, Wu J (2002) Cooling abolishes neuronal network synchronization in rat hippocampal slices. Epilepsia 43:574–580
75. Motamedi GK, Salazar P, Smith EL, Lesser RP, Webber WR, Ortinski PI, Vicini S, Rogawski MA (2006) Termination of epileptiform activity by cooling in rat hippocampal slice epilepsy models. Epilepsy Res 70:200–210
76. Wu J, Javedan SP, Ellsworth K, Smith K, Fisher RS (2001) Gamma oscillation underlies hyperthermia-induced epileptiform-like spikes in immature rat hippocampal slices. BMC Neurosci 2:18
77. Esmann M, Skou JC (1988) Temperature-dependencies of various catalytic activities of membrane-bound Na^+/K^+-ATPase from ox brain, ox kidney and shark rectal gland and of C12E8-solubilized shark Na^+/K^+-ATPase. Biochim Biophys Acta 944:344–350
78. Moser EI, Mathiesen LI (1996) Relationship between neuronal activity and brain temperature in rats. Neuroreport 7:1876–1880
79. Andersen P, Moser EI (1995) Brain temperature and hippocampal function. Hippocampus 5:491–498
80. Rothman SM (2009) The therapeutic potential of focal cooling for neocortical epilepsy. Neurotherapeutics 6:251–257
81. Rothman SM, Smyth MD, Yang XF, Peterson GP (2005) Focal cooling for epilepsy: an alternative therapy that might actually work. Epilepsy Behav 7:214–221
82. Johnson EW, Olsen KJ (1960) Clinical value of motor nerve conduction velocity determination. J Am Med Assoc 172:2030–2035
83. Pinto DJ, Patrick SL, Huang WC, Connors BW (2005) Initiation, propagation, and termination of epileptiform activity in rodent neocortex in vitro involve distinct mechanisms. J Neurosci 25:8131–8140
84. Hajos N, Ellender TJ, Zemankovics R, Mann EO, Exley R, Cragg SJ, Freund TF, Paulsen O (2009) Maintaining network activity in submerged hippocampal slices: importance of oxygen supply. Eur J Neurosci 29:319–327

Chapter 9

Living Human Brain Slices: Network Analysis Using Voltage-Sensitive Dyes

Tilman Broicher and Erwin-Josef Speckmann

Abstract

The study of diseases of the nervous system relies heavily on the use of animal models which try to replicate the human condition via various methods. The underlying assumption is that the root causes of the impairment are similar in the animal model and the human patient, thus validating the findings in the animal model as relevant for the human condition. This assumption is difficult to verify, as experiments analogous to those being performed in the animal model obviously cannot be performed in humans. In severe cases of epilepsy and during treatment of brain tumors, surgical removal of diseased brain tissue is the last available option. The extracted tissue may be used to study the disease directly using in vitro methodology, circumventing the need for an animal model. During the surgical procedure, the epileptic "focus" is removed from the temporal lobe. The dissected tissue is usually composed of neocortex, hippocampus, and amygdala. This chapter describes the preparation of brain slices from the tissue excised during the surgery and the use of voltage-sensitive dyes to investigate network activity in the human brain slices.

Key words: Epilepsy, Human brain slice, Membrane potential, Neuronal population, Voltage-sensitive dye

1. Introduction

The study of diseases of the nervous system relies heavily on the use of animal models which try to replicate the human condition via various methods. The underlying assumption is that the root causes of the impairment are similar in the animal model and the human patient, thus validating the findings in the animal model as relevant for the human condition. This assumption is difficult to verify, as experiments analogous to those being performed in the animal model obviously cannot be performed in humans. In some cases of epilepsy, pharmacological treatment is not successful.

Klaus Ballanyi (ed.), *Isolated Central Nervous System Circuits*, Neuromethods, vol. 73,
DOI 10.1007/978-1-62703-020-5_9, © Springer Science+Business Media New York 2012

These forms of epilepsy are referred to as pharmacoresistant epilepsies. If medication fails and the frequency and severity of seizures endanger the life of the patient or considerably impair the quality of life, surgical removal of diseased brain tissue is the last available option. This procedure is only feasible if the site of the origin of seizures (often called the "focus") can be identified and does not lie within regions indispensable for speech, motor, or sensory function. This is most often the case in pharmacoresistant temporal lobe epilepsy. Given the patient's consent, the tissue extracted during this surgical procedure may be used to study the disease directly using in vitro methodology, circumventing the need for an animal model.

During the surgical procedure, the epileptic focus is removed from the temporal lobe. The dissected tissue is usually composed of neocortex, hippocampus, and amygdala. Once tissue from human patients is available for study, appropriate methods to address the given question have to be selected. In the case of tissue extracted from epileptic patients, the investigation of neuronal and network activity and excitability are of primary interest. These questions may be adequately addressed in brain slices, which leave the local neuronal networks sufficiently intact and allow experimental access to single neurons. In principle, all the advantages and limitations of animal brain slices apply to slices from human brain tissue. The major advantages of this technique are good control over the extracellular milieu, including the ability to apply defined concentrations of pharmacological agents, as well as easy access to individual anatomical subregions and cells within the slice during the experiment (see also, e.g., Chaps. 1-Trapp and Ballanyi (1), 2- Ruangkittisakul et al. (2), 3-Moore et al. (3), 8-Sanchez-Vives (4), 10-Luhmann and Kilb (5), and 12-De Curtis et al. (6)).

Several recording methods are currently available to study neural network activity, including electrical and optical techniques. Voltage-sensitive dyes (VSDs) are an excellent tool in this regard due to their fast response properties and the possibility to simultaneously record from a large number of neurons (7–9) (see also Chap. 2-Ruangkittisakul et al. (2)). The principle of VSD imaging is the incorporation of optically active compounds into the cell membrane, which change their optical properties as a function of the membrane potential. The two major types are fluorescent and absorption dyes. Fluorescent dyes change the amount of emitted light in response to changes in membrane potential, while absorption dyes change the amount of absorbed light. The emitted/passed light is usually recorded with a charge-coupled device (CCD) camera or an array of photodiodes attached to a microscope. Changes in emitted/passed light are directly proportional to changes in membrane potential (7–12).

2. Materials

This section provides brief reference to equipment needed for VSD imaging in brain slices and the commercial sources. For more detailed information regarding equipment for electrophysiological and optical analysis of isolated neural networks, see other chapters (see also, e.g., Chaps. 1-Trapp and Ballanyi (1), 2- Ruangkittisakul et al. (2), 3-Moore et al. (3), 8-Sanchez-Vives (4) and 12 De Curtis et al. (6)).

2.1. Necessary Equipment

The core of the recording setup is a microscope with a submerged-type recording chamber to hold the slice during the experiment. The microscope has to be equipped with an appropriate filter set, a light source, and a recording device (CCD camera or photodiode array). In addition, a computer with appropriate recording software is needed. A vibration isolated table should be used to support the recording setup. To ensure a sufficient flow of solution through the recording chamber, either a peristaltic pump or gravitational flow system can be used. If electrophysiological recordings or (electrical) stimulations are to be performed, the setup has to be equipped with manipulators for positioning the electrodes. Furthermore, amplifiers, stimulus generators/isolators, and a computer with the appropriate recording software are necessary. To prepare brain slices, a vibratome and a holding chamber for the slices are needed. The solutions used in physiological experiments involving brain slices have to be gassed. Specifically, depending on the pH buffer used (usually CO_2/HCO_3 or HEPES), it is necessary to have access to either O_2 or carbogen (95% O_2, 5% CO_2) in the laboratory.

2.2. Manufacturers

The microscopes, filter sets, light sources, recording software, and cameras for imaging experiments can be obtained, e.g., from Carl Zeiss (www.zeiss.com), Nikon (www.nikoninstruments.com), Leica (www.leica-microsystems.com), or Olympus (www.olympus-global.com). Optical filters for imaging experiments are also offered by Chroma (www.chroma.com) and Semrock (www.semrock.com). A separate source for camera systems and recording software is RedShirtImaging (www.redshirtimaging.com). Vibration isolated tables are produced either by Technical Manufacturing Corporation (TMC www.techmfg.com) or Newport (www.newport.com). Fluid pumps can be obtained from Masterflex (www.masterflex.com) or Gilson (www.gilson.com). Slice recording chambers and manipulators can be obtained from Luigs und Neumann (www.luigs-neumann.com), Sutter Instruments (www.sutter.com), or Warner Instruments (www.warneronline.com). Amplifiers for electrophysiological recordings are manufactured by

Axon Instruments (www.moleculardevices.com), HEKA (www.heka.com), or Warner Instruments (www.warneronline.com). These companies also supply recording software. Vibratomes are sold by Leica (www.leica-microsystems.com) or Warner Instruments (www.warneronline.com). The ingredients for the solutions used in brain slice experiments can be obtained from Sigma-Aldrich (www.sigmaaldrich.com). Pharmacological agents can be obtained from Sigma-Aldrich or Tocris (www.tocris.com). Dyes can be obtained from Invitrogen (www.invitrogen.com).

3. Methods

3.1. Human Brain Slice Preparation

Once the surgeon has performed the resection, the excised tissue should be submerged in 2–5°C cold carbogenated (Sect. 2) artificial cerebrospinal fluid (ACSF). It is important to process the resected tissue as quickly as possible, because it deteriorates after being cut off from oxygen and glucose supply. The composition of ACSF used by us is (in mM) 124 NaCl, 4 KCl, 2 $CaCl_2$, 1.24 NaH_2PO_4, 1.3 $MgSO_4$, 26 $NaHCO_3$, and 10 D-glucose. The pH should be kept between 7.35 and 7.45 and depends on factors such as the exact CO_2 content in (analytical) carbogen and temperature. The excised tissue block, which may vary in size from one to several cm^3, should be trimmed prior to the slicing procedure to remove any parts which are not of interest. After the tissue has been trimmed, it is ready to be sliced using a conventional vibratome (see Chaps. 1-Trapp and Ballanyi (1) and 2-Ruangkittisakul et al. (2)). To foster tissue survival, the slicing procedure should be performed as soon as possible after the resection. For this reason, it is advisable to slice the tissue in a location as close as possible to the site of the surgery. The slicing procedure itself takes ~10 min and should be performed in 2–5°C cold, carbogenated ACSF. Once the slices are prepared, they have to be transferred into a holding chamber filled with carbogenated ACSF at 30°C (13, 14). Within the holding chamber, the slices should rest upon a mesh of inert material (e.g., nylon) to allow both sides to be in contact with the ACSF. After the slicing procedure, the brain slices should be allowed to rest for at least 60 min before commencing recordings of any kind.

Often, the sites of surgery and in vitro experimentation are separated by larger distances (see also Chap. 3-Moore et al. (3)). If this is the case, it is necessary to construct a portable holding chamber for the brain slices. The portable chamber should be composed of a crate allowing for a carbogen supply, a power supply, a heating system, and a holding chamber. Figure 1 shows a schematic drawing of the portable holding chamber used by our group.

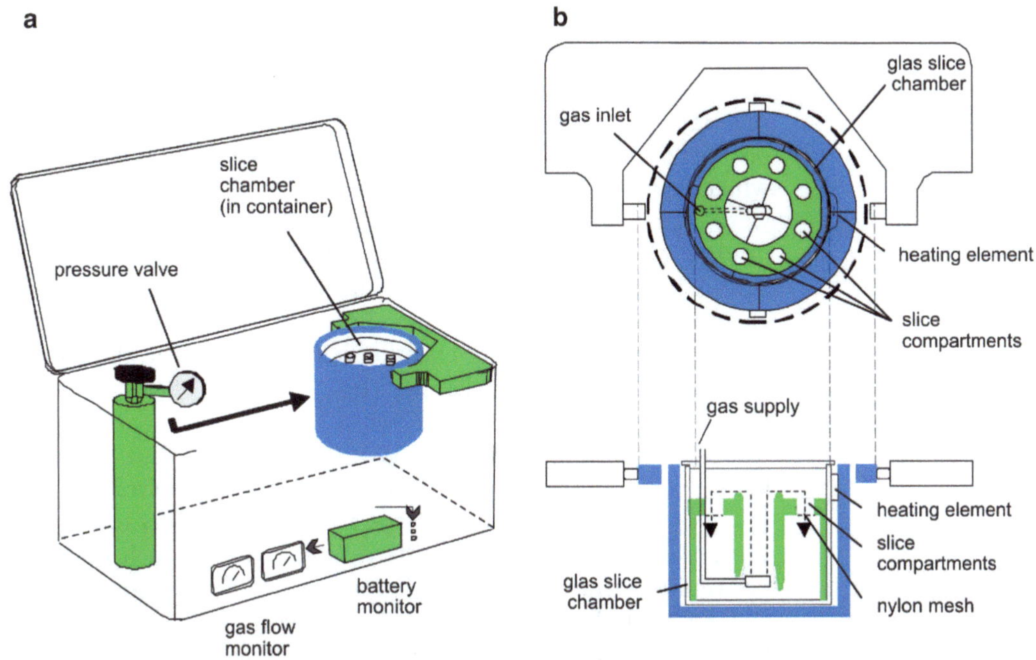

Fig. 1. *Schematic drawing of a portable human brain slice holding chamber*. (**a**) Schematic drawing of the crate containing the holding/slice chamber including a heating system, a gas supply, and a battery. (**b**) The holding/slice chamber viewed from the top (*upper panel*) and from the side (*lower panel*). The holding/slice chamber is filled with carbogenated and heated (30°C) artificial cerebrospinal fluid (ACSF). Slices rest on nylon meshes in individual compartments and are guarded against floating out of their compartments by additional nylon mesh placed about 1 cm above the first. For a more detailed description of the crate and chamber, the reader is referred to Köhling et al. (15) (Reprinted with permission from (15)).

Using this device, it was possible to transport the brain slices for distances of up to 100 km without noticeable adverse effects on slice quality. A detailed description of a similar device is given in Köhling et al. (15).

3.2. Voltage-Sensitive Dye Imaging

3.2.1. Staining of Brain Slices

Prior to optical recording, the brain slices have to be stained with the VSD. The description given here applies to the fluorescent VSD RH795 (Invitrogen). A series of pilot experiments were performed to determine a good staining parameter set. Initial experiments for finding optimal dye staining parameters should not be performed on human tissue. Initially, our group used adult (6 months or older) Long-Evans rats for this purpose. Our experiments with RH795 indicate that staining parameters for human neocortex, hippocampus, and amygdala are similar to those in adult Long-Evans rats and C57BL/6 mice. Though, this may not be true for other dyes and/or tissues. Furthermore, staining parameters yielding optimal results as defined by reproducibility of optical signals, phototoxicity, and bleaching/washout will vary with dyes and setups (7, 10–12). Thus, optimal parameters have to be found prior to experimentation. For population recordings, the entire slice should be stained with dye. To minimize the amount of dye needed per slice, we used a staining chamber consisting of wells each holding a single slice. These wells each had their own carbogen supply, were kept at 30°C, and were gently shaken to prevent dye aggregation. Each well was filled with 1 ml of ACSF in which RH795 was dissolved at a dilution of 12.5 μg/ml. The slices were stained for 60 min before being transferred back into the holding chamber containing standard ACSF (also kept at 30°C). To wash out unbound dye, the slices were kept in the holding chamber for at least 60 min before recording commenced. Furthermore, the holding chamber should be kept in the dark to minimize dye bleaching (13, 14, 16).

3.2.2. Optical Recording Setup

The VSD recording setup consists of a light source (usually a xenon lamp); an optical filter set including a dichroic mirror, a light microscope, and a submerged-type recording chamber to hold the slice; and a device to record the optical signals, which may be a photodiode array or a fast CCD camera (see also Chaps. 1-Trapp and Ballanyi (1) and 8-Sanchez-Vives (5)). To take advantage of the rapid kinetics of the VSD, an image sampling rate of ~1 kHz or more is desirable. Recording software for optical data is usually supplied by the company providing the CCD camera or the photodiode array. In our laboratory, we use Neuroplex (RedShirtImaging) [for other systems, see Chaps. 1-Trapp and Ballanyi (1) and 8-Sanchez-Vives (5) or Sect. 2 in this chapter]. Either upright or inverted microscopes may be incorporated into the setup. If an upright microscope is used, immersion objectives have to be employed to avoid artifacts caused by differences in the refractive index of air and the ACSF and scattering artifacts caused by ACSF surface curvature. Furthermore, placing stimulation or recording electrodes inside the imaged area might lead to obstruction in case of light-impermeable electrodes or scattering in case of glass electrodes. If an inverted microscope is used, these problems do not apply, but optical recordings will have to be obtained from the lower side of the slice, while electrical stimulation and electrophysiological recording will originate from the upper side of the slice.

Figure 2 shows a schematic drawing of a filter set, which can be used for imaging with RH795, which has an excitation maximum at 530 nm and an emission maximum at 712 nm. The optical setup is designed to reduce the spectral composition of the excitation light and guide this light onto the stained slice, while guiding the emitted light onto a photodiode array which records the VSD signal. In the setup described here, the spectral composition of the excitation light generated by the xenon lamp is reduced by an optical band-pass filter (546 ± 20 nm). The band-pass filtered light is then led onto a dichroic mirror, which reflects the excitation wavelength but allows the emission wavelength to pass (with a cutoff wavelength of 580 nm). The excitation light then passes through

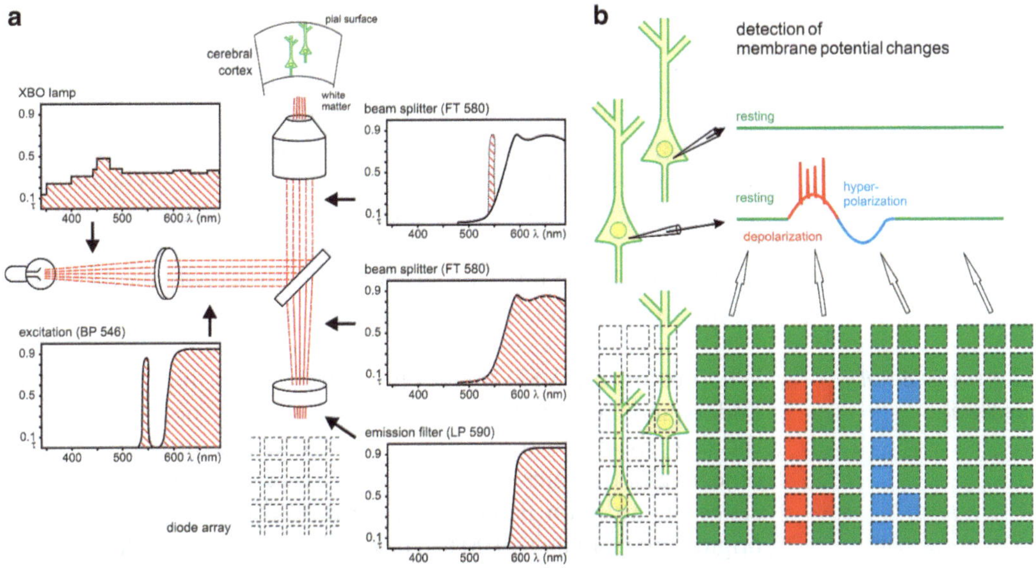

Fig. 2. *Schematic drawing of an optical recording setup using the voltage-sensitive fluorescent dye RH795 (excitation maximum 530 nm, emission maximum 712 nm) and an example for the detection of membrane potential changes by a photodiode array.* (**a**) The optical recording setup. Plots showing the spectral composition of the excitation/emission wavelengths are displayed at each stage, as indicated by *arrows*. The excitation light is generated by a xenon lamp (XBO lamp) and is led through an optical band-pass filter (BP, 546 nm) to reduce spectral composition. The excitation light is then led onto a dichroic mirror (beam splitter), with a cutoff wavelength of 580 nm. The excitation light is then passed through the objective of a light microscope onto the stained tissue. The emitted light travels through the objective, passes the dichroic mirror, and is high-pass filtered at 590 nm (emission filter) before hitting the photodiode array (diode array). (**b**) Example detection of voltage-sensitive dye (VSD)-imaged membrane potential changes. Consider two neurons in the field of view. One neuron experiences no membrane potential change (*upper panel*, upper neuron), while the other experiences depolarization followed by hyperpolarization (*upper panel*, lower neuron). If the two neurons map onto the photodiode array as shown in the leftmost graph of the lower panel, a signal as shown in the four right graphs of the *lower panel* will be detected (*white arrows* indicate temporal correlation between the membrane potential and the signal recorded by the photodiode array): Initially, none of the cells change their membrane potential, leading to no change in the detected signal (2nd graph from the left, baseline levels of fluorescence, shown in *green*). Next, the lower neuron depolarizes (3rd graph from the left). This leads to a decrease in emitted light and a change in the signal detected by the photodiodes onto which the neuron maps (*change from green to red*). Following the depolarization, the neuron is hyperpolarized, which leads to an increase in emitted light (shown in *blue*), which is again detected by the photodiodes onto which the neuron maps (4th graph from the left). Finally, the neuron returns to rest, and the emitted light returns to baseline levels (shown in green, 5th graph from the left).

the objective before reaching the slice. The emitted fluorescence travels back through the objective of the microscope, passes the dichroic mirror, and is low-pass filtered (590 nm) before arriving at a photodiode array, which records the VSD signal. For details on optical imaging, see the textbook by Yuste and Konnerth (17).

The imaged area, and thus the spatial resolution, is controlled by the objective. This allows for a range of spatial resolutions to be used during the experiment. For example, a 4× objective used with a photodiode array composed of 464 individual photodiodes will cover a total area of 12.02 mm^2, with a spatial resolution of 161 × 161 μm per diode, while a 20× objective will cover a total area of

0.464 mm^2 with a spatial resolution of 32 × 32 μm per diode. The objective magnification should be chosen according to the size of the structure to be imaged. Often low-magnification objectives will have a lower numerical aperture and therefore a lower signal-to-noise ratio. Thus, it is advisable to use the objective with the highest numerical aperture possible. When imaged activity is stereotypical, e.g., a response to an (electrical) stimulus, high-resolution/small-area images taken from successive recordings may be combined into one larger image, thus enabling the experimenter to image a larger structure with a high spatial resolution.

3.2.3. Voltage-Sensitive Dye Signals

When recording VSD images from brain slices, it is important to understand the nature of the recorded signal. The surface area of the slice, from which signals will be detected, is determined by the objective of the microscope (Sect. 3.2.2). Due to the optical properties of the brain tissue and VSD diffusion characteristics, light originating from up to 100 μm below the slice surface will contribute to the recorded signal. Thus, VSD signals recorded from completely stained slices correspond to average potential changes in columns of tissue extending ~100 μm into the slice (7–13, 18). Furthermore, optical signal amplitudes are directly proportional to the magnitude of voltage deflections as well as the amount of cell membrane experiencing a voltage change (19, 20). This must be taken into consideration when imaging areas with heterogeneous density of neuronal membrane. In addition, non-neuronal sources (e.g., glia) may contribute to the optical signal (Sect. 5.2).

Due to phototoxicity and dye bleaching, VSD imaging has to be limited to trials of a few-second duration (7, 10, 11). After a single session of VSD imaging, the slices should be allowed to recover for an amount of time depending on the length of the recording. The maximal recording length and separation time depend on the type of brain tissue and dye and have to be found empirically. For example, in human brain slices, RH795 allows recording for 1–4 s with trial separations of 3–5 min (13, 14). A wider range of time spans has been reported in animal studies (21–27). In order to directly relate a change of an electrophysiologically recorded potential to an optical signal, calibration procedures are necessary. This is feasible when single cells are analyzed and membrane voltage can be measured directly, but much less so when network activity is investigated. Thus, VSD imaging of network activity reports relative changes in membrane potential.

VSD signals are often expressed as relative changes of the resting fluorescence: $(F_{rest}-F)/F_{rest}$, where F_{rest} corresponds to the resting fluorescence and F corresponds to the recorded signal. $F_{rest}-F$ is often abbreviated as ΔF. The resting fluorescence should be recorded prior to each run of data acquisition. This method prevents errors caused by inhomogeneous staining of the slice. One problem associated with recording of VSD signals is the low signal-to-noise ratio. In our experience, the optimal signal-to-noise ratio for RH795 in brain slices is ~10:1 when a strong stimulus is used or a robust spontaneous event is imaged.

A common approach to circumvent this problem is to record multiple trials of each experimental condition, followed by averaging of the recorded traces, e.g., 3 trials averaged in (21) and 8 trials averaged in (26) (see also Chap. 2-Ruangkittisakul et al. (2)). Final image analysis is then performed on the averaged signals. Obviously, this approach is only feasible when individual trials are sufficiently similar, e.g., responses to stimuli.

One intuitive way of displaying the extent of network activity is via color-coded activity maps. This is done by calibrating a color scale ranging, e.g., from blue to red, to the maxima and minima of the recorded VSD signals and subsequently assigning the corresponding color to each data point recorded by each pixel or photodiode. A map is then generated when color-coded data points from all CCD camera pixels or photodiodes are plotted for a given point in time (Fig. 2). This color map needs to be aligned with the slice. In case of CCD camera recording, the color map can be superimposed with a transmitted light image from the slice. In case of a photodiode array, the position of the array relative to the slice has to be determined. This can be done by incorporating light-emitting diodes into the array, which indicate the position of defined points of the photodiode array.

4. Results

4.1. Optical Recording in Human Brain Slices: Spontaneous Activity

Slices obtained after surgical treatment of temporal lobe epilepsy are prepared and stained as described above (Sect. 3.1), see also (13, 14). Based on the work by Köhling and colleagues, spontaneous VSD-imaged activity is observed in ~40% of cortical slices originating from the temporal lobe of patients with pharmacoresistant temporal lobe epilepsy (13, 28, 29). Importantly, this form of activity is absent in slices obtained from tumor resectates of non-epileptic patients (13). Spontaneously active slices generate interictal-like activity at frequencies around 1 Hz (average 0.7 Hz), as assessed with field potential recordings (28). Examples of spontaneous depolarizing events recorded using VSD imaging and field potentials are shown in Fig. 3. To localize the site of spontaneous activity within the slice, field potential recordings should be employed. Once the site of spontaneous activity has been localized, VSD imaging can commence. Due to the limited recording duration of each run of VSD signal acquisition (1–4 s), the likelihood of recording or "catching" a spontaneous event depends on their frequency and duration. As trace averaging is not possible when

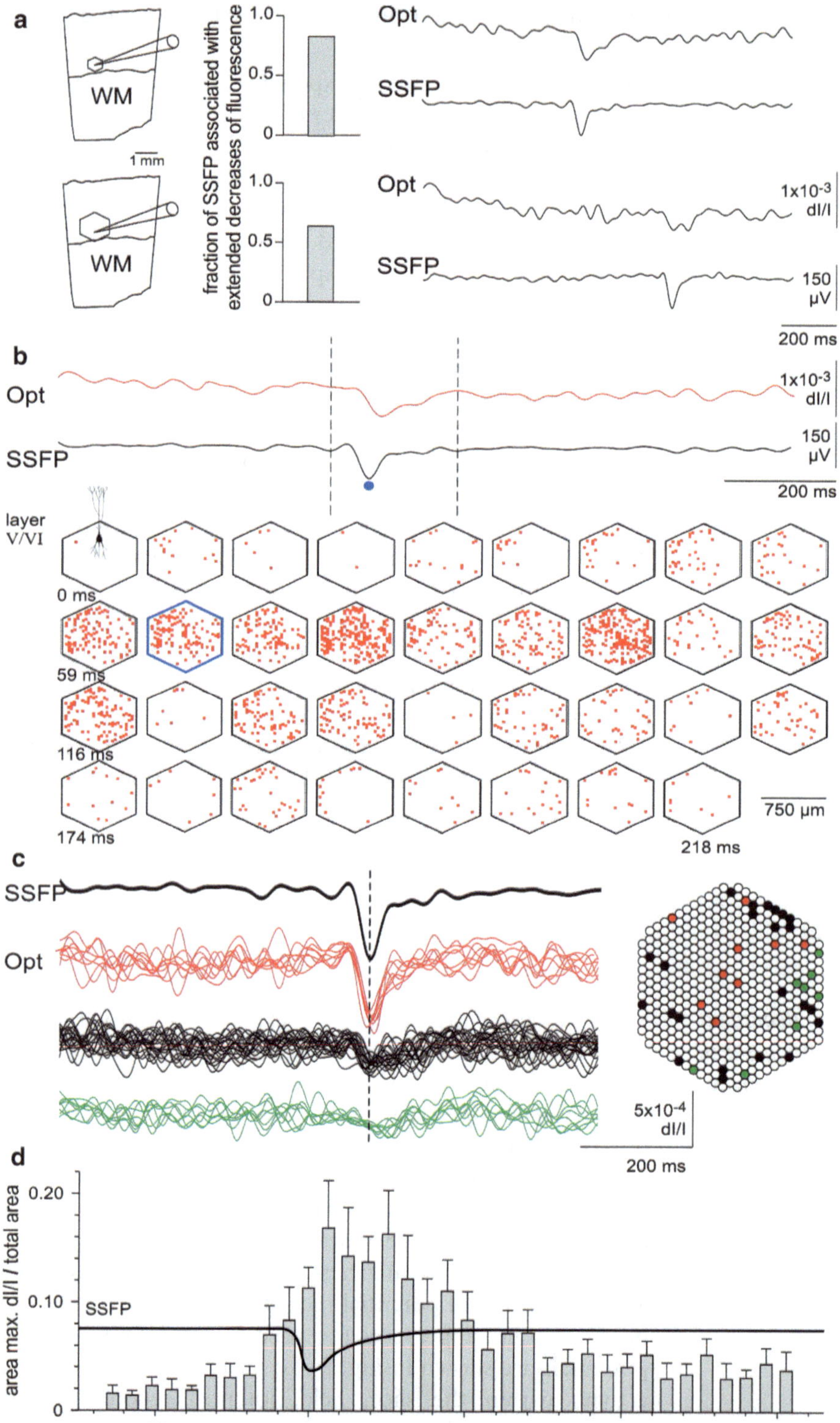

Fig. 3. *Spontaneous optical and concomitant electrical activities in neocortical slice from an epilepsy patient.* (**a**) Synchronous sharp field potentials (SSFP) occurring spontaneously and concomitant neuronal depolarizations detected by VSD imaging (Opt). Recording positions of field potential electrode and optical recording fields (via 20× and 10× objective lenses) are given in the diagram. WM: white matter. The fraction of SSFP associated with extended decreases of

investigating spontaneous activity, a high-magnification/numerical aperture objective should be employed to improve the signal-to-noise ratio. Köhling et al. reported a drop from 83% of field potential deflections co-occuring with a clear depolarization in the VSD signal with a 20× objective to 64% of field potential deflections co-occuring with a depolarization in the VSD signals with a 10× objective (Fig. 3a) (13). Spontaneous interictal-like activity has been described in different brain regions resected from pharmacoresistant temporal lobe epilepsy patients including the cortex (13, 28), the subiculum (30), and the CA2 region of the hippocampus (31). According to these studies, spontaneous activity is reliably observed (in >50% of slices) using field potential recordings (28, 30, 31). The frequency of these events appears to be similar in the cortex and the subiculum (~1 Hz) and slower in the CA2 region (~0.5 Hz). The spontaneous activity in the cortex and the subiculum is sensitive to blockade of ionotropic glutamate and A-type γ-aminobuturic acid ($GABA_A$) receptors, while this has (to the best of our knowledge) not been investigated in the CA2 region (28, 30, 31). In the subiculum, depolarizing GABAergic responses, caused by a depolarizing shift in the Cl^- reversal potential, have been observed in a minority of pyramidal cells. This has been suggested to contribute to the generation of the spontaneous interictal-like activity in this region (30, 32). This mechanism does not seem to be present in the CA2 region however. Here, an imbalance between excitation and inhibition could be involved in the generation of abnormal network activity (31).

Importantly, care has to be taken when interpreting in vitro recordings of spontaneous activity in epileptic human material, because neither recordings of physiological in vitro activity in healthy human tissue nor in vivo extracellular recordings of epileptic tissue are available for comparison.

4.2. Optical Recordings in Living Human Brain Slices: Evoked Activity

Slices obtained after surgical treatment of temporal lobe epilepsy are prepared and stained as described above (Sect. 3.1), see also (13, 14). Evoked activity is recorded after electrical stimulation of the slice. The structure to be stimulated depends on the question to be addressed and the brain region the slice is taken from. Resectates from temporal lobe epilepsy patients have primarily been used to study the hippocampus and the cortex. Commonly

Fig. 3. (continued) fluorescence (neuronal depolarizations) differed depending on the magnification. (**b**) Spatial spread of neuronal depolarizations associated with SSFP; single traces as in (**a**). Optical signals at different time points (6.4 ms interframe-interval; 64 ms before and 160 ms after SSFP peak marked by *blue dot* and frame) covering a field of 750-mm diameter. *Red dots* represent relative minima of fluorescence (upper 10% of fluorescence changes). (**c**) Single SSFP and single optical traces (superimposed). Traces with largest (*red*), substantially (<50%) smaller (*black*), and missing (*green*) optical peaks. The location of the diodes within the array is marked by corresponding colors. (**d**) Fraction of total recording area (750-μm diameter) with maximal fluorescence changes at different time points as in (**b**). Mean ± standard error of 13 events (Reproduced with permission from (13)).

used stimulation targets in hippocampal preparations are the mossy fibers (input to CA3), the Schaffer collaterals (input to CA1), and the perforant pathway (input to the dentate gyrus). In neocortical slices, commonly used stimulation targets are the white matter below layer 6 (activating ascending and descending fiber tracts), layer 1 (horizontal cortical connections), and layer 4 (recipient of subcortical inputs). Figure 4 shows VSD responses of cortical slices from the temporal lobe after stimulation of the border between the white matter and layer 6 and stimulation of layer 1. The traces shown in Fig. 4 are averages from 3 successive recordings separated by 3 min. The increased signal-to-noise ratio allows the use of lower-magnification objectives to cover a larger area of the slice. (The recordings shown in (Fig. 4) were done using a 5× objective.)

5. Limitations and Pitfalls

5.1. Human Brain Slices

A major problem using brain slices from human patients is the great heterogeneity of the origin of the isolated tissue. In case of removal of the epileptic "focus" in temporal lobe epilepsy, the "foci" will be distributed over the entire temporal lobe. Moreover, all patients undergoing surgery have long medication histories. This long exposure to pharmacological agents may influence the properties of nervous tissue. Adding to this complexity is the difference in medication history between patients, which may lead to different pharmacological adaptation processes depending on the individual and the drugs used. Furthermore, patients undergoing surgery may differ substantially in age. This may lead to confounds based on different maturational stages of neuronal tissue, which can be obtained from children as well as aging patients. Finally, the interpretation of findings on tissue from human patients is limited by the absence of healthy control material. A further issue arises due to the size of human material. Most equipment for in vitro neurophysiology is designed for rodent brain slices. The homologous structures in humans are several times larger, which may cause problems due to equipment of insufficient size.

5.2. Voltage-Sensitive Dye Signals

As stated above (Sect. 3.2.3), VSD signals correspond to average membrane potential changes in tissue columns extending ~100 μm into the slice. The amplitudes of VSD signals are proportional to the magnitude of voltage deflections and the amount of membrane generating a potential change. This results in a strong influence of dendritic potential changes and may lead to the detection of the strongest signals in a different area than the somata of

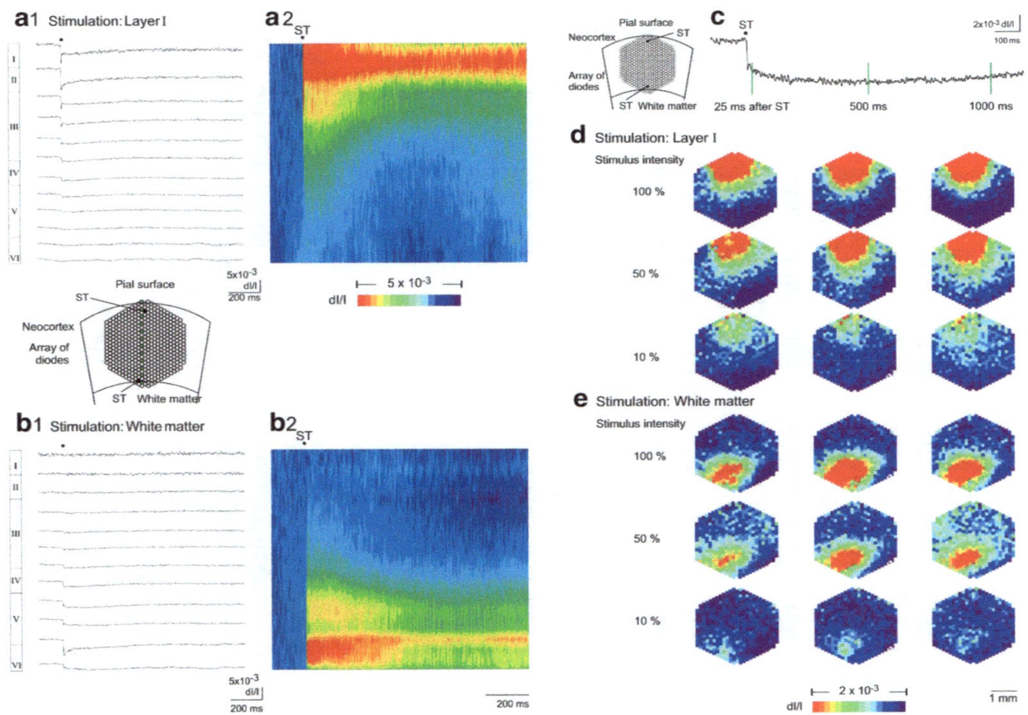

Fig. 4. *VSD imaging of stimulus-induced activity in a line-time plot (*a, b*) and an area plot (*c, d, e*) in a human neocortical slice from an epilepsy patient.* Insets: schematic drawings of the slice and the location of diode array, stimulation electrodes (ST). (**a**, **b**) Line-time plots. The diodes from which optical signals in the line-time plots were generated are shown in the *left inset*. Time of stimulation is indicated by *black dots*. ($\mathbf{a}_1$, $\mathbf{b}_1$) Optical responses of selected diodes (see inset) following stimulation in layer I ($\mathbf{a}_1$) and in white matter ($\mathbf{b}_1$). The corresponding cortical depth is indicated on the *left* side. ($\mathbf{a}_2$, $\mathbf{b}_2$) Pseudocolor line-time plot of fluorescence changes along the stimulation axis following layer 1 ($\mathbf{a}_2$) and white matter ($\mathbf{b}_2$) stimulation. All diodes of the column (see inset) are included. Color calibration for fluorescence intensity is indicated at the *bottom* of ($\mathbf{a}_2$). Cortical depth as indicated in ($\mathbf{a}_1$) and ($\mathbf{b}_1$). (**c**, **d**, **e**) Area plots. (**c**) Fluorescence signal of a single diode positioned in supragranular layers after stimulation of layer 1. The *green lines* mark the time points after stimulation (25 ms, 500 ms, 1,000 ms) at which area plots were constructed (**d**, **e**). Time of stimulation (ST) is indicated by *black dot*. (**d**, **e**) Pseudocolor area plots of fluorescence changes at 25 ms (*left*), 500 ms (*middle*), and 1,000 ms (*right column*) following stimulation in layer 1 (**d**) and white matter (**e**) at stimulus intensities of 100% (*top row*), 50% (*middle row*), and 10% (*bottom row*). Calibration of relative fluorescence changes (dI/I) is given at the *bottom* of (**e**) (Reprinted with permission (14)).

excited neurons are located in (19, 20). Further, amplitudes of VSD signals recorded in areas with a high density of neuronal membrane are easily overestimated when compared to amplitudes recorded in lower-density areas. As VSD imaging on stained slices does not resolve individual cells, a contribution of non-neuronal sources such as astrocytic or oligodendrocytic glial cells to the recorded signal is difficult to rule out. Although optical signals originating from glia have been described to have much slower time courses than neuronal signals, this issue needs to be kept in mind (33, 34).

5.3. Combining Voltage-Sensitive Dye Imaging and Electrophysiology

Due to the low signal-to-noise ratio and the limited recording duration of VSD, it is advisable to combine VSD imaging with electrical field potential or intracellular recordings. Field potential recordings are highly valuable for investigations of spontaneous activity as the experimenter can continuously monitor the ongoing activity and can use the electrical signals to guide optical recordings spatially (rough localization) and temporally (e.g., by triggering optical recordings from electrical events). Combining VSD imaging and intracellular recordings will be most easily applicable to investigation of evoked activity because of the limited duration of intracellular recordings. When investigating spontaneous activity, the duration of the intracellular recording will set a time window within which a spontaneous event has to be recorded or "caught" using VSD imaging.

6. Concluding Remarks

In our experience, recordings from human brain slices are comparable to recordings from rodent brain slices in terms of slice survivability (8–10 h) and quality of signals. This is likely due to the immediate processing of the human tissue after the resection. We have not noticed major differences in the difficulty of obtaining good field potential recordings or VSD signals between rodent and human material. Intracellular (both patch-clamp and sharp microelectrode) recordings are feasible in human brain slices (see Chap. 3-Moore et al. (3)). Patch-clamp recordings tend to be more difficult in adult nervous tissue, while sharp microelectrode recordings in human material are of the same difficulty as in animal preparations. Combining intracellular recordings of spontaneous activity with a VSD signal of the same event can be complicated by the limited duration of intracellular electrophysiological recordings.

References

1. Trapp S, Ballanyi K (2012) Autonomic nervous system in vitro: studying tonically active neurons controlling vagal outflow in rodent brainstem slices. In Isolated Central Nervous System Circuits (Ed K Ballanyi), Neuromethods Series Vol. 73 (Ed W Walz). Springer Science+Business Media, LLC, New York, NY, pp 1–59
2. Ruangkittisakul A, Panaitescu B, Secchia L, Bobocea N, Kantor C, Kuribayashi J, Iizuka M, Ballanyi K (2012) Isolated brainstem respiratory centers from perinatal rodents. In Isolated Central Nervous System Circuits (Ed K Ballanyi), Neuromethods Series Vol. 73 (Ed W Walz). Springer Science+Business Media, LLC, New York, NY, pp 61–124
3. Moore AR, Zhou WL, Jakovcevski I, Zecevic N, Antic SD (2012) Physiological properties of human fetal cortex in vitro. In Isolated Central Nervous System Circuits (Ed K Ballanyi), Neuromethods Series Vol. 73 (Ed W Walz). Springer Science+Business Media, LLC, New York, NY, pp 125–158
4. Sanchez-Vives MV (2012) Spontaneous rhythmic activity in the adult cerebral cortex in vitro. In Isolated Central Nervous System Circuits (Ed K Ballanyi), Neuromethods Series Vol. 73 (Ed W Walz). Springer Science+Business Media, LLC, New York, NY, pp 263–284
5. Luhmann HJ, Kilb W (2012) Intact in vitro preparation of the neonatal rodent cerebral cortex-analysis of cellular properties and network activity. In Isolated Central Nervous System Circuits (Ed K Ballanyi), Neuromethods Series Vol. 73 (Ed W Walz). Springer Science+Business Media, LLC, New York, NY, pp 301–314
6. De Curtis M, Lilbrizzi L, Uva L, Gnatkovsky V (2012) Neuronal networks in the in vitro isolated guinea pig brain. In Isolated Central Nervous System Circuits (Ed K Ballanyi), Neuromethods Series Vol. 73 (Ed W Walz). Springer Science+Business Media, LLC, New York, NY, pp 357–383
7. Cohen LB, Salzberg BM (1978) Optical measurement of membrane potential. Rev Physiol Biochem Pharmacol 83:35–88
8. Ebner TJ, Chen G (1985) Use of voltage-sensitive dyes and optical recordings in the central nervous system. Prog Neurobiol 46:463–506
9. Grinvald A, Hildesheim R (2004) VSDI: a new era in functional imaging of cortical dynamics. Nat Rev Neurosci 5:874–885
10. Baker BJ, Kosmidis EK, Vucinic D, Falk CX, Cohen LB, Djurisic M, Zecevic D (2005) Imaging brain activity with voltage- and calcium-sensitive dyes. Cell Mol Neurobiol 25: 245–282
11. Cohen LB, Salzberg BM, Grinvald A (1978) Optical methods for monitoring neuron activity. Annu Rev Neurosci 1:171–182
12. Wu JY, Cohen LB (1993) Fast multisite optical measurement of membrane potential. In: Mason WT (ed) Fluorescent and luminescent probes for biological activity. Academic, London, pp 389–404
13. Köhling R, Hohling JM, Straub H, Kuhlmann D, Kuhnt U, Tuxhorn I, Ebner A, Wolf P, Pannek HW, Gorji A, Speckmann EJ (2000) Optical monitoring of neuronal activity during spontaneous sharp waves in chronically epileptic human neocortical tissue. J Neurophysiol 84:2161–2165
14. Straub H, Kuhnt U, Hohling JM, Köhling R, Gorji A, Kuhlmann D, Tuxhorn I, Ebner A, Wolf P, Pannek HW, Lahl R, Speckmann EJ (2003) Stimulus-induced patterns of bioelectric activity in human neocortical tissue recorded by a voltage sensitive dye. Neuroscience 121:587–604
15. Köhling R, Lucke A, Straub H, Speckmann EJ (1996) A portable chamber for long-distance transport of surviving human brain slice preparations. J Neurosci Methods 67: 233–236
16. Köhling R, Reinel J, Vahrenhold J, Hinrichs K, Speckmann EJ (2002) Spatio-temporal patterns of neuronal activity: analysis of optical imaging data using geometric shape matching. J Neurosci Methods 114:17–23
17. Yuste R, Konnerth A (2005) Imaging in neuroscience and development. Cold Spring Harbor Laboratory Press, Cold Spring Harbor
18. Salzberg BM, Grinvald A, Cohen LB, Davila HV, Ross WN (1977) Optical recording of neuronal activity in an invertebrate central nervous system: simultaneous monitoring of several neurons. J Neurophysiol 40:1281–1291
19. Cinelli AR, Salzberg BM (1990) Multiple site optical recording of transmembrane voltage (MSORTV), single-unit recordings, and evoked field potentials from the olfactory bulb of skate (Raja erinacea). J Neurophysiol 64: 1767–1790
20. Cinelli AR, Salzberg BM (1992) Dendritic origin of late events in optical recordings from salamander olfactory bulb. J Neurophysiol 68:786–806
21. Broicher T, Bidmon HJ, Kamuf B, Coulon P, Gorji A, Pape HC, Speckmann EJ, Budde T (2010) Thalamic afferent activation of supra-

granular layers in auditory cortex in vitro: a voltage sensitive dye study. Neuroscience 165: 371–385

22. Contreras D, Llinas R (2001) Voltage-sensitive dye imaging of neocortical spatiotemporal dynamics to afferent activation frequency. J Neurosci 21:9403–9413
23. Kubota M, Sugimoto S, Horikawa J, Nasu M, Taniguchi I (1997) Optical imaging of dynamic horizontal spread of excitation in rat auditory cortex slices. Neurosci Lett 237:77–80
24. Laaris N, Carlson GC, Keller A (2000) Thalamic-evoked synaptic interactions in barrel cortex revealed by optical imaging. J Neurosci 20:1529–1537
25. Llinas RR, Leznik E, Urbano FJ (2002) Temporal binding via cortical coincidence detection of specific and nonspecific thalamocortical inputs: a voltage-dependent dye-imaging study in mouse brain slices. Proc Natl Acad Sci USA 99:449–454
26. Sato H, Shimanuki Y, Saito M, Toyoda H, Nokubi T, Maeda Y, Yamamoto T, Kang Y (2008) Differential columnar processing in local circuits of barrel and insular cortices. J Neurosci 28:3076–3089
27. Yuste R, Tank DW, Kleinfeld D (1997) Functional study of the rat cortical microcircuitry with voltage-sensitive dye imaging of neocortical slices. Cereb Cortex 7:546–558
28. Köhling R, Lucke A, Straub H, Speckmann EJ, Tuxhorn I, Wolf P, Pannek H, Oppel F (1998) Spontaneous sharp waves in human neocortical slices excised from epileptic patients. Brain 121:1073–1087
29. Köhling R, Qu M, Zilles K, Speckmann EJ (1999) Current-source-density profiles associated with sharp waves in human epileptic neocortical tissue. Neuroscience 94:1039–1050
30. Cohen I, Navarro V, Clemenceau S, Baulac M, Miles R (2002) On the origin of interictal activity in human temporal lobe epilepsy in vitro. Science 298:1418–1421
31. Wittner L, Huberfeld G, Clemenceau S, Eross L, Dezamis E, Entz L, Ulbert I, Baulac M, Freund TF, Magloczky Z, Miles R (2009) The epileptic human hippocampal cornu ammonis 2 region generates spontaneous interictal-like activity in vitro. Brain 132: 3032–3046
32. Huberfeld G, Wittner L, Clemenceau S, Baulac M, Kaila K, Miles R, Rivera C (2007) Perturbed chloride homeostasis and GABAergic signaling in human temporal lobe epilepsy. J Neurosci 27:9866–9873
33. Garaschuk O, Milos RI, Grienberger C, Marandi N, Adelsberger H, Konnerth A (2006) Optical monitoring of brain function in vivo: from neurons to networks. Pfluger's Arch 453:385–396
34. Konnerth A, Obaid AL, Salzberg BM (1987) Optical recording of electrical activity from parallel fibres and other cell types in skate cerebellar slices in vitro. J Physiol 393:681–702

Chapter 10

Intact In Vitro Preparations of the Neonatal Rodent Cortex: Analysis of Cellular Properties and Network Activity

Heiko J. Luhmann and Werner Kilb

Abstract

In vitro preparations of one whole hippocampus or one entire neocortical hemisphere (so-called intact preparations) combine the technical possibilities of conventional in vitro slice preparations with the advantage of a preserved intrinsic connectivity of a defined brain region. The initiation and propagation of spontaneous or experimentally induced network activity can be studied with various electrophysiological techniques from the single cell level to large neuronal networks. The good control of the extracellular milieu allows in-depth neuropharmacological analyses on the molecular and cellular mechanisms underlying physiological or pathophysiological network activity. This chapter aims to give an overview on the prospects and limitations of intact in vitro preparations of the newborn rodent cortex.

Key words: Cerebral cortex, Development, Electrophysiology, Epileptiform activity, Hippocampus, Intact preparation, In vitro, Network activity, Neuronal network, Rodents, Temporal lobe

1. Introduction

The ideal model to study the anatomical and physiological properties of any given neural structure should have the following properties: (1) fully preserved and functionally intact afferent, intrinsic and efferent connections, including inputs from afferent neuromodulatory systems, (2) in the case of a sensory brain region, functional connectivity from the peripheral sensory organ, (3) easy access to visually identifiable neurons and their dendritic and axonal compartments by multiple recording devices, (4) easy, reliable and reversible application of pharmacological agents at defined concentrations, (5) an extracellular milieu that can be controlled experimentally, (6) possibility for temporally and spatially precise stimulation of presynaptic and/or postsynaptic elements and

Klaus Ballanyi (ed.), *Isolated Central Nervous System Circuits*, Neuromethods, vol. 73,
DOI 10.1007/978-1-62703-020-5_10, © Springer Science+Business Media New York 2012

(7) sufficient stability of anatomical and physiological properties to study the question(s) of interest. Obviously, such a model does not exist, although the in vitro isolated brain preparation of the guinea pig (1, 2, see also Chap. 12-De Curtis et al. (3)), new optogenetic tools (4–6) and other recent genetic and molecular biological technical developments come close to this 'neurophysiologist's dream' (7).

In the last decades, many attempts were made to expand in vitro electrophysiological investigations to neural tissue of enhanced functional integrity to approach some of the properties listed above. A first step was the introduction of acute brain slices (8). The isolated spinal cord of immature and adult rodents, as a thin linear structure, was successfully maintained under submerged conditions for electrophysiological examinations ((8, 9, reviewed in (10)) (see also Chaps. 2-Ruangkittisakul et al. (11), 14-Biggs et al. (12) and 15-Mandadi et al. (13)). Later, it was shown that brainstem-cerebellar preparations, the isolated hypothalamus and even the whole brain (1, 14, 15) can be maintained in vitro by a technically challenging arterial perfusion. A seminal step was the introduction of the intact hippocampus in vitro preparation of the newborn rat by Khalilov and co-workers (16). It was subsequently shown that also two interconnected hippocampi (17, 18), the complete cortico-hippocampal formation (19–21), or intact neocortical hemispheres (22) of newborn rodents remain functional for many hours in submerged-type recording chambers. These hippocampal and neocortical preparations thus allow the investigation of isolated parts of the brain with intact connectivity under reproducible conditions. In the following paragraphs, we will describe the generation of and applications for these intact cortical in vitro models, focusing on the intact neocortical hemisphere, the intact hippocampus and the intact cortico-hippocampal formation. We use the term 'intact' to emphasize the preservation of intrinsic (i.e. intracortical) connections in these in vitro preparations.

Intact in vitro models offer substantial advantages such as (1) the intrinsic connectivity is well preserved, (2) superficially located neurons and their dendritic compartments can be visually identified by infrared differential interference contrast (IR-DIC) optics (23, 24) for subsequent electrophysiological recordings or cellular imaging after filling with a fluorescent dye, (3) relatively good control of the extracellular milieu (although not as good as in brain slices) and (4) easy (and to some extent reversible) application of pharmacological agents at defined concentrations. In comparison to brain slices, which for technical reasons represent strongly reduced models with typically, but not always, limited preservation of the intrinsic neural network (25, 26) (see also Chaps. 2-Ruangkittisakul et al. (11), 8-Sanchez-Vives (27) and 11-Kantor et al. (28)), intact in vitro models are especially well suited for the structural and functional analyses of properties such as intrinsic axonal projections, spontaneous and evoked network oscillations or epileptiform activity (29–38). In addition, intact in vitro preparations can be used for various imaging techniques (39).

2. Materials

Materials for standard electrophysiological and imaging analyses in brain slices are listed in other chapters of this volume (e.g. Chaps. 1-Trapp and Ballanyi (40), 2-Ruangkittisakul et al. (11), 3-Moore et al. (41), 4-Fish et al. (42), 5-Nakamura et al. (43), 6-Stachniak et al. (44), 7-McKay et al. (45), 9-Broicher and Speckmann (46) and 11-Kantor et al. (28)).

3. Methods

3.1. Intact Cerebral Cortex

Neonatal mice or rats between postnatal day (P) 0 to ~ P10 (day of birth = P0) are deeply anaesthetized by hypothermia and decapitated (22, 47). The brain is rapidly removed and transferred to carbogenated (95% O_2/5% CO_2) ice-cold (2–5°C) artificial cerebrospinal fluid (ACSF) having, for example, the following composition (in mM): 124 NaCl, 5 KCl, 1.6 $CaCl_2$, 1 $MgSO_4$, 26 $NaHCO_3$, 1.25 NaH_2PO_4 and D-glucose 10 (pH 7.4). A special composition of the ACSF (low Na^+/Ca^{2+}, elevated Mg^{2+} or glutamate receptor antagonists) is not necessary (see also chapter 1-Trapp and Ballanyi (40)). The subsequent preparation is performed in carbogenated ice-cold ACSF and lasts 10–15 min. The two hemispheres are separated by a scalpel cut through the midline. In one hemisphere, all subcortical structures including the hippocampus can be removed. Subsequently, the leptomeninges are carefully removed. For the preparation of early postnatal mice cortices, it is helpful to glue the isolated brain on a spatulum (using the cerebellum to fix the brain) for easier handling. For these small cortices, it is also recommended to remove the leptomeninges before separating the hemispheres. The intact hemisphere of the cerebral cortex is transferred into a submerged-type acrylic recording chamber (volume 2.2–2.5 ml) (see Chaps. 1-Trapp and Ballanyi (40) and 2-Ruangkittisakul et al. (11)) superfused with ACSF at 32–33°C at a rate of 4–10 ml/min (22). The tissue is placed on a plastic mesh with the medial side down and fixed to the Sylgard® bottom using three entomological needles with a diameter of 0.1 mm (Fig. 1a). Experiments usually start ~1 h after preparation. Using the methods described above, the cortical hemisphere can be studied for at least 8 h.

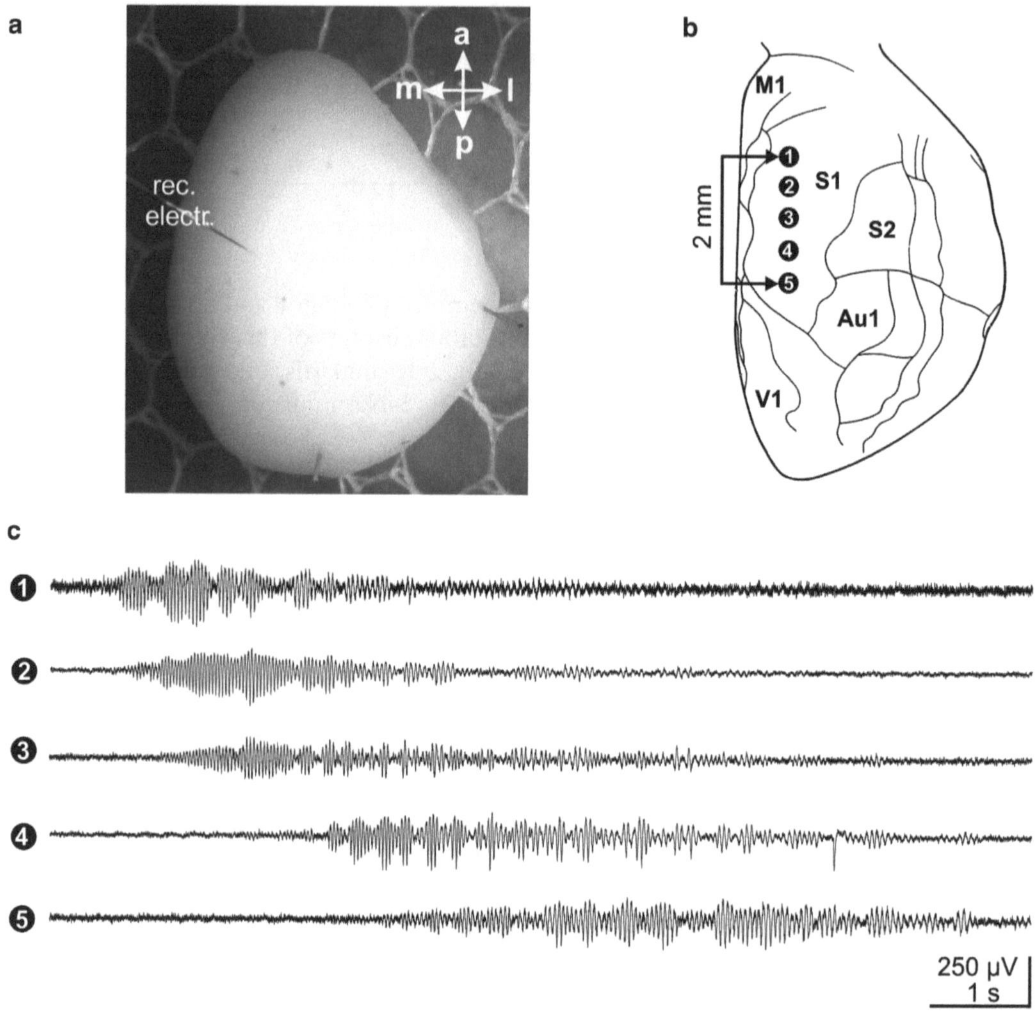

Fig. 1. *The intact whole neocortical hemisphere preparation in vitro.* (**a**) Photograph of a lateral view of the right cortical hemisphere from a newborn rat. Tungsten electrode (rec. electr.) for recording extracellular activity is inserted in the primary somatosensory cortex. (**b**) Multi-site extracellular recordings in primary somatosensory cortex (S1) with five tungsten electrodes (no. 1–5) positioned in rostro-caudal direction with a tip separation of 0.5 mm. Schematic illustration of the electrode positions in a lateral view of the intact cortex with locations of the primary motor (M1), S1, primary auditory (Au1), primary visual (V1) and secondary somatosensory (S2) cortex according to Zilles and Wree (71). (**c**) Bath application of the acetylcholine receptor agonist carbachol (≥30 μM) elicits large-scale oscillatory network activity in the cerebral cortex of a postnatal day (P) 5 rat. Note propagation of activity from the most rostral electrode (no. 1) towards the caudal pole at a velocity of approximately 0.7 mm/s (Reproduced with permission from (22)).

Extracellular multi-site recordings can be performed with tungsten 4–5 MΩ microelectrodes (FHC, Brunswick, USA, www.fh-co.com) positioned in different cortical areas (Fig. 1b, c). Field potentials are recorded with AC-coupled differential amplifiers, low-pass filtered at 3 kHz, and are sampled into a computer for off-line analysis using specific software (see Chap. 1-Trapp and Ballanyi (40)). DC recordings with suitable electrodes and

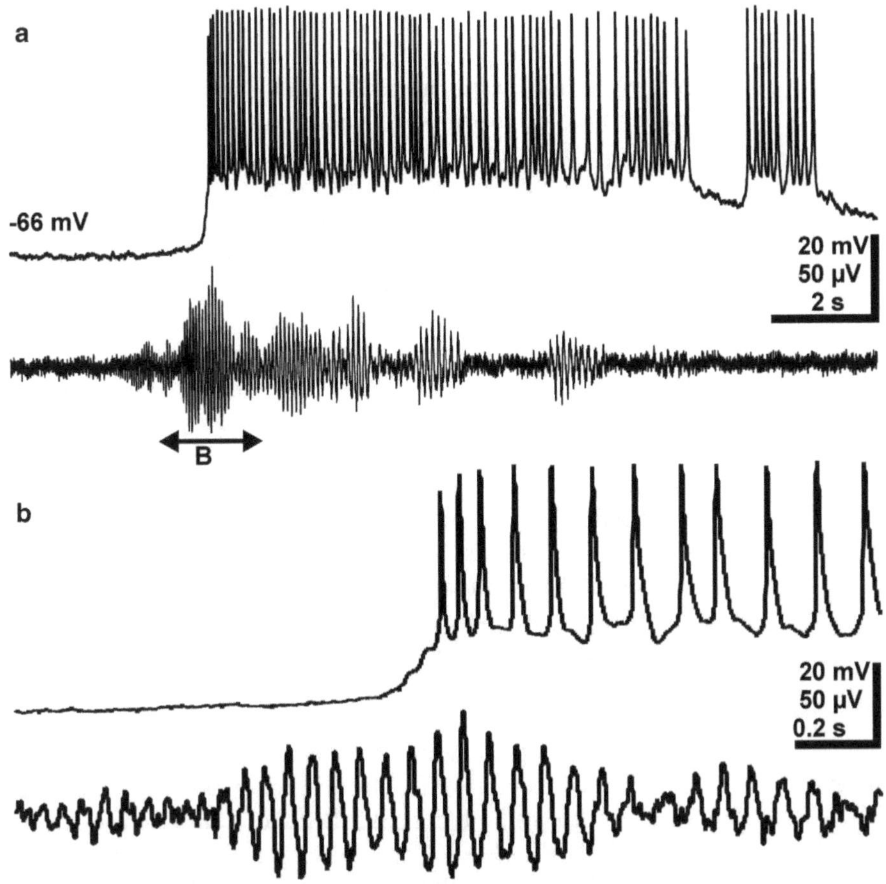

Fig. 2. *Simultaneous whole-cell patch-clamp membrane potential and extracellular field potential recordings in the intact in vitro neocortical preparation.* (**a**) Patch-clamp recording was obtained from a layer II/III pyramidal neuron (*upper trace*) and nearby field potential was recorded with an extracellular electrode (*lower trace*). Bath application of 100 μM carbachol induces a membrane depolarization with repetitive action potential discharge in the pyramidal neuron and a transient oscillatory field potential. (**b**) Same recordings as in (**a**), but at higher temporal resolution (interval marked with B) (Reproduced with permission from (22)).

amplifiers can be used to analyse slow DC shifts, for example, evoked by spreading depression, anoxic depolarizations or slow network oscillations (48). The preparation is also well suited to combine extracellular measurements with 'blind' whole-cell patch-clamp recordings of membrane potentials or currents, for example, in the supragranular layers of the intact cortex (Fig. 2) according to the methods described by Blanton et al. (49). For this purpose, recording pipettes with DC resistances of 6–12 MΩ are filled with appropriate pipette solutions that include 0.5% biocytin for later cell identification (50) (see also Chaps. 1-Trapp and Ballanyi (40), 2-Ruangkittisakul et al. (11) and 3-Moore et al. (41)). For experiments involving whole-cell recordings, extracellular $MgSO_4$ was substituted by $MgCl_2$.

An intact neocortex of the newborn rodent is thin and transparent enough to allow visually guided patch-clamp experiments using IR-DIC optics (23, 24). This technique has been used by us to record from visually identified Cajal-Retzius cells (Fig. 3a). Cajal-Retzius cells are located in the most superficial layer of the immature cerebral cortex, the marginal zone or layer I, and are oriented parallel to the pial surface (51, 52). They reveal characteristic active membrane properties (53, 54), like a prominent hyperpolarization-activated cation current (I_h), repetitive discharge of action potentials with a relatively small amplitude of 30 ± 2.1 mV and a rather long duration of 4 ± 0.6 ms at half-maximal amplitude (Fig. 3b). After intracellular labelling with biocytin and histological processing of the tissue (50), Cajal-Retzius cells can be analysed regarding their somato-dendritic morphology (Fig. 3c) and axonal projections (e.g. Fig. 1c in (55)). For such histological analyses, the cortices should be fixed by immersion in 4% paraformaldehyde solution for at least 24 h, sectioned in series of thin slices and stained using a standard biocytin-based staining protocol (21, 51). In principle, also somata of neurons in the upper cortical layers are visible using IR-DIC optics; however, their identification is impeded by the vertical view on these neurons.

3.2. Intact Hippocampus

The intact hippocampus preparation of the newborn rat has been initially described by Khalilov and co-workers (16). In subsequent studies, this model has also been established for the newborn mouse (30, 32, 34–36, 56). Since the mouse brain is considerably smaller than that of the rat at comparable ages, the mouse is even better suited to obtain intact in vitro preparations, because of better oxygen and glucose supply in particular to deeper tissue layers. Accordingly, viable intact hippocampal preparations can be made from up to P28 mice (57).

Newborn rats or mice are deeply anaesthetized by hypothermia. The brain is quickly removed and immersed for 2–3 min in ACSF (Sect. 3.1). For isolation of the hippocampi, the subcortical areas are thoroughly abscised and the hemispheres divided. The hippocampi are now clearly visible 'inside' the cortical hemispheres and can be carefully isolated from both hemispheres of the cerebral cortex. The preparation is performed in carbogenated ice-cold ACSF in a similar way as described above for the cerebral cortex and lasts ~15 min. The isolated hippocampi are fixed with entomological needles (Sect. 3.1) in a submerged-type chamber where they are superfused with ACSF at a rate of 4–5 ml/min at 31°C. Experiments started normally ~1 h after preparation. Most studies were performed with extracellular electrodes, but Zhang and co-workers also performed whole-cell patch-clamp recordings from visually identified hippocampal neurons in this preparation (32). The intact hippocampus preparation proved to be most valuable to

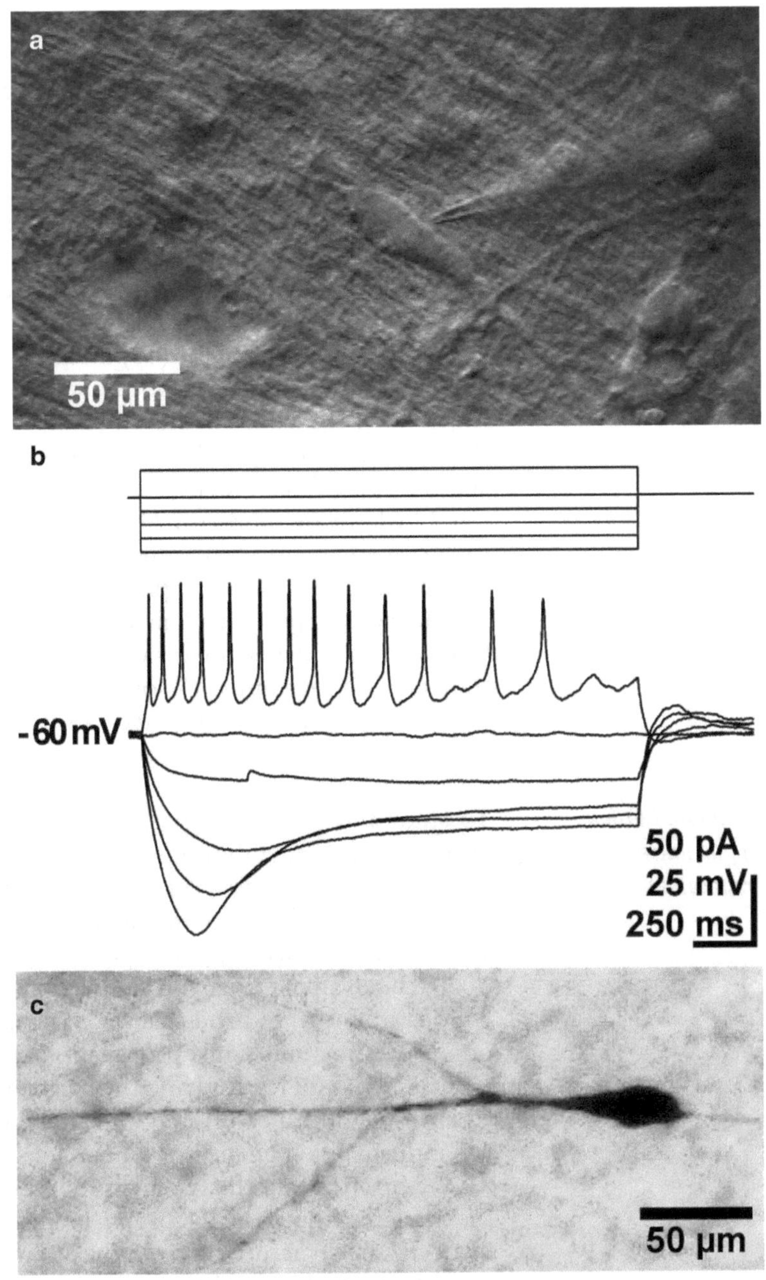

Fig. 3. *Single cell identification and recording in the intact whole neocortical hemisphere preparation in vitro.* (**a**) Visual identification of a Cajal-Retzius neuron in the intact neocortex preparation of a P1 mouse using infrared differential interference contrast (IR-DIC) optics. Note the patch pipette close to the soma of the cell. (**b**) Whole-cell patch-clamp recording from a Cajal-Retzius cell in a whole neocortical preparation. Membrane potential recordings show responses to injection of de- and hyperpolarizing DC current pulses (upper traces) at a holding potential of −60 mV. (**c**) Photomicrograph of the cell recorded in (**b**) after intracellular filling with biocytin and histological processing.

study the expression, mechanisms and propagation of epileptiform activity within the hippocampus (58). Accordingly, this preparation has frequently been used to unravel pharmacological influences on epileptiform activity, to compare the properties of different in vitro epilepsy models, to investigate mechanisms of epileptogenesis and to characterize molecular changes induced by epileptiform activity (30, 35, 36, 56). In addition, the intact hippocampus preparation was used to characterize the connectivity between visually identified interneurons (32) or the mechanisms underlying theta oscillations (37).

Khazipov and colleagues (18) described also a three-compartment chamber and a dual hippocampal preparation to study the function of the commissural fibres in vitro. In this preparation, the two interconnected hippocampi and their commissural fibres are placed in three independent compartments, which are separated by latex membranes. This chamber allows the perfusion of each compartment with a different solution. This intact in vitro preparation can, for example, be utilized for the analysis of a secondary epileptogenic mirror focus formation (after application of convulsive agents to only one hippocampus) in the contralateral hippocampus (17) or for the analysis of pharmacological differences between primary and secondary epileptogenic foci (38).

3.3. Intact Cortico-hippocampal Preparation

To study the interaction between different brain regions, or the propagation to neighbouring areas, cortico-hippocampal preparations are most valuable (Fig. 4). These intact in vitro preparations have been described for the hippocampus-entorhinal cortex (19) and for the cortico-hippocampal formation (20, 31, 33). They offer the advantage that the connections between different brain regions are preserved and that several stages of neuronal information processing and signal propagation can be investigated. These models are useful for studying the mechanisms of seizure initiation and propagation at early developmental stages and to screen potential anticonvulsant agents for human infantile epilepsies.

After separating the two hemispheres by a scalpel cut through the midline, the frontal cortex, brainstem, cerebellum and all diencephalic structures are removed, and the leptomeninges are carefully peeled off. The preparation is performed in carbogenated ice-cold ACSF in a similar way as described above for the cerebral cortex. The hippocampus-entorhinal cortex or the cortico-hippocampal formation is transferred to a submerged-type recording chamber, superfused with ACSF at 31°C at a flow rate of ~5 ml/min. The preparation of the intact hippocampus-entorhinal cortex or the cortico-hippocampal formation takes 15–20 min after some training.

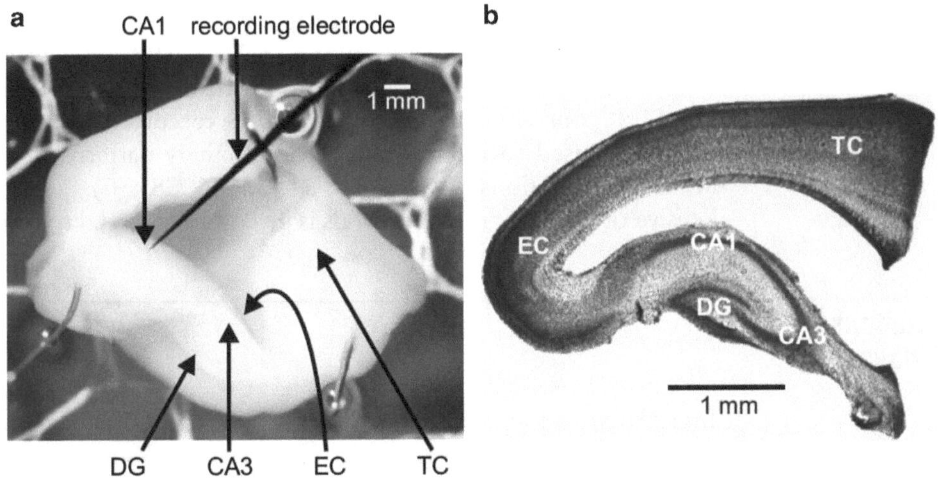

Fig. 4. *The intact cortico-hippocampal formation in vitro.* (**a**) Photograph of the P4 mouse cortico-hippocampal formation in a submerged-type recording chamber. One tungsten recording electrode is located in the hippocampal CA1 region. The other four recording sites in dentate gyrus (DG), CA3, entorhinal cortex (EC) and temporal cortex (TC) are marked by *arrows*. (**b**) Photograph of a Nissl-stained horizontal section of the intact cortico-hippocampal formation of a P4 mouse (Reproduced with permission from (33)).

4. How to Obtain Spontaneously Active Intact In Vitro Preparations

The intact in vitro hippocampus preparation generates spontaneous theta activity, which is a clear evidence that endogenous hippocampal theta oscillators mediate this particular periodic activity essential for spatial memory formation (37). In contrast, intact in vitro cerebral cortex preparations and also acute cortical slices are mostly electrically silent when compared to in vivo recordings [e.g. see (59) for spontaneous activity in the newborn rat somatosensory cortex in vivo]. Extracellular solutions which more closely mimic the ionic composition of brain interstitial fluid in vivo (e.g. 1–1.2 mM Ca^{2+}, 1–2 mM Mg^{2+} and 3–4 mM K^{+}) can induce physiologically relevant activity patterns in vitro (60) (see also Chaps. 1-Trapp and Ballanyi (40), 2-Ruangkittisakul et al. (11) and 11-Kantor et al. (28)).

Similar activities can be induced by (transient) bath application of neuromodulatory agents. Carbachol (>30 μM), a cholinergic agonist, induces oscillatory network activity in the intact in vitro neocortex preparation (22, 61) (Fig. 1), which resembles in many aspects the in vivo pattern (59). Oscillatory network activity can be also induced by transient bath application of the non-specific group I/II mGluR agonist (1 S, 3R)-1-aminocyclopentane-1,3-dicarboxylic acid (ACPD) (47).

As in slices, bath application of agents like 4-amino pyridine or antagonists of A-type γ-aminobutyric acid ($GABA_A$) receptors, as well as modification of ACSF composition (e.g. reduction in the Mg^{2+} concentration to boost NMDA receptor activation), is frequently used to induce epileptiform activity patterns in intact in vitro preparations (17, 18, 20, 31, 33–36, 38, 56). A comprehensive overview of these methods is given by Schwartzkroin (62).

5. Limitations and Pitfalls

Intact in vitro preparations have one major limitation: only intact preparations from brain regions of immature and small mammals, preferentially small rodents, can be maintained in a close-to-physiological state for up to 10–12 h or even longer, for up to 24 h, according to Quilichini and co-workers (20). Due to the thickness of the intact hippocampus or whole cortical hemisphere preparations (exceeding in some cases 1 mm), the oxygen and glucose supply of the tissue in the centre of the preparation may be very low. In preparations from mature mammals, this hypoxic-ischaemic condition causes neuronal dysfunction and cell death. In contrast, intact in vitro preparations from immature animals are more resistant to this hypoxia/ischaemia (63) (see also Chap. 2-Ruangkittisakul et al. (11)). Several factors contribute to the resistance of intact in vitro preparations from immature brains: (1) Immature neurons have smaller demands for energy-consuming processes (64–67) such as intracellular transport mechanisms and ion pump activity to preserve intracellular ion homeostasis. The smaller somatic size, axonal arborization and dendritic branching of immature neurons would support this hypothesis. (2) Under hypoxic/ischaemic conditions, the immature brain is capable of maintaining sufficiently high concentrations of intracellular energy resources via anaerobic glycolysis (66, 67). (3) The larger volume fraction and smaller tortuosity in the extracellular compartment of the immature neocortex and hippocampus facilitate diffusion of glucose into the preparation (68–70). The age limit in rats for functionally intact in vitro preparations of the cerebral cortex and the hippocampus is in the range of P10–12, corresponding to the cortex of a full-term newborn human baby (71). In species with smaller brains (e.g. mice) and under some conditions, this age limit may be considerably expanded (57).

The successful use of intact in vitro preparations requires the anatomical and physiological integrity throughout the preparation and experiment. Mechanical and biochemical stress should thus be minimized during the preparation procedure as well as in the submerged recording chamber. Otherwise no reliable recordings (or no signals at all) can be obtained from such large structures.

Therefore, the following aspects should be considered: (1) The preparation must be performed without inducing lesions to the cortex/hippocampus. Any direct contact with the cortical/hippocampal surface should be minimized. (2) A sufficiently low temperature and high oxygen supply of the chamber must be maintained throughout the preparation. (3) The fixation of the isolated hippocampi and cortices must be obtained without introducing tension to the tissue. (4) A relatively high perfusion rate of the recording chamber is necessary to maintain stable responses; the preparation should be in direct contact with the solution from all sides. (5) The recording electrodes should be placed at appropriate locations and depths. The determination of the exact depth may be complicated by the opaqueness of the tissue.

Acknowledgements

This work was supported by grants of the Deutsche Forschungsgemeinschaft to the authors and by the EC contract LSH-CT-2006-037315 (EPICURE).

References

1. Mühlethaler M, De Curtis M, Walton K, Llinás RR (1993) The isolated and perfused brain of the guinea-pig in vitro. Eur J Neurosci 5:915–926
2. Biella G, Panzica F, De Curtis M (1996) Interactions between associative synaptic potentials in the piriform cortex of the in vitro isolated guinea pig brain. Eur J Neurosci 8:1350–1357
3. De Curtis M, Lilbrizzi L, Uva L, Gnatkovsky V (2012) Neuronal networks in the in vitro isolated guinea pig brain. In Isolated Central Nervous System Circuits (Ed K Ballanyi), Neuromethods Series Vol. 73 (Ed W Walz). Springer Science+Business Media, LLC, New York, NY, pp 357–383
4. Sohal VS, Zhang F, Yizhar O, Deisseroth K (2009) Parvalbumin neurons and gamma rhythms enhance cortical circuit performance. Nature 459:698–702
5. Miller G (2006) Optogenetics. Shining new light on neural circuits. Science 314:1674–1676
6. Ernst OP, Sanchez Murcia PA, Daldrop P, Tsunoda SP, Kateriya S, Hegemann P (2008) Photoactivation of channelrhodopsin. J Biol Chem 283:1637
7. Berndt A, Yizhar O, Gunaydin LA, Hegemann P, Deisseroth K (2009) Bi-stable neural state switches. Nat Neurosci 12:229–234
8. Yamamoto C, Mcilwain H (1966) Electrical activities in thin sections from mammalian brain maintained in chemically-defined media in vitro. J Neurochem 13:1333–1343
9. Bagust J, Kelly MEM, Kerkut GA (1985) An isolated mammalian brain-stem spinal-cord preparation suitable for the investigation of descending control of motor-activity. Brain Res 327:370–374
10. Kerkut GA, Bagust J (1995) The isolated mammalian spinal-cord. Progr Neurobiol 46: 1–48
11. Ruangkittisakul A, Panaitescu B, Secchia L, Bobocea N, Kantor C, Kuribayashi J, Iizuka M, Ballanyi K (2012) Isolated brainstem respiratory centers from perinatal rodents. In Isolated Central Nervous System Circuits (Ed K Ballanyi), Neuromethods Series Vol. 73 (Ed W Walz). Springer Science+Business Media, LLC, New York, NY, pp 61–124
12. Biggs JE, Lu VB, Kim H, Lai A, Todd KG, Ballanyi K, Colmers WF, Smith PA (2012) Defined medium organotypic cultures of spinal cord put 'pain in a dish'. In Isolated Central Nervous System Circuits (Ed K Ballanyi), Neuromethods Series Vol. 73 (Ed W Walz). Springer Science+Business Media, LLC, New York, NY, pp 405–435

13. Mandadi S, Nakanishi ST, Han P, Humphreys J, Whelan PJ (2012) The use of rodent isolated spinal cord preparations to examine motor output. In Isolated Central Nervous System Circuits (Ed K Ballanyi), Neuromethods Series Vol. 73 (Ed W Walz). Springer Science+Business Media, LLC, New York, NY, pp 437–464
14. Llinas R, Mühlethaler M (1988) An electrophysiological study of the in vitro, perfused brain-stem cerebellum of adult guinea-pig. J Physiol 404:215–240
15. Bourque CW, Renaud LP (1984) Activity patterns and osmosensitivity of rat supraoptic neurons in perfused hypothalamic explants. J Physiol 349:631–642
16. Khalilov I, Esclapez M, Medina I, Aggoun D, Lamsa K, Leinekugel X, Khazipov R, Ben-Ari Y (1997) A novel in vitro preparation: the intact hippocampal formation. Neuron 19:743–749
17. Khalilov I, Holmes GL, Ben-Ari Y (2003) In vitro formation of a secondary epileptogenic mirror focus by interhippocampal propagation of seizures. Nat Neurosci 6(10):1079–1085
18. Khazipov R, Desfreres L, Khalilov I, Ben-Ari Y (1999) Three-independent-compartment chamber to study in vitro commissural synapses. J Neurophysiol 81:921–924
19. Luhmann HJ, Dzhala VI, Ben-Ari Y (2000) Generation and propagation of 4-AP-induced epileptiform activity in neonatal intact limbic structures in vitro. Eur J Neurosci 12: 2757–2768
20. Quilichini PP, Diabira D, Chiron C, Ben-Ari Y, Gozlan H (2002) Persistent epileptiform activity induced by low Mg^{2+} in intact immature brain structures. Eur J Neurosci 16:850–860
21. Horikawa K, Armstrong WE (1988) A versatile means of intracellular labeling: injection of biocytin and its detection with avidin conjugates. J Neurosci Methods 25:1–11
22. Kilb W, Luhmann HJ (2003) Carbachol-induced network oscillations in the intact cerebral cortex of the newborn rat. Cereb Cortex 13:409–421
23. Stuart GJ, Dodt H-U, Sakmann B (1993) Patch-clamp recordings from the soma and dendrites of neurons in brain slices using infrared video microscopy. Pflüger's Arch 423: 511–518
24. Dodt H-U, Zieglgänsberger W (1990) Visualizing unstained neurons in living brain slices by infrared DIC-videomicroscopy. Brain Res 537:333–336
25. Steriade M (2001) The intact and sliced brain. MIT Press, Cambridge
26. Stepanyants A, Martinez LM, Ferecskó AS, Kisvárda ZF (2009) The fractions of short- and long-range connections in the visual cortex. Proc Natl Acad Sci USA 106:3555–3560
27. Sanchez-Vives MV (2012) Spontaneous rhythmic activity in the adult cerebral cortex in vitro. In Isolated Central Nervous System Circuits (Ed K Ballanyi), Neuromethods Series Vol. 73 (Ed W Walz). Springer Science+Business Media, LLC, New York, NY, pp 263–284
28. Kantor C, Panaitescu B, Kuribayashi J, Ruangkittisakul A, Jovanovic I, Leung V, Lee TF, MacTavish D, Jhamandas JH, Cheung PY, Ballanyi K (2012) Spontaneous neural network oscillations in hippocampus, cortex and *locus coeruleus* of newborn rat and piglet brain slices. In Isolated Central Nervous System Circuits (Ed K Ballanyi), Neuromethods Series Vol. 73 (Ed W Walz). Springer Science+Business Media, LLC, New York, NY, pp 315–356
29. Chang SL, LoTurco JJ, Nisenbaum LK (2000) In vitro biocytin injection into perinatal mouse brain: a method for tract tracing in developing tissue. J Neurosci Methods 97:1–6
30. Li J, Shen H, Naus CCG, Zhang L, Carlen PL (2001) Upregulation of gap junction connexin 32 with epileptiform activity in the isolated mouse hippocampus. Neuroscience 105: 589–598
31. Quilichini PP, Diabira D, Chiron C, Milh M, Ben-Ari Y, Gozlan H (2003) Effects of antiepileptic drugs on refractory seizures in the intact immature cortico-hippocampal formation in vitro. Epilepsia 44:1365–1374
32. Zhang XL, Zhang L, Carlen PL (2004) Electrotonic coupling between stratum oriens interneurones in the intact in vitro mouse juvenile hippocampus. J Physiol 558:825–839
33. Moser J, Kilb W, Werhahn KJ, Luhmann HJ (2006) Early developmental alterations of low-Mg^{2+}-induced epileptiform activity in the intact corticohippocampal formation of the newborn mouse in vitro. Brain Res 1077:170–177
34. Derchansky M, Rokni D, Rick JT, Wennberg R, Bardakjian BL, Zhang L, Yarom YA, Carlen PL (2006) Bidirectional multisite seizure propagation in the intact isolated hippocampus: the multifocality of the seizure "focus". Neurobiol Dis 23:312–328
35. Kilb W, Sinning A, Luhmann HJ (2007) Model-specific effects of bumetanide on epileptiform activity in the in-vitro intact hippocampus of the newborn mouse. Neuropharmacology 53:524–533
36. Derchansky M, Jahromi SS, Mamani M, Shin DS, Sik A, Carlen PL (2008) Transition to seizures in the isolated immature mouse hippocampus: a switch from dominant phasic inhibition to dominant phasic excitation. J Physiol 586:477–494

37. Goutagny R, Jackson J, Williams S (2009) Self-generated theta oscillations in the hippocampus. Nat Neurosci 12(12):1491–1493
38. Nardou R, Ben-Ari Y, Khalilov I (2009) Bumetanide, an NKCC1 antagonist, does not prevent formation of epileptogenic focus but blocks epileptic focus seizures in immature rat hippocampus. J Neurophysiol 101(6): 2878–2888
39. Davies ML, Kirov SA, Andrew RD (2007) Whole isolated neocortical and hippocampal preparations and their use in imaging studies. J Neurosci Methods 166:203–216
40. Trapp S, Ballanyi K (2012) Autonomic nervous system *in vitro*: studying tonically active neurons controlling vagal outflow in rodent brainstem slices. In Isolated Central Nervous System Circuits (Ed K Ballanyi), Neuromethods Series Vol. 73 (Ed W Walz). Springer Science+Business Media, LLC, New York, NY, pp 1–59
41. Moore AR, Zhou WL, Jakovcevski I, Zecevic N, Antic SD (2012) Physiological properties of human fetal cortex in vitro. In Isolated Central Nervous System Circuits (Ed K Ballanyi), Neuromethods Series Vol. 73 (Ed W Walz). Springer Science+Business Media, LLC, New York, NY, pp 125–158
42. Fish KN, Gonzales-Burgos G, Zaitsev AV, Lewis DA (2012) Histological characterization of physiologically determined fast spiking interneurons in slices of the primate dorsolateral prefrontal cortex. In Isolated Central Nervous System Circuits (Ed K Ballanyi), Neuromethods Series Vol. 73 (Ed W Walz). Springer Science+Business Media, LLC, New York, NY, pp 159–181
43. Nakamura TJ, Michel S, Block GD, Colwell CS (2012) Neural Circuits underlying Circadian Oscillations in Mammals: Clocks in a Dish. In Isolated Central Nervous System Circuits (Ed K Ballanyi), Neuromethods Series Vol. 73 (Ed W Walz). Springer Science+Business Media, LLC, New York, NY, pp 183–210
44. Stachniak TS, Sudbury JR, Trudel E, Choe KY, Ciur S. Bourqu CW (2012) Osmoregulatory circuits in slices and en bloc preparations of rodent hypothalamus. In Isolated Central Nervous System Circuits (Ed K Ballanyi), Neuromethods Series Vol. 73 (Ed W Walz). Springer Science+Business Media, LLC, New York, NY, pp 211–231
45. McKay BE, Tadayonnejad R, Anderson DM, Engbers JDT, Fernandez FR, Iftinca M, Turner RW (2012) Establishing in vivo like activity in rat cerebellar cells maintained in vitro. In Isolated Central Nervous System Circuits (Ed K Ballanyi), Neuromethods Series Vol. 73 (Ed W Walz). Springer Science+Business Media, LLC, New York, NY, pp 233–262
46. Broicher T, Speckmann EJ (2012) Living human brain slices: network analysis using voltage sensitive dyes. In Isolated Central Nervous System Circuits (Ed K Ballanyi), Neuromethods Series Vol. 73 (Ed W Walz). Springer Science+Business Media, LLC, New York, NY, pp 285–300
47. Wagner J, Luhmann HJ (2006) Activation of metabotropic glutamate receptors induces propagating network oscillations in the intact cerebral cortex of the newborn mouse. Neuropharmacology 51:848–857
48. Luhmann HJ, Kral T (1997) Hypoxia-induced dysfunction in developing rat neocortex. J Neurophysiol 78:1212–1221
49. Blanton MG, LoTurco JJ, Kriegstein AR (1989) Whole cell recording from neurons in slices of reptilian and mammalian cerebral cortex. J Neurosci Methods 30:203–210
50. Schröder R, Luhmann HJ (1997) Morphology, electrophysiology and pathophysiology of supragranular neurons in rat primary somatosensory cortex. Eur J Neurosci 9: 163–176
51. Frotscher M (1998) Cajal-Retzius cells, Reelin, and the formation of layers. Curr Opin Neurobiol 8:570–575
52. Mienville JM (1999) Cajal-Retzius cell physiology: just in time to bridge the 20th century. Cereb Cortex 9:776–782
53. Kilb W, Luhmann HJ (2000) Characterization of a hyperpolarization-activated inward current in Cajal-Retzius cells in rat neonatal neocortex. J Neurophysiol 84:1681–1691
54. Luhmann HJ, Reiprich RA, Hanganu I, Kilb W (2000) Cellular physiology of the neonatal rat cerebral cortex: intrinsic membrane properties, sodium and calcium currents. J Neurosci Res 62:574–584
55. Kilb W, Ikeda M, Uchida K, Okabe A, Fukuda A, Luhmann HJ (2002) Depolarizing glycine responses in Cajal-Retzius cells of neonatal rat cerebral cortex. Neuroscience 112:299–307
56. Derchansky M, Shahar E, Wennberg RA, Samoilova M, Jahromi SS, Abdelmalik PA, Zhang L, Carlen PL (2004) Model of frequent, recurrent, and spontaneous seizures in the intact mouse hippocampus. Hippocampus 14:935–947
57. Wu CP, Shen H, Luk WP, Zhang L (2002) A fundamental oscillatory state of isolated rodent hippocampus. J Physiol 540:509–527
58. Khalilov I, Dzhala V, Medina I, Leinekugel X, Melyan Z, Lamsa K, Khazipov A, Ben-Ari Y (1999) Maturation of kainate-induced epileptiform activities in interconnected intact neonatal limbic structures in vitro. Eur J Neurosci 11: 3468–3480

59. Yang JW, Hanganu-Opatz IL, Sun JJ, Luhmann HJ (2009) Three patterns of oscillatory activity differentially synchronize developing neocortical networks in vivo. J Neurosci 29:9011–9025
60. Sanchez-Vives MV, McCormick DA (2000) Cellular and network mechanisms of rhythmic recurrent activity in neocortex. Nat Neurosci 3:1027–1034
61. Dupont E, Hanganu IL, Kilb W, Hirsch S, Luhmann HJ (2006) Rapid developmental switch in the mechanisms driving early cortical columnar networks. Nature 439:79–83
62. Schwartzkroin PA (2009) Encyclopedia of basic epilepsy research. Elsevier, Amsterdam
63. Ballanyi K, Onimaru H, Homma I (1999) Respiratory network function in the isolated brainstem-spinal cord of newborn rats. Prog Neurobiol 59:583–634
64. Duffy TE, Kohle SJ, Vannucci RC (1975) Carbohydrate and energy metabolism in perinatal rat brain: relation to survival in anoxia. J Neurochem 24:271–276
65. Kawai S, Yonetani M, Nakamura H, Okada Y (1989) Effects of deprivation of oxygen and glucose on the neural activity and the level of high energy phosphates in the hippocampal slices of immature and adult rat. Dev Brain Res 48:11–18
66. Hansen AJ, Nordstrom C-H (1979) Brain extracellular potassium and energy metabolism during ischemia in juvenile rats after exposure to hypoxia for 24 h. J Neurochem 32:915–920
67. Hansen AJ (1985) Effect of anoxia on ion distribution in the brain. Physiol Rev 65:101–148
68. Kilb W, Dierkes PW, Syková E, Vargová L, Luhmann HJ (2006) Hypoosmolar conditions reduce extracellular volume fraction and enhance epileptiform activity in the CA3 region of the immature rat hippocampus. J Neurosci Res 84:119–129
69. Lehmenkühler A, Syková E, Svoboda J, Zilles K, Nicholson C (1993) Extracellular space parameters in the rat neocortex and subcortical white matter during postnatal development determined by diffusion analysis. Neuroscience 55:339–351
70. Sykova E, Nicholson C (2008) Diffusion in brain extracellular space. Physiol Rev 88:1277–1340
71. Romijn HJ, Hofman MA, Gramsbergen A (1991) At what age is the developing cerebral cortex of the rat comparable to that of the full-term newborn human baby? Early Hum Dev 26:61–67
72. Zilles K, Wree A (1985) Cortex: areal and laminar structure. In: Paxinos G (ed) The rat nervous system, vol 1. Academic, Sydney, pp 375–415

Chapter 11

Spontaneous Neural Network Oscillations in Hippocampus, Cortex, and Locus Coeruleus of Newborn Rat and Piglet Brain Slices

Chase Kantor, Bogdan Panaitescu, Junya Kuribayashi, Araya Ruangkittisakul, Igor Jovanovic, Victoria Leung, Tze-Fun Lee, Dave MacTavish, Jack H. Jhamandas, Po-Yin Cheung, and Klaus Ballanyi

Abstract

Rises of cytosolic Ca^{2+} (Ca_i) associated with early network oscillations (ENOs) are important for brain maturation. Thus, developing neural networks are often studied by combining Ca_i imaging with electrophysiological recording of extracellular activity and/or intracellular "patch-clamp" analysis. At birth, some nervous systems such as medullary respiratory networks are functional while cortical circuits are yet quite immature. Here, we summarize our experimental approaches to investigate how both mature and developing neuron-glia networks in newborns generate spontaneous synchronized bursting and how such activity is modulated by (pharmacological) experimental manipulation mimicking neurological diseases or their treatment. For this, we studied ENOs in cortex and hippocampus of newborn rat and piglet brain slices, whereas ENO-like bursting in *locus coeruleus* was only analyzed in rat slices. All these activities are stable for several hours in superfusate of close-to-physiological ion content. Similar to isolated inspiratory network bursting, ENOs depend on a "Ca^{2+}/K^{+} antagonism" meaning that depressed bursting in elevated superfusate Ca^{2+} is countered by raised K^{+}. As further example for our findings, anoxia abolishes ENOs and bursting in *locus coeruleus*, whereas μ-opioid receptor activation blocks bursting, transforms burst pattern, or has no clear effect in *locus coeruleus*, hippocampus, and cortex, respectively. Multiphoton Ca_i imaging reveals different responses to neuromodulators in neurons *versus* neighboring astrocytic glia which forms the basis for their further discrimination via morphological fluorescence imaging of sulforhodamine-101 or glial acidic fibrillary protein. Our findings indicate that "electrophysiological imaging" in brain slices from neonatal mammals is a potent tool for studying spontaneously active (developing) central neuron-glia networks.

Key words: Brain slice, Calcium, Cortex, Hippocampus, Locus coeruleus, Rhythm

---

Chase Kantor and Bodgan Panaitescu contributed equally to this study.

Klaus Ballanyi (ed.), *Isolated Central Nervous System Circuits*, Neuromethods, vol. 73,
DOI 10.1007/978-1-62703-020-5_11, © Springer Science+Business Media New York 2012

1. Introduction

Spontaneous electrical activity that is often tagged "early network oscillations" (ENOs) is pivotal for development of neural structures that are comprised of mutually interacting neurons and glial cells (1–4). A major cellular second messenger influenced by ENOs is a rise in the free concentration of cytosolic Ca^{2+} (Ca_i) that is primarily due to Ca^{2+} influx through voltage-activated Ca^{2+} channels and ligand-activated cation channels (1, 3, 5–7). In addition, Ca_i can increase via Ca^{2+} release from intracellular stores in response to activation of metabotropic receptors by neurotransmitters or other neuroactive substances. Particularly, glial cells, which typically lack voltage-activated Ca^{2+} channels, mainly use such metabotropic "Ca^{2+} excitability" for communicating with each other and with neurons or microglial cells that are of non-neural origin (8–10). Because of the important mutual interrelations of ENOs with Ca_i dynamics, developing neural networks are often analyzed by combining optical Ca_i imaging with electrophysiological recording of neural population activity and/or single-cell "patch-clamp" recording of biophysical membrane properties (1–5, 7, 11–17).

At a given stage, neural networks differ substantially in their degree of maturation. For example, the rhythmogenic pre-Bötzinger complex (preBötC) inspiratory center in the lower brainstem (*medulla oblongata*) has to already function robustly at birth (18–20). In contrast, circuits in cortical areas including hippocampus are rather immature during this time period (2, 4, 14–17, 21). Thus, medication administered to stabilize breathing in early postnatal life may impair developing cortical circuits. For example, not yet fully developed respiratory networks in (preterm) infants are successfully stimulated in clinical practice with caffeine and theophylline to counter spontaneous apneas of prematurity that eventually lead to low, that is, hypoxic, arterial oxygen levels (18–20, 22, 23). These methylxanthines are thus potent respiratory stimulants but also exert adverse effects including nausea, hyperglycemia, tachycardia, tachypnea, and seizures (23). Therefore, despite their beneficial effects on respiratory networks, methylxanthines may notably perturb developing cortical circuits from which neonatal seizures originate (14, 15, 24).

In this context, the Ballanyi laboratory studies cellular mechanisms by which opioids depress the preBötC, how methylxanthines antagonize such inhibition, and how this inspiratory center is inhibited by experimentally evoked hypoxia that may cause transient oxygen depletion (anoxia) in the brain (18, 19, 25). For this, we use "en bloc" newborn rat brainstem-spinal cord preparations and transversal brainstem slices with a functional preBötC (20, 22, 25, 26) (see also Chap. 2-Ruangkittisakul et al. (27)). Our recent analyses using these "respiratory networks in the dish" revealed, for example, that

(1) apparently similar slowing of preBötC bursting due to anoxia or activation of μ-opioid receptors with [D-Ala2, N-MePhe4, Gly-ol]-enkephalin (DAMGO) is caused by different mechanisms (28), (2) methylxanthines may stimulate breathing by reversing in preBötC neurons (opioid-evoked) depletion of cyclic adenosine monophosphate which is presumably a major stimulus to these rhythmogenic cells (18, 19, 22, 23, 25), (3) inhibition of preBötC neurons by opioids or extracellular Ca^{2+} levels above the physiological range (1–1.2 mM) seems to involve primarily presynaptic mechanisms (22, 28, 29), and (4) in context of the latter finding, preBötC activity depends on a "Ca^{2+}/K^{+} antagonism" meaning that depressed preBötC bursting in raised Ca^{2+} is countered by also increasing K^{+} from physiological 3 mM to 7–9 mM (30, 31).

Based on these findings, we started to use newborn rat brain slices to investigate how opioids, methylxanthines, or anoxia affect spontaneous activity in neural circuits that are more or less immature at birth and whether a Ca^{2+}/K^{+} antagonism determines also their excitability. Firstly, we used horizontal brain slices for which it was shown with multiphoton (2-photon) population Ca_i imaging that the entorhinal cortex generates ENOs at a rate of <0.05 Hz, whereas hippocampal ENOs occur faster, at 0.1–0.5 Hz (6). The latter study also demonstrated that cortical ENOs propagate in a wavelike fashion at ~2 mm/s from the posterior to the anterior end of the cortex and require endogenous activation of "AMPA" and "NMDA" subtypes of ionotropic glutamate receptors. In contrast, the authors reported in that study that hippocampal ENOs do not propagate and instead depend on A-type γ-aminobutyric acid ($GABA_A$) receptors. Involvement of a depolarizing and excitatory action of $GABA_A$ receptors in hippocampal ENOs has been elucidated previously in a milestone study by (21) who labeled these events as giant depolarizing potentials (GDPs). Now, it is established that spontaneous GABA depolarization is typical primarily (but not exclusively) for developing neurons and that it requires an inwardly directed Cl^- transporter, NKC1 (14, 32–34). During neuronal maturation, a switch occurs toward the hyperpolarizing and inhibitory action of these receptors in association with expression of an outwardly directed Cl^- transporter, KCC2 (14, 34). Cortical and hippocampal ENOs are considered as in vitro correlates of spontaneous bursting in these brain areas in intact newborn rodents and humans (2, 4, 14, 15, 35–37). When we established horizontal newborn rat slices for studying cortical and hippocampal ENOs, we also monitored ENO-like spontaneous bursting in the area of the *locus coeruleus* which provides noradrenergic input to most brain regions (38, 39) and is known to generate tonic (synchronized) discharge in vitro (40, 41). Apart from one study revealing tonic activity in the *locus coeruleus* of anesthetized rat pups (42), it is not clear yet to which extent ENO-like bursting in these circuits reflects in vivo activity. Of interest to our research is the finding that a prominent opioid-evoked hyperpolarization of *locus coeruleus* neurons involves a G protein-mediated cellular signaling cascade leading to activation of postsynaptic inwardly rectifying K^+ channels (43). Moreover, some classes of *locus coeruleus* neurons are respiratory modulated (41) but are not part of a rhythmogenic respiratory group (10, 20, 25, 28). It is likely that *locus coeruleus* networks in newborn rats are more mature than cortical structures because noradrenergic neurons are already functional in the early postnatal period and can actively support other developing networks (44, 45). Accordingly, we hypothesize that spontaneous *locus coeruleus* bursts in the slices are not genuine ENOs, so we refer to them here as ENO-like events.

In this chapter, we summarize our approaches for studying ENOs in cortex and hippocampus of brain slices from newborn rats and piglets and ENO-like bursting in *locus coeruleus* of newborn rat brain slices. Our (preliminary) findings revealed that (1) all types of these events are stable for several hours in superfusate of close-to-physiological ion content, (2) similar to inspiratory network bursting, their excitability depends on a Ca^{2+}/K^+ antagonism, (3) anoxia blocks all spontaneous bursts, whereas μ-opioid receptor activation abolishes bursting, transforms burst pattern, or has no clear effect in *locus coeruleus*, hippocampus, and cortex, respectively, (4) pharmacological blockade of $GABA_A$ receptors as a commonly used epilepsy model (46) evokes seizure-like bursting in cortex and hippocampus but not in *locus coeruleus*, and (5) multiphoton/confocal Ca_i imaging is most suitable to discriminate responses to neuromodulators in neurons *versus* neighboring (astrocytic) glia. The latter cells can be further identified by fluorescence imaging of sulforhodamine-101 (SR-101) and, at least for newborn rat hippocampus and cortex, glial acidic fibrillary protein (GFAP) staining (8, 9, 16, 47–49). In summary, our findings indicate that “electrophysiological imaging” in brain slices from neonatal mammals is a potent tool for studying how pharmacological or metabolism-related perturbation of spontaneously active central neural circuits affects their (developing) function.

2. Materials

Some materials used by us for studying ENOs in brain slices are identical with those for analyzing dorsal vagal neurons or respiratory neurons in isolated brainstem models. We refer to them here in abbreviated form because these materials, and rationales for their use, are described in the original papers on these systems and in corresponding chapters of this book (Chaps. 1-Trapp and Ballanyi (50) and 2-Ruangkittisakul et al. (27)).

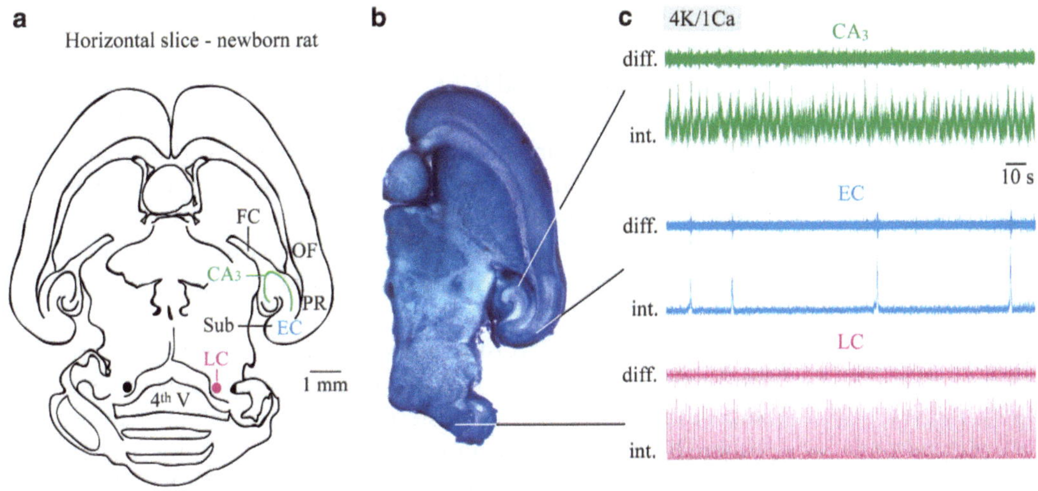

Fig. 1. *Simultaneous suction electrode recording of spontaneous early network oscillations (ENOs) in horizontal newborn rat brain slices.* (**a**) Schema of a horizontal brain section at a level that enables simultaneous recording of ENOs in both entorhinal cortex (EC) and the CA_3 hippocampal area (CA_3) plus ENO-like bursting in the brainstem area of the *locus coeruleus* (LC). Other abbreviations: FC fornix commissure, OF orbifrontal cortex, PR perirhinal cortex, Sub subiculum, 4th V fourth ventricle. (**b**) Detail from a paraformaldehyde-fixed and thionin-stained horizontal slice from a postnatal day (P) 1 old Wistar rat including the LC, EC, and CA_3 regions. (**c**) Simultaneous suction electrode recordings from a 400-µm-thick P1 Sprague-Dawley rat slice kept in superfusate containing (among other components) 4 mM K^+ and 1 mM Ca^{2+} (4K/1Ca) plus 20 mM glucose. Recordings of ENO (-like) bursting in the CA_3, EC, and LC regions show the "raw" differentially amplified (×10k) and band-pass filtered (0.3–3 kHz) extracellular neural population signal before (upper traces) and after (lower traces) integration at a time constant of 20 ms using a "Moving Averager." Note that ENOs in the EC are typically the slowest events with a rate of 1–4 bursts/min, whereas ENOs in the CA_3 area occur at 10–40 bursts/min, and ENO-like bursting in the LC has a frequency of 0.5–3 Hz.

2.1. Animal Models

ENOs, also called GDPs (14, 15, 21), are typically analyzed in brain slices from early postnatal rats and mice. Our group has so far only studied ENOs in slices from Sprague-Dawley or Wistar rats of postnatal day (P) 0–12 obtained from Charles River Laboratory Inc. (Wilmington, MA, USA; www.criver.com) and P1-4 Yorkshire-Landrace piglets weighing 1.4–2.2 kg (51) obtained from the Swine Research and Technology Centre of University of Alberta.

2.2. Solutions

In the following, we refer to compositions of solutions for (1) superfusing brain slices, (2) filling patch electrodes for "whole-cell" recording (from *locus coeruleus* neurons), and (3) histology.

2.2.1. Superfusates

ENOs in hippocampus, cortical regions, and *locus coeruleus* (Fig. 1) are principally stable in superfusate that we also recommend for studying preBötC slices with a rostrocaudal thickness >600 µm (see also Chap. 2-Ruangkittisakul et al. (27)). Because the frequently used term "artificial cerebrospinal fluid" ("ACSF") is not fully appropriate for reasons outlined in the latter chapter, we prefer to use the terms "(standard) superfusate" or "standard/

control solution." Standard superfusate has the following ionic composition (in mM): 120 NaCl, 3 KCl, 1 $CaCl_2$, 2 $MgSO_4$, 1.25 NaH_2PO_4, and 26 $NaHCO_3$. The pH of this solution is adjusted to 7.4 by ongoing gassing with carbogen, which is a mixture of 95% O_2 and 5% carbon dioxide (CO_2). Such "carbogenation" is, for example, achieved with a gas dispersion tube, also termed "micro filter candle" (Fisher Scientific, Canada; wwwfisher.ca or Robu, Hattert, Germany; www.robuglas.com). Standard superfusate also contains typically 10 or 20 mM D-glucose ("dextrose") (Sects. 3.1 and 4.1).

For studying ENO-like bursting in *locus coeruleus*, horizontal newborn rat slices can be cut at room temperature in standard superfusate and directly transferred to the recording chamber. The other slice types are cut in ice-cold "sucrose" solution containing (in mM) 234 sucrose, 4 KCl, 8 $MgSO_4$, 26 $NaCO_3$, 1 NaH_2PO_4, and 10–20 glucose to prevent excitotoxicity and preserve spontaneous neuronal activity later during recording in standard solution (6, 52) (see also Chaps. 1-Trapp and Ballanyi (50) and 2-Ruangkittisakul et al. (27)). Following cutting, slices were directly transferred to the recording chamber in a limited series of experiments.

For the majority of experiments, slices were stored on a nylon net in a 150-ml beaker filled with "recovery" solution to allow ongoing superfusion and subfusion of carbogenated solution and kept at 25–27°C by placing the beaker in a water bath (Isotemp102S, Fisher). Recovery solution is identical to standard solution, except K^+ varies between 3 and 4 mM and Ca^{2+} between 1 and 1.2 mM. For investigating Ca^{2+}/K^+ antagonism of ENOs (Sect. 1), superfusate Ca^{2+} was varied between 0 and 5 mM and K^+ between 3 and 8 mM (Sect. 4) (Fig. 2).

2.2.2. Electrode Solutions

The Ballanyi laboratory currently uses two types of patch electrode solutions for whole-cell patch-clamp recording (Sect. 3.2.2), that is, (1) standard ("low Cl^-") pipette solution containing (in mM) 120 K-gluconate, 1 NaCl, 1 $CaCl_2$, 1 $MgCl_2$, 11 K_4-BAPTA, 0.5 Na_2-ATP, and 10 Hepes with pH adjusted to 7.4 with KOH and (2) "high Cl^-" solution of identical composition except that 120 mM KCl replaces 120 mM K-gluconate. Properties of *locus coeruleus* neurons were studied using low Cl^- patch solution (Sects. 3.2.2 and 4.1.1). Rationales for particular compositions of patch electrode solutions are given elsewhere (53, 54) (see also Chaps. 1-Trapp and Ballanyi (50), 2-Ruangkittisakul et al. (27) and 3-Moore et al. (55)).

2.2.3. Histological Solutions

For histological analysis based on thionin staining of nervous structures such as layers or nuclei of neuron somata (Fig. 1) (30, 56, 57) (see also Chap. 2-Ruangittisakul et al. (27)), slices are fixed in 4% paraformaldehyde (Sigma) in phosphate buffer, which is a ~1:2

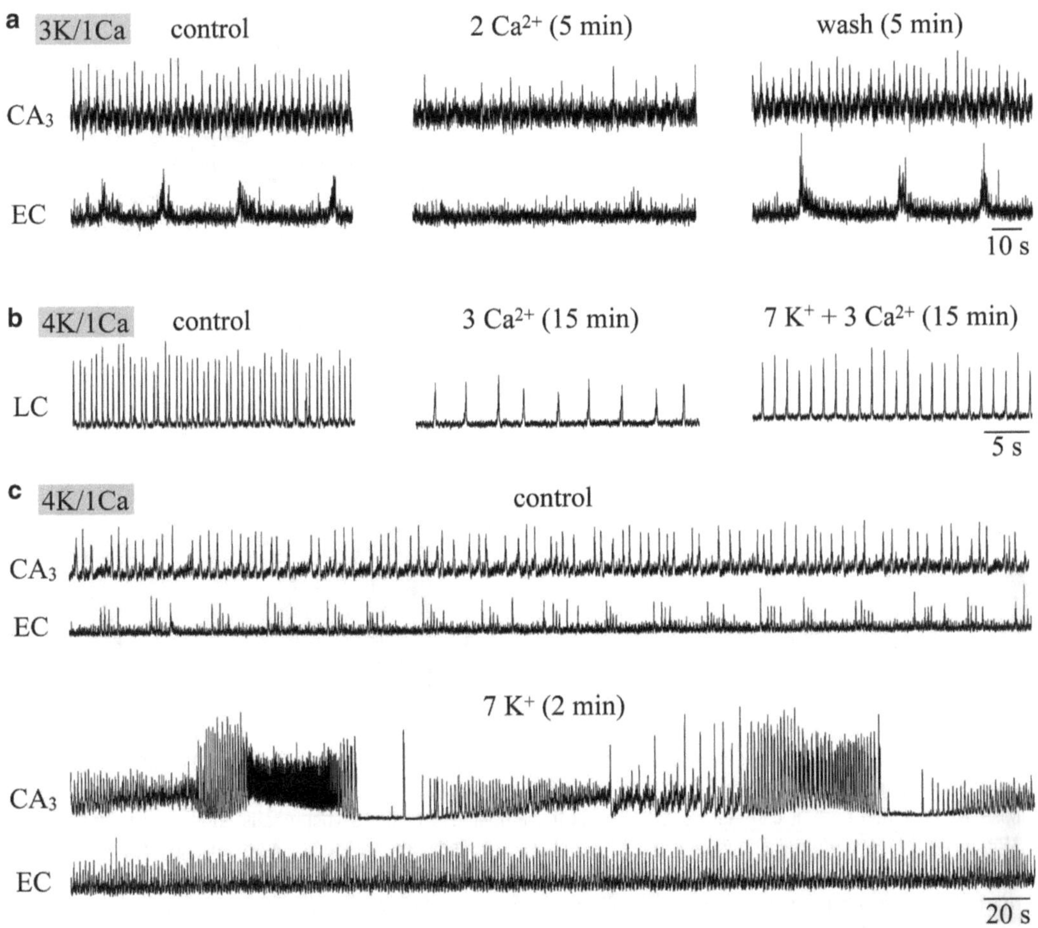

Fig. 2. *Ca^{2+}/K^+ antagonism of ENOs in horizontal newborn rat brain slices.* (**a**) Simultaneous recording of ENOs in CA_3 and EC areas of a 500-μm-thick P6 Wistar rat slice shows regular bursting in 3K/1Ca solution with 10 mM glucose. Within 5 min after start of bath-application of 2 mM Ca^{2+} in solution of otherwise identical composition, both types of ENOs were greatly depressed after 5 min but recovered within 5 min after start of return to control superfusate ("wash"). (**b**) ENO-like LC bursting in a 400-μm-thick slice from a P1 Wistar rat was inhibited by increasing the Ca^{2+} content of 4K/1Ca superfusate with 10 mM glucose to a total of 3 mM. This depression was partially reversed by also raising K^+ to a total of 7 mM. (**c**) In a 400-μm-thick P10 Sprague-Dawley rat slice in 4K/1Ca solution with 20 mM glucose, raising K^+ to a total of 7 mM transformed the regular ENO pattern in the CA_3 region to large amplitude seizure-like bursts interrupted by silent periods while ENOs in EC became faster and more uniform in amplitude.

mixture of 0.1 M NaH_2PO_4 + 0.1 M Na_2HPO_4 in H_2O, resulting pH 7.2. Thionin staining solution consists of 0.1 M Na^+ acetate trihydrate + 0.1 M acetic acid + 1% thionin acetate dissolved in distilled water. For live cell labeling of neurons and glia (Fig. 3), slices are incubated for 1–1.5 h in carbogenated standard solution with SR-101 (Sigma) at 165 μM 20–23°C or 1 μM at 34°C, respectively. Immunohistochemical identification of GFAP in astrocytes of horizontal newborn rat brain slices (Sect. 1) was performed on slices fixed with 4% paraformaldehyde for at least 2 h. Subsequently, phosphate buffer solution is used with either 4% paraformaldehyde or

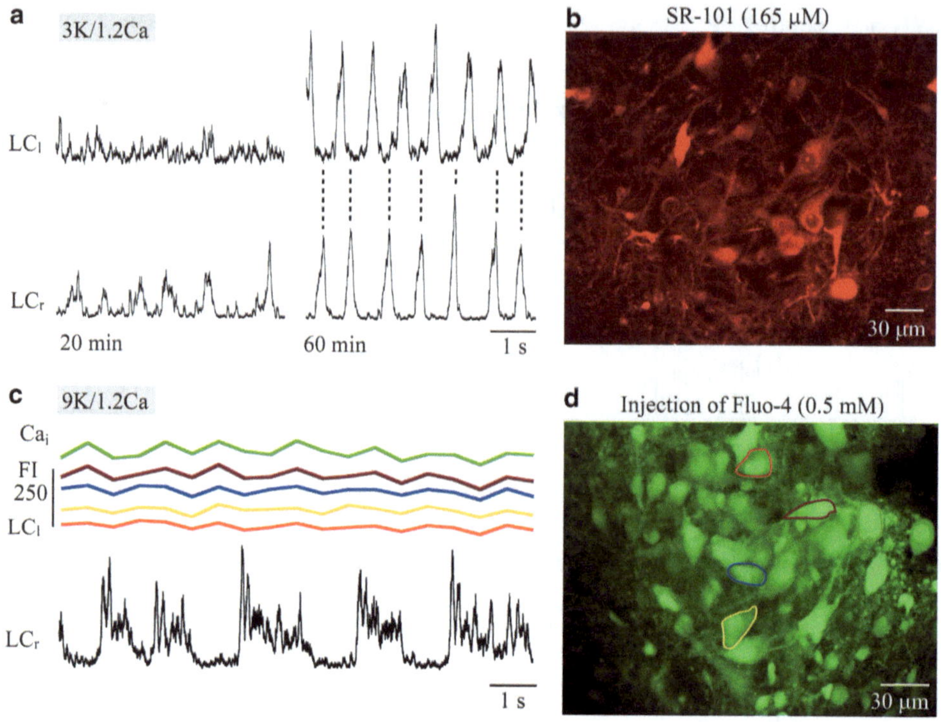

Fig. 3. *Imaging of ENO-like bursts in the bilaterally organized LC of horizontal newborn rat slices.* (**a**) Simultaneous suction electrode recording of rhythms in both LC aspects of a 400-μm-thick P2 Wistar rat slice in 3K/1.2Ca solution with 10 mM glucose. As indicated here, it can last up to 1 h until the amplitudes of the asynchronous bursts reach steady state. (**b**) Morphology of the neuron-glia network in one LC aspect as revealed upon loading of a 400-μm-thick P2 Wistar rat slice that was incubated for 1 h at 22°C in standard superfusate containing 165 μM of the red fluorescent dye sulforhodamine-101 (SR-101). Cells with a soma diameter of 20–35 μm likely represent neurons, whereas cells with a soma diameter of ~10 μm are presumably astrocytic glial cells. Thus, note that SR-101 does not show preferential staining of neurons versus glia in LC. (**c**) LC bursting often occurs at rates of >1 Hz, which is beyond the binding/unbinding kinetics of commonly used Ca^{2+}-sensitive dyes such as Fluo-8L-AM (see also Figs. 7, 8, and 10). Therefore, burst-related rises of the free cytosolic Ca^{2+} concentration (Ca_i) are typically of small amplitude in standard 3K/1.2Ca solution. Lowering superfusate Ca^{2+} to 0.5 mM or raising K^+ to 7–9 mM may transform the pattern of the ENO-like rhythm to longer lasting bursts that can be reflected by periodic Ca_i increases as labeled in relative units of fluorescence intensity (FI) here. Each Ca_i trace was measured in neuronal soma regions that are outlined in the image in (**d**) by regions of interest (ROIs) using Fluoview software of the FV300 Olympus confocal laser scan imaging system. Cells were loaded with the membrane-permeant (AM) form of Ca^{2+}-sensitive Fluo-4 via pressure injection into the LC using a broken patch electrode. The "sharp" shape of the Ca_i peaks is due to the slow scanning rate of 0.58 s/frame. Note that this or slower scanning rates are sufficient to resolve the peak of rhythmic Ca_i rises in the EC or CA_3 areas (see Figs. 7, 8, and 10). (**d**) Shows the imaged area from which the traces in (**c**) were derived. The image represents a summated "z-stack" between slice surface and 50 μm obtained after bath-application of 0.5 mM glutamate for 10 min using the confocal laser scan Olympus system for a better visualization of neuronal morphology.

0.5% Triton X (Sigma) added. As described under (Sect. 3.3), slices are ultimately exposed via incubation to a rabbit GFAP primary antibody and, 1 day later, to an Alexa 488 goat anti-rabbit secondary antibody both from Invitrogen (Carlsbad, CA, USA; www.invitrogen.com) (Sect. 3.3). SR-101 and/or GFAP labeled slices are inspected using the multiphoton/confocal imaging stations (Sect. 2.8).

2.3. Devices for Generating Brain Slices

Under anesthesia using isoflurane (Baxter Corp., Deerfield, IL, USA; www.baxter.com) (Sect. 3.1), the brain of newborn rats is dissected with fine forceps (e.g., Dumont #5, straight tip [0.05 × 0.02 mm] or Dumont #5SF [0.025 × 0.005 mm]) and (iris) scissors (e.g., Vannas-style, straight tip angle [length 80 mm, 2.5-mm blade] or Vannas-Tübingen [length 85 mm, 5-mm blade]), supplied, for example, by Fine Science Tools Inc. (Vancouver, BC, Canada; www.finescience.com).

For isolating the piglet brain (Sect. 3.1.2), an animal is anesthetized with 2–5% isoflurane (Benson Medical Industries Inc., Markham, ON, Canada; www.bensonmedical.com) through a gas mask. A catheter (5 Fr Argyle TM, Sherwood Medical Co., St. Louis, MO, USA; www.epa.gov) is placed at the femoral artery for systemic pressure measurement and medication administration. Using a pulse oximeter (Nellcor, Hayward, CA, USA; www.respiratorysolution.covidien.com) and a HP78833B monitor (Hewlett Packard Co., Palo Alto, CA, USA; www.hp.com), O_2 saturation, heart rate, and blood pressure are continuously monitored. Following surgical instrumentation, the piglet is allowed to stabilize for 30 min before euthanization using an intra-arterial bolus of 120 mg/kg sodium pentobarbital (Bimeda-MTC, Cambridge, ON, Canada; www.bimeda.com). The skull is opened with an electric saw (e.g., Dremel MiniMite Cordless Rotary Tool, Racine, WI, USA; www.dremel.com) and then dissected with fine forceps (see above), large scissors (e.g., Student Surgical Scissors [length 14.5 cm, 4.5-cm blade], Fine Science Tools Inc.) or a razor blade (Double Edge Stainless Steel Razor Blade [1.0 mm thickness], Electron Microscopy Sciences, Hatfield, PA, USA; www.emsdiasum.com), and a pair of large forceps/tongs (e.g., Multi-Use Tongs FSE [length 20 cm, 2.5-cm stirrup jaws], Fisher Scientific) for stability.

The entire newborn rat brain or a piglet brain block (Sect. 3.1) is transferred to the vise of a vibratome such as the VT1000S (Leica Microsystems, Nussloch, Germany; www.leica-microsystems.com) or the 5000mz (Campden Instruments Ltd, Loughborough, UK; www.campdeninstruments.com). A less expensive device, such as the "McIlwain" tissue chopper (Warner Instruments, Hamden, CT, USA; www.warneronline.com), may cause more damage to superficial slice layers but can be used if recording approaches are restricted to extracellular or "blind" patch-clamp techniques in thicker slices (Sect. 3.2).

2.4. Recording Chamber and Superfusion System

As reviewed in (22, 52, 58) and outlined and exemplified with figures elsewhere in this book (see Chaps. 1-Trapp and Ballanyi (50) and 2-Ruangkittisakul et al. (27)), a commercial submerged-type acrylic recording chamber with glass bottom, for example, from Warner Instruments, is used by us for experiments involving visualization of cells using an upright microscope. A custom-made acrylic chamber is used for "blind" recording experiments, in which basic slice structures and electrode positions are recognizable using a low magnification stereomicroscope (Sect. 2.8). The latter type of chamber requires a cold-light source with flexible glass fiber guides, for example, from World Precision Instruments (WPI, Sarasota, FL, USA; www.wpiinc.com) or Zeiss (Oberkochen, Germany; www.zeiss.com). The recording chambers are continuously perfused with carbogenated standard solution (Sect. 2.2) that is prewarmed in a water bath (DC-10, Thermo Haake, Sigma-Aldrich, Canada; www.sigmaaldrich.com) and administered via a peristaltic ("roller") pump such as the Watson-Marlow Alitea (Sin-Can, Calgary, AB, Canada; www.sincaninc.com). Superfusate passes through an acrylic custom-made bubble trap before entering the recording chamber that is kept at 26–28°C via a temperature control system such as the TC-324B (Harvard Apparatus, Saint-Laurent, QC, Canada; www.sincaninc.com). Superfusate is removed from the recording chamber via a hypodermic needle (gauge 18 G1½") connected to either (another channel of) the peristaltic pump or a vacuum system, for example, an aquarium pump. Tygon® or stainless steel tubing (e.g., Fisher Scientific, Canada; www.fishersci.ca) should be used for the superfusion system because it is impermeable to O_2 and CO_2 thus preventing the loss of these or other gases such as nitrogen for eliciting hypoxia during transport of fluid to the recording chamber. Optimal flow rates are at least 5 ml/min for recording chambers with a volume of ~1 ml and should be adapted correspondingly to provide similar fluid exchange rates if chambers of different volumes need to be used (for rationales, see Chaps. 1-Trapp and Ballanyi (50) and 2-Ruangkittisakul et al. (27)). In the majority of our published studies, we applied drugs by addition to the superfusate ("bath-application").

For imaging experiments or patch-clamp recording from visualized superficial cells (6, 22, 52, 58), a brain slice is transferred to the recording chamber and held in place with nylon strings. The strings are fixed to a frame made from platinum, or tantalum as a cheaper alternative, that are both supplied by Goodfellow (Huntington, United Kingdom; www.goodfellow.com) (see Fig. 4 in Chap. 1-Trapp and Ballanyi (50)). In recording chambers with sylgard-184® ("KWIKGARD," WPI) added, slices can alternatively be fixed with insect pins ("minutien pins," stainless steel, length 10 mm, tip diameter 0.02 mm, base diameter 0.2 mm; Fine Science Tools Inc.). Suction electrode(s) for electrophysiological recording (Sect. 2.5) provides further mechanical stability. Depending on the angle of the light guide from the cold-light source, carbon powder may be added while preparing the initially fluid sylgard-184® for better visualization of electrode positions or slice details.

2.5. Patch and Suction Electrodes

For electrophysiological recording of extracellular neural population activity and biophysical membrane properties of single cells, we use suction electrodes and patch electrodes, respectively (see also Chaps. 1-Trapp and Ballanyi (50) and 2-Ruangkittisakul et al. (27)). Both types of electrodes are pulled from borosilicate glass capillaries (GC-150TF-10; 1.5 mm outer diameter × 1.17 mm inner diameter, Harvard Apparatus). The tip of suction electrodes can be grinded down ("beveled") to a desired size (50–250 μm) with sand paper or whetstone ("True hard Arkansas whetstone," Fine Science Tools). Then they are fire-polished using a commercial lighter while blowing air through the electrode to avoid closure by melting. To minimize tissue damage particularly for recording under visual control using a short distance high magnification objective (6, 22, 52, 58), electrodes are beveled at an angle (30–45°) to produce a cut tip surface that is parallel to the slice surface. Alternatively, a straight tip electrode is heated and bent down ~45° at its shank to obtain a parallel tip surface and allow maneuvering under the objective. Suction and patch electrodes are inserted into a holder system (ESP-M15-P and MHH-25, Warner Instrument Corp.) with an outlet for connecting a Tygon® tube that is attached for suction to a 5-ml standard medical syringe. Fine-tuning of suction is achieved with a manually tunable syringe (e.g., Gilmont micrometer syringe; www.vwrsp.com). Suction electrodes and patch electrodes are fabricated using microelectrode pullers such as the P-1000 (Sutter Instrument Company, Novato, CA, USA; www.sutter.com) or the DMZ (Dagan Corporation, Minneapolis, MI, USA; www.dagan.com). The PC-10 (Narishige International Inc., East Meadow, NY, USA; www.usa.narishige-group.com) is a less expensive alternative but is not designed for pulling sharp microelectrodes.

2.6. Micromanipulators and Antivibration Tables

As illustrated in Fig. 3 in Chap. 2-Ruangkittisakul et al. (27), patch electrodes are positioned with either an MP-285 (Sutter Instrument Company) or an MX7600 stepmotor, both (potentially) equipped with digital display of recording depth (Siskiyou Corporation, Grants Pass, OR, USA; www.siskiyou.com). Cheaper non-electronic manipulators, such as the MM33 (Märzhäuser, Wetzlar, Germany; www.marzhauser.com) or the MMN-1 (Narishige International Inc., East Meadow, NY, USA; www.narishige.co.jp), are sufficient for positioning extracellular electrodes. For positioning electrodes on the xy-stage of an upright microscope (Sect. 2.8), You-2 (Narishige) or Kleindiek Nanotechnik (Reutlingen, Germany; www.nanotechnik.com) miniature manipulators are well suited. Manual or motorized xy-stages can, for example, be obtained from Siskiyou Design Inc. or Luigs-Neumann (Ratingen, Germany; www.Luigs-neumann.com). Mechanically unstable and electrically noisy laboratories require use of an antivibration table in combination with a Faraday cage both, for example, provided by Newport (Irvine, CA, USA; www.newport.com) or Kinetic Systems (Boston, MA, USA; www.kineticsystems.com). The Faraday cage is mounted on the antivibration table and thus electrically shields the microscope, experimental chamber, and manipulators that are positioned on this table. A simple and less expensive (but often sufficient) system uses a stone or metal plate resting on rubber or silicone dampers while a metal (tool) cabinet serves as a Faraday cage.

2.7. Amplifiers, Stimulators, and Computer-Based Data Acquisition

For intracellular electrophysiological recording, we use patch-clamp amplifiers from HEKA (Lambrecht, Germany, www.heka.com) or "sample-and-hold" SEC amplifiers from npi electronic (Tamm, Germany; www.npielectronic.com). For suction electrode recording, AC-coupled differential amplifiers (gain 1–10 k, band-pass filtering 0.3–3 kHz) are needed such as the Model-1700 from AM-Systems (Sequim, WA, USA; www.am-systems.com) or the DAM-50 from WPI. For better resolution particularly of signals with an amplitude <100 μV, integration of the output signal at a time constant of 20–50 ms is recommended using, for example, the MA-821/RSP "Moving Averager" (CWE; www.cwe-inc.com) (see Fig. 4 in Chap. 2-Ruangkittisakul et al. (27)). Signals are sampled into a computer via an interface and acquisition software at rates of 1–10 kHz for suction electrode signals and 4–20 kHz for membrane potential (or current) recordings. Convenient hardware/software packages are supplied by HEKA, npi electronic, or Axon Instruments (Molecular Devices; Sunnyvale, CA, USA; www.moleculardevices.com). Commonly used multichannel electrophysiological recording systems are "Powerlab" (ADInstruments, Colorado Springs, CO, USA; www.adinstruments.com) or "Spike-2" (Cambridge Electronic Design Ltd., Cambridge, UK; www.ced.co.uk). In most of these systems, software options are implemented for electrical stimulation in combination with an isolation unit, for example, from WPI, AM-Systems, or A.M.P.I. (Jerusalem, Israel; www.ampi.co.il).

2.8. Microscopes, Imaging Devices, and Fluorescent Dyes

For optic control or mechanic fixation of a preparation and subsequent microelectrode positioning for blind recording, a stereomicroscope is used such as the Stemi-2000 (Zeiss), SZXZ (Olympus, Center Valley, PA, USA; www.olympusamerica.com) or the ELZ (Scienscope, Chino, CA, USA; www.scienscope.com). Visualization is facilitated by attaching a simple video camera, for example, from WPI, to the stereomicroscope which can also be fully substituted by a video system equipped with zoom optics. The video signal is displayed on a separate monitor or, using a frame grabber, as an inset on the computer monitor used for controlling the (intracellular) recording (see also Chap. 3-Moore et al. (55)). Recording under visual control from superficial neurons in slices (6, 22, 52, 58) is done using an upright microscope such as the Zeiss Axioskop-2FS (or successor) or the Olympus BX61WI equipped with a water immersion lens (usually 40× or 60× magnification) with a working distance of at least ~2 mm to facilitate electrode positioning. The microscope should be equipped with differential interference contrast (DIC) optics for approaching individual cells which then appear almost in 3D-like fashion (7) (see also Chaps. 1-Trapp and Ballanyi (50), 4-Fish et al. (59), 5-Nakamura et al. (60) and 9-Broicher and Speckmann (61)). Cells located up to 80 μm below the slice surface can be visualized for targeted patch-clamp recording by attaching a charge-coupled device (CCD) camera with infrared (IR) sensitivity to the microscope and displaying the resulting IR-DIC image on a monitor. Convenient CCD cameras are the ImageM or Orca-R2 (Hamamatsu, Bridgewater, NJ, USA; www.hamamatsu.com) or Coolsnap (Röper, Ottobrunn, Germany; www.roperscientific.de). For patch-clamping cells that are fluorescence-tagged, for example, with green fluorescent protein, an epifluorescence system is required. Details of such systems using a CCD camera or photomultiplier are described elsewhere (7) (see also Chaps. 1-Trapp and Ballanyi (50), 3-Moore et al. (55), 4-Fish et al. (59), 5-Nakamura et al. (60) and 9-Broicher and Speckmann (61)).

The Ballanyi laboratory is equipped with confocal and multiphoton microscopes for imaging the activity and morphology of neurons showing ENOs and neighboring (astrocytic) glia. Similar to our approaches for loading respiratory neurons and glia in preBötC slices (56, 57) (see also Chap. 2-Ruangkittisakul et al. (27)), we load rhythmically active neurons and glia in horizontal brain slices with the membrane-permeant acetoxymethyl (AM) forms of the fluorescent Ca^{2+}-sensitive dyes Fluo-4 or Fluo-8L (TEFLabs, Austin, TX, USA; www.teflabs.com or Invitrogen) (Fig. 3). These dyes are firstly dissolved at 5 mM in dimethyl sulfoxide, containing 20% pluronic acid (Invitrogen) and then further diluted with standard superfusate (Sect. 2.2.1) to 0.5 mM for tissue injection or to 10 μM for loading via incubation (Sect. 3.3). For dye incubation, a slice is transferred to a 1.8-ml cryotube (Nunc A/S, Roskilde, Denmark; www.nuncbrand.com) with a hole in the lid for insertion of a Tygon® tube used for carbogenation of dye-containing standard superfusate (Sect. 2.4). Direct contact of the tube inlet with the solution needs to be avoided to prevent mechanic stress of the slice by perturbation of the fluid due to bubble/foam formation originating from pluronic acid. The cryotube is inserted into a heating unit (Isotemp-125D, Fisher) and the slice is exposed to the dye for 30 min at 30°C. After entering the cells, the AM component of the Ca^{2+} dyes is dissociated by esterases thus enabling Fluo-4 or Fluo-8L molecules to interact with Ca_i (6, 7, 62–64). These and other fluorescent dyes are light-sensitive and exposure to room light should thus be kept to a minimum.

Fluorescence signals are monitored using either a FV300 laser scanning confocal microscope and software (488-nm argon laser, 20× XLUMPlanF1, NA 0.95 objective; Olympus FV300, Fluoview software) or a FV300 connected with an infrared titanium-sapphire laser (10 W Mira/Verdi, Coherent, Santa Clara, California, USA; www.coherent.com) for multiphoton imaging (7, 22, 56–58, 62). The latter microscopes and software are also used by us for imaging of fluorescent SR-101 and Alexa Fluor 488 tagged to GFAP (Sect. 1) using techniques that we have adapted to further improve the resolution of morphological features of both neurons and glia in slice regions showing ENOs (Sects. 3.3 and 4).

For histological analysis of thionin-stained morphological structures (Fig. 1), slices are either transferred on a "hanging drop" glass slide (Fisher Scientific) to an upright microscope (Zeiss Standard 16, 32× magnification) connected to a PL-A642-1.3-Megapixel CCD camera (PixeLINK, Ottawa, ON, Canada; www.pixelink.com) or to a Petri dish under a stereo microscope (Zeiss SR15, 32× magnification) connected to a PL-A686-6.6-Megapixel CCD camera (PixeLINK).

2.9. Agents

Most agents for our studies on ENOs are obtained from Sigma-Aldrich except salts for superfusate, Na^+ acetate trihydrate, or acetic acid (Fisher Scientific). Most other agents, such as neurotransmitters and corresponding (ant)agonists, are purchased from Tocris Biosciences (Ellisville, Missouri, USA; www.tocris.com), MP Biomedicals (www.mpbio.com), TEFLabs Inc. (Austin, Texas, USA; www.teflabs.com), or Alomone Labs Ltd. (Jerusalem, Israel; www.alomone.com). Fluorescence dyes and immunohistochemical agents are purchased from Invitrogen, ATT Bioquest (Sunnyvale, CA, USA; www.attbio.com), and TEF Labs Inc.

3. Methods

In this section, techniques are described for generating newborn rat and piglet brain slices in which ENOs are generated spontaneously by different neural networks for several hours in superfusate with close-to-physiological ion content. Reference is made here also to microelectrode-based recording of extracellular neuronal population activity, membrane potential fluctuations, or Ca_i rises at cellular resolution. Finally, histological approaches are mentioned for discriminating neurons from glia in these slices.

3.1. Generation of Brain Slices

Initially, we used horizontal brain slices from newborn rats (Fig. 1) to establish recording of ENOs from hippocampus and cortex (studied in P0-P12 animals) and/or *locus coeruleus* (studied in P0-P4 pups). After conditions were determined for obtaining long-term recording of ENOs in rat slices, horizontal and coronal slices were generated from the brains of newborn piglets that are used in the Cheung laboratory as an in vivo model for systemic effects of asphyxia in human infants (51, 65). Basically, the same procedures are used for slicing, except that the piglet brain is substantially larger that the newborn rat brain and therefore needs to be cut into smaller blocks prior to starting the slicing procedure.

3.1.1. Horizontal Slices from Newborn Rats

All procedures were approved by the University of Alberta Animal Care and Use Committee and carried out in compliance with guidelines of the Canadian Council for Animal Care. Animals are anesthetized with 100–200 μl isoflurane until the paw withdrawal reflex disappears and then decapitated above the shoulder. To visualize the skull, the cranium is pinched with fine forceps and the skin plus nasal cavity is cut away with small scissors (Sect. 2.3). The preparation is then transferred to a customized acrylic chamber with sylgard-184® (WPI) at its base and secured in place with two 21-gauge "PrecisionGlide" needle tips (Becton Dickinson, Franklin Lakes, NJ; www.bd.com). The chamber is quickly filled with 0–4°C cold continuously carbogenated sucrose solution (Sect. 2.2.1) For initial experiments, horizontal brain sectioning was performed at 19–22°C in standard superfusate (Sect. 2.2.1) similar to preparation of newborn rat preBötC slices (22, 30, 56, 57). In our hands, slices generated with that approach only show minor spontaneous hippocampal and cortical activities, whereas stable ENOs in these areas are revealed following sectioning in ice-cold sucrose solution (Sect. 2.2.1). With the preparation kept submerged, a sagittal incision is made down the midline of the skull starting from the nasal cavity and terminating immediately posterior to the cerebellum with spring scissors (Sect. 2.3). Next, two lateral incisions are made to the left and right of the posterior end of this opening. Two additional lateral cuts are made just anterior to the olfactory bulb. To expose the brain, the skull is removed by connecting longitudinal incisions that are made between the anterior and posterior lateral cuts. The brain is isolated from the spinal cord with a coronal section through the pons, immediately posterior to the cerebellum and another section through the olfactory bulbs. Using a small spatulum, the brain is lifted slightly to enable cutting of remaining arteries and cranial nerves on the ventral surface with microscissors (Sect. 2.3) followed by removal of the brain including cerebellum from the cranial cavity.

Next, the brain is glued with cyanoacrylate glue on its ventral surface to a metal cutting plate that is then inserted into the vise of the Leica vibratome (Sect. 2.3). Subsequently, the vise is filled with ice-cold carbogenated sucrose solution. The brain is sectioned horizontally at maximal frequency (about 100 Hz) of lateral blade movement, initially in 0.5–1 mm steps in posterior-to-anterior direction until the fornix commissure appears thin and elongated. Once the desired anatomical section level is reached (Fig. 1), the speed for forward movement of the blade is reduced to achieve a smooth slice surface with minimized damage of superficial cell layers. Depending on the size of the recording chamber (Sect. 2.4) as well as the age of the animal and slice thickness (typically 200–500 μm), horizontal slices may be hemisected along the midline, potentially generating 2–10 rhythmic preparations (Fig. 1). With practice, the entire procedure takes 20–25 min. Complete or hemisected horizontal slices are placed on a net in a bath incubator and stored at 27°C for at least 1 h in continuously carbogenated recovery solution (Sects. 2.2.1 and 2.3). A single slice is then transferred to the recording chamber and fixed with insect pins to the sylgard-184® layer on the bottom of the recording chamber and the superfusion system is activated (Sect. 2.4).

Generation of horizontal rat brain slices containing the *locus coeruleus* is technically very similar to generating slices for recording of ENOs in hippocampus and cortex, except as follows. For one, caudal separation of the brain from the rest of the brainstem/spinal cord must be further posterior to the cerebellum, eventually including some anterior portion of the spinal cord, thus preserving a complete fourth ventricle. It is not necessary to isolate and slice the brain in sucrose solution because spontaneous bursting in *locus coeruleus* is very similar regardless of whether dissection is done with ice-cold sucrose solution or in standard superfusate at 19–22°C (Sect. 2.2.1). However, if simultaneous recording in cortex, hippocampus, and *locus coeruleus* in a single slice is intended for direct comparison of drug effects on these events (Fig. 1), ice-cold sucrose solution needs to be used to increase the probability for recording stable hippocampal and cortical ENOs. As described above, the brain is glued on its ventral surface and transferred to the Leica vibratome. Unlike the previous method described, it is sectioned in anterior-to-posterior direction to avoid tearing of the fourth ventricle. Serial sections are cut at maximal frequency for lateral blade movement, initially at 400–600 μm steps until the apex of the fourth ventricle appears. At this point, slice thickness is reduced to 100 μm when approaching the *locus coeruleus* with forward cutting speed reduced. Each slice is inspected "online" using a microscope (Sect. 2.8). Once the *locus coeruleus* starts to appear in the 100 μm thin section as a dark oval area close to the lateral border of the fourth ventricle and the fornix commissure appears extremely thin and elongated if not altogether absent, one (typically 400 μm thick) brain slice is cut. Hemisection of this slice provides two independently rhythmic *locus coeruleus* aspects with each a diameter of 200–400 μm in all dimensions (Fig. 1). These procedures take 25–30 min.

3.1.2. Horizontal and Coronal Slices from Newborn Piglets

After instrumentation, a newborn piglet is stabilized for 30 min before harvesting the brain with tools described above (Sect. 2.3). Immediately after euthanization, the skin above the cranium is removed to allow access to the skull. An electric, rotating circular saw is used to cut away the dorsal portion of skull covering the cortex. It is essential that the window is laterally wide enough for easy removal of the brain, which takes between 1 and 3 min. A large spatulum is used for cutting through the olfactory bulb and isolating the brain from the rest of the spinal cord by cutting anterior to the cerebellum as well as removing the whole brain from the intracranial space. Extra care must be taken at this step to ensure that the spatulum does not penetrate brain tissue and potentially destroys areas to be studied. The isolated brain is immediately placed in a 100-ml Pyrex beaker (Corning Life Sciences, Wilkes Barre, PA, USA; www.corning.com) containing 50 ml of 0–4°C cold carbogenated sucrose solution. With a pair of large forceps/tongs holding the brain in place, a pair of large scissors is used to vertically hemisect the brain. An additional coronal cut is made to separate the frontal lobes from the rest of the brain and yielding four similarly sized tissue blocks, which can cool down more quickly than the whole brain. One posterior tissue block containing the somatosensory cortex is then transferred to the dissecting chamber used for generating rat slices and filled with ice-cold continuously carbogenated sucrose solution. A coronal ~5–10-mm-thick slab containing the somatosensory cortex is cut (66), and the ventral half of the slab, which contains most of the caudate and putamen, is removed manually with a razor blade. Next, the tissue is glued either on its posterior or anterior surface with cyanoacrylate glue to the vibratome vise. Using a fine-tipped forceps (Sect. 2.3), extra care should be given when removing the arachnoid membrane covering the cortex to ensure complete sectioning of slices. Coronal slices, 250–500 μm thick, are cut using Leica vibratome in the inferior-to-superior direction and stored as described above.

The ventral half of the posterior brain quarter contains the hippocampus and most of the limbic system. The tissue block is glued with dorsal cut surface down to the vibratome vise as described above. The arachnoid membrane surrounding the gyri is then carefully removed with fine-tipped forceps to facilitate smooth sectioning of tissue. Horizontal slices are cut in posterior-to-anterior direction. Roughly 1–1.5 mm of tissue must be sectioned relatively quickly before the hippocampal formation can be identified and 6–8 slices of 250–500 μm thickness can be generated which are stored as described above (Sect. 2.2.1).

3.2. Electrophysiological Recording

We found that suction electrodes used for electrophysiological recording of extracellular activity in isolated respiratory networks (see also Chap. 2-Ruangkittisakul et al. (27)) are also very useful for recording ENOs and ENO-like bursting in the *locus coeruleus* (Sects. 3.2.1 and 4). Our experience with patch recording in the slices of the present study is restricted to *locus coeruleus* neurons. For this, we use the same electrodes as for recording from medullary dorsal vagal neurons or respiratory neurons (see also Chaps. 1-Trapp and Ballanyi (50), 2-Ruangkittisakul et al. (27) and 3-Moore et al. (55)).

3.2.1. Suction Electrode Recording

Extracellular population activity recorded with a suction electrode that is connected to a differential amplifier for AC-coupled recording typically has an amplitude in the low millivolt range but is quite often as small as 10–100 μV (Sect. 2.5). These signals are very sensitive to electrical noise, which may require use of a Faraday cage. It is also recommended to record and display them as a moving average (integrated) trace for better signal-to-noise ratio (Sect. 2.7) (Fig. 1). Suction electrodes of apparently identical shape can differ notably in the capability of recording small signals. Prior to starting a series of experiments, it is thus beneficial to prepare a set of about ten such electrodes and record ENOs in a test slice for choosing the most responsive electrodes. These electrodes can be used for up to several months, provided that they are cleaned with distilled water after each experiment.

Following positioning of these electrodes on the slice surface, the initial (modest) suction may be relieved by disconnecting the syringe from the suction tube (Sect. 2.5). The electrode should only make contact with the superficial tissue layer, particularly in <300-μm thin slices. Though, when placed on the slice surface, gentle suction using a micrometer syringe (Sect. 2.5) may improve the signal notably. In case of a "leaky" recording with steadily decreasing signal amplitude, it may be necessary to reestablish modest suction every 10–20 min. In particular, in hippocampal slice areas, suction electrodes record tonic neuronal population discharge in addition to phasic ENO bursting. This is most evident in the integrated signal (Sect. 2.7) when blockade of such activity with anoxia or high superfusate Ca^{2+} causes a sometimes pronounced decrease of the baseline of this signal (22, 29) (see also Chap. 2-Ruangkittisakul et al. (27)).

3.2.2. Patch-Clamp Recording

Similar to our experience with dorsal vagal or respiratory brainstem neurons (see Chaps. 1-Trapp and Ballanyi (50) and 2-Ruangkittisakul et al. (27)), blind whole-cell recordings from *locus coeruleus* neurons are stable for up to several hours (mostly 30–60 min) (Fig. 4). For such recordings, the electrode is advanced perpendicularly using an MX7600 motor-driven micromanipulator with an added display of recording depth (Sect. 2.6). While advancing the electrode through the tissue, positive pressure (~30 mmHg) is applied to avoid clogging which is evident from an increase in electrical resistance (26, 29). Augmentation by up to 300% of a voltage signal in response to repetitive (~25 Hz) injection of negative DC current pulses (−1 nA, 10 ms) indicates that the electrode is close to a neuronal cell membrane. The probability for a GΩ seal, which is necessary for every mode of patch recording (53, 54), correlates

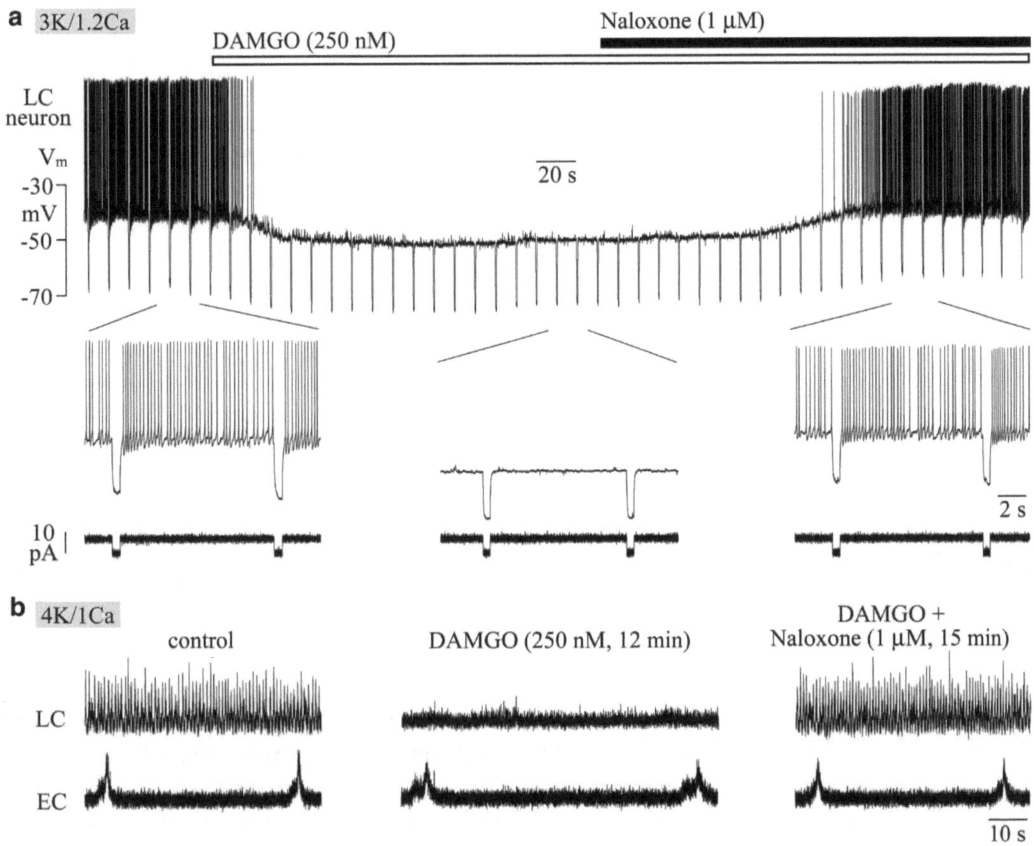

Fig. 4. *Differential effects of opioids on cortical ENOs and ENO-like bursting in LC of horizontal newborn rat slices.* (**a**) "Blind" whole-cell patch-clamp recording from a LC neuron in a 400-µm-thick P2 Wistar rat brain slice in 3K/1.2Ca superfusate with 10 mM glucose. A persistent membrane potential (V_m) hyperpolarization in response to bath-application of the µ-opioid receptor agonist [D-Ala2, N-MePhe4, Gly-ol]-enkephalin (DAMGO) was not accompanied by a major change in membrane resistance measured upon repetitive injection of an 8 pA hyperpolarizing DC current pulse (see lower trace of insets at higher time resolution). The DAMGO-evoked hyperpolarization and resulting inhibition of spontaneous "tonic" action potential discharge were reversed by adding the opioid receptor antagonist naloxone. (**b**) ENO-like LC bursting in 4K/1Ca solution with 20 mM glucose in a 400-µm-thick P2 Sprague-Dawley rat slice was abolished in naloxone-sensitive fashion by DAMGO, whereas simultaneously recorded ENOs in the EC of the same slice were not profoundly affected.

with the occurrence of electrode potential oscillations that resemble sine waves. At this instance, the patch electrode is moved upward and downward by 1–5 µm for several times until both the amplitude and speed of such electrode potential oscillations becomes maximal. At this point, the amplitude of the stimulation pulse is decreased to 0.1 nA and positive pipette pressure is released. In ~30% of cases, a GΩ seal develops spontaneously within <3 s as evident from an increase in the amplitude of the current-induced electrode potential change. If electrode potential does not start to continuously increase upon removal of positive pressure during this time period, negative pressure of up to −50 mmHg is applied. After formation of a seal >1 GΩ, the whole-cell configuration is established by abrupt and brief suction (~100 mmHg). If patch electrodes are not "perfect," for example, due to use-related corrosion of the electrode puller filament (Sect. 2.5), suction may have to be substituted by (or combined with) 1 ms 500–1,000 nA pulse using the "ZAP" function of the amplifier/software, if implemented. We do not correct membrane potential values for liquid-junction potentials that presumably range from 10 to 20 mV (53, 54) (for rationale, see Chap. 2-Ruangkittisakul et al. (27)). Once membrane potential stabilizes within several minutes after establishing whole-cell recording, membrane resistance is monitored via injection of hyperpolarizing DC current pulses (40–250 pA, 0.5–0.7 s) at intervals of 5–15 s (Fig. 4).

3.3. Multiphoton/ Confocal Imaging

Because neural activity is associated with cytosolic Ca_i rises (Sect. 1), fluorescence imaging using Ca^{2+}-sensitive dyes is widely used for monitoring both activity and morphology of brain cells (3, 6–10, 22, 56, 57, 63). As exemplified in the latter (review) articles, imaging technologies changed over the past 10 years from primarily CCD camera to multiphoton/confocal approaches, with a concomitant increase in the number of optical studies on in vivo neural systems. One major drawback of Ca_i imaging is related to the fact that fluorescent Ca^{2+} dyes non-differentially stain neurons, astrocytes, and oligodendrocytes (plus microglia of non-neural origin) particularly in immature or cultured neural structures (10, 63, 64). For discriminating between neurons and glia, it has been proposed based on both in vivo (48) and in vitro findings (16) that the fluorescence dye SR-101 selectively stains glia. However, according to other reports the same dye selectively stains neurons (47, 49).

In the following, we describe our approach for Ca_i imaging in bursting neurons and adjacent (astrocytic) glia in brain slices and also refer to our novel distinct protocols for loading either primarily neurons or astrocytes with SR-101. Moreover, we outline an established immunohistochemical approach for fluorescence labeling with GFAP in fixed slices for ultimately attributing Ca_i recordings to astrocytes (8, 9, 16, 48). For our multiphoton/confocal Ca^{2+} imaging approach of ENOs in brain slices, we use two procedures for loading cells with either Fluo-4 or Fluo-8L via injection into a specific target area or incubation of the entire slice, respectively (Sect. 2.8). For incubation with dye at 30°C, a slice is transferred for 30 min to a cryotube containing continuously carbogenated standard superfusate with 10 μM Fluo-8L-AM added (Sect. 2.8). This results in almost equal loading of neurons and glial cells in rat slices whereas primarily glial cells are loaded in hippocampus of piglet slices (Sect. 4.2.1). For focal dye application (Fig. 3), 0.5 mM Fluo-4-AM is backfilled into a broken patch pipette with an outer diameter of 5–10 μm (Sect. 2.8) and subsequently pressure-injected (25–50 mmHg) for 10 min (56, 57, 62) at 30–100 μm depth into the spontaneously active slice area of interest. This leads to staining of cells located between the surface and a depth of at least 100 μm. While imaging, rhythmic neuronal population activity can principally be recorded with a suction electrode (Sects. 2.5 and 3.2.1) from a spot in the vicinity of the area to which Ca^{2+} dye has been applied (Sect. 6).

Fluorescence signals in tissue depths of up to 60 μm and 90 μm are measured using either an Olympus FV300-based laser scanning confocal microscope or a custom-made multiphoton system based on a further FV300 system connected to an IR titanium-sapphire laser (Sect. 2.8) (Fig. 3). The Ca^{2+} dye–stained region is mostly monitored using a 20× objective with 2–3× digital zoom at reduced settings for y-axis scanning. Compared to full frame acquisition at a resolution of 512 × 512 pixels, such "clipped mode" imaging samples an area of 512 × 80–220 pixels thus enabling scan rates of 1.25–3 Hz which are normally sufficient to detect 70–100% of the peak of spontaneous Ca_i rises. An exception are ENO-like rhythms in *locus coeruleus* which occur at a rate of 1–4 Hz and are thus normally too fast to be resolved with our imaging approach (Sect. 6). However, oscillations can be slowed down pharmacologically in *locus coeruleus* for Ca_i imaging (Sect. 4.1.1) (Fig. 3). The morphology of cell bodies and primary dendrites is best revealed by combining "z-stack" image series, using single z-steps sizes of 0.05–1 μm, with 3D animation using Fluoview software (Sect. 2.8). Fluo-4 or Fluo-8L fluorescence emission is weak at low cytosolic Ca^{2+} levels (7, 62–64). Therefore, to improve morphological reconstruction, slices are exposed at the end of Ca_i imaging experiments to glutamate (0.25–1 mM, 5–10 min) for generation of z-stacks (Fig. 3). Glutamate-induced (persistent) fluorescence signals are 3–10 times larger than those associated with spontaneous activities.

For visualizing cell layers or nuclei in newborn rat and piglet brain slices with thionin (Fig. 1), slices are fixed after experiments in 4% paraformaldehyde for at least 15 min, typically overnight. For subsequent staining with thionin, slices are "washed" in phosphate buffer for 2 min, then immersed for 90 s in thionin solution (Sect. 2.2.3), and sequentially washed in phosphate buffer (2 min), in 50% ethanol (4 min), and finally in phosphate buffer (2 min).

The following procedures are used for immunohistochemical identification of GFAP in astrocytes of horizontal newborn rat brain slices (see also (16)). After electrophysiological recording, the (mostly 400 μm thick) slices are fixed and stored at 4°C in phosphate buffer with 4% paraformaldehyde for at least 2 h, optimally > 1 day. Fixed slices are then washed 3 times for 10 min each in phosphate buffered solution. Slices are subsequently incubated in methanol for 10 min followed immediately by wash with 3% hydrogen peroxide in phosphate buffer for 10 min and washed 3 times in this solution each time for 10 min. They are then incubated overnight at 4°C in rabbit GFAP antibody diluted 1:500 in phosphate buffer with 0.5% Triton X. The following day, slices are washed 3 times for 10 min each in phosphate buffer and then incubated at room temperature for 2 h in Alexa 488 goat anti-rabbit diluted 1:500 in phosphate buffer with 0.5% Triton X. Finally, sections are washed 3 times each for 10 min in phosphate buffer and then stored in this solution with 4% paraformaldehyde added until visual inspection using the confocal/multiphoton microscopes (Sect. 2.8).

3.4. Data Analysis

No quantitative data are provided in the examples for this article. However, the following sections describe methods used for quantification of data. ENOs are continuously recorded at a sampling rate of 1–10 kHz (Sect. 2.7). Burst rate is averaged over 2 min time windows. Burst duration is defined using Clampfit software (Molecular Devices Corporation, Chicago, Il, USA; www.moleculardevices.com) as the time interval from when the signal increases above and decreases below a threshold set at 10% of the peak amplitude for that burst. Pharmacologically altered rhythms are described by averaging burst rates over a 2 min time period at steady state. Values are means ± SEM. Significance ($P < 0.05$) is assessed by Student's *t* test or one-way ANOVA using Prism (GraphPad Software Inc., La Jolla, CA, USA; www.graphpad.com).

Imaging analysis is performed by drawing with Fluoview software (Olympus) regions of interest (ROIs) around cell somata and obtaining a time-series analysis of fluorescent intensity changes for each region of interest (Sect. 2.8). Fluorescence intensity data are transferred to Excel for further computation and plotting of graphs. 3D reconstruction is obtained from z-stack series (Sect. 2.8). As typical 3D settings, we use an x-y plane with rotation in x and y axes to give a 3D impression of cell morphology and x-z plane with y-axis rotation to reveal the depth of fluorescence signal. For generating figures for publication, fluorescence intensities are plotted in graphs using both Excel (Microsoft office) and Illustrator (Adobe; www.adobe.com). For generation of movies, large files are compressed using Premiere Pro CS3 software (Adobe).

4. Results

In the following, some of our "electrophysiological imaging" findings are exemplified for ENOs in horizontal newborn rat slices and horizontal plus coronal newborn piglet slices. A detailed description will follow in original reports that are currently being prepared for publication.

4.1. ENO(-Like) Rhythms in Horizontal Newborn Rat Brain Slices

Here, properties of robust ENO-like activity in *locus coeruleus* are compared with those of developmentally related ENOs (Sect. 1) in entorhinal cortex and hippocampus in the same slice type.

4.1.1. ENO-Like Bursting in Locus Coeruleus

ENO-like bursting in *locus coeruleus* consists of particularly robust bursts with very good signal-to-noise ratio, which do thus not require use of "quiet" suction electrodes (Sect. 3.2.1) (Figs. 1, 2, and 3). Because both *locus coeruleus* aspects are not coupled, bilateral bursting is asynchronous (Fig. 3). Accordingly, a horizontal slice can be sectioned in the midline for studying both *locus coeruleus* aspects separately. In both complete and split slices, ENO-like rhythms are revealed right at the start of recording at normal rate while burst amplitude may take 0.5–1 h to reach steady state. Bursting in *locus coeruleus* is stable for >5 h in standard superfusate with physiological Ca^{2+} (1–1.2 mM) and K^+ content (3 mM).

Similar to bursting of the isolated preBötC (Sect. 1) (22, 30, 31), *locus coeruleus* activity is determined by a Ca^{2+}/K^+ antagonism. Specifically, supraphysiological superfusate Ca^{2+} levels depress the rate (and amplitude) of ENO-like rhythm which ceases upon raising Ca^{2+} to >3 mM. This Ca^{2+} depression of *locus coeruleus* rhythm is reversed by raising K^+ to 6–8 mM (Fig. 2). The blocker of voltage-activated Na^+ channels tetrodotoxin (TTX) abolishes bursting within <10 min at concentrations >10 nM. Riluzole, which is an inhibitor of the "persistent" subtype of such channels that mediate intrinsic bursting (15, 20, 22), does not fully block rhythm within 20 min at 50 μM which is already beyond the level at which the drug is specific (Fig. 5). Bursting in *locus coeruleus* is also depressed following activation of ionotropic $GABA_A$ and glycine receptors with muscimol (2.5 μM) and glycine (1 mM), respectively. This depression is countered by the $GABA_A$ or glycine receptor antagonists gabazine (1–5 μM) and strychnine (5 μM), respectively. Moreover, >50 nM DAMGO inhibits *locus coeruleus* rhythm via a pronounced hyperpolarization, and this depression is antagonized by either >250 nM of the opioid receptor antagonist naloxone (Fig. 4) or >1 mM methylxanthine. The latter observations are in line with our findings on the isolated preBötC (23). Also chemical anoxia evoked by bath-application of cyanide (1 mM) blocks these bursts after more than 5 min, whereas rhythm persists, albeit with a transformed pattern, in response to 0.01–5 mM adenosine (Fig. 6). Neither gabazine (Fig. 6), strychnine, nor methylxanthines disrupted *locus coeruleus* bursting when added to standard superfusate.

In control, Ca_i fluctuations are detectable in most *locus coeruleus* cells although their amplitudes are often rather small (Sect. 6). Injection of Ca^{2+} dye or SR-101 into the *locus coeruleus* reveals basic morphological features of several cell classes (Sect. 3.3) (Fig. 3). Presumptive neurons have a soma diameter of 20–35 μm,

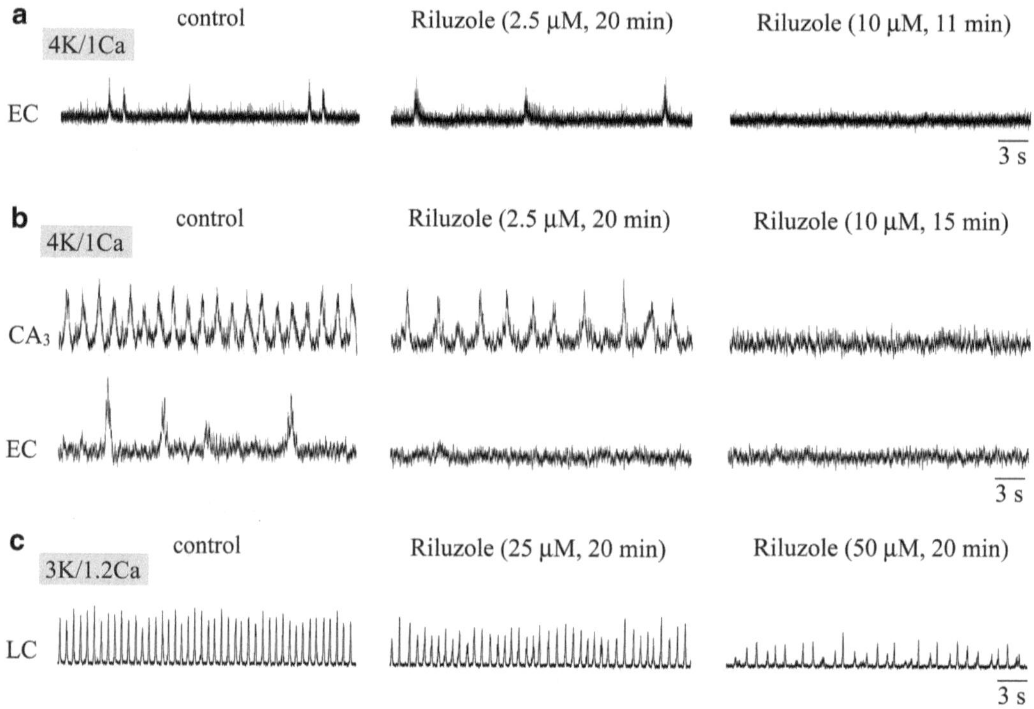

Fig. 5. *Involvement of persistent Na+ channels in ENOs of horizontal newborn rat brain slices.* (**a**) ENOs were recorded in 4K/1Ca superfusate with 10 mM glucose in the EC region of a 500-μm-thick P0 Wistar rat slice. These events were not affected by bath-application of 2.5 μM of riluzole, a blocker of the "persistent" subtype of voltage-activated Na^+ channels, but were abolished within 11 min following bath-application of the agent at 10 μM. (**b**) In a 500-μm-thick slice from a P7 Wistar rat, ENOs in 4K/1Ca superfusate with 20 mM glucose in the EC region were abolished within 20 min by application of 2.5 μM riluzole, whereas 10 μM of the drug was necessary to block ENOs in the CA_3 region of the same slice. (**c**) Rhythmic bursting in 3K/1.2Ca superfusate with 10 mM glucose in the LC region of a 400-μm-thick P1 Wistar rat slice was not affected by 25 μM riluzole while bursting was notably depressed 20 min after bath-application of 50 μM of the agent. Note that riluzole effects were typically irreversible.

whereas most cells with a soma diameter of ~10 μm most likely represent astrocytes (8–10). In response to either lowering Ca^{2+} to 0.5 mM or raising K^+ to 7–9 mM in the superfusate, the rate of ENOs decreases to <0.3 bursts/s while single burst duration increases, often to >1 s. In this situation, larger cells show synchronous Ca_i rises, whereas presumptive astrocytes remain silent (Fig. 3). Bath-application of glutamate (0.5–1 mM) evokes a prominent Ca_i rise that persists in both cell types for >5 min. Therefore, this feature can be used for z-stack imaging of their morphology (Sect. 3.3) (Fig. 3). In contrast, the metabotropic glutamate receptor agonist trans-1-aminocyclopentane-1,3-dicarboxylic acid (t-ACPD, 25 μM) causes primarily (oscillatory) Ca_i rises in presumptive astrocytes. This is in line with the view that their excitability is primarily comprised of store-mediated Ca^{2+} release (Sect. 1) (8–10). Bath-application at 20–23°C of 165 μM of the morphological dye SR-101 improves the resolution of

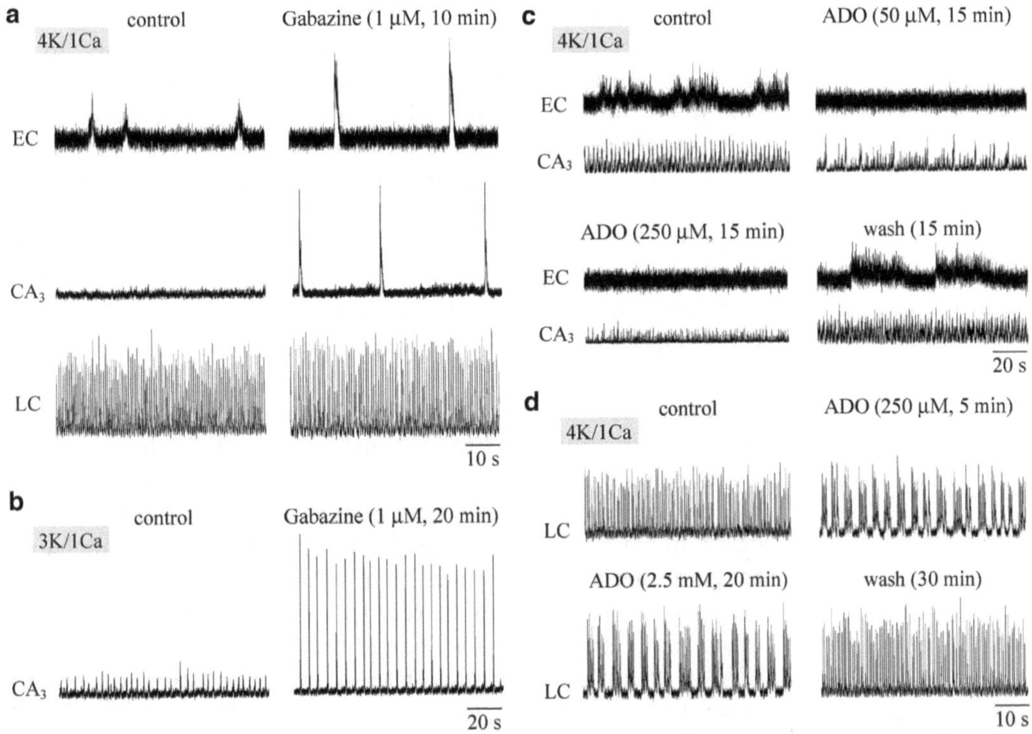

Fig. 6. *Differential modulation by (anti)convulsant agents of bursting in CA_3, EC, and LC regions of horizontal newborn rat slices.* (**a**) Bath-application in 4K/1Ca solution with 10 mM glucose of the pharmacological "seizure model" gabazine, an antagonist of A-type γ-aminobutyric acid ($GABA_A$) receptors, caused large amplitude bursting in both CA_3 and EC areas of a 400-μm-thick P4 Wistar rat slice. Contrary, LC bursting in the same slice was not affected. (**b**) In the CA_3 region of a 500-μm-thick P6 Wistar rat slice in 3K/1Ca superfusate with 10 mM glucose, ENOs were transformed by gabazine into seizure-like large amplitude bursts of similar rate. (**c**) ENOs in both the CA_3 and EC area of a 400-μm-thick P8 Wistar rat slice in 4K/1Ca superfusate with 20 mM glucose were reversibly abolished upon bath-application of the "endogenous anticonvulsant" adenosine (ADO). Note that hippocampal ENOs were more resistant to ADO than ENOs in the EC region in this experiment. (**d**) In a 400-μm-thick P1 Wistar rat slice in 4K/1Ca superfusate with 20 mM glucose, LC bursting was not abolished even by 2.5 mM ADO while the agent notably transformed burst pattern already at 250 μM.

morphological details in *locus coeruleus* but does not appear to primarily stain mainly neurons as in hippocampus (Sect. 4.1.2) (Fig. 3). Conversely, 1 μM SR-101 at 34°C results in (fainter) staining of all cell types contrary to astrocyte labeling in newborn rat hippocampus (16) (Sect. 4.1.2).

4.1.2. ENOs in Hippocampus

The capability of hippocampal networks in the CA_3 area or neighboring regions (Fig. 1) for generating long-term stable ENOs appears to depend critically on a recovery period of at least 1 h (Sects. 2.2.1 and 3.1). The rate of these ENOs ranges between 4 and 30 bursts/min (mostly 10–20 bursts/min), and single burst duration is typically 0.5–2 s. In ~20% of slices, ENOs of slightly higher rate and shorter single burst duration ("spikes") can occur in addition to the typical ENOs. Slices that have been stored in recovery solution for several hours can mostly generate robust ENOs in the recording chamber for further several hours. In our initial experiments performed in spring/summer of 2010, we used 20 mM glucose as in other studies (6) (see also Chap. 2-Ruangkittisakul et al. (27)). Changing to 10 mM glucose later did not seem to influence the probability for occurrence (or basic properties) of stable hippocampal ENOs. However, success rates of obtaining stable ENOs decreased in the following fall/winter. We have resumed using 20 mM glucose in spring/summer of 2011 and currently (summer/fall 2011) have a success rate of >70% for simultaneous recording of robust persistent hippocampal and cortical rhythms.

Regardless of whether analysis is done on slices from a single pup or across multiple litters, hippocampal ENOs that persist for >3 h are seen in ~40% of slices. Similar to ENO-like bursting in *locus coeruleus* (Sect. 4.1.1), hippocampal ENOs typically show a good signal-to-noise ratio (Figs. 1, 2, 5, and 6). In our hands, they are rarely seen in slices from <P3 rats and their amplitude seems to be larger at more regular rate in slices from >P5 animals. Hippocampal ENOs depend presumably on a depolarizing GABA action that switches to a hyperpolarization during the second postnatal week (Sect. 1) (14, 15, 34). However, in our hands hippocampal ENOs are also typical for slices from P12 rats which are the oldest animals tested so far by us. We have not studied yet whether hippocampal ENOs depend on $GABA_A$ receptors at that age.

Hippocampal ENOs in newborn rats show a pronounced Ca^{2+}/K^+ antagonism (Fig. 2). In ~50% of slices, raising Ca^{2+} to 2 mM abolishes the typical type of bursting which is restored by raising K^+ from 3 to 6–8 mM while spike-like ENOs are less sensitive to high Ca^{2+} similar to ENO-like *locus coeruleus* rhythm (Sect. 4.1.1). Raising K^+ to >6 mM in 1 mM Ca^{2+} hyperexcites the hippocampal circuits which can profoundly perturb hippocampal ENOs (Fig. 2). In ~50% of slices, ENOs are not robust in 3 K/1Ca, but bursting is enhanced and stabilized by raising K^+ to 4–5 mM. TTX (at >10 nM) and riluzole (5–20 μM) (Fig. 5) abolish both types of ENOs within <20 min. Bursting is also abolished during 5–10-min time periods of chemical anoxia or bath-applied adenosine (10–250 μM) (Fig. 6). Gabazine (1–5 μM) often transforms hippocampal ENOs into large amplitude seizure-like events that occur at a rate of 0.5–5 bursts/min (Fig. 6). Occasionally, a similar transformation of burst pattern is seen in one population of slices upon application of >100 nM DAMGO. In other slices, DAMGO transforms ENOs into rhythmic events with a rate of 0.2–1 bursts/min and a single burst duration of up to 1 min.

Suction electrode recorded hippocampal ENOs are principally well reflected by Ca_i oscillations in CA_3 pyramidal layer neu-

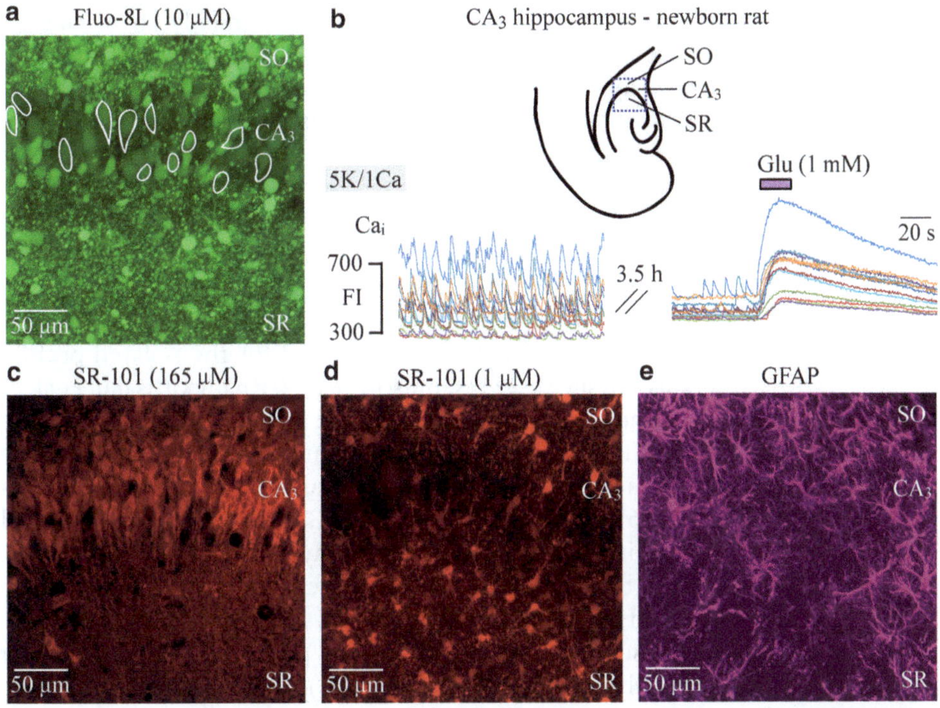

Fig. 7. *Neuron-glia imaging in CA_3 hippocampal area of horizontal newborn rat slices.* (**a**) Incubation of a 400-μm-thick P3 Wistar rat slice with 10 μM of the membrane-permeant AM form of the Ca^{2+}-sensitive dye Fluo-8L resulting in loading of neurons in the CA_3 layer and (astrocytic) glia that are densely packed particularly in *stratum radiatum* (SR) and *stratum oriens* (SO) (see also **c–e**). The image represents a summated z-stack of images at 0.5 μm steps in a tissue depth between 0 and 50 μm below the slice surface (see also Fig. 3). (**b**) Colored traces show the kinetics of spontaneous, mostly synchronous, Ca_i rises in 4 K/1Ca solution 10 mM glucose in somatic areas of CA_3 neurons indicated by the ROIs in (**a**). Traces in the right part illustrate that Ca_i baseline is still low after 3.5 h of (noncontinuous) imaging and that the amplitude of a Ca_i rise associated with bath-application of glutamate (Glu) for 20 s is notably larger than spontaneous Ca_i increases. (**c**) Morphology of CA_3 neurons revealed following incubation of a 400-μm-thick P3 Wistar rat slice for 1 h at 20–23°C with 165 μM of SR-101. (**d**) Incubation for 0.5 h with 1 μM SR-101 at 34°C resulted in staining of presumptive astrocytes in a 400-μm-thick P4 Wistar rat slice. (**e**) In a 400-μm-thick P10 Wistar rat slice, astrocytes were identified by immunohistochemical staining (details in methods) for glial fibrillary acidic protein (GFAP). Movies for (**a**, **c–e**) are provided online (see Movies 1–7).

rons with a soma diameter of 15–25 μm (Fig. 7). Because incubation of slices with Fluo-8L-AM stains both neurons and glia, injection of dye is not necessary for loading neurons as in isolated newborn rodent respiratory networks in our hands (Sect. 3.3) (56, 57) (see also Chap. 2-Ruangkittisakul et al. (27)). Synchronized bursting seems to be more prominent in slice layers deeper than 60 μm as evident in ~50% of cases from "diffuse" Ca_i increases originating from cells that cannot be visualized at sufficient resolution at such depths. Ca_i rises in a notable portion of superficial neurons are synchronized with the "background" Ca_i oscillations while neighboring neurons may show out-of-phase oscillations. Sparsely distributed cells with a soma diameter of ~10 μm in the pyramidal CA_3 or CA_1 cell layer (*stratum pyramidale*) and numerous cells of similarly small size in the adjacent (dendritic) areas of the *stratum radiatum* and *stratum oriens*, both likely representing mostly astrocytes, do not show synchronized spontaneous Ca_i rises. Glutamate notably increases Ca_i in both types of cells and enables their morphological characterization (Sect. 3.3) (Fig. 7), whereas t-ACPD and adenosine triphosphate (ATP) often induce Ca_i oscillations in presumptive astrocytes with minor effects on Ca_i in neurons.

The latter finding shows that profiling of neuromodulator-evoked Ca_i responses is one tool for discriminating small neurons from astrocytes as we have described for the isolated preBötC in brainstem slices (22) (see also Chap. 2-Ruangkittisakul et al. (27)). A further approach in this regard involves fluorescence imaging using SR-101. In initial experiments, we found that incubation of living horizontal newborn rat brain slices at room temperature with 165 μM of the dye (Sect. 3.3) primarily stains neuronal cell bodies in the CA_3 pyramidal cell layer (Fig. 7). These findings are similar to previous reports on selective neuronal staining in other isolated nervous tissues (47, 49). More recently, we have incubated such slices at 34°C with 1 μM SR-101 which appears to primarily label astrocytes (Fig. 7). This is similar to what was observed in newborn rat hippocampal slices (16) and similar to selective astrocyte staining by this dye in vivo (48). It thus seems that these different parameters used for loading cells with SR-101 are suitable to discriminate neurons from glia. For unknown reasons, it seems unfeasible to incubate slices with Fluo-8L-AM plus 165 μM SR-101 for simultaneous Ca_i imaging and neuron labeling, respectively. Contrary, dual Ca_i imaging and morphological labeling is possible in glia of slices exposed to 10 μM Fluo-8L-AM and subsequent loading of 1 μM SR-101 at high temperature (Fig. 7). It should be considered that SR-101 may induce hyperexcitability of neural networks as suggested by findings in a recent study (67).

Our very recent findings indicate that it is possible to stain slices that were chemically fixed after optical Ca_i recording with SR-101 for morphological analysis. Also in chemically fixed slices, we have applied an established immunohistochemical approach (16) for staining astrocytes with GFAP. While staining can be very pronounced in slices that have only been used in standard superfusate for time periods <1.5 h (Fig. 7), it is not clear yet whether double labeling of GFAP with Fluo-8L (as shown by Kafitz and colleagues (16)) or triple labeling with Fluo-8L and SR-101 can be done routinely in our hands in slices used for electrophysiological imaging for several hours.

4.1.3. ENOs in (Entorhinal) Cortex

Similar to hippocampal ENOs, spontaneous synchronized bursting for several hours in (entorhinal) cortex is primarily observed in slices that have been stored in recovery solution for 1 to >5 h

(Sect. 2.2.1). In spring/summer of 2010, "simple" ENOs in entorhinal cortex were revealed in almost all slices studied in superfusate with either 10 or 20 mM glucose. These cortical ENOs consisted mostly of 0.3–1.5 s lasting events that occurred at a rate of 0.2–2 bursts/min and thus resemble very much ENOs shown in the original paper on these rhythms (6). In ~50% of slices, the signal-to-noise ratio was very good as for hippocampal ENOs or ENO-like bursting in *locus coeruleus*, whereas in the other cases, amplitudes of signals could be <100 μV thus requiring selected suction electrodes (Sect. 3.2.1). Since winter 2010, we observe also other patterns of cortical ENOs in a majority of slices. In addition to slow single peak events described above, multipeak bursts can occur at either slow (<1 burst/min) or faster rates (>4 bursts/min). Moreover, (multipeak) events lasting several seconds are observed that either represent the only type of rhythm in a given slice or can be intermingled with one or several of the other patterns. Independent on the types of ENO pattern in entorhinal cortex, rhythmic bursting in this region and neighboring areas (Fig. 1) shows pronounced Ca^{2+}/K^{+} antagonism and is abolished by TTX, riluzole, adenosine, and anoxia with a sensitivity similar to that of hippocampal ENOs (Figs. 2, 5, and 6). Gabazine evokes large amplitude seizure-like rhythmic bursting in (entorhinal) cortex, similar to that seen in hippocampus (Fig. 6).

So far, we have not systematically studied Ca_i dynamics in cortical areas. But, preliminary findings indicate that our multiphoton/confocal imaging approach using loading of cells via incubation of slices with Fluo-8L-AM is feasible for Ca_i imaging in both neurons and glia (Fig. 8). Similarly, morphological imaging with high or low SR-101 doses (plus GFAP staining) seems principally possible for identifying cortical neurons *versus* astrocytes (Fig. 8).

4.2. ENOs in Newborn Piglet Brain Slices

During the past decade, piglets are increasingly used as an experimental model in neuroscience (51, 65, 68). These studies employ various techniques such as behavioral testing, electroencephalogram recording, functional in vivo imaging such as positron emission tomography, and also molecular biology, for studying brain development and establishing models for various brain disorders. However, we are not aware of previous electrophysiological (imaging) studies to identify functional neural circuits in newborn piglet brain slices. In summer 2010, we have established techniques corresponding to those applied to newborn rat slices to detect and characterize for the first time ENOs in piglet slices.

4.2.1. ENOs in Hippocampus and Cortex of Horizontal Slices

Following recovery of slices for 1 to >5 h (Sect. 2.2.1), suction electrode recording in both hippocampus and (entorhinal) cortex reveals spontaneous activity in >80% of cases that can last for

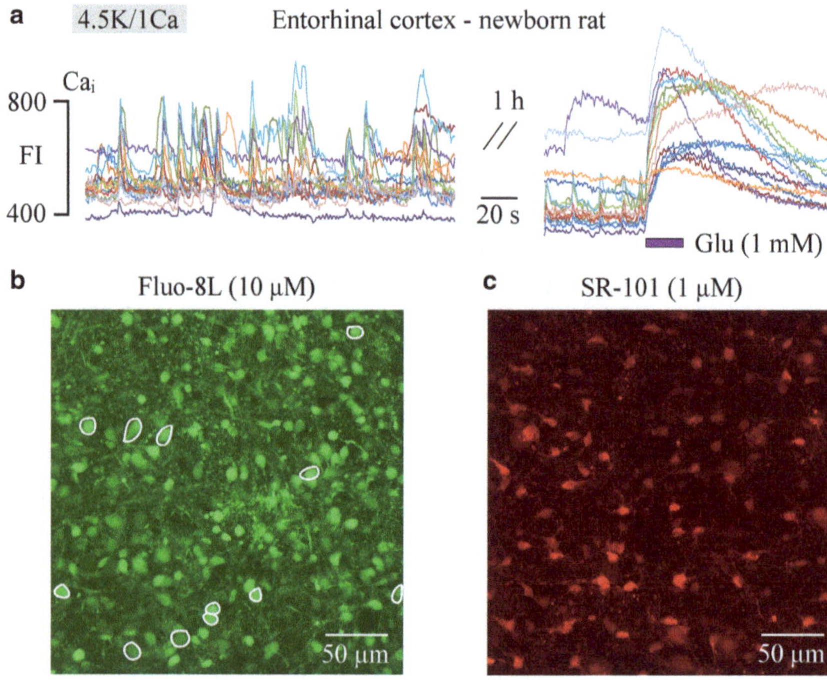

Fig. 8. *Neuron-glia imaging in EC of horizontal newborn rat slices.* (**a**) Colored traces show the kinetics of spontaneous Ca_i rises in 4.5K/1Ca solution with 20 mM glucose in the soma regions (indicated by ROIs in the image of **b**) of neurons and glial cells located 15 μm below the slice surface of a 400-μm-thick P1 Wistar rat slice loaded with Fluo-8L by incubation. As is more evident from the movie, seen online, these Ca^{2+} rises occur (synchronously) in small clusters of up to several dozens of cells. Traces in the right show that cells are still active and responsive to bath-applied Glu after long-term (non-continuous) recording. The movie, seen online, illustrates that almost all cells in the imaged area respond to Glu. (**b**) Shows the area imaged in (**a**) at the peak of the Glu response, with ROIs around the somata of cells whose Ca_i responses are shown in (**a**) (see also Figs. 3 and 7). (**c**) In a different 400-μm-thick P7 Wistar rat slice, loading with 1 μM SR-101 at 34°C resulted in staining of presumptive astrocytes. Movies for (**a–c**) are shown online (see Movies 8–10).

several hours. Contrary to ENOs in rat slices, very regular large amplitude bursting is detected in only <20% of cases in the CA_3 area of hippocampus. Instead, this area exhibits sharp activity peaks with a duration of 0.1–0.5 s. Such bursting occurs at a rate of 2–20 bursts/min with intermittent periods of up to 30 s during which the baseline of the integrated signals shows fast low amplitude fluctuations (Fig. 9). Both large peaks and baseline fluctuations reflect synchronized neural bursting because they disappear in response to >0.1 μM TTX plus >50 μM of the voltage-activated Ca^{2+} channel blocker Cd^{2+}. In ~60% of slices, spontaneous activity is depressed by raising Ca^{2+} from 1 to 2–3 mM or by addition of adenosine (10–250 μM) (Fig. 9). Chemical anoxia abolishes bursting in all cases, whereas gabazine (1–5 μM) evokes mostly large amplitude seizure-like bursting in ~40% of slices. Principally similar spontaneous activity and effects of the above drugs are revealed in entorhinal cortex of horizontal slices (Fig. 9).

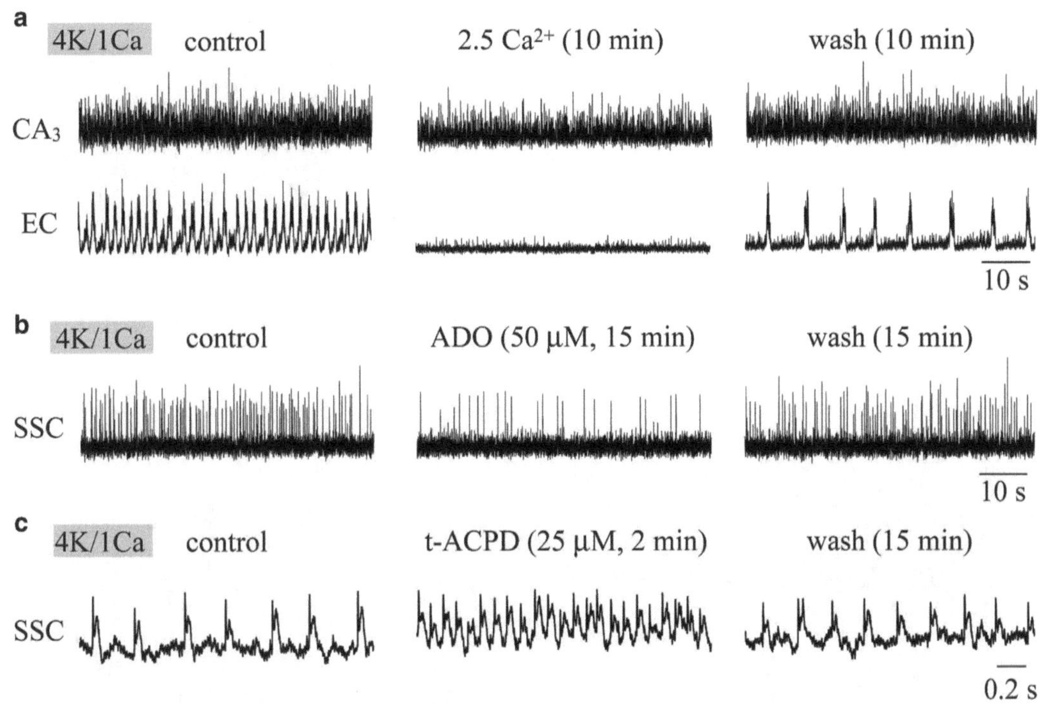

Fig. 9. *ENOs in horizontal and coronal slices from newborn piglet.* (**a**) ENOs in 4K/1Ca superfusate with 10 mM glucose were simultaneously recorded with suction electrodes in the CA_3 and EC region of a 400-μm-thick horizontal P4 piglet slice. Raising Ca^{2+} by 1.5 mM to a total of 2.5 mM reversibly abolished ENOs in the EC, whereas CA_3 bursting was only attenuated. (**b**) ENOs in 4K/1Ca superfusate with 10 mM glucose in the somatosensory cortex (SSC) of a 300-μm-thick coronal P4 piglet slice were reversibly attenuated by ADO. (**c**) ENOs in 4K/1Ca solution with 20 mM glucose in the SSC region of 500-μm-thick coronal P3 piglet slice were notably accelerated upon bath-application of the metabotropic glutamate receptor agonist trans-1-aminocyclopentane-1,3-dicarboxylic acid (t-ACPD).

Incubation of slices with Fluo-8L-AM seems to be less effective compared to injection of Fluo-4-AM which stains both neurons and presumptive astrocytes in entorhinal cortex (Fig. 10). In only few cases, spontaneous Ca_i rises occur primarily in astrocytes while glutamate evokes a notable Ca_i increase in both cell types. In hippocampal regions, only astrocytes appear to respond to either glutamate or t-ACPD with a Ca_i rise. Neurons in that area do not show a change in fluorescence intensity in response to glutamate although the dye can enter the cells and reveal their morphology quite well in some cases. Upon staining with 165 μM SR-101, morphological features of neurons in both hippocampus and cortex can be resolved well (Fig. 10).

4.2.2. ENOs in Somatosensory Cortex of Coronal Slices

Newborn pigs show great similarities to human infants in regard to gross anatomy and brain development (68), making them suited as models for neurological diseases. Of particular interest to us is the somatosensory cortex because significant histological and

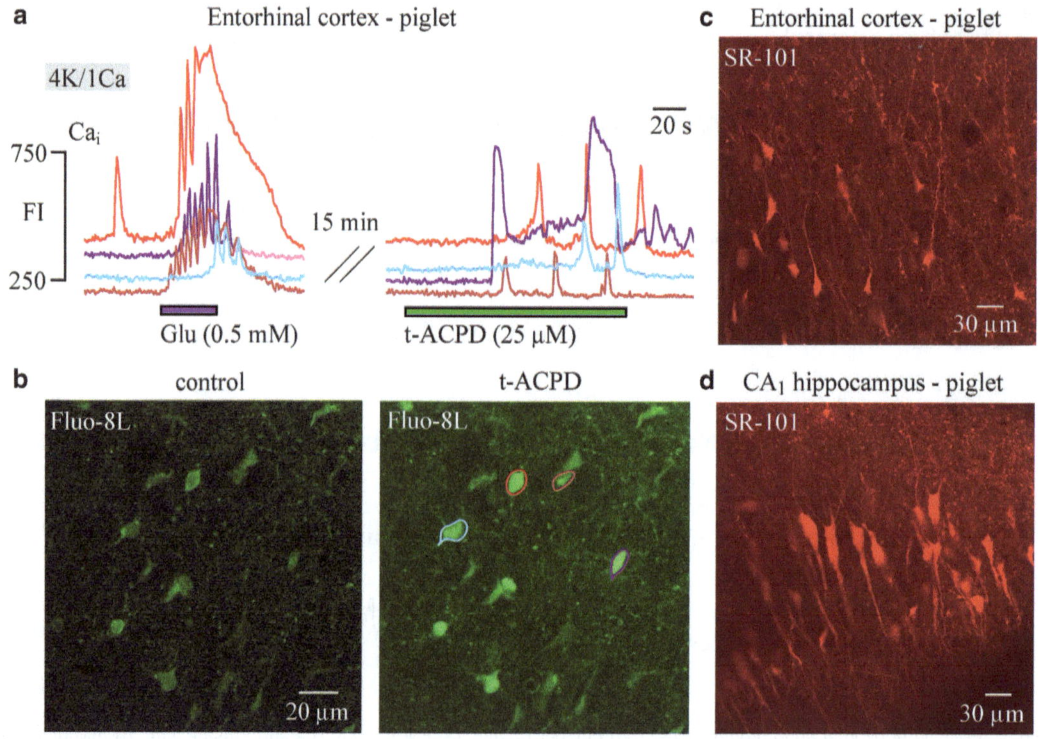

Fig. 10. *Ca_i dynamics and morphological cellular features in horizontal newborn piglet slices.* (**a**) Bath-application of Glu for 30 s led to prominent Ca_i increase in cells in the EC region of a 500-μm-thick horizontal P2 piglet slice loaded via incubation with 10 μM Fluo-8L. Exposure to 2 min of t-ACPD 15 min later evoked partly oscillatory Ca_i rises. (**b**) Shows in the right image the area imaged in (**a**) with ROIs around the somata of the cells whose Ca_i responses are shown in (**a**). The image in the left shows baseline fluorescence in control, while the right image shows the summated image (over 100 s) during the peak of t-ACPD response. (**c**) In a 500-μm-thick horizontal P3 piglet slice, incubation with 165 μM SR-101 at 22°C revealed the morphology of neurons in the EC region. (**d**) In a 500-μm-thick horizontal P3 piglet slice, corresponding incubation with SR-101 results in staining of neurons in the CA_1 area. Movies for (**a**) are shown online (see Movies 11 and 12).

biochemical injuries have been demonstrated in this area of newborn pig after hypoxic-ischemic insults (69). We therefore started to investigate ENOs in this region using coronal slice and to develop this model for future electrophysiological imaging study of hypoxia/ischemia effects in the newborn brain (Sect. 1). Similar to horizontal piglet slices, coronal somatosensory cortical slices were stored for 1 to >5 h in recovery solution. In >70% of cases, we observed robust ENOs occurring at a regular rate of 0.2–1 Hz and having a single burst duration of 0.1–0.3 s (Fig. 9). These ENOs are depressed by TTX (plus Cd^{2+}) and (mostly) also by raised Ca^{2+} or adenosine (Fig. 9) at concentrations similar to those used in horizontal piglet or rat slices. Chemical anoxia seems to abolish these events while gabazine can evoke seizure-like bursting. Upon loading cells in these slices with Fluo-4 via focal injection, multi-photon/confocal Ca^{2+} imaging is feasible in neurons and glia and reveals glutamate-evoked Ca_i rises in both types of cells. Also SR-101 imaging can be applied successfully for characterizing the morphology of neurons in somatosensorty cortex while glial labeling has not been tested yet.

5. Discussion

Findings from in vitro studies suggest that ENOs are pivotal for development of neural structures (1–4) (Sect. 1). There is steadily increasing evidence that ENOs in isolated hippocampal and cortical circuits are similar to spontaneous activities in these brain regions in vivo (2, 4, 14, 15, 35–37). Contrary, so far only one study assessed in vivo activity in neonatal *locus coeruleus* (42). Specifically, the authors found that synchronized tonic activity is typical for the majority of *locus coeruleus* neurons in newborn rat pups in vivo several days after birth, whereas such discharges occur only sporadically in some of these cells immediately after birth. However, the animals in that study were anesthetized, and it is established that this can notably depress spontaneous bursting (70). In contrast to these in vivo findings, tonic activity seems to be typical for *locus coeruleus* neurons in slices from newborn and juvenile rodents (40). In line with this, synchronized *locus coeruleus* neural networks in newborn rat brainstem-spinal cords are either tonically active or show rhythmic bursts of activity while some cells in both classes receive additional phasic synaptic input from respiratory networks (41). Besides this discrepancy between in vivo and in vitro findings, it is not clear whether *locus coeruleus* networks in newborn rats are immature as cortical structures or rather (almost) mature as the preBötC. There is though evidence from studies in newborn rats that noradrenergic neurons are already functional in the early postnatal period and can support other developing networks (44, 45). Based on these considerations, we hypothesized (Sect. 1) that *locus coeruleus* circuits in the newborn rat slices are functional, yet still developing, and thus generate non-developmentally related ENOs that are likely also present in intact unanesthetized animals.

ENOs in hippocampus, cortex, and *locus coeruleus* rhythms share several common features. Firstly, a Ca^{2+}/K^{+} antagonism determines bursting in these isolated brain regions, in both rats and newborn piglets, in a fashion resembling that in the isolated preBötC and respiratory networks in vivo (30) (see also Chap. 2-Ruangkittisakul et al. (27)). In the latter studies, we hypothesized that depression of preBötC bursting by modestly elevated extracellular Ca^{2+} is primarily of presynaptic origin and stated that the underlying molecular mechanism is yet unknown (22, 29, 31). It should be noted that raised Ca^{2+} also depresses spontaneous activity in cortical slices from mature mammals (Chap. 8-Sanchez-Vives (71)). All this indicates that Ca^{2+}/K^{+} antagonism is a general characteristic feature of spontaneously active neural networks. As a further common feature of these networks, chemical anoxia abolishes all three types of spontaneous bursting in newborn rats plus hippocampal and cortical ENOs in piglets. The finding that this effect is stable for time periods >10 min without occurrence of a major tonic discharge indicates lack of profound perturbation of ion homeostasis. This is similar to our findings on a high tolerance to anoxia of neonatal respiratory neurons and adult dorsal vagal neurons, which is contrary to the fact that most adult central neurons respond to anoxia within few minutes with irreversible depolarization and Ca^{2+} deregulation (18, 19, 25) (see also Chaps. 1-Trapp and Ballanyi (50) and 12-De Curtis et al. (72)). Also adenosine, which accumulates in the extracellular space of (isolated) brain tissues during normal activity and particularly during hypoxia/anoxia (18, 19, 23, 25, 52), depresses ENOs in neonatal hippocampus as previously reported (73) and also in cortex. In contrast, ENO-like rhythm in *locus coeruleus* seems to be resistant to this neuromodulator similar to minor effects on the isolated preBötC at least in our hands (18, 19, 23, 25). Also similar to the preBötC (18, 19, 23, 25), neural networks in *locus coeruleus* are depressed by μ-opioid receptor activation, which is intensely studied since several decades in *locus coeruleus* slices (e.g., (40)). In contrast, μ-opioids either transform burst patterns or have minor effects on ENOs generated by hippocampal and cortical networks, respectively. Furthermore, seizure-like activity associated with blockade of $GABA_A$ receptors can severely perturb ENOs in hippocampus and cortex, whereas rhythmogenic circuits in *locus coeruleus* are resistant to this manipulation. Finally, our findings on riluzole support observations by others (12, 13, 15) suggesting that generation of hippocampal and cortical ENOs depends on persistent Na^{+} channels. While these channels also contribute to intrinsic bursting in newborn rat respiratory networks (20, 25), ENOs in *locus coeruleus* seem to be quite independent on these ion channels based on lack of effect of riluzole at doses <50 μM as shown by us here.

In summary, these (preliminary) results show that rhythmic bursting in isolated neural networks of the preBötC, hippocampus, cortex, and *locus coeruleus* shares various common properties albeit their varying degree of network maturation. Nevertheless, each of these structures has also characteristic features enabling their pharmacological discrimination, for example, based on different responses to riluzole or adenosine. This indicates that the activity of each of these neural networks can potentially be targeted by specific novel clinically relevant agents that may not have an adverse effect on the other structures. We also found that electrophysiological imaging by complementing suction electrode or patch-clamp based observations with those revealed upon fluorescent Ca_i

and morphological imaging is a potent tool to analyze the role of both neurons and neighboring astrocytes in the generation of these spontaneous activities and their modulation by pharmacological agents or experimental strategies that mimic neurological diseases or are candidates for their treatment.

6. Limitations and Perspectives

In the following, we refer to (potential) limitations of studying ENOs in newborn rat and piglet brain slices. We focus this discussion on factors that may be important for long-term stable ENOs in hippocampus and cortex and the feasibility of (combined) electrophysiological imaging.

6.1. Stability and Variability of ENOs

Spontaneous hippocampal and cortical ENOs are studied since >20 and >10 years, respectively (2, 4, 6, 12–17, 21, 74–76). References in the latter articles document the notably different experimental conditions among research groups. Major factors in this respect are the immediate use of slices after cutting *versus* storing them (on a net) in "recovery solution" of particular composition and use of different types of "cerebrospinal" superfusates (Sect. 2.2.1). Most relevant for the latter aspect are differing concentrations for superfusate Ca^{2+}, K^{+}, and glucose. Both, Ca^{2+} and K^{+} have a major influence on neuronal excitability (22, 52, 58, 77) and our findings of pronounced Ca^{2+}/K^{+} antagonism for all three types of ENOs strongly support this view. While hippocampal and cortical ENOs can be stable in our hands for at least 5 h in 3 K/1Ca solution, elevation of K^{+} to 4–5 mM can have a stabilizing effect on weak bursting in these areas. In contrast, ENO-like bursting in *locus coeruleus* is extremely robust and long lasting in 3 K/1Ca. *Locus coeruleus* rhythms seem to be similar regardless of whether slices are directly used for recording or firstly stored in recovery solution. Contrary, both hippocampal and cortical ENOs apparently depend greatly on such preincubation for recovery in our hands.

Important for the practicability of experiments, slices from both rats and piglets that have been stored for several hours can still generate all types of ENOs in the recording chamber for several more hours. This indicates that neural circuits in slices from newborn mammals remain functional for >10 h contrary to the view that (adult) brain slices should not be used for >5 h because of limited viability of neurons (52, 58). In fact, it is possible that the neonatal ENOs persist for >24 h in acute slices similar to inspiratory rhythm in preBötC slices (Chap. 2-Ruangkittisakul et al. (27)) and spontaneous diurnally related activity in slices from the suprachiasmatic nucleus (Chap. 5-Nakamura et al. (60)). If this were the case, effects could be studied of "slow" neuromodulators such as neurotrophic factors or responses of these networks to long-lasting application of opioids or methylxanthines to test whether this clinical scenario affects their bursting. However, it needs to be considered that some properties of neural circuits in slices may reconfigure within several hours after their isolation (52). One obvious example for this is the "washout" of preBötC rhythm in transversal medullary slices after several hours in 3 K/1Ca solution. preBötC rhythm can be reactivated for multiple hours by raising of K^+ to 6–9 mM. However, high K^+ changes the sensitivity of the isolated inspiratory center to neuromodulators such as opioids or to anoxia (22).

Regarding superfusate glucose, hippocampal ENOs are possibly more robust in 20 *versus* 10 mM glucose. But, we did not test this systematically, and other experimental factors may also have changed at the same time. Regarding the most appropriate superfusate glucose content, it needs to be considered that glucose in brain interstitial space ranges between 0.1 and 2 mM, mostly depending on methods used for its determination (76) (for other references, see Chap. 1-Trapp and Ballanyi (50)). Accordingly, it has been proposed that brain slices of newborn mammals should be studied in close-to-physiological glucose, that is, ~1–2 mM, and/or in solution in which glucose is supplemented or fully substituted by energy substrates that are more relevant for newborns, specifically ketone bodies, lactate, and pyruvate (76). In the latter review article, the authors refer to their related work showing that hippocampal and cortical ENOs are inhibited using close-to-physiological energy substrate levels in the superfusate. However, findings from other recent studies suggest that (10 mM) glucose provides an appropriate metabolic fuel for (newborn rodent) slices (74, 75). Obviously, this deserves further investigation.

6.2. Feasibility of Electrophysiological Imaging

One limitation that is common to the analysis of ENOs in all regions and slice types described above refers to simultaneous electrophysiological imaging. This approach is particularly feasible for studying inspiratory networks in newborn rodent brainstem slices because population activity can be recorded with suction electrodes in one lateral aspect of the bilaterally organized networks. Concurrently, the contralateral preBötC aspect or inspiratory active XII motor nucleus can be loaded with Ca^{2+} and/or morphological dyes for imaging which can be combined with patch-clamp recording (22, 63, 64) (see also Chap. 2-Ruangkittisakul et al. (27)).

Simultaneous electrophysiological imaging is principally also applicable for analyzing ENOs. For *locus coeruleus*, however, this bilaterally organized network does not generate synchronized activity, at least not in vitro under our experimental conditions. This does not allow activity phase analyses between extracellular recording via suction electrode in one aspect and imaging and/or intracellular patch-clamp recording in the contralateral aspect. Nevertheless, (pharmacological) imaging findings in one *locus coeruleus* aspect can be supported by results from electrophysiological recording in the contralateral aspect (Fig. 3.). As a major limiting factor, Ca_i imaging in *locus coeruleus* is mostly not feasible in superfusate of close-to-physiological ion content. This is because ENOs occur at a rate of ~1 Hz while recovery from a single burst-related Ca_i increase in cortical structures or the isolated preBötC lasts typically >1 s using conventional Ca^{2+} dyes like Fluo-4 or Fluo-8L (5, 6, 22, 63, 64) (see also Chap. 2-Ruangkittisakul et al. (27)). Accordingly, *locus coeruleus* rhythm needs to be slowed down for Ca_i imaging, for example, by raising superfusate K^+ or Ca^{2+} (Sect. 4.1.1). As a further limitation, SR-101 imaging using either low or high concentrations of this dye does not seem to discriminate between neurons and astrocytes in *locus coeruleus*, whereas Ca_i imaging of responses to particular "gliomodulators" such as t-ACPD or ATP seems to be useful. Also, there is some evidence that SR-101 induces hyperexcitability in isolated (hippocampal) neural networks (67) while it is established that loading of cell with Ca^{2+}-sensitive dyes affects their intrinsic buffering capability (7).

In contrast to *locus coeruleus*, the entire set of Ca_i and morphological imaging tools described above (Sects. 3.3.and 4) can be used for analyzing hippocampal ENOs, at least in newborn rat slices. This will likely allow distinguishing different contributions of neuron (subtypes) and neighboring glia to processes involved in generation of ENOs. Moreover, it may be possible to study whether hampered interactions between these brain cell types are causally related to perturbation of ENOs by anoxia and opioids (and possibly also methylxanthines), which is a primary research interest for our group (Sect. 1). To elucidate the roles of glia, further optical tools need to be applied for specifically discriminating astrocytes from oligodendrocytic glia and similarly small microglia of non-neural origin (8–10, 64). Our established Ca_i and morphological imaging approaches are likely also suitable for analyzing neuron *versus* glia roles in cortical ENOs in newborn rat and piglet slices although we have not explored this in detail yet. For simultaneous electrophysiological imaging in different areas of (entorhinal) cortex, it needs to be considered that these ENOs propagate in a wavelike fashion (6). In hippocampus, ENOs are typically most stable in the CA_3 area, which covers an area that extends for <500 μm. However, sometimes, ENOs and related Ca_i oscillations are more prominent (and stable) in CA_1 regions. Consequently, it is difficult to position in hippocampus, cortex, and *locus coeruleus* suction electrodes that have a short shank and outer diameters of 50–250 μm (Sects. 2.5 and 3.2.1) next to a patch electrode or a fluorescent dye-loaded area. While we found that suction electrode recording provides signals with excellent signal-to-noise ratio, much finer microelectrodes or slightly broken patch electrodes are traditionally used to record "field" potentials which provide a suitable alternative to record population activity next to imaged or patch-clamped cells (2, 4, 5, 12–16, 21, 35).

Our preliminary findings indicate that whole-cell patch-clamp recording is feasible for time periods of up to several hours (mostly 0.5–1.5 h) from bursting *locus coeruleus* neurons as has also been reported by others (40, 41). It seems to be easier to initiate a whole-cell recording compared to neonatal respiratory neurons or dorsal vagal neurons in vitro (see Chaps. 1-Trapp and Ballanyi (50) and 2-Ruangkittisakul et al. (27)), but >50% of recordings are stable for only <10 min. Nevertheless, the success rate for >1 h whole-cell recording in *locus coeruleus* neurons is quite high overall. We have not yet established intracellular sharp microelectrode, whole-cell, or perforated patch-clamp recording for analyzing hippocampal and cortical ENOs in the slices. However, all these approaches are feasible as indicated by extensive literature in that regard (2, 4, 5, 12–16, 21).

6.3. Species Differences for Hippocampal and Cortical ENOs

We found that ENOs are less robust in hippocampus and entorhinal cortex of brain slices from newborn piglets compared to newborn rats. However, it needs to be considered that our data are based on preliminary findings from <10 animals and it took some effort to optimize all procedures between anesthetizing the animals and putting the cut slices into recovery solution. Accordingly, it is likely that with new series of experiments, a higher percentage of horizontal slices would show more robust ENOs in the hippocampus and cortex. However, it is possible that ENOs in newborn piglets, which are more mature at birth and thus more resemble human newborns regarding brain development (68), are principally less regular than corresponding events in newborn rats. There may also be differences between brain regions regarding stability of ENOs because regular bursting was revealed in most somatosensory cortex slices. Moreover, Ca_i and morphological imaging seems feasible in somatosensory cortex, whereas we have yet no explanation why neuronal Ca_i imaging seems more difficult in hippocampus of piglet slices. Possibly, also this may improve with further technical refinement.

Acknowledgments

Work contributing to this study was supported by the Alberta Heritage Foundation for Medical Research (AHFMR), Alberta Innovates Health Solutions (AIHS), Hotchkiss Brain Institute, the Canada Foundation for Innovation (CFI-ASRIP), and operating plus training grants from the Canadian Institutes of Health Research (CIHR).

References

1. Spitzer NC (2006) Electrical activity in early neuronal development. Nature 444:707–712
2. Khazipov R, Luhmann HJ (2006) Early patterns of electrical activity in the developing cerebral cortex of humans and rodents. Trends Neurosci 29:414–418
3. O'Donovan MJ, Bonnot A, Mentis GZ, Arai Y, Chub N, Shneider NA, Wenner P (2008) Imaging the spatiotemporal organization of neural activity in the developing spinal cord. Dev Neurobiol 68:788–803
4. Allene C, Cossart R (2010) Early NMDA receptor-driven waves of activity in the developing neocortex: physiological or pathological network oscillations? J Physiol 588:83–91
5. Leinekugel X, Medina I, Khalilov I, Ben-Ari Y, Khazipov R (1997) Ca^{2+} oscillations mediated by the synergistic excitatory actions of GABA(A) and NMDA receptors in the neonatal hippocampus. Neuron 18:243–255
6. Garaschuk O, Linn J, Eilers J, Konnerth A (2000) Large-scale oscillatory calcium waves in the immature cortex. Nat Neurosci 3:452–459
7. Yuste R, Konnerth A, Masters B (2006) Imaging in neuroscience and development, a laboratory manual. J Biomed Opt 11:19902
8. Agulhon C, Petravicz J, McMullen AB, Sweger EJ, Minton SK, Taves SR, Casper KB, Fiacco TA, McCarthy KD (2008) What is the role of astrocyte calcium in neurophysiology? Neuron 59:932–946
9. Wang DD, Bordey A (2008) The astrocyte odyssey. Prog Neurobiol 86:342–367
10. Ballanyi K, Panaitescu B, Ruangkittisakul A (2010) Control of breathing by nerve glue. Sci Sig 3: pe41
11. Metzger F, Klapproth N, Kulik A, Sendtner M, Ballanyi K (2005) Optical assessment of motoneuron function in a 'twenty-four-hour' acute spinal cord slice model from fetal rats. J Neurosci Methods 141:309–320
12. Sipilä ST, Huttu K, Soltesz I, Voipio J, Kaila K (2005) Depolarizing GABA acts on intrinsically bursting pyramidal neurons to drive giant depolarizing potentials in the immature hippocampus. J Neurosci 25:5280–5289
13. Sipilä ST, Huttu K, Voipio J, Kaila K (2006) Intrinsic bursting of immature CA3 pyramidal neurons and consequent giant depolarizing potentials are driven by a persistent Na^{+} current and terminated by a slow Ca^{2+} -activated K^{+} current. Eur J Neurosci 23:2330–2338
14. Ben-Ari Y, Gaiarsa JL, Tyzio R, Khazipov R (2007) GABA: a pioneer transmitter that excites immature neurons and generates primitive oscillations. Physiol Rev 87:1215–1284
15. Sipilä ST, Kaila K (2008) GABAergic control of CA3-driven network events in the developing hippocampus. Results Probl Cell Differ 44:99–121
16. Kafitz KW, Meier SD, Stephan J, Rose CR (2008) Developmental profile and properties of sulforhodamine 101-Labeled glial cells in acute brain slices of rat hippocampus. J Neurosci Methods 169:84–92
17. Bonifazi P, Goldin M, Picardo MA, Jorquera I, Cattani A, Bianconi G, Represa A, Ben-Ari Y, Cossart R (2009) GABAergic hub neurons orchestrate synchrony in developing hippocampal networks. Science 326:1419–1424
18. Ballanyi K (2004) Protective role of neuronal K_{ATP} channels in brain hypoxia. J Exp Biol 207:3201–3212
19. Ballanyi K (2004) Neuromodulation of the perinatal respiratory network. Curr Neuropharmacol 2:221–243
20. Feldman JL, Del Negro CA (2006) Looking for inspiration: new perspectives on respiratory rhythm. Nat Rev Neurosci 7:232–242
21. Ben-Ari Y, Cherubini E, Corradetti R, Gaiarsa JL (1989) Giant synaptic potentials in immature rat CA3 hippocampal neurones. J Physiol 416:303–325
22. Ballanyi K, Ruangkittisakul A (2009) Structure-function analysis of rhythmogenic inspiratory pre-Botzinger complex networks in 'calibrated' newborn rat brainstem slices. Respir Physiol Neurobiol 168:158–178
23. Ruangkittisakul A, Ballanyi K (2010) Methylxanthine reversal of opioid-evoked inspiratory depression via phosphodiesterase-4 blockade. Respir Physiol Neurobiol 172:94–105
24. Lombroso CT (2007) Neonatal seizures: gaps between the laboratory and the clinic. Epilepsia 48:83–106
25. Ballanyi K, Onimaru H, Homma I (1999) Respiratory network function in the isolated brainstem-spinal cord of newborn rats. Prog Neurobiol 59:583–634
26. Smith JC, Ellenberger HH, Ballanyi K, Richter DW, Feldman JL (1991) Pre-Bötzinger complex: a brainstem region that may generate respiratory rhythm in mammals. Science 254:726–729
27. Ruangkittisakul A, Panaitescu B, Secchia L, Bobocea N, Kantor C, Kuribayashi J, Iizuka M, Ballanyi K (2012) Anatomically 'calibrated' isolated respiratory networks from newborn

rodents. In Isolated Central Nervous System Circuits (Ed K Ballanyi), Neuromethods Series Vol. 73 (Ed W Walz). Springer Science+Business Media, LLC, New York, NY, pp 61–124

28. Ballanyi K, Ruangkittisakul A, Onimaru H (2009) Opioids increase and anoxia decreases delay of rhythmogenic pre-inspiratory (pFRG) and inspiratory (preBötC) network bursts in newborn rat brainstems. Eur J Physiol (Pflüger's Archives) 458:571–587
29. Panaitescu B, Ruangkittisakul A, Ballanyi K (2009) Silencing by raised extracellular Ca^{2+} of pre-Bötzinger complex neurons in newborn rat brainstem slices without change of membrane potential or input resistance. Neurosci Lett 456:25–29
30. Ruangkittisakul A, Secchia L, Bornes TD, Palathinkal DM, Ballanyi K (2007) Dependence on extracellular Ca^{2+}/K^{+} antagonism of inspiratory centre rhythms in slices and *en bloc* preparations of newborn rat brainstem. J Physiol 584:489–508
31. Ruangkittisakul A, Panaitescu B, Ballanyi K (2011) K^{+} and Ca^{2+} dependence of inspiratory-related rhythm in novel "calibrated" mouse brainstem slices. Respir Physiol Neurobiol 175:37–48
32. Ballanyi K, Grafe P (1985) An intracellular analysis of γ-aminobutyric-acid-associated ion movements in rat sympathetic neurones. J Physiol 365:41–58
33. Kulik A, Nishimaru H, Ballanyi K (2000) Role of bicarbonate and chloride in GABA- and glycine-induced depolarization and $[Ca^{2+}]_i$ rise in fetal rat motoneurons in situ. J Neurosci 20:7905–7913
34. Blaesse P, Airaksinen MS, Rivera C, Kaila K (2009) Cation-chloride cotransporters and neuronal function. Neuron 61:820–838
35. Leinekugel X, Khazipov R, Cannon R, Hirase H, Ben-Ari Y, Buzsáki G (2002) Correlated bursts of activity in the neonatal hippocampus in vivo. Science 296:2049–2052
36. Milh M, Kaminska A, Huon C, Lapillonne A, Ben-Ari Y, Khazipov R (2007) Rapid cortical oscillations and early motor activity in premature human neonate. Cereb Cortex 17:1582–1594
37. Brockmann MD, Pöschel B, Cichon C, Hanganu-Opatz IL (2011) Coupled oscillations mediate directed interactions between prefrontal cortex and hippocampus of the neonatal rat. Neuron 71:332–347
38. Aston-Jones G, Cohen JD (2005) An integrative theory of locus coeruleus-norepinephrine function: adaptive gain and optimal performance. Annu Rev Neurosci 28:403–450
39. Van Bockstaele EJ, Reyes BA, Valentino RJ (2010) The locus coeruleus: a key nucleus where stress and opioids intersect to mediate vulnerability to opiate abuse. Brain Res 1314:162–174
40. Christie MJ, Williams JT, North RA (1989) Electrical coupling synchronizes subthreshold activity in locus coeruleus neurons in vitro from neonatal rats. J Neurosci 9: 3584–3589
41. Ballantyne D, Andrzejewski M, Mückenhoff K, Scheid P (2004) Rhythms, synchrony and electrical coupling in the locus coeruleus. Respir Physiol Neurobiol 143:199–214
42. Nakamura S, Kimura F, Sakaguchi T (1987) Postnatal development of electrical activity in the locus coeruleus. J Neurophysiol 58: 510–524
43. Ocaña M, Cendán CM, Cobos EJ, Entrena JM, Baeyens JM (2004) Potassium channels and pain: present realities and future opportunities. Eur J Pharmacol 500:203–219
44. Felten DL, Hakan H, Jonsson G (1982) Evidence for a neurotrophic role of noradrenaline neurons in the postnatal development of rat cerebral cortex. J Neurocytol 11: 119–135
45. Moriceau S, Roth TL, Sullivan RM (2010) Rodent model of infant attachment learning and stress. Dev Psychobiol 52:651–660
46. Jefferys JG (1994) Experimental neurobiology of epilepsies. Curr Opin Neurol 7:113–122
47. Keifer J, Vyas D, Houk JC (1992) Sulforhodamine labeling of neural circuits engaged in motor pattern generation in the in vitro turtle brainstem-cerebellum. J Neurosci 12:3187–3199
48. Nimmerjahn A, Kirchhoff F, Kerr JN, Helmchen F (2004) Sulforhodamine 101 as a specific marker of astroglia in the neocortex in vivo. Nat Methods 1:31–37
49. Mandal R, Anderson CW (2009) Anatomical organization of brainstem circuits mediating feeding motor programs in the marine toad. Bufo marinus Brain Res 1298:99–110
50. Trapp S, Ballanyi K (2012) Autonomic nervous system in vitro: studying tonically active neurons controlling vagal outflow in rodent brainstem slices. In Isolated Central Nervous System Circuits (Ed K Ballanyi), Neuromethods Series Vol. 73 (Ed W Walz). Springer Science+Business Media, LLC, New York, NY, pp 1–59
51. Jantzie LL, Cheung PY, Johnson ST, Bigam DL, Todd KG (2010) Cerebral amino acid profiles after hypoxia-reoxygenation and N-acetylcysteine treatment in the newborn piglet. Neonatology 97:195–203

52. Ballanyi K (1999) Isolated tissues: in vitro preparations. In: Windhorst U, Johansson H (eds) Modern techniques in neuroscience research. Springer, Heidelberg, p 307
53. Neher E, Sakmann B (2009) Single-channel recording, 2nd edn. Springer, New York/Dordrecht/Heidelberg/London
54. Walz W (2009) Patch-clamp analysis: advanced techniques, 2nd edn. Humana, Totowa/New Jersey
55. Moore AR, Zhou WL, Jakovcevski I, Zecevic N, Antic SD (2012) Physiological properties of human fetal cortex in vitro. In Isolated Central Nervous System Circuits (Ed K Ballanyi), Neuromethods Series Vol. 73 (Ed W Walz). Springer Science+Business Media, LLC, New York, NY, pp 125–158
56. Ruangkittisakul A, Schwarzacher SW, Secchia L, Poon BY, Ma Y, Funk GD, Ballanyi K (2006) High sensitivity to neuromodulator-activated signaling pathways at physiological [K^+] of confocally imaged respiratory center neurons in on-line-calibrated newborn rat brainstem slices. J Neurosci 26:11870–11880
57. Ruangkittisakul A, Schwarzacher SW, Secchia L, Ma Y, Bobocea N, Poon BY, Funk GD, Ballanyi K (2008) Generation of eupnea and sighs by a spatiochemically organized inspiratory network. J Neurosci 28:2447–2458
58. Ballanyi K, Ruangkittisakul A (2009) Brain slices. In: Binder M, Hirokawa N, Windhorst U (eds) Encyclopedia of neuroscience. Springer, Heidelberg/New York/Tokyo, pp 483–490
59. Fish KN, Gonzales-Burgos G, Zaitsev AV, Lewis DA (2012) Histological characterization of physiologically determined fast spiking interneurons in slices of the primate dorsolateral prefrontal cortex. In Isolated Central Nervous System Circuits (Ed K Ballanyi), Neuromethods Series Vol. 73 (Ed W Walz). Springer Science+Business Media, LLC, New York, NY, pp 159–181
60. Nakamura TJ, Michel S, Block GD, Colwell CS (2012) Neural circuits underlying circadian oscillations in mammals: clocks in a dish. In Isolated Central Nervous System Circuits (Ed K Ballanyi), Neuromethods Series Vol. 73 (Ed W Walz). Springer Science+Business Media, LLC, New York, NY, pp 183–210
61. Broicher T, Speckmann EJ (2012) Living human brain slices: network analysis using voltage sensitive dyes. In Isolated Central Nervous System Circuits (Ed K Ballanyi), Neuromethods Series Vol. 73 (Ed W Walz). Springer Science+Business Media, LLC, New York, NY, pp 285–300
62. Stosiek C, Garaschuk O, Holthoff K, Konnerth A (2003) In vivo two-photon calcium imaging of neuronal networks. Proc Natl Acad Sci USA 100:7319–7324
63. Ruangkittisakul A, Ballanyi K (2009) Neuron-glia-imaging. In: Binder M, Hirokawa N, Windhorst U (eds) Encyclopedia of neuroscience. Springer, Heidelberg/New York/Tokyo, pp 2756–2764
64. Ruangkittisakul A, Okada Y, Oku Y, Koshiya N, Ballanyi K (2009) Fluorescence imaging of active respiratory networks. Respir Physiol Neurobiol 168:26–38
65. Lingwood BE, Healy GN, Sullivan SM, Pow DV, Colditz PB (2008) MAP2 provides reliable early assessment of neural injury in the newborn piglet model of birth asphyxia. J Neurosci Methods 171:140–146
66. Craner SL, Ray RH (1991) Somatosensory cortex of the neonatal pig: I. Topographic organization of the primary somatosensory cortex (SI). J Comp Neurol 306:24–38
67. Kang J, Kang N, Yu Y, Zhang J, Petersen N, Tian GF, Nedergaard M (2010) Sulforhodamine 101 induces long-term potentiation of intrinsic excitability and synaptic efficacy in hippocampal CA1 pyramidal neurons. Neuroscience 169:1601–1609
68. Lind NM, Moustgaard A, Jelsing J, Vajta G, Cumming P, Hansen AK (2007) The use of pigs in neuroscience: modeling brain disorders. Neurosci Biobehav Rev 31:728–751
69. Martin LJ, Brambrink A, Koehler RC, Traystman RJ (1997) Primary sensory and forebrain motor systems in the newborn brain are preferentially damaged by hypoxic-ischemia. J Comp Neurol 377:262–285
70. Teppema LJ, Baby S (2011) Anesthetics and control of breathing. Respir Physiol Neurobiol 177:80–92
71. Sanchez-Vives MV (2012) Spontaneous rhythmic activity in the adult cerebral cortex in vitro. In Isolated Central Nervous System Circuits (Ed K Ballanyi), Neuromethods Series Vol. 73 (Ed W Walz). Springer Science+Business Media, LLC, New York, NY, pp 263–284
72. De Curtis M, Lilbrizzi L, Uva L, Gnatkovsky V (2012) Neuronal networks in the in vitro isolated guinea pig brain. In Isolated Central Nervous System Circuits (Ed K Ballanyi), Neuromethods Series Vol. 73 (Ed W Walz). Springer Science+Business Media, LLC, New York, NY, pp 357–383
73. Safiulina VF, Kasyanov AM, Giniatullin R, Cherubini E (2005) Adenosine down-regulates giant depolarizing potentials in the developing rat hippocampus by exerting a negative control on glutamatergic inputs. J Neurophysiol 94:2797–2804

74. Ruusuvuori E, Kirilkin I, Pandya N, Kaila K (2010) Spontaneous network events driven by depolarizing GABA action in neonatal hippocampal slices are not attributable to deficient mitochondrial energy metabolism. J Neurosci 30:15638–15642

75. Tyzio R, Allene C, Nardou R, Picardo MA, Yamamoto S, Sivakumaran S, Caiati MD, Rheims S, Minlebaev M, Milh M, Ferré P, Khazipov R, Romette JL, Lorquin J, Cossart R, Khalilov I, Nehlig A, Cherubini E, Ben-Ari Y (2011) Depolarizing actions of GABA in immature neurons depend neither on ketone bodies nor on pyruvate. J Neurosci 31:34–45

76. Zilberter Y, Zilberter T, Bregestovski P (2010) Neuronal activity in vitro and the in vivo reality: the role of energy homeostasis. Trends Pharmacol Sci 31:394–401

77. Somjen GG (2002) Ion regulation in the brain: implications for pathophysiology. Neuroscientist 8:254–267

Chapter 12

Neuronal Networks in the In Vitro Isolated Guinea Pig Brain

Marco de Curtis, Laura Librizzi, Laura Uva, and Vadym Gnatkovsky

Abstract

Neuronal networks can be properly analyzed in experimental preparations that preserve the largest connectivity within and between brain structures. The isolated whole guinea pig brain maintained in vitro by arterial perfusion is ideal for this purpose. The methods and techniques utilized to study brain networks in this preparation are reviewed in this chapter. The advantages, potentials, and limitations for the study of pathological activity in models of acute (and chronic) cerebral diseases, such as epilepsy and ischemia, are also discussed.

Key words: Blood-brain barrier, Cerebral ischemia, Epilepsy, In vitro, Isolated guinea pig brain, Limbic system, Olfactory cortex

List of Abbreviations

4AP	4-aminopyridine
BBB	Blood-brain barrier
CSD	Current source density
CNS	Central nervous system
EC	Entorhinal cortex
IPSP	Inhibitory postsynaptic potentials
ID	Ischemic depolarizations
LOT	Lateral olfactory tract
MCA	Medial cerebral artery
OB	Olfactory bulb
PID	Peri-infarctual depression
PRC	Perirhinal cortex
PC	Piriform cortex
VSD	Voltage-sensitive dye

Klaus Ballanyi (ed.), *Isolated Central Nervous System Circuits*, Neuromethods, vol. 73,
DOI 10.1007/978-1-62703-020-5_12, © Springer Science+Business Media New York 2012

1. Introduction

The technical constraints imposed by in vivo experiments and the compression of the synaptic circuitry dictated in most cases by the use of the in vitro slice preparation represent serious limitations to the functional study of brain networks. These restrictions can be partially overcome by isolating en bloc parts of the cerebrum, with the aim to preserve connectivity between distant brain areas. One of the most complete of these preparations is the isolated brain of the guinea pig maintained in vitro by arterial perfusion (1–4) and extension of the isolated guinea pig brainstem and cerebellum (5).

The isolated guinea pig brain has the obvious advantages of any in vitro preparation. It shows mechanical stability, easy access for pharmacological manipulations, direct accessibility to brain tissue, and facilitated control of the extracellular solutions. In addition, this preparation represents so far a unique model in which neuronal connectivity between remote brain regions is completely preserved and available for neurophysiological and imaging studies. In particular, ventral surface areas, such as the piriform cortex (PC) or the entorhinal cortex (EC), that are impracticable in vivo can be easily approached in the isolated brain to perform neurophysiological and imaging studies, as detailed in Sect. (4). The stereotaxic references are obviously missing in this preparation, and the access to deep brain regions is based on the use of external reference points and the guinea pig brain atlas by Luparello (6). A fundamental guide to reach deep structures is the assistance of stimulus-evoked field responses, as illustrated for the hippocampus (Sect. 4).

One of the distinctive features of the isolated brain preparation in comparison to other in vitro preparations is the preservation of the cerebral vascular system in functional contiguity to brain structures (2, 7, 8) and the accessibility of both neuronal and vascular compartments. The guinea pig brain can be successfully isolated and studied in vitro because of the peculiar arrangement of the communication between the posterior (vertebrobasilar) and the anterior (carotid) arterial systems that form the Willis circle. The basilar artery in the guinea pig divides into two large diameter posterior communicating arteries, from which the posterior cerebral arteries originate. The guinea pig is the only animal analyzed so far, in which the arrangement of the Willis circle is compatible with perfusion of the entire brain via the basilar artery. In other animal species, such as the rat and the mouse, the small diameter of the posterior communicating arteries does not allow good brain perfusion in vitro when the basilar artery is cannulated (9).

This peculiar anatomical aspect opens a number of possibilities to study of the changes in vascular tone and neurovascular interaction at large (Sects. 3). The facilitated access to this preparation is ideal to induce acute changes that mimic neurological disorders, such as seizures, acute ischemia, brain inflammation, and others (Sects. 5 and 6). Finally, the integrity of the blood-brain barrier (BBB) was demonstrated for the isolated guinea pig brain (7, 8). Therefore, the preparation is suitable for studies on the role of BBB in pathophysiology of brain dysfunction (e.g., seizures, ischemia, inflammation) and for screening studies on BBB permeability of new drugs and neuroactive compounds.

2. Materials

The equipment required to isolate and maintain the isolated guinea pig preparation in vitro is detailed here. Intracardiac perfusion is performed with a peristaltic pump (Gilson Minipulse 3, France) connected to a set of tygon tubings that end with a metal cannula (2 mm internal gauge) inserted in the aorta after torachotomy. Standard surgical instruments are utilized to prepare the intracardiac perfusion and to isolate the brain. Of particular relevance is the use of an iris microscissor (Boneschi, Italy) to cut arteries and cranial nerves at their emergence from the foramens of the base of the skull. The dissected brain is transferred to a custom-made incubation chamber that is internally coated with a 0.5–1-mm-thick silicon elastomer film (Sylgard, Dow-Corning). The chamber was machine-carved into an aluminum block, in which holes are drilled to allow the passage of the perfusion and the suction tubing and the temperature probe. The temperature of both the chamber and the solution are regulated by heat exchanges (Peltier devices) connected to a RG-01 temperature controller (npi electronic, Tamm, Germany; www.npielectronics.com). The perfusion of the brain is sustained by the flow of the solution through tygon and polyethylene tubings connected in line with the Minipulse-3 peristaltic pump, a bubble trap, and a model 800 pressure monitor device (Bentley Trantec).

Equipment for neurophysiological and imaging recordings depends on the planned experiment and will not be detailed here (for this, see e.g., Chaps. 1-Trapp and Ballanyi (10), 2-Ruangkittisakul et al. (11) and 3-Moore et al. (12)).

3. Methods

3.1. Generation of Isolated Arterially Perfused Guinea Pig Brain

The technique for isolating the guinea pig brain is described in detail elsewhere (2–4). In brief, brains are isolated from young adult animals weighing 150–200 g. After intraperitoneal barbiturate anesthesia (thiopental, 120 mg/kg), the animal undergoes intracardiac perfusion using a peristaltic pump via the aorta with a cold carbogenated complex saline (10°C, pH = 7.1) of composition indicated below. This internal perfusion serves to reduce brain metabolism and preserve viability of the brain tissue during the dissection. After the animal is decapitated, the brain is carefully removed from the skull and transferred to the incubation chamber where it is positioned on its dorsal surface. The olfactory bulbs and the cervical spinal cord are held down by two silk treads for mechanical stabilization (Fig. 1a). Under a stereomicroscope, the *dura mater* that enfolds the basilar artery is gently removed, and a polyethylene cannula (terminal gauge 250 μm) is inserted into the basilar artery (Fig. 1a). The cannula is tied to the basilar artery with a silk thread, and the arterial perfusion is started at a rate of 7 mL/min. The temperature of the perfusion chamber is increased from 15°C to 32°C at steps of 0.2°C per minute with a ramp commanded by a thermostat (npi). The carotid arteries and hypophyseal arteries are tied with silk knots to reestablish the physiological cerebral vascular flow through the Willis circle (see below) without major leaks. The ionic composition of the intra-arterial solution is (in mM) 126 NaCl, 3 KCl, 1.2 KH_2PO_4, 1.3 $MgSO_4$, 2.4 $CaCl_2$, 26 $NaHCO_3$, 15 D-glucose, 3% dextran (M.W. 70,000), carbogenated with a 95% O_2–5% CO_2 gas mixture (pH 7.3). To avoid pharmacological manipulations that might interfere with the experiments, we utilized low temperature as a substitute for anesthesia. The temperature of the preparation is maintained at 32°C.

3.2. Electrophysiological Techniques

3.2.1. Extracellular Recording and Stimulation

The isolated guinea pig brain is ideal to record activity from different and remote brain structures with multiple electrodes, the limit being the number of recording channels available in the laboratory equipment. As illustrated in Fig. 1b, c, extracellular field potentials and their main current sink/source generators can be analyzed by recording with either single electrodes, multichannel silicon probes (NeuroNexus Technologies, Ann Arbour, MI, USA; www.neuronexus.com), or multichannel arrays (FHC, Bowdoin ME, USA; www.fhc-co.com). Extracellular recording electrodes (glass micropipettes or metal electrodes) are used to study the propagation of responses evoked by electrical stimulation and to analyze the distribution pattern of spontaneous and pharmacologically induced cortical activity (Fig. 1b) (13–16). When the aim of the study is the reconstruction of local neuronal networks, the use of multichannel silicon probes is advantageous. The multichannel electrodes feature multiple recording sites (up to 16) separated by 50 or 100 μm on a single shaft (Figs. 1a and 4c). Simultaneous field potential recordings with narrow-spaced silicon probes inserted in a laminar structure, such as a cortex, generate laminar depth profiles (Fig. 1c). The study of the laminar profiles with a current source density (CSD) analytical approach (17, 18) allows a detailed reconstruction of the laminar distribution and time course

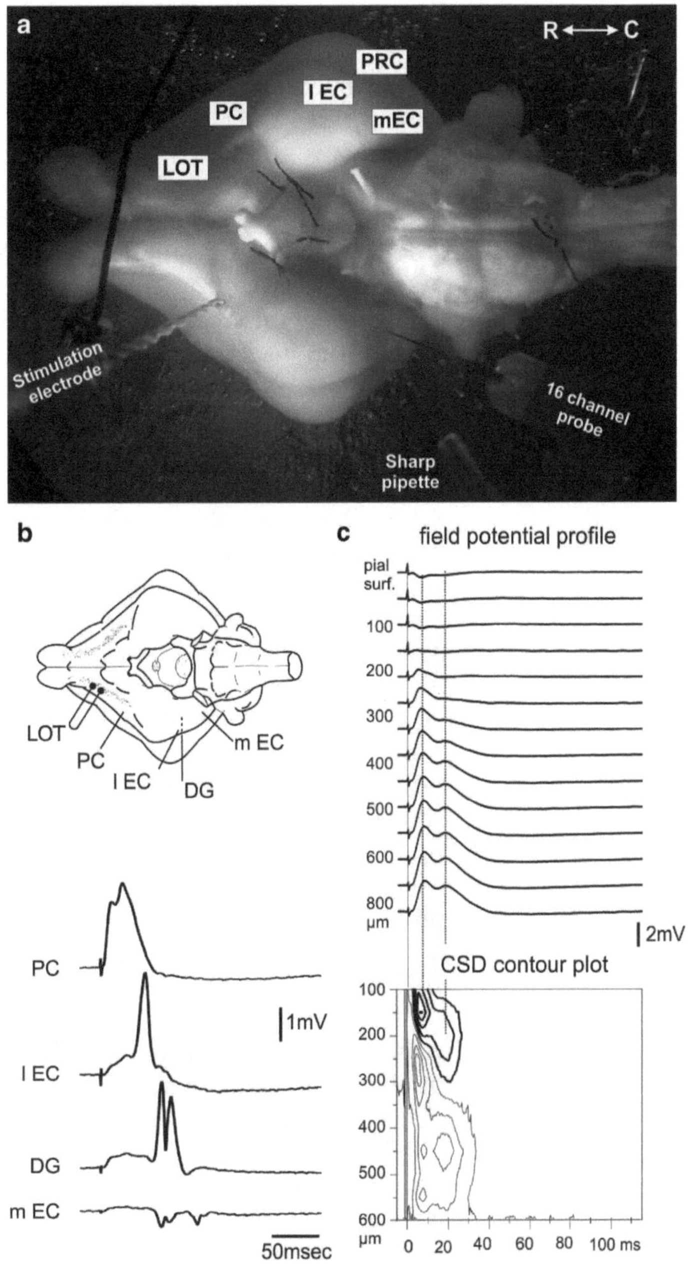

Fig. 1. Evoked laminar profiles in the isolated guinea pig brain in vitro. (**a**) Photograph of ventral view of an isolated young adult guinea pig brain. Recording and stimulating electrodes are illustrated, calibration bar = 2.5 mm. (**b**) Schematic representation of the guinea pig brain showing the position of the recording electrodes (*upper panel*) utilized to record simultaneously field potentials in piriform cortex (PC), lateral entorhinal cortex (lEC), dentate gyrus (DG), and medial entorhinal cortex (mEC) after single electrical shock stimulation of the lateral olfactory tract (LOT). Traces demonstrate that LOT stimulation induces a sequential activation from PC to mEC. (**c**) An example of a field potential profile recorded with a multichannel silicon probe (illustrated in Figs. 1a and 4c) positioned in PC after LOT stimulation. Tissue depths of recording leads of the silicon probe are shown on the *left* in the profile. The corresponding current source density (CSD) contour plot illustrated in the *lower part* of the panel shows the local sink/sources dipole at different depths in the PC. The sinks are represented with *thick lines* and sources with *thin lines*, respectively.

of the extracellular currents that underlie the different field potential components (lower panel in Fig. 1c) (4.). Moreover, with CSD analysis, it is possible to distinguish the locally generated components of field potentials from far-field components generated elsewhere and passively conducted within the tissue where the recording is done. The exact position of the CSD sinks/sources dipoles is then reconstructed from the histological identification of the electrode track in the recorded structure. Stimuli were delivered either with custom-made twisted Teflon-coated silver wires or with commercially available stainless steel electrodes (FHC).

3.2.2. Intracellular Microelectrode Recordings

Characterization of single neurons in brain circuits is one of the main tools to study basic functions in the central nervous system (CNS). Most available in vitro preparations permit intracellular recordings of membrane potentials and underlying ion currents. Intracellular recording during in vivo experiments is feasible but demanding. Limited access to brain structures in ventral brain aspects is an obvious obstacle in that regard. Contrary, the isolated guinea pig brain enables simultaneous intra- and extracellular recordings even from remote brain areas including structures close to the ventral surface. Intracellular recordings are successfully performed in the isolated guinea pig brain with sharp glass microelectrode that are, for example, filled with 2 M K^+ acetate and 1.5% of the histological marker dye biocytin and have a DC resistance of 60–120 MΩ. As for brain slices, the cell position and morphology can be revealed after the electrophysiological recording with standard biocytin-horseradish peroxidase visualization protocols (e.g., ABC kit, Vector Laboratories, Burlingame, CA; www.vectorlabs.com). The examples in Fig. 2 illustrate these advantages of the isolated brain preparation. Intracellular recordings can be performed simultaneously in more neurons at a distance; in Fig. 2a, two intracellular recordings were simultaneously performed in the EC of both hemispheres during a spontaneous event. New information on how the intracellularly recorded neuron contributes to the information flow and to the local network can be gained by implementing simultaneous intracellular recordings together with CSD analysis of laminar profiles, as illustrated for the EC in Fig. 2b.

3.2.3. Ion-Sensitive Microelectrodes

Neuronal activity is associated with functionally significant changes in both extracellular and intracellular concentrations of pH and ions such as K^+, Ca^{2+}, Na^+, and Cl^- which can be monitored with ion-sensitive microelectrodes (see also Chaps. 1-Trapp and Ballanyi (10) and 2-Ruangkittisakul et al. (11)). Double- or triple-barreled glass pipettes were utilized by us in the isolated guinea brain to simultaneously record field potentials and interstitial K^+ or pH at the same cortical site (7). Measurements of extracellular Ca^{2+} or other ions are, in principle, also possible. Moreover, we performed simultaneous recordings of all three parameters during periodic bicuculline-induced interictal spikes and seizure-like events in this

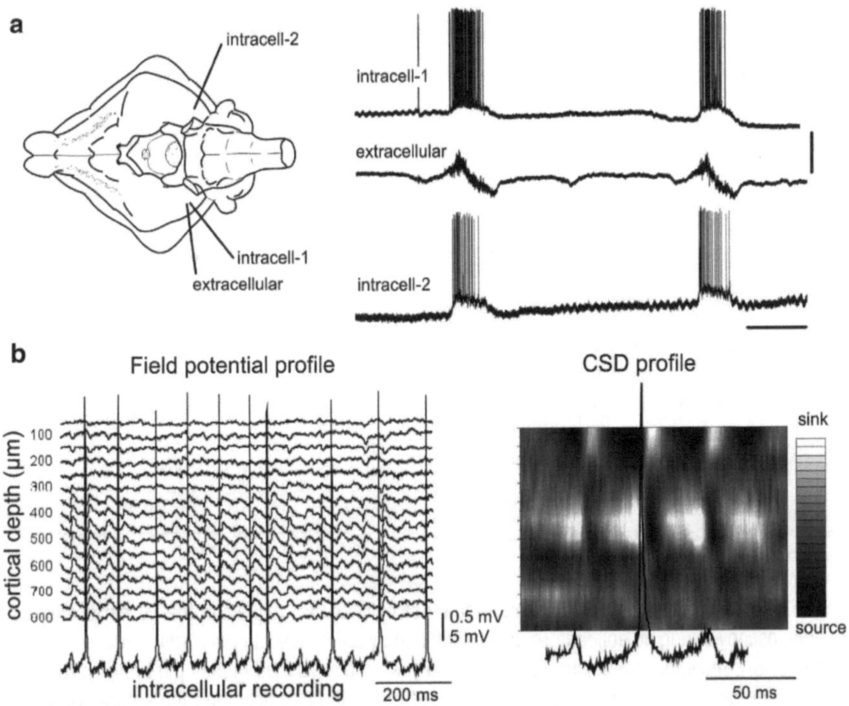

Fig. 2. Rhythmic activity patterns in the isolated guinea pig brain. (**a**) Activity of two principal EC neurons simultaneously recorded in both hemispheres during a spontaneous "up" deflection recorded with the extracellular pipette (*middle trace*) next to one intracellular recording site in the isolated guinea pig brain. The position of the electrodes is shown in the schema on the *left*. (**b**) Intracellular recording of membrane potential from a layer II EC neuron superimposed on the field potential depth profile simultaneously recorded with a 16-channel linear silicon probe in the EC during gamma oscillations induced by arterial perfusion of 100 μM carbachol, an agonist of acetylcholine receptors. CSD analysis performed on the multichannel depth profile (*right panel*) shows the sink and source sequences that represent local generator of the recorded oscillation. This pattern correlates with action potentials and excitatory postsynaptic potentials in the neuron recorded simultaneously close to the multichannel probe.

model (2, 7) and also analyzed extracellular K^+ dynamics to evaluate changes in the permeability of the BBB. We found, for example, that brief intra-arterial administration of a bolus of high K^+ solution did not change interstitial K^+ unless the BBB was damaged (7, 19). Recently, we used ion-sensitive microelectrodes to analyze changes of the extracellular cortical microenvironment caused by fictive ischemia (20) (Sect. 6).

3.3. Optical Imaging Techniques

3.3.1. Voltage-Sensitive Dye Imaging

Optical recording with voltage-sensitive dye (VSD) is a method to simultaneously investigate the propagation of the neuronal activity throughout different cerebral areas. In the last years, the technique has greatly improved in terms of temporal and spatial resolution and signal-to-noise ratio (see also Chaps. 2-Ruangkittisakul et al. (11) and 9-Broicher and Speckmann (21)). We have successfully adapted VSD imaging techniques to the isolated guinea pig brain preparation to study the propagation of neural signal throughout cortical structures at the

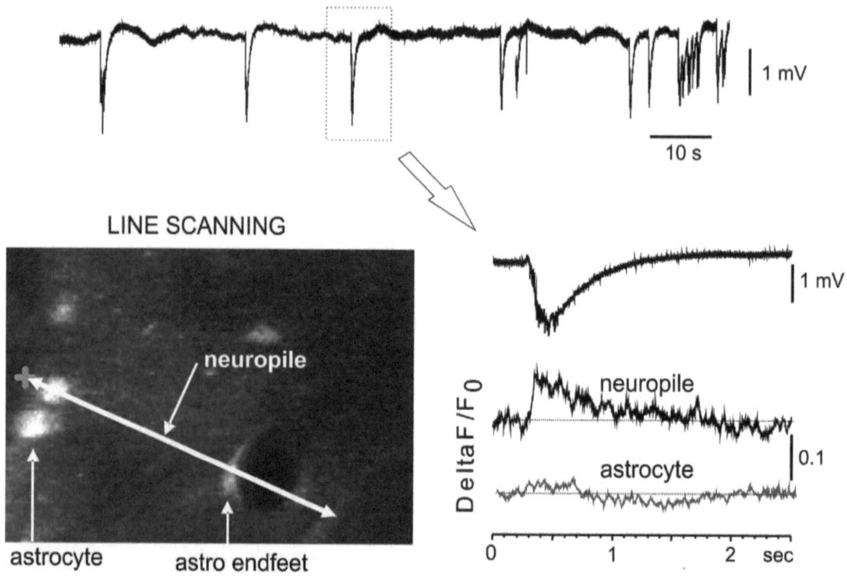

Fig. 3. Imaging of cytosolic Ca^{2+} signals. Line-scan imaging of cellular Ca^{2+} signals in the EC performed using 2-photon laser scanning microscopy during an epileptiform spike (outlined by dotted box in *upper trace*) induced in the EC by arterial application of the $GABA_A$ receptor antagonist bicuculline (50 μM) in the isolated guinea pig brain. In the *lower left panel*, Ca^{2+} changes are analyzed along line scans that cross an astrocyte, the neuropile and the endfoot of an astrocyte, after loading the EC with the Ca^{2+}-sensitive fluorescent dye Oregon Green 488 BAPTA-1 AM by local injection. As shown on the *right*, the interictal spike (trace expanded from the *upper panel*) corresponds to a Ca^{2+} increase in the neuropile (*black line*) but not in the astrocyte (*grey line*).

ventrolateral surface of the brain (22–24). For such experiments, brains are perfused via the arterial system with perfusate containing 0.36 M of the VSD di-2-ANEPEQ (Invitrogen, Carbad, CA, USA; www.invitrogen.com). A single staining of the brain allows recording of optical signals for 3–4 h with a MiCAM02 system (Brain Vision, Tsukuba, Japan; www.scimedia.com). With different lens sets, exposed brain areas may vary from 3.2 × 4.8 to 10.2 × 7.6 mm with a spatial resolution of 192 × 128 pixels. Our studies demonstrated the propagation of neuronal activity, evoked by lateral olfactory tract (LOT) stimulation, to the PC, to the lateral EC, and, with a delay due to propagation of activity within the hippocampal loop, to the medial EC.

3.3.2. Ca^{2+} Imaging

Recently, we studied changes in the free concentration of cytosolic Ca^{2+} in neurons and astrocytes during epileptiform activity induced in the isolated guinea pig brain by acute pharmacological manipulation (25). For this approach, 20 μM of the Ca^{2+}-sensitive dye Oregon Green BAPTA-1 AM was locally injected into the EC (see also Chaps. 2-Ruangkittisakul et al. (11) and 11-Kantor et al. (26)). Changes in cytosolic Ca^{2+} were imaged using 2-photon laser scanning microscopy in collaboration with Gian Michele Ratto and

Giorgio Carmignoto (27). Figure 3 illustrates findings from one of these experiments, in which "line-scanning" was performed in the EC between an astrocyte, the neuropile, and a perivascular endfoot of an astrocyte during an interictal epileptiform discharge.

3.4. Recording of Vascular Tone

Changes in the vascular resistance to perfusion could be recorded in the isolated guinea pig brain by interposing a model 800 pressure transducer along the perfusion line connected to the peristaltic pump that supplies the arterial solution (2. The pressure values are converted in voltage values that represent an indirect measurement of the vessel resistance, i.e., of the tone of the vascular system. Activity-dependent vasodilation and vasoconstriction could be recorded with this method during epileptiform discharges and during spreading depression-like phenomena (2).

3.5. Technical Notes and Pitfalls

There are some critical points during the process of transferring the preparation in vitro that should be considered. The isolation of the carotid arteries during the brain dissection may be problematic since these vessels can be easily pulled from the base of the skull and can be damaged. To avoid this, the artery should be cut just after its passage in the skull without excessively lifting the brain. The flow rate must be always calibrated before the brain is positioned into the recording chamber. A drop in the flow rate during the experiment can be due to rupture of the tubing damaged by the rotors of the peristaltic pump. Increases and decreases in flow rate and pressure can be easily verified by controlling the level of the solution into the bubble trapper placed between the peristaltic pump and the filter or by monitoring the pressure of the system on a chart recorder.

If carefully isolated, the in vitro guinea pig brain shows close-to-physiological activity for 6–8 h. Though, after 5 h in vitro, a certain amount of vasogenic edema develops, and this hampers stable intracellular recordings.

4. Results: Study of Local Neural Networks

The viability of the olfactory-limbic system in the isolated guinea pig brain preparation has been extensively analyzed by us over the last 20 years (4, 28–30). This revealed the potentials of the preparation for the evaluation of system physiology of this brain region and confirmed that this preparation retains close-to-physiological properties that resemble in vivo conditions. Some of these results and new findings, that are not achievable with other preparations, will be referred to in the following. Other systems were also investigated in the isolated guinea pig whole brain (31) and isolated brainstem/cerebellum models (5, 32).

Electrophysiological and optical imaging approaches have been used to test and analyze synaptic network activation via electrical stimulation. All areas of the olfactory-limbic circuit retain their polysynaptic intrinsic properties. Indeed, following a single LOT stimulus, field potentials were observed in the olfactory bulbs (OB) (33), PC (28, 34), EC (35), hippocampal subfields in the dentate gyrus (DG), CA3 and CA1, and parahippocampal cortices (30, 35, 36). Olfactory cortex pyramidal neurons project through a rostrally to caudally directed associative fiber system all over the PC and the lateral EC. EC neurons receive the associative input from the PC and project to the hippocampus via the perforant path.

4.1. Olfactory Bulbs

The *OBs* are the first site within the brain to receive odor information. The axons of the odor receptors target the glomeruli located superficially in the OBs. Moving into depth, the external plexiform layer, the mitral cells layer that receives the information from the glomeruli, the internal plexiform layer, and the granular cell layer with small inhibitory cells are encountered. The pattern of mitral cell discharge is regulated by the synaptic OB circuit (37, 38). The granule cells are responsible for the recurrent and lateral inhibition onto the mitral cells regulating their firing rate through dendrodendritic synapses (38, 39).

In the isolated guinea pig brain, electrical stimulation of the LOT, which is constituted by axons of the mitral cells, induces a response in OBs corresponding to a sequence of sinks/sources dipoles as demonstrated by the CSD analysis of field potential profiles recorded with the multichannel probe (see above). High-frequency (100 Hz) stimulation and pharmacological manipulation via blocking kainate/AMPA- and NMDA-type ionotropic glutamate receptor-mediated as well as A-type γ-aminobutyric acid ($GABA_A$) receptor-mediated fast synaptic transmission using CNQX plus APV and bicuculline, respectively, enabled the identification of each component. The first evoked component is an antidromic population spike corresponding to a sink located in the mitral cell layer, with the corresponding source in the external plexiform layer. This is followed by a distributed sink in the external plexiform layer with its source in the mitral cell layer/granular cell layer. This component has been identified with the dendritic activation of the mitral cells triggered by the back-propagation of the antidromic spike. The antidromic spike induces a complex sink in the external plexiform layer with a source in external plexiform layer/mitral cell layer that reflects the synaptic dendritic activation of the granule cells and is abolished by CNQX and by high-frequency stimulation. Finally, a late response with a sink spanning the mitral cell layer and the superficial granular cell layer is evoked which is abolished by CNQX and high-frequency stimulation but enhanced by bicuculline. A late component is also elicited when PC is stimulated after surgical section of LOT; such a response disappears when both PC and LOT are cut. Taken those observations together, the late component could correspond to the excitation coming either from the PC or other caudal structure (33).

4.2. Piriform Cortex

The *PC* represents the first associative cortical station of afferent input generated in the olfactory bulb. Anatomical studies have revealed a diffusely arranged network of associative connections in the PC (40). Local circuit associative fibers run in layer Ib and III and synapse respectively on the apical and basal dendrites of superficial and deep layers of pyramidal cells (41, 42). The PC is uniformly activated in response to olfactory afferent input (43–45). When the LOT is electrically stimulated, the apical dendrites of the PC neurons are monosynaptically activated. Because of the diffuse distribution of intrinsic associative connections, each PC neuron receives a large number of associative inputs. In line with in vivo studies by Haberly's group (46), laminar profiles recorded with multichannel silicon probes in the isolated guinea pig brain and CSD analysis showed that LOT-evoked field potentials are composed of a mono- and a disynaptic response followed by a slow surface-positive, depth-negative component (29, 34) (Fig. 1c). The early monosynaptic sink is localized in the superficial layer Ia. The disynaptic associative input is represented by two sinks at the cortical depth of layer Ib and III (29). The delayed low-amplitude and long-lasting sink localized in the layer II follows the disynaptic response. This sink is commonly associated with the activation of inhibitory postsynaptic potentials (IPSPs). The associative connections within the PC are organized in a continuous set of local-remote projective fibers directed caudally that contact pyramidal neurons in the PC and in the lateral EC at variable distances from the neurons of origin.

A set of synaptic events with a laminar distribution similar to that described in the PC is revealed in the *lateral EC* (Fig. 4b) (35, 36, 47). CSD analysis performed on laminar field responses demonstrates that the mono- and polysynaptic responses in lateral EC are generated by current sink located in the superficial layers. The short latency small-amplitude field responses recorded in medial EC represent far fields generated in the lateral EC and passively volume-conducted through the tissue (medial EC in Fig. 4b) (35). The propagation of LOT-induced activity to the PC and to the lateral EC was also confirmed by the analysis of signals generated by VSD (23, 48). Optical imaging of LOT-evoked events also included the characterization of responses in the olfactory tubercle (24). These findings demonstrated that the lateral EC, but not medial, can be regarded as an associative olfactory area (49).

4.3. Entorhinal Cortex

The *EC* is important for the processing of inputs coming from the olfactory areas, the hippocampus, and the associative polymodal neocortical areas. It is one of the brain regions involved in spatial

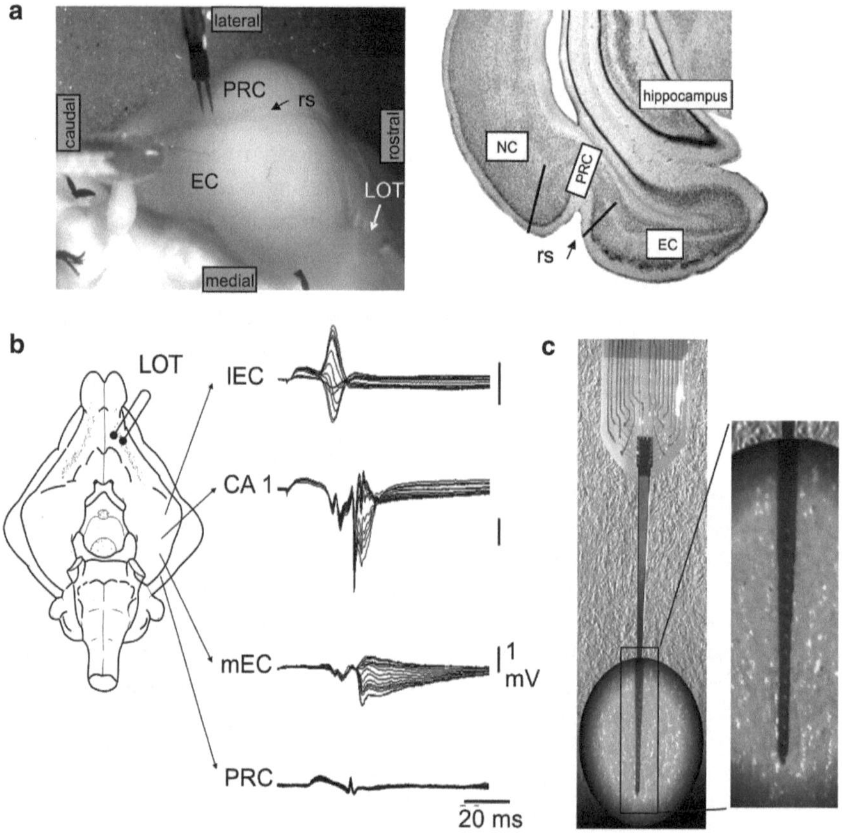

Fig. 4. Laminar profiles performed in the temporal lobe with linear multichannel silicon probes. (**a**) Different areas of the ventral part of isolated guinea pig brain accessible for electrophysiological recordings. In the photomicrograph on the *left* and on the coronal slice on the *right*: LOT, EC, perirhinal cortex (PRC), neocortex (NC), rhinal sulcus (rs). (**b**) LOT-evoked field potential depth profiles in lateral EC, medial EC, PRC, and hippocampus (CA1 region) with a 16-channel silicon probe (**c**) inserted in four different positions (scheme on the *left* in **b**). In each profile obtained in the four positions, superimposed traces simultaneously recorded at 16 different depths with the 16-channel probe are shown. Polysynaptic phase reversal responses presented in the medial EC, lateral EC, and hippocampus. The depth profile recorded in the PRC shows no activity reversal, suggesting volume conduction from neighboring regions. (**c**) A 16-channel silicon probe by NeuroNexus Technologies. The 16 recording leads magnified under the water drop are illustrated in right insert.

learning and formation of the declarative memory. In the isolated guinea pig brain, stimulation of LOT fibers induces a direct mono- or disynaptic response in the lateral EC (Fig. 4) but not in the medial EC (23, 48). Field potential analysis and optical imaging of VSD signals demonstrated that local medial EC activation is induced by repetitive LOT stimulation at a delta-theta frequency (2–8 Hz; see below). During this patterned stimulation, the lateral EC response is followed by a potential in the hippocampal DG. The DG potential is followed by activation of the tri-synaptic hippocampal pathway that involves subfields CA3, CA1, and subiculum, and feeds back into the EC (50). The largest potentials (and optical VDS signal) were observed in the medial EC (23, 35). CSD

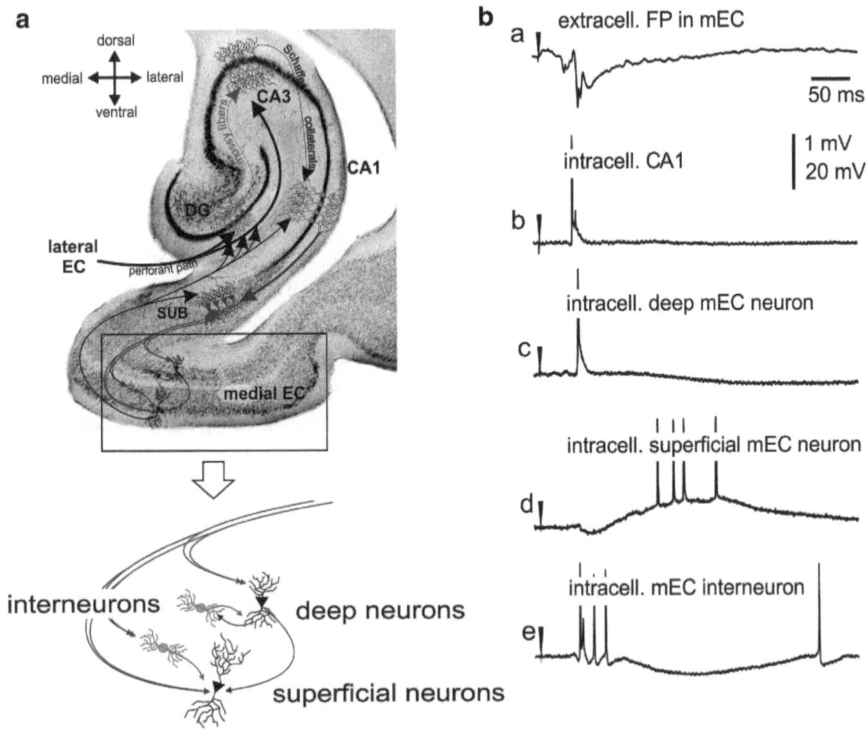

Fig. 5. Network interactions between hippocampus and entorhinal cortex analyzed by combined intra- and extracellular recordings. (**a**) Schema of interconnections between the hippocampus and medial EC based on presented data. A simplified schema of the local network in the mEC is illustrated in the *lower part* of the panel. (**b**) Extracellular (*upper trace* a) and intracellular potentials (traces b–e) evoked by single-shock LOT stimulation in different types of neurons in the medial EC and CA1 recorded non-simultaneously in the same experiment. The EC field potential response (***a***) was preceded by activation of CA1 hippocampal neuron (***b***) and correlated with excitation of deep layer neurons (***c***) and with feed-forward inhibition of superficial layer neurons (***d***). LOT-evoked firing in local EC interneurons (***e***) correlated with the hyperpolarizing phase of the superficial neuron (***d***).

analysis of laminar profiles demonstrated that the output from the hippocampus mediated by the LOT stimulation induced a small-amplitude sink in superficial layers of the lateral EC, whereas in the medial EC, a large sink was observed also in the deep layers (23). These findings demonstrated the existence of a stronger output to the medial EC compared to the lateral EC which further supports the concept that these two EC regions are functionally different also with respect to the hippocampal afferent input (49, 51).

To better evaluate the networks involved in the medial EC output response to the hippocampus, we performed simultaneous intra- and extracellular recordings in the medial EC following stimulation of the LOT (Fig. 5). LOT stimulation induced activation of CA1 neurons (Fig. 5b*b*) followed by excitation of deep layer (V and VI) neurons of the medial EC (Fig. 5b*c*) and feed-forward inhibition of superficial layer neurons (layers II and III) (Fig. 5b*d*) mediated by local inhibitory EC interneurons (Fig. 5b*e*). This novel finding (23), achievable exclusively on a preparation such as the isolated guinea pig brain, completely revised the understanding of the role played by the hippocampal projection to the EC. It was, indeed, demonstrated that the afferent CA1 input to the medial EC is mainly inhibitory. These data complete the study of network interactions that rule the electroresponsiveness of olfactory-evoked output of the hippocampus in the medial EC (see scheme in Fig. 5a).

LOT stimulation also activates a response in the contralateral hemisphere. CSD analysis demonstrated that contralateral LOT stimulation induces a sink in the outer molecular layer of the DG, which is not followed by a response in CA1. Since there are no fibers connecting CA1 to contralateral CA1, this demonstrated that inter-hemispheric hippocampal propagation of the olfactory input is mediated by CA3/contralateral CA3 fibers (50).

4.4. Parahippocampal Region

The *parahippocampal region*, which is crucial for learning and memory formation and retrieval, mediates the reciprocal interaction between the associative polymodal cortical areas and the hippocampus processes (52–54). It includes the EC, the perirhinal cortex (PRC), the postrhinal cortex, and the retrosplenial region. Our studies performed on the isolated guinea pig brain demonstrated that propagation of activity from the EC to the PRC/postrhinal cortex and vice versa is under a powerful inhibitory control that regulates the bidirectional transfer of information between neocortex and hippocampus (36, 48), reviewed in (55). The existence of this "wall of inhibition" was confirmed by in vitro and in vivo work performed in different animal species (55–58). Our VSD imaging experiments revealed a clear block of activity propagation from the PRC to the EC (Fig. 6a). Signals evoked by local stimuli were recorded in different portions of the PRC; postrhinal and retrosplenial cortices never propagated across the rhinal sulcus. Combined stimulation of the lateral amygdala and either PRC or neocortex also did not facilitate the transfer of information in this area. Experiments with local neocortical stimulation 2 mm away from the rhinal sulcus (Fig. 6a) showed how the locally generated activity invades the PRC/EC border and stops there. Neither high/low frequency patterned stimulation nor paired associative stimuli facilitated the propagation of activity. Similar stimulation during cholinergic/adrenergic neuromodulation was also ineffective in overcoming inhibition that controls propagation to the EC.

Reciprocal synaptic connections from neocortex to hippocampus are anatomically arranged in a sequential cascade of excitatory synapses (59, 60). Along this pathway, feed-forward inhibition modulates the reciprocal propagation of activity between the PRC and EC. The blockade of inhibitory networks at the border between the PRC and EC, induced by local application of bicuculline in area 35 and in the adjacent EC, represents the permissive conditions for a massive transfer of neural activity between these two regions. During local bicuculline application, propagation is released in both

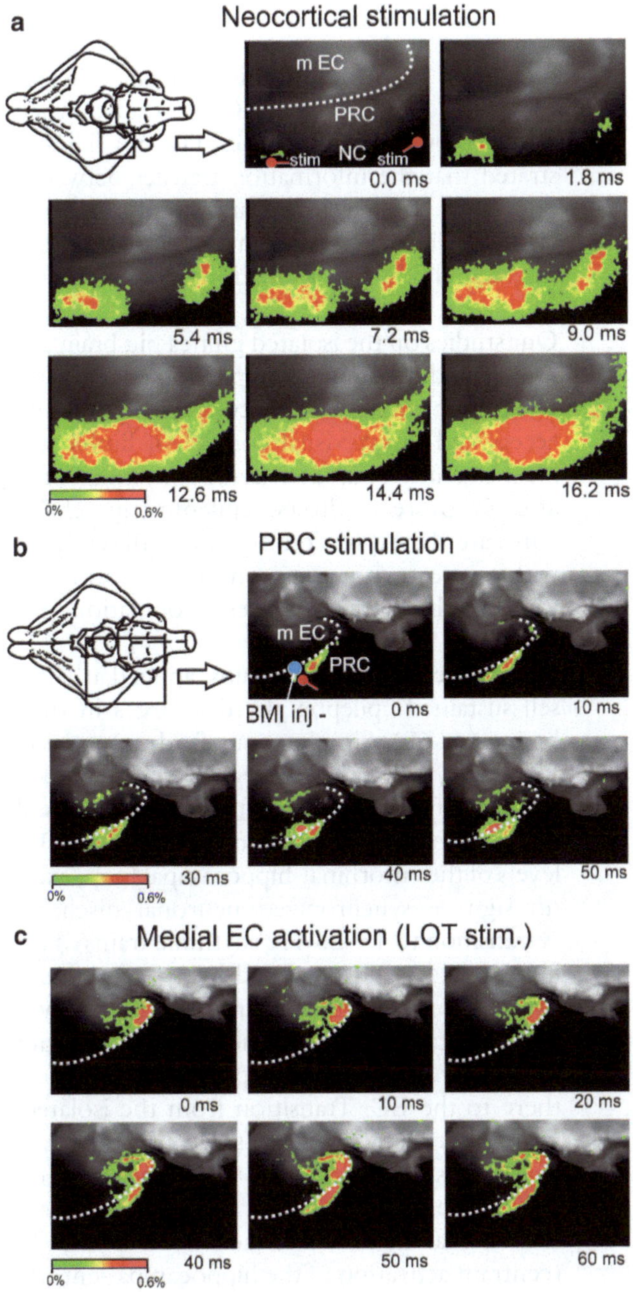

Fig. 6. Optical signals induced in the parahippocampal region by tissue application of the voltage-sensitive dye 2-ANEPS. (**a**) In the *upper left panel*, the area covered by the recording charge-coupled device video camera is illustrated on a schema of the ventral surface of the isolated guinea pig brain. Photomicrographs of imaged area with stimulation electrode positions in the NC (*white dots*) are shown in the *middle panel* of the *upper row*. The sequence of panels illustrates how combined stimulation of two NC areas induces large activation of the NC that propagates to the PRC at different times after the stimulus. No activity propagation across rhinal sulcus into the mEC was observed. (**b**) Optical imaging of the evoked activity recorded in the EC and PRC following local injection of 1 mM bicuculline at the border between PRC and mEC (*blue dot*). A field larger than in A is covered by the large-field camera utilized in this experiment. Stimulation applied to the PRC (*red dot*) induced a stronger activation of the PRC, and the voltage signal spreads into the mEC across the rhinal sulcus close to the area of bicuculline injection. Frame at 30 ms represents the time point when the evoked activity crosses the rhinal sulcus. (**c**) Optical imaging of LOT-evoked activity recorded in mEC and PRC following local bicuculline injection (location as above). LOT stimulation after injection induces a strong activation of the mEC (at 10 ms) followed by voltage signal spread to the PRC, across the rhinal sulcus (40 ms).

directions: when the PRC is stimulated, activity propagates to the EC (Fig. 6b) and the PRC is invaded when the medial EC is activated by LOT stimulation (Fig. 6c). These experiments demonstrated that the information transfer between neocortex and hippocampus is finely tuned and highly regulated. It remains to be clarified which processes in the parahippocampal region regulate the natural flow of information during cognitive functions.

4.5. Study of Epileptic Networks

Our studies on the isolated guinea pig brain also contributed to the understanding of epileptic phenomena (61–63). The preservation of synaptic connectivity between temporal lobe areas and the easy access and direct visual control of limbic cortical structures are ideal to study brain activity in conditions of hyperexcitability and after exposure to diverse epileptogenic challenges. Epileptic seizures are the expression of many different pathological conditions with one feature in common: the activation of an excessive, paroxysmal discharge that involves a population of CNS cells.

4.5.1. Seizures Induced by High-Frequency Stimulation

Repeated electrical pulse trains, applied at 100 Hz every 2 s, induce self-sustained epileptic afterdischarges in the isolated guinea pig brain (64, 65). Simultaneous field potential recordings along the DG, CA3, CA1, and EC regions of the hippocampus showed two patterns of rhythmic epileptiform discharges. An early phase characterized by quasi-rhythmic spikes at 2–3 Hz was recorded at all levels of the entorhinal-hippocampal loop, except in the DG, where no sign of synchronized neuronal discharge has been found. A transition to 15–25-Hz rhythmic trains of population spikes lasting 0.3–0.4 s and separated by 1–2 s was recorded 30–90 s after the beginning of the discharge onset at every levels of the entorhinal-hippocampal circuit. The earliest population spikes occurred in the CA3 region and then propagated to the CA1 area and from there to the EC. Transition from the isolated CA3-CA1 population spikes characteristic of stage 1 to the rhythmic 15–25-Hz discharge of stage 2 coincides with appearance of a prominent, sharp population spike in the DG (64). Thus, these experiments demonstrated that massive DG activation is the determinant event in reentrant activation of the hippocampal-entorhinal loop during the late phase of these seizure-like events.

4.5.2. Pharmacologically Evoked Seizures

Recently, the isolated guinea pig brain has been used to analyze interictal and ictal (seizure-like) activity induced by acute application of pro-convulsive compounds. These drugs applied either locally in the brain parenchyma or by systemic perfusion allow the analysis of network interactions during generation of seizures (ictogenesis). Brief focal ejection of bicuculline (1 mM) in the anterior PC is able to transform physiological network organization to generate spontaneous interictal potentials showing characteristics of interictal spikes (10, 29, 61, 66, 67). The first event observed after

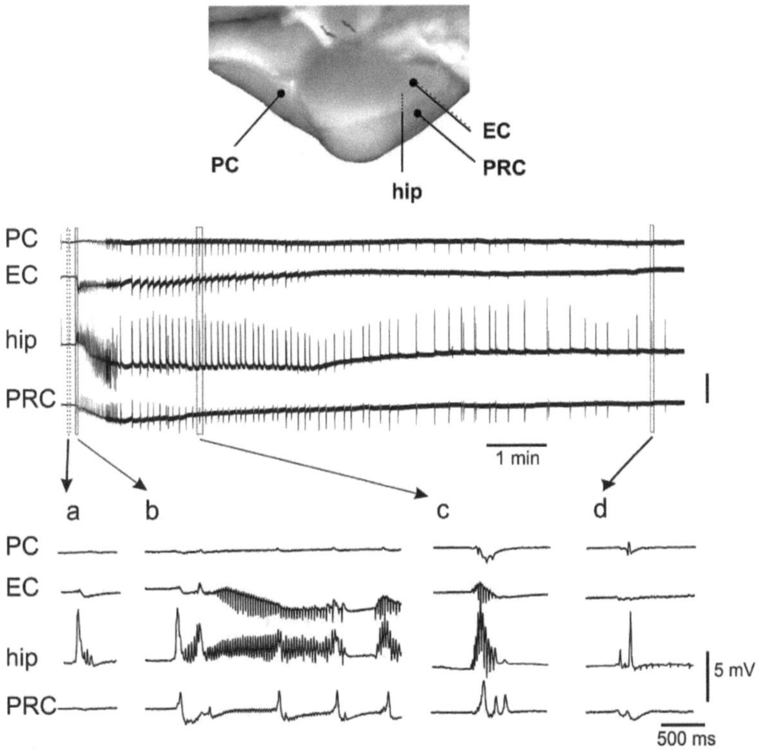

Fig. 7. Pharmacologically induced seizure-like activity in the olfactory-limbic cortex. Photograph of the brain showing recording electrode positions in PC, hippocampus (hip), PRC, and EC. Interictal and seizure-like activity induced by 3-min perfusion of bicuculline (50 μM) that is recorded simultaneously in the four positions is represented. Enlargements of the periods outlined by the dotted lines in the upper traces are illustrated below: (***a***) preictal spikes, (***b***) ictal onset, (***c***) afterdischarges, (***d***) postictal spikes.

the onset of bicuculline ejection is an enlargement of the disynaptic response in the LOT-evoked field response, immediately followed by a rhythmic (100–200 Hz) short-lasting (30–50 ms) burst-like discharge (66) that is also observed in spontaneous interictal spikes. Laminar profile and CSD analysis demonstrated that interictal spikes correlated with a sink in layer II due to the synchronous activation of inward currents entering the soma of superficial neurons during bursting. The early sink was followed by a large sink in layer Ib mediated by the activation of a recurrent associative circuit (30, 68). Intracellular recordings from PC neurons and simultaneous CSD analysis of laminar profiles were employed to further characterize the network determinants of the interictal spikes (66, 67).

Local application of bicuculline seldom induced seizure-like activity. Arterial application for 3 min of 50 μM bicuculline diluted in the perfusion solution is able to induce interictal spikes in the PC-EC, followed by ictal events that typically originate in the EC/hippocampus and do not propagate diffusely into the brain (Fig. 7). Epileptiform activity initiates with isolated complex interictal spikes in the PC. Just before the onset of the ictal event, the interictal focus shifts to the medial EC/hippocampus where independent and asynchronous interictal spikes are activated (50). The hippocampal preictal spikes increase in amplitude and progressively entrain a sequence of fast rhythmic activity at 25–30 Hz that correlates with the onset of the all-or-none seizure discharge (Fig. 7*b*) (50, 51). Such fast recruiting activity is followed by large amplitude afterdischarges that propagate to the EC and secondarily invade the PC only when no interictal spiking is present in this region (10).

Within 30 s from onset of ictal activity, medial EC/hippocampal afterdischarges initiated and gradually increased in amplitude. At the end of the ictal discharges, postictal depression ensued. The appearance of periodic interictal spikes and ictal-like events is associated with simultaneous changes in extracellular ion concentration and pH (see also Fig. 9). Spontaneous interictal spikes recorded extracellularly in the PC consistently correlate with a rapid increase in extracellular K^+ concentration by 0.5–1 mM. The recovery to control values coincided with the activation of next interictal spike. Similarly, a fast-rising extracellular alkaline shift (ranging between 0.05 and 0.08 pH units) was observed during the inter-spike period (69). A slow increase in extracellular K^+ associated with a progressive decrease of fast IPSPs in superficial neurons is also associated with fast activity (20–30 Hz) characterizing the initial phase of seizure-like discharge recorded in the medial EC after 3 min of arterial bicuculline perfusion (Fig. 8). During the transition toward an epileptiform ictal event, extracellular K^+ increases further while the initial alkalosis turns into a large and long-lasting extracellular acidosis (Fig. 8). When extracellular K^+ reaches a plateau value, irregular firing initiates in extracellular recording. Extracellular K^+ and acidosis gradually recover to control with concomitant attenuation of the amplitude and frequency of seizure-like event (62).

In correlation with the appearance of fast activity (20–30 Hz) at seizure onset, superficial layers of principal cells stopped spontaneous firing. When neurons were depolarized by intracellular injection of positive current above firing threshold before seizure onset, action potential generation ceased and inhibitory postsynaptic rhythmic potentials appeared during the extracellular fast oscillation (Fig. 9a, b and upper trace in Fig. 8). These hyperpolarizing fast potentials correlated with fast extracellular activity. Neuronal membrane potential showed an abrupt hyperpolarization followed by a slow depolarization. Three to five seconds later, in coincidence with fragmentation of fast activity, irregular firing resumed in superficial neurons in parallel with the emergence of irregular bursting in the extracellular trace (Fig. 9b*a*). Within few seconds, bursts of action potentials superimposed on a paroxysmal depolarizing shift appeared in coincidence with occurrence of regular extracellular bursts. (Fig. 9a). These bursts became progressively more robust and less frequent and were followed by postictal

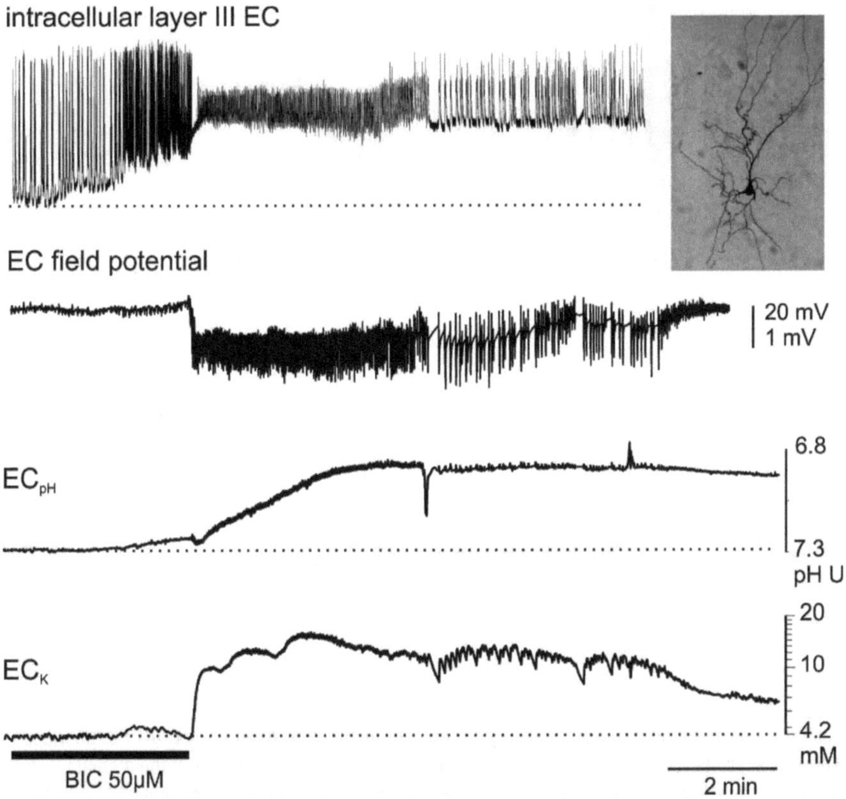

Fig. 8. Changes in electrical activities and extracellular milieu during seizure in medial EC due to arterial bicuculline perfusion. Simultaneous recordings of membrane potential in principal layer III neuron (*top trace*; reconstruction on the right after biocytin injection), extracellular field potential (second trace), and changes in extracellular pH (EC_{pH}; third trace) and K^+ (EC_{K+}; *bottom trace*) during an ictal event. Resting membrane potential of the neuron was −51 mV.

depression. The intracellular correlates of both preictal spikes and repetitive small-amplitude potentials coupled with the fast activity are IPSPs. A gradual reduction of the fast IPSPs paralleled the spontaneous membrane depolarization that occurred at the onset of seizure (64) in correlation with the increase in extracellular K^+. The increase in K^+ could be responsible for the switch from fast oscillation to recovery of neuronal firing that promotes the progression of the seizure (Fig. 8) (62).

Our intracellular recordings demonstrated that also deep EC layer principal cells did not generate neuronal firing at seizure onset. Unlike superficial neurons, no fast IPSPs were observed in deep principal cells. Neuronal firing reappeared in deep neurons during the extracellular burst activity that developed later on during the seizure. Putative interneurons recorded in superficial layers of the medial EC generated a barrage of high-frequency action potentials at more than 150 Hz either at the onset or few milliseconds ahead of the seizure. Continuous firing evolved into phasic bursting that gradually became time locked with the

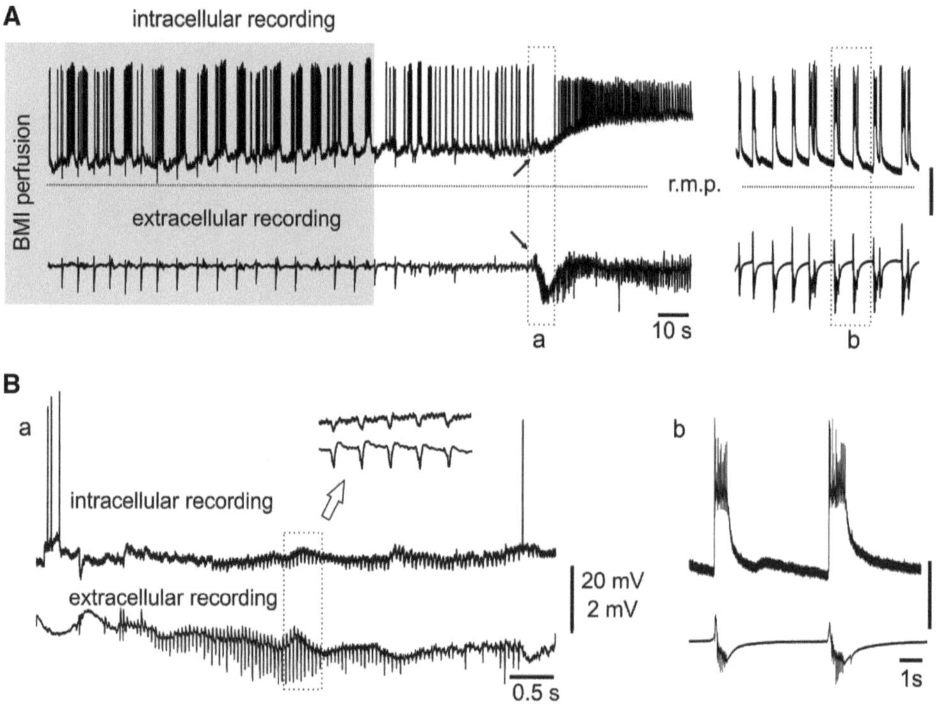

Fig. 9. Intra-extracellular recordings during seizure-like activity in the entorhinal cortex. (**A**) Simultaneous extracellular (*lower trace*) and intracellular recordings from a principal neuron of the superficial layers of the medial EC (*top trace*) during the transition into seizure-like discharge induced by the systemic perfusion of bicuculline (50 μM, 3 min). Recording segments of the ictal onset (***a***) and postictal bursts (***b***) are expanded in **B**. Trace interrupted for 1 min during the recordings illustrated in **A. (B)** At the onset of seizure (***a***), an abrupt depolarization at the seizure onset (*arrows*) correlated with extracellular field potential oscillations at 20–25 Hz (insets in **b**, *left panel*). Regular bursting (***b***) followed during the seizure, as illustrated in the right traces in **B**. Note spontaneous, physiological up-down events during artificial depolarization before bicuculline effect.

extracellular discharge pattern (62). These experiments demonstrate that inhibitory networks are crucial for the initiation of a seizure in the EC. Since fast activity at 20–30 Hz was observed also in intracranial human recordings during temporal lobe seizures, our data obtained on the isolated guinea pig brain model may be of general relevance for the understanding of human ictogenesis in the temporal lobe (63).

Seizure-like activity could also be induced in the isolated guinea pig brain by application via the perfusate of the nonselective muscarinic acetylcholine receptor agonist pilocarpine (12). Epileptiform discharges could be achieved exclusively when the permeability of the BBB was enhanced by bradykinin (0.6–2.5 μM) or histamine (100 μM) (12, 70). The seizure pattern observed in these conditions was similar to that observed after bicuculline application. Also the K^+ channel blocker 4-aminopyridine (4AP, 100 μM) induced epileptiform activity in the isolated guinea pig brain (13). It has been demonstrated in experimental models that 4AP stimulates the generation of GABAergic spikes (71–73). These potentials, elicited in the isolated guinea pig brain by co-perfusion of 4AP with blockers of ionotropic glutamate receptors, have the capability to propagate at a distance and bilaterally in the hippocampal-entorhinal region and are sustained by local circuits, as shown by the CSD analysis. Intracellular recordings in the EC (13) demonstrated that the correlate of the GABAergic spikes consists in a hyperpolarizing potential, with the typical reversal of $GABA_A$ receptor-mediated IPSPs, interrupted by a series of spikes on the hyperpolarizing component of the IPSP. Those spikes show the features of ectopic spikes and are possibly induced by the extracellular K^+ change associated with the massive GABA release (74). Other non-synaptic mechanisms could also sustain ectopic spikes, such as pH modification due to HCO_3^- efflux during IPSPs. The ectopic spikes could then work for the further recruitment of interneurons, by means of gap junctions between principal cells and interneurons or by other non-synaptic mechanisms. The propagation of the $GABA_A$ receptor-mediated potentials could be mediated by the same non-synaptic mechanisms.

The role of astroglia has been investigated in the generation and maintenance of epileptiform activity. According to the traditional perspective, glial cells play structural and metabolic roles. Nowadays, other functions have been attributed to glia, such as regulation of the synaptic strength, synaptogenesis, and neurogenesis (75–77). Indeed, it is well known that glutamate released from activated astrocytes (78) binds to the neuronal extrasynaptic NMDA receptors and can synchronize neuronal firing through a non-synaptic mechanism (79–81). As mentioned above, 2-photon laser scanning microscopy in the guinea pig whole brain maintained in vitro offers a useful opportunity to investigate the relationship between neuronal and astrocytic activities (82). With this method, the activity of superficial layers of the cortex (up to 300 μm depth) could be probed. Astrocytes in the medial EC (and also neurons) can be visualized after local loading of membrane-permeant fluorescent Ca^{2+} dye, and the temporal correlation between the astrocytic, neuronal, and neuropile Ca^{2+} elevations and the electrophysiological signal can be analyzed during seizures induced by arterial application of bicuculline (50 μM). Activation of neuropile and neurons was observed during the interictal spikes as well as during the ictal activity (Fig. 3). On the contrary, since astrocyte involvement was detected only during the seizure-like discharges, it is suggested that astrocytes play a role in the evolution of a seizure epileptiform discharge.

4.6. Network Changes in Focal Ischemia

Recently, we have developed a model of acute focal ischemia and reperfusion for the isolated guinea pig brain preparation (17, 83, 84). In this preparation, it is possible to induce transitory focal ischemia under microscopic control by tying one of the medial cerebral

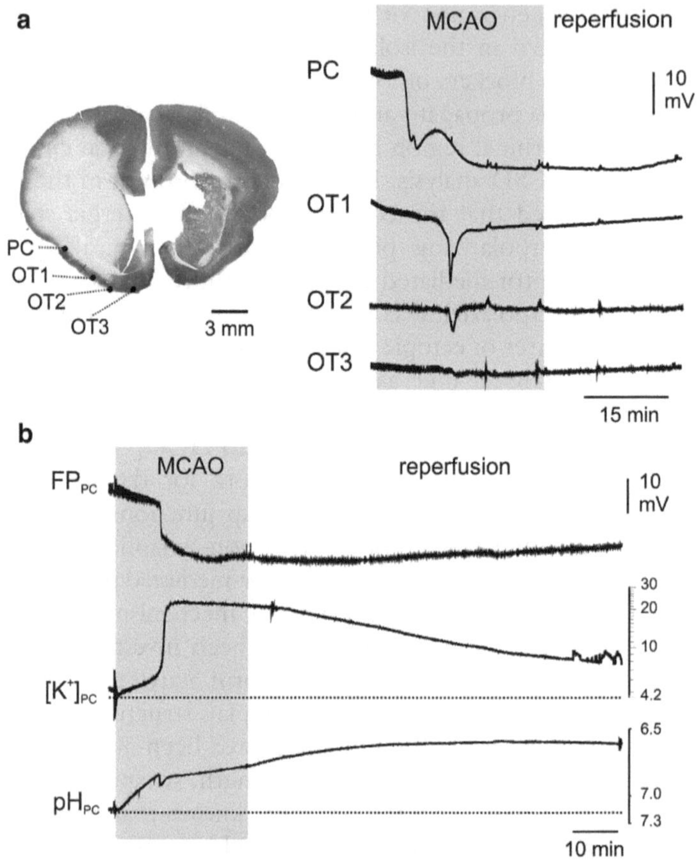

Fig. 10. Effect of 30-min occlusion of medial cerebral arteries (MCA) on electrical activity and microenvironment in isolated guinea pig brain. (**a**) The position of the recording electrodes is illustrated in the low power images of a microtubule-associated protein 2-immunoreacted coronal section cut at the rostro-caudal position in which recordings were performed. On the right, simultaneous recordings performed in PC, lateral (OT1 and OT2) and medial olfactory tubercle (OT3) during ipsilateral MCA occlusion (MCAO; grey-shaded area), and after MCA reperfusion are illustrated. MCAO induced high-amplitude ischemic depolarization (ID) in the PC and fast depression called peri-infarctual depolarization (PID) in OT1 and OT2 ipsilateral to the MCA occlusion. (**b**) Changes in extracellular K^+ ($[K^+]_{PC}$) and pH (pH_{PC}) in the PC induced by ipsilateral MCA occlusion and reperfusion. Occlusion induced an enhancement of $[K^+]$ associated with a rapid metabolic acidification during the ischemic depolarization (ID). The field response (FP) recorded with the conventional barrel is also shown (*top trace*).

arteries (MCA) for different time periods, while brain activity is recorded with multiple electrodes positioned in brain areas within and outside MCA perfusion territory. MCA occlusion induces large and persistent negative-going voltage deflections defined as ischemic depressions or depolarizations (IDs) (85) in PC and EC of the hemisphere ipsilateral to the occluded MCA (Fig. 10).

Multisite extracellular recordings demonstrated prolonged ID and the abolition of evoked responses in the PC served by MCA, but not in the olfactory tubercle, supplied by the anterior cerebral artery. Here, evoked responses were preserved and brief peri-infarctual depressions (PID) could be observed during occlusion (Fig. 10a). Simultaneous extracellular pH and K^+ measurements performed in the PC of the ischemic hemisphere showed a gradual K^+ increase associated with a rapid metabolic acidification interrupted by a transient and mild ID-associated alkalosis (Fig. 10b). The presence of post-ischemic changes in those areas, in which ID occurred, was further confirmed by histological evaluation at the end of the electrophysiological experiments. Microtubule-associated protein-2 is confirmed to be a good marker of the dendro-somatic neuronal injury that defines and localizes the ischemic MCA territory (left panel in Fig. 10a). Microtubule-associated protein-2 staining was markedly reduced or absent in the MCA territory of the guinea pig (9), whereas intense immunostaining was observed in large dendrites of superficial and deep layer neurons of neocortex in the control hemisphere and in nonischemic regions ipsilateral to the occluded MCA (83). Magnetic resonance imaging of these changes can be performed on fixed brains in which MCA occlusion was performed in vitro (84). Our preliminary findings using this optical imaging approach suggest that the correlation between data from electrophysiology, histology, and such imaging helps to identify and characterize the core and penumbra region in the ischemic hemisphere.

5. Conclusions

This chapter provides an overview of the use of the in vitro isolated guinea pig brain preparation for the study of network activity in close-to-physiological conditions and in models of diseases. The versatility of this preparation is captivating. Neurophysiological and imaging studies that cannot be performed in other experimental conditions are possible using this model. In spite of the aura of complexity casting this preparation, it is not too difficult to learn how to master the isolation technique. Its use has been proven beneficial to study both, system and cellular brain physiology. Moreover, this preparation is ideal to study interactions between neuronal and vascular compartments of the brain, a novel approach that has not been yet appropriately exploited.

Acknowledgments

Experiments described in this chapter are the result of ~20 years of research sponsored by the Italian Health Ministry and by grants of the Human Frontiers Science Program Organization, the Mariani Foundation, the CARIPLO Foundation, and the European Community (VSAMUEL and EPICURE).

References

1. Llinás R, Yarom Y, Sugimori M (1981) Isolated mammalian brain in vitro: new technique for analysis of electrical activity of neuronal circuit function. Fed Proc 40:2240–2245
2. de Curtis M, Biella G, Buccellati C, Folco G (1998) Simultaneous investigation of the neuronal and vascular compartments in the guinea pig brain isolated in vitro. Brain Res Protoc 3:221–228
3. Mühlethaler M, de Curtis M, Walton K, Llinas R (1993) The isolated and perfused brain of the guinea-pig in vitro. Eur J Neurosci 5:915–926
4. de Curtis M, Pare D, Llinas RR (1991) Electrophysiology of the olfactory-hippocampal circuit in the isolated and perfused adult mammalian brain in vitro. Hippocampus 1:341–354
5. Llinas R, Mühlethaler M (1988) An electrophysiological study of the in vitro, perfused brain stem-cerebellum of adult guinea-pig. J Physiol 404:215–240
6. Luparello TJ (ed) (1967) Stereotaxic atlas of the forebrain of the guinea pig. Krager, Basel
7. Librizzi L, Janigro D, De Biasi S, de Curtis M (2001) Blood brain barrier preservation in the *in vitro* isolated guinea -pig rain preparation. J Neurosci Res 66:289–297
8. Mazzetti S, Librizzi L, Frigerio S, de Curtis M, Vitellaro-Zuccarello L (2004) Molecular anatomy of cerebral microvessels in the isolated guinea-pig brain. Brain Res 999:81–90
9. Librizzi L, Biella G, Cimino C, de Curtis M (1999) Arterial supply of limbic structures in the guinea pig. J Comp Neurol 411:674–682
10. Trapp S, Ballanyi K (2012) Autonomic nervous system *in vitro*: studying tonically active neurons controlling vagal outflow in rodent brainstem slices. In Isolated Central Nervous System Circuits (Ed K Ballanyi), Neuromethods Series Vol. 73 (Ed W Walz). Springer Science+Business Media, LLC, New York, NY, pp 1–59
11. Ruangkittisakul A, Panaitescu B, Secchia L, Bobocea N, Kantor C, Kuribayashi J, Iizuka M, Ballanyi K (2012) Isolated brainstem respiratory centers from perinatal rodents. In Isolated Central Nervous System Circuits (Ed K Ballanyi), Neuromethods Series Vol. 73 (Ed W Walz). Springer Science+Business Media, LLC, New York, NY, pp 61–124
12. Moore AR, Zhou WL, Jakovcevski I, Zecevic N, Antic SD (2012) Physiological properties of human fetal cortex in vitro. In Isolated Central Nervous System Circuits (Ed K Ballanyi), Neuromethods Series Vol. 73 (Ed W Walz). Springer Science+Business Media, LLC, New York, NY, pp 125–158
13. Librizzi L, de Curtis M (2003) Epileptiform ictal discharges are prevented by periodic interictal spiking in the olfactory cortex. Ann Neurol 53:382–389
14. Uva L, Librizzi L, Wendling F, de Curtis M (2005) Propagation dynamics of epileptiform activity acutely induced by bicuculline in the hippocampal-parahippocampal region of the isolated Guinea pig brain. Epilepsia 46:1914–1925
15. Uva L, Librizzi L, Marchi N, Noe F, Bongiovanni R, Vezzani A, Janigro D, de Curtis M (2008) Acute induction of epileptiform discharges by pilocarpine in the in vitro isolated guinea-pig brain requires enhancement of blood-brain barrier permeability. Neuroscience 151:303–312
16. Uva L, Avoli M, de Curtis M (2009) Synchronous GABA-receptor-dependent potentials in limbic areas of the in-vitro isolated adult guinea pig brain. Eur J Neurosci 29:911–920
17. Mitzdorf U (1985) Current source-density method and application in cat cerebral cortex: investigation of evoked potentials and EEG phenomena. Physiol Rev 65:37–100
18. Freeman WJ, Skarda CA (1985) Spatial EEG patterns, non-linear dynamics and perception: the neo-Sherringtonian view. Brain Res 357:147–175
19. Kuhlmann CR, Librizzi L, Closhen D, Pflanzner T, Lessmann V, Pietrzik CU, de Curtis M, Luhmann HJ (2009) Mechanisms of C-reactive protein-induced blood-brain barrier disruption. Stroke 40:1458–1466
20. Pastori C, Librizzi L, Breschi GL, Regondi C, Frassoni C, Panzica F, Frigerio S, Gelati M, Parati E, De Simoni MG, de Curtis M (2008) Arterially perfused neurosphere-derived cells distribute outside the ischemic core in a model of transient focal ischemia and reperfusion in vitro. PLoS One 3:e2754
21. Broicher T, Speckmann EJ (2012) Living human brain slices: network analysis using voltage sensitive dyes. In Isolated Central Nervous System Circuits (Ed K Ballanyi), Neuromethods Series Vol. 73 (Ed W Walz). Springer Science+Business Media, LLC, New York, NY, pp 285–300
22. de Curtis M, Takashima I, Iijima T (1999) Optical imaging of cortical activity after in vitro perfusion of cerebral arteries with a voltage-sensitive dye. Brain Res 837:314–319

23. Gnatkovsky V, de Curtis M (2006) Hippocampus-mediated activation of superficial and deep layer neurons in the medial entorhinal cortex of the isolated guinea pig brain. J Neurosci 26:873–881
24. Carriero G, Uva L, Gnatkovsky V, de Curtis M (2009) Distribution of the olfactory fibre input into the olfactory tubercle of the in vitro isolated guinea pig brain. J Neurophysiol 101: 1613–1619
25. Fellin T, Gomez-Gonzalo M, Gobbo S, Carmignoto G, Haydon PG (2006) Astrocytic glutamate is not necessary for the generation of epileptiform neuronal activity in hippocampal slices. J Neurosci 26:9312–9322
26. Kantor C, Panaitescu B, Kuribayashi J, Ruangkittisakul A, Jovanovic I, Leung V, Lee TF, MacTavish D, Jhamandas JH, Cheung PY, Ballanyi K (2012) Spontaneous neural network oscillations in hippocampus, cortex and *locus coeruleus* of newborn rat and piglet brain slices. In Isolated Central Nervous System Circuits (Ed K Ballanyi), Neuromethods Series Vol. 73 (Ed W Walz). Springer Science+Business Media, LLC, New York, NY, pp 315–356
27. Ratto GM, Brondi M, Carmignoto G, de Curtis M, Solis-Satu S, Uva L (2008) A window with a view: structure and function in the cortex seen through the two photon microscope. Abstract V meeting molecular mechanism neuroscience, Milan
28. Biella G, de Curtis M (1995) Associative synaptic potentials in the piriform cortex of the isolated guinea-pig brain in vitro. Eur J Neurosci 7:54–64
29. Forti M, Biella G, Caccia S, de Curtis M (1997) Persistent excitability changes in the piriform cortex of the isolated guinea-pig brain after transient exposure to bicuculline. Eur J Neurosci 9:435–451
30. Biella G, Uva L, de Curtis M (2001) Network activity evoked by neocortical stimulation in area 36 of the guinea pig perirhinal cortex. J Neurophysiol 86:164–172
31. Mühlethaler M, Serafin M (1990) Thalamic spindles in an isolated and perfused preparation. Brain Res 524:17–21
32. Llinas R, Mühlethaler M (1988) Electrophysiology of guinea-pig cerebellar nuclear cells in the in vitro brain stem-cerebellar preparation. J Physiol 404:241–258
33. Uva L, Strowbridge BW, de Curtis M (2006) Olfactory bulb networks revealed by lateral olfactory tract stimulation in the in vitro isolated guinea-pig brain. Neuroscience 142:567–577
34. Biella G, Panzica F, de Curtis M (1996) Interactions between associative synaptic potentials in the piriform cortex of the in vitro isolated guinea pig brain. Eur J Neurosci 8:1350–1357
35. Biella G, de Curtis M (2000) Olfactory inputs activate the medial entorhinal cortex via the hippocampus. J Neurophysiol 83:1924–1931
36. Biella G, Uva L, de Curtis M (2002) Propagation of neuronal activity along the neocortical-perirhinal-entorhinal pathway in the guinea pig. J Neurosci 22:9972–9979
37. Shepherd GM (1972) Synaptic organization of the mammalian olfactory bulb. Physiol Rev 52:864–917
38. Shipley MT, Ennis M (1996) Functional organization of olfactory system. J Neurobiol 30:123–176
39. Phillips CG, Powell TP, Shepherd GM (1963) Responses of mitral cells to stimulation of the lateral olfactory tract in the rabbit. J Physiol 168:65–88
40. Haberly LB, Bower JM (1984) Analysis of association fiber system in piriform cortex with intracellular recording and staining techniques. J Neurophysiol 51:90–112
41. Haberly L, Behan M (1983) Structure of the piriform cortex of the opossum. III. Ultrastructural characterization of synaptic terminals of association and olfactory bulb afferent fibers. J Comp Neurol 219:448–460
42. Haberly LB, Presto S (1986) Ultrastructural analysis of synaptic relationships of intracellularly stained pyramidal cell axons in piriform cortex. J Comp Neurol 248:464–474
43. Ketchum KL, Haberly LB (1993) Synaptic events that generate fast oscillations in piriform cortex. J Neurosci 13:3980–3985
44. Takagi SF (1984) The olfactory nervous system of the old world monkey. Jpn J Physiol 34:561–573
45. Cattarelli M (1989) Selective olfactory pathway lesions differentially affect runway behavior of the rat. Physiol Behav 46:393–396
46. Ketchum KL, Haberly LB (1993) Membrane currents evoked by afferent fiber stimulation in rat piriform cortex. I. Current source-density analysis. J Neurophysiol 69:248–260
47. Biella G, Forti M, de Curtis M (1996) Propagation of epileptiform potentials in the guinea-pig piriform cortex is sustained by associative fibres. Epilepsy Res 24:137–146
48. Biella GR, Gnatkovsky V, Takashima I, Kajiwara R, Iijima T, de Curtis M (2003) Olfactory input to the parahippocampal region of the isolated guinea pig brain reveals weak entorhinal-to-perirhinal interactions. Eur J Neurosci 18:95–101
49. Sewards TV, Sewards MA (2003) Input and output stations of the entorhinal cortex:

superficial vs. deep layers or lateral vs. medial divisions? Brain Res Rev 42:243–251
50. Uva L, de Curtis M (2005) Polysynaptic olfactory pathway to the ipsi- and contralateral entorhinal cortex mediated via the hippocampus. Neuroscience 130:249–258
51. Gnatkovsky V, Wendling F, de Curtis M (2007) Cellular correlates of spontaneous periodic events in the medial entorhinal cortex of the in vitro isolated guinea pig brain. Eur J Neurosci 26:302–311
52. Eichenbaum H, Yonelinas AP, Ranganath C (2007) The medial temporal lobe and recognition memory. Annu Rev Neurosci 30:123–152
53. Squire LR, Wixted JT, Clark RE (2007) Recognition memory and the medial temporal lobe: a new perspective. Nat Rev Neurosci 8:872–883
54. Braak H, Braak E (1993) Entorhinal-hippocampal interaction in mnestic disorders. Hippocampus 3:239–246
55. de Curtis M, Pare D (2004) The rhinal cortices: a wall of inhibition between the neocortex and the hippocampus. Prog Neurobiol 74:101–110
56. Martina M, Royer S, Paré D (2001) Propagation of neocortical inputs in the perirhinal cortex. J Neurosci 21:2878–2888
57. Pelletier JG, Apergis J, Pare D (2004) Low-probability transmission of neocortical and entorhinal impulses through the perirhinal cortex. J Neurophysiol 91:2079–2089
58. Kajiwara R, Takashima I, Mimura Y, Iijima T (2003) Amygdala input promotes spread of excitatory neural activity from perirhinal cortex to the entorhinal-hippocampal circuit. J Neurophysiol l89:2176–2184
59. Witter MP, Wouterlood FG, Naber PA, Van Haeften T (2000) Anatomical organization of the parahippocampal-hippocampal network. Ann NY Acad Sci 911:1–24
60. Burwell R (2000) The parahippocampal region: corticocortical interconnectivity. In: Scharfman HE, Witter MP, Schwarcz R (eds) The parahippocampal region, vol 2. New York Academy of Science, New York, pp 25–42
61. de Curtis M, Avanzini G (2001) Interictal spikes in focal epileptogenesis. Prog Neurobiol 63:541–567
62. Gnatkovsky V, Librizzi L, Trombin F, de Curtis M (2008) Fast activity at seizure onset is mediated by inhibitory circuits in the entorhinal cortex in vitro. Ann Neurol 64:674–686
63. de Curtis M, Gnatkovsky V (2009) Re-evaluating the mechanisms of focal ictogenesis: the role of low-voltage fast activity. Epilepsia 7:1–12
64. Pare D, de Curtis M, Llinas R (1992) Role of the hippocampal-entorhinal loop in temporal lobe epilepsy: extra- and intracellular study in the isolated guinea pig brain in vitro. J Neurosci 12:1867–1881
65. de Curtis M, Llinas RR (1993) Entorhinal cortex long-term potentiation evoked by theta-patterned stimulation of associative fibers in the isolated in vitro guinea pig brain. Brain Res 600:327–330
66. de Curtis M, Biella G, Forti M, Panzica F (1994) Multifocal spontaneous epileptic activity induced by restricted bicuculline ejection in the piriform cortex of the isolated guinea pig brain. J Neurophysiol 71: 2463–2475
67. de Curtis M, Radici C, Forti M (1999) Cellular mechanisms underlying spontaneous interictal spikes in an acute model of focal cortical epileptogenesis. Neuroscience 88:107–117
68. de Curtis M, Biella G, Forti M (1996) Epileptiform activity in the piriform cortex of the in vitro isolated guinea pig brain preparation. Epilepsy Res 26:75–80
69. de Curtis M, Manfridi A, Biella G (1998) Activity-dependent pH shifts and periodic recurrence of spontaneous interictal spikes in a model of focal epileptogenesis. J Neurosci 18:7543–7551
70. Marchi N, Oby E, Batra A, Uva L, De Curtis M, Hernandez N, van Boxel-Dezaire A, Najim I, Janigro D (2007) In vivo and in vitro effects of pilocarpine: relevance to ictogenesis. Epilepsia 48:1934–1946
71. Michelson HB, Wong RK (1994) Synchronization of inhibitory neurones in the guinea-pig hippocampus in vitro. J Physiol 477:35–45
72. Avoli M, Barbarosie M, Lücke A, Nagao T, Lopantsev V, Köhling R (1996) Synchronous GABA-mediated potentials and epileptiform discharges in the rat limbic system in vitro. J Neurosci 16:3912–3924
73. Lopantsev V, Avoli M (1998) Participation of GABAA-mediated inhibition in ictallike discharges in the rat entorhinal cortex. J Neurophysiol 79:352–360
74. Barolet AW, Morris ME (1991) Changes in extracellular K^+ evoked by GABA, THIP and baclofen in the guinea-pig hippocampal slice. Exp Brain Res 84:591–598
75. Nedergaard M, Ransom B, Goldman SA (2003) New roles for astrocytes: redefining the functional architecture of the brain. Trends Neurosci 26:523–530

76. Zonta M, Carmignoto G (2002) Calcium oscillations encoding neuron-to-astrocyte communication. J Physiol Paris 9:193–198
77. Verkhratsky A, Orkand RK, Kettenmann H (1998) Glial calcium: homeostasis and signaling function. Physiol Rev 78:99–141
78. Parpura V, Basarsky TA, Liu F, Jeftinija K, Jeftinija S, Haydon PG (1994) Glutamate-mediated astrocyte-neuron signalling. Nature 369:744–747
79. Angulo MC, Kozlov AS, Charpak S, Audinat E (2004) Glutamate released from glial cells synchronizes neuronal activity in the hippocampus. J Neurosci 24:6920–6927
80. Fellin T, Carmignoto G (2004) Neurone-to-astrocyte signalling in the brain represents a distinct multifunctional unit. J Physiol 559:3–15
81. Perea G, Araque A (2005) Properties of synaptically evoked astrocyte calcium signal reveal synaptic information processing by astrocytes. J Neurosci 25:2192–2203
82. Gomez-Gonzalo M, Zonta M, Losi G, Chiavegato A, Brondi M, Vetri F, Uva L, Pozzan T, de Curtis M, Ratto GM, Carmignoto G (2010) A recurrent excitatory loop with astrocytes contributes to drive neurons to seizure threshold. PLoS Biol 8:1–19
83. Pastori C, Regondi MC, Librizzi L, de Curtis M (2007) Early excitability changes in a novel acute model of transient focal ischemia and reperfusion in the in vitro isolated guinea pig brain. Exp Neurol 204:95–105
84. Breschi GL, Librizzi L, Pastori C, Zucca I, Cattalini A, Regondi MC, de Curtis M (2010) Functional and structural correlates of MRI patterns in a model of focal cerebral ischemia in vitro. Neurobiol Dis 39:181–191
85. Somjen GG (2001) Mechanisms of spreading depression and hypoxic spreading depression-like depolarization. Physiol Rev 81: 1065–1096

Chapter 13

The Study of Respiratory Chemoreflexes Using a Novel Dual-Perfused Rodent Preparation

Trevor A. Day and Richard J.A. Wilson

Abstract

Artificially perfused brainstem preparations have several advantages over both in vivo and superfused en bloc preparations for studying respiratory circuits. They allow the analysis of neuronal function without confounding influences of anesthesia and major peripheral feedback loops (chemoreceptor, vagal, and hormonal) while maintaining the brainstem in an oxygenated state and preserving many of the salient features of the normal respiratory motor pattern. Here we have summarized the development, characterization, and utility of a novel "dual-perfused preparation": a vagotomized, decerebrate, in situ juvenile rat brainstem preparation in which central and peripheral chemoreceptors are perfused separately with defined media containing precisely controlled gas concentrations. This preparation has been instrumental in recent studies of time-dependent effects of chemoreflexes in response to intermittent chemostimuli as well as in demonstrating a hypoadditive interaction between central and peripheral chemoreceptors in the absence of cortical and vagal afferents (1–3).

Key words: Artificially perfused, Brainstem, Carotid body, Chemoreceptors, In situ, Respiration

1. Introduction

In 2005, we introduced a dual-perfused, vagotomized, decerebrate rodent preparation (DPP) developed to study various aspects of respiratory chemoreflexes (1). What follows is a brief review of some reduced (en bloc and in vitro slice) brainstem preparations used in respiratory neurobiology and a description of the development, characterization, methods, and utility of the DPP, which overcomes many of the limitations of these preparations. The chapter will end with a discussion of the advantages and disadvantages of the DPP as a model system for investigating central and peripheral chemoreflexes in respiratory control.

Klaus Ballanyi (ed.), *Isolated Central Nervous System Circuits*, Neuromethods, vol. 73,
DOI 10.1007/978-1-62703-020-5_13, © Springer Science+Business Media New York 2012

1.1. En Bloc Superfused Preparations

Until the second half of the last century, most investigations into the neuronal control of breathing were centered around in vivo preparations. Recording from single cells was possible but difficult owing to movement associated with pulsatile blood flow and ventilation. Further, the interpretation of data from in vivo preparations is complicated by the complexity of autonomic regulation that includes a number of powerful feedback control loops involving peripheral chemoreceptors, pulmonary stretch receptors, and cardiovascular, sympathetic, thermal, and hormonal regulations. In addition, most experiments were performed on large mammals, making experiments time intensive and costly.

In 1984, Suzue introduced a novel en bloc superfused brainstem-spinal cord preparation to study respiratory rhythm and pattern generation in neonatal rats (4). Respiratory-related neural output can be recorded from cranial and spinal nerves. While this preparation has been an instrumental tool for investigating mammalian respiratory rhythm generation (e.g., (5, 6)), only the superficial 500–700 μm of tissue are oxygenated; the core is both anoxic and acidic (7, 8). Some groups have proposed that the synchronized decrementing hypoglossal and phrenic bursts produced by these preparations are similar in waveform to gasping, a motor pattern indicative of brainstem tissue hypoxia (9). However, others argue that the inspiratory-related motor patterns are more similar to those of vagotomized neonates at reduced temperature (10, 11). Regardless of how well the en bloc preparations capture the motor pattern and underlying neuronal activity of normal "eupneic" breathing, there can be little doubt that the work to determine the mechanisms responsible for the inspiratory-like activity has greatly enhanced our understanding of the brainstem respiratory circuit.

Through a series of transverse sections, Smith and colleagues (6) demonstrated that a ~200-μm rostrocaudal brainstem region they called the pre-Bötzinger complex is necessary for inspiratory-related rhythm in the en bloc brainstem preparation. When isolated in a transverse brainstem slice and superfused, this area continues to produce rhythmic inspiratory-related activity. In intact adult animals, functional or anatomical ablation of the pre-Bötzinger complex has a devastating impact on ventilation, suggesting this nucleus is vital for inspiratory rhythm generation (12). Onimaru and colleagues identified another region, the parafacial respiratory group (pFRG) rostral to the pre-Bötzinger complex, that contains pacemaker-like neurons and may participate in generation of expiratory-related activity (5). In fact, recent data suggest that under some conditions the pFRG is active before, and excites, the pre-Bötzinger complex (13, 14). These studies are dealt with in more detail elsewhere in this book (Chap. 2-Ruangkittisakul et al. (15)). Along with comparative studies in Amphibia, these studies have lead to the hypothesis that respiratory rhythms are produced in large part by two coupled oscillators, one for inspiration and one

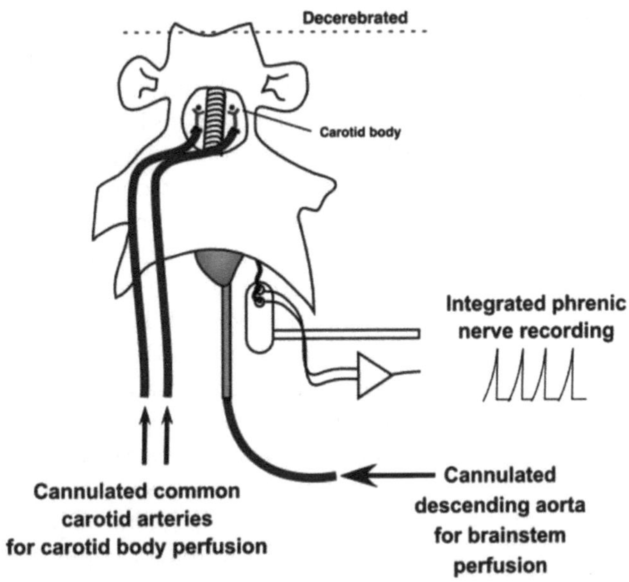

Fig. 1. *Schematic of the in situ dual-perfused preparation (DPP)*. The descending aorta in juvenile rats is cannulated and perfused retrogradely, with rostral flow towards the aortic arch and vertebral arteries for brainstem perfusion. The common carotid arteries are cannulated above the clavicles, proximal to the carotid bodies. The phrenic nerve neurogram is recorded, rectified, and integrated, as illustrated.

for expiration (16). Key questions resulting from this hypothesis are how these oscillators are integrated within the larger medullary-pontine respiratory network during normal breathing and how the motor patterns this network produces are shaped by tonic and phasic sensory inputs. To address these questions necessitates the use of more intact preparations that retain sensory inputs and replicate the essence of the motor pattern that occurs in vivo.

1.2. Arterially Perfused Brainstem Preparations

Perfused preparations, where brainstems are artificially perfused via the vasculature (e.g., basilar, vertebral, descending, or ascending aorta) or cerebral ventricles, were created to provide a more intact model of respiratory rhythm and pattern generation without the need for anesthetics or resorting to the difficulties and complexities of in vivo preparations. Many artificial perfusion models have been developed and utilized, including artificial perfusion of (1) the dog or goat cerebral ventricles with artificial cerebral spinal fluid (in vivo systemic circulation left intact through the animal's own cardiorespiratory system (17–19)), (2) the dog or goat peripheral chemoreceptors via cannulation of carotid artery (in vivo brainstem perfusion left intact through the animal's own cardiorespiratory system (20, 21)), (3) the cat vertebral artery (22), (4) the guinea pig ascending aorta (23, 24), (5) the guinea pig basilar artery (25), (6) the rat abdominal aorta (26), and (7) the mouse or rat descending aorta (27) (see Fig. 1 above).

1.3. The Working Heart-Brainstem Preparation (WHBP)

Of the arterially perfused preparations, the recently developed in situ working heart-brainstem preparation (WHBP) has been used most widely (27). In this preparation, a rodent is heavily anesthetized, decerebrated, and transected below the diaphragm (the caudal half is discarded). The descending aorta is rapidly cannulated and perfused in the retrograde direction with anesthetic-free physiological saline to oxygenate the brainstem and regulate PCO_2 (for detailed description, see below). The dissection takes ~15 min, and the preparation remains viable for several hours.

Paton introduced the WHBP in 1996 using the mouse, but it has been used most widely as a technique for artificial vascular perfusion of the decerebrate juvenile rat brainstem (9, 28–31). The brainstem of this preparation is well oxygenated (tissue PO_2 ~200 mmHg when perfused with saline having a PO_2 over 600 mmHg), has a tissue pH that matches the perfusate, retains vasoreactivity to CO_2 and O_2, and displays a ramping phrenic neurogram, indicative of eupnea (9, 31). The WHBP is a strong candidate for studying respiratory chemosensitivity, being chemoresponsive to both CO_2 and O_2 perturbations (9) and can be studied with and without intact vagi and carotid sinus nerves.

1.4. The In Situ Dual-Perfused Preparation (DPP)

To facilitate the study of central and peripheral respiratory chemoreceptor contribution to the control of breathing, including chemoreceptor interaction, we adapted the WHBP to create the DPP. Our modifications to the WHBP include (1) separating the perfusion of central and peripheral chemoreceptor compartments, (2) clamping of descending aorta perfusion pressures at ~90 mmHg using an automated feedback controller, (3) midcervical vagotomy, and (4) use of (close to) normokalemic perfusate (4 mM vs. 6.25 mM) (9, 27) (compare Chap. 2-Ruangkittisakul et al. (15)).

In the WHBP, only the descending aorta is cannulated, so the brainstem and carotid bodies are perfused by the common perfusate, which is hyperoxic (>600 mmHg) (9, 27). The hyperoxia likely reduces activity in the carotid sinus nerve, blunting important chemical afferents contributing to eupneic ventilation, but may not entirely eliminate the contribution of the carotid bodies (32–34). However, in the DPP, by perfusing the carotid bodies and brainstem separately with defined media, we can target chemostimuli or pharmacological perturbations to either compartment independently (Fig. 1). Thus, experimental stimuli acting on either central or peripheral chemoreceptors can be regulated precisely.

While other investigators utilize the WHBP at perfusion pressures of ~60 mmHg (Drs. Julian Paton and Jeffrey Potts, personal communications), we chose to use ~90 mmHg, more closely matching the mean arterial blood pressure of an intact rat. As a result, the mean flow rate through the central circuit in the DPP is ~50 ml/min (unpublished measurements), significantly higher than that used by other investigators. Further, while other investigators attempt to maintain the perfusion pressure by increasing or decreasing flow manually, we clamp the pressure using a computer-controlled proportional integral derivative (PID) feedback system such that the speed of the pump supplying perfusate to the preparation is adjusted automatically. This helps maintain a constant perfusion pressure despite transient changes in vascular resistance in response to changes in perfusate gas concentrations.

The vagus nerve (cranial nerve X) carries visceral afferents from organs in the abdominal and thoracic cavities to the *nucleus tractus solitarius* in the brainstem, including tonic and phasic feedback from pulmonary stretch receptors (30, 35), and afferents from the aortic baroreceptors and chemoreceptors (36, 37). Midcervical vagotomy eliminates these afferents, allowing the contribution of brainstem and carotid body chemoreceptors to be isolated without the confounding influence of pulmonary stretch and aortic arch baro- and chemoafferent input.

2. Materials

Special materials required for this preparation include a custom-made dish with ear and snout bars and a dual-line catheter for simultaneous retrograde perfusion of the descending aorta and arterial pressure measurement. A pressure transducer and amplifier, for example, from CWE (USA; www.cwe-inc.com/default.htm) are also needed as are two high-quality peristaltic pumps (e.g., Minipulse 5, Gilson Incorporation, Middleton, WI, USA; www.gilson.com). One pump, supplying the descending aorta, is servo-controlled using the arterial pressure measurement. This is achieved using custom software that incorporates a PID feedback. Other equipment includes water baths, heat exchangers (custom made), filter holders (VWR Incorporation, Radnor, PA, USA; www.vwr.com), 25-μm filters (Advantec MFS Incorporation, Dublin, CA, USA; wwwadvantecmfs.com), and bubble traps (custom made; one for the peripheral perfusion circuit, one for central). In addition to the above, we also employ a custom-built tonometer system, having a common reservoir that receives and distributes recirculated perfusion to four subchambers. Solution is pulled from the bottom of the subchambers, each of which contains a bubble stone for independent equilibration of solutions. This tonometer prevents mixing of equilibrated perfusate in the four subchambers. We tested this system within our experimental setup using a Clark-style polarographic PO_2 electrode (Diamond General Inc.) to ensure efficient equilibration without cross contamination between reservoirs (unpublished measurements).

We employ a custom-made computer-controlled system to mix gases, based around a bank of mass flow controllers (MKS Instruments Inc.). Gas from this system was sampled to ensure accuracy using a Medical Gas Analyzer LB-2 (SensorMedics, Milano, Italy) and a PA-1B O_2 analyzer (Sable Systems International, Las Vegas, NV, USA; www.sablesystems.com); both calibrated daily. Gas concentrations reported here are accurate to within 0.1% (~1.3 mmHg in Calgary). In addition to the above specialized equipment, standard electrophysiological equipment is needed to record respiratory motor activity from phrenic and cranial nerves including extracellular amplifiers (e.g., AM Systems 1700, Sequim, WA, USA; www.a-msystems.com), integrators (CWE Inc.), and a data acquisition system (e.g., Digidata board, Molecular Devices, Sunnyvale, CA, USA; www.moleculardevices.com). Details in this regard are provided in other sections of this book (e.g., Chaps. 1-Trapp and Ballanyi (38), 2-Ruangkittisakul et al. (15), and 12-De Curtis et al. (39)).

3. Methods

3.1. Dissection and Perfusion System

All procedures described below were approved by the University of Calgary Animal Care Committee and are in accordance with national guidelines. To prepare the DPP, we use juvenile male Sprague-Dawley albino rats (Charles River, Quebec, Canada). We find that success rate is optimal with animals between 120 and 150 g, that is, ~4–6 weeks old: below 120 g, animals are very small and blood vessels are fragile and difficult to cannulate; above 150 g, animals become too large to provide adequate perfusion.

Rats are heparinized (intraperitoneal, 1,500 units) ~15 min prior to being anesthetized for dissection. Deep anesthesia is then induced using inhaled halothane. Adequate anesthesia is assessed by testing for absence of response to noxious tail pinch. In rapid succession, rats are then transected subdiaphragmatically and decerebrated at the midcollicular level (approximate level of lambda, the more caudal of coronal sutures on the dorsal aspect of the rat skull). All tissues rostral to the decerebration and all remaining cortex dorsal to the colliculi are removed. This eliminates the circle of Willis (confirmed by visual inspection), removing anastomoses between the carotid and basilar circuits. These procedures are performed in cold saline (5–8°C, volume: 500 ml) containing (in mM) 115 NaCl, 24 $NaHCO_3$, 4 KCl, 2 $CaCl_2$, 1 $MgSO_4$, 1.25 NaH_2PO_4 and 10 dextrose, equilibrated ("carboxygenated") with 95% O_2–5% CO_2. Next, the skin is removed, the preparation placed in an experimental chamber and secured with snout and ear bars. The descending aorta is cannulated with a double-lumen catheter.

One lumen of the catheter is connected to a peristaltic pump (Sect. 2) and used to perfuse the descending aorta retrogradely with room temperature saline containing 1.33% Ficoll Type-70 (a nonionic polymer of sucrose, molecular weight 70,000, Sigma-Aldrich F-2878) equilibrated with 40 mmHg PCO_2 in O_2. In the absence of plasma, Ficoll is utilized as an oncotic agent to minimize preparation swelling resulting from artificial vascular perfusion. 10 mM sucrose can be used in place of Ficoll without any apparent detriment. The other lumen of the catheter is connected to a pressure transducer. Once cannulated, perfusion pressure is increased to 60 mmHg over the first few minutes using a custom-built, computer-controlled feedback system that modulates pump speed. A bilateral vagotomy is then performed at the midcervical level to remove afferent input from collapsed lungs. Both common carotid arteries are tied off above the clavicles and cannulated downstream so as to provide perfusion to the carotid bodies via a separate peristaltic pump at ~15–20 ml/min/carotid. Up to this stage in the dissection, central and peripheral perfusions are from the same container (~300 ml).

Perfusion is then switched to fresh Ficoll/sucrose-containing media, pulled from two different subchambers of the custom-built tonometer, allowing independent perfusion of central (descending aorta) and peripheral (carotid arteries) circuits. Between the tonometer subchambers and the preparation, the central and peripheral perfusate passes through a 25-µm filter, bubble trap, and heat exchanger. Perfusate leaks from cut vessels in the preparation and returns to the common return of the tonometer system where it is recycled and re-equilibrated (total volume of perfusate: 750 ml). When independent perfusion is initiated, the perfusate is virtually blood free. Subsequently, the central and peripheral perfusates are equilibrated with 40 mmHg PCO_2 in O_2 and 40 mmHg PCO_2 and 100 mmHg PO_2 in N_2, respectively. Over the next 30 min, the temperature of the preparation is increased to 32–33°C, and the central perfusion pressure is increased to ~90 mmHg. Protocols commence 70–75 min after central cannulation. Note that neuromuscular blockers may have potential confounding effects on peripheral chemosensitivity and are thus not used (40).

3.2. Electrophysiology

Respiratory activity is recorded from the left phrenic nerve using a small custom-made Plexiglass recording chamber that incorporates a pair of silver bipolar extracellular electrodes. The recording chamber was designed to (1) eliminate the electrocardiogram signal in the phrenic recording, (2) protect the nerve from dehydration, and (3) allow stable recordings for several hours. The neurogram is amplified, band-pass filtered (0.3–5 KHz), rectified, and integrated and then computer archived at a sampling rate of 50 Hz and analyzed off-line.

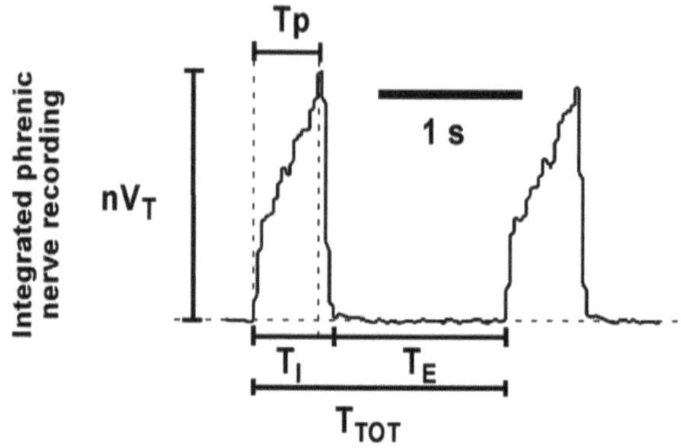

Fig. 2. *Quantification of respiratory variables from phrenic neurogram.* The shape of the integrated phrenic nerve bursts here is representative of this preparation. Note that respiratory variables shown are used to derive the following: respiratory frequency (f_R), $60/T_{TOT}$; neural minute ventilation (nV_E), $f_R \times nV_T$; and eupneic index (EI), Tp/T_I. These variables were calculated on a burst-by-burst basis (Reproduced with kind permission from (1)).

3.3. Data Analysis

Analysis of the integrated phrenic neurogram is performed using custom software (1) (Fig. 2). This software quantifies the following respiratory variables: period (T_{TOT}), respiratory frequency (f_R, 60 times the inverse of the period), inspiratory duration (T_I), expiratory duration (T_E), T_I:T_E ratio, neural tidal volume (nV_T, the peak phrenic amplitude), neural minute ventilation (nV_E, the product of f_R and nV_T), inspiratory ramp (IR, nV_T divided by time to peak, Tp), and eupneic index (EI, Tp divided by T_I). Typically, we reject data from preparations that have a T_I in excess of 1 sec and an inspiratory duty cycle >50%; these preparations are considered apneustic.

Although peak phrenic amplitude is highly correlated with tidal volume (41), the units of this measurement are arbitrary. Thus, in order to make meaningful comparisons between animals when measuring changes in respiratory variables that include peak amplitude (e.g., nV_T, nV_E, and IR), we routinely normalized phrenic amplitude from each animal to a common condition that all animals experienced (e.g., baseline). Furthermore, displaying normalized data has the advantage of illustrating percentage changes from a baseline and eliminating interanimal variability. Raw data could reliably be used in measurements of timing, such as T_{TOT}, f_R, T_I, and T_E.

4. Results: Preparation Characterization

Before utilizing the DPP for studies of chemoreflexes, we characterized the stability of the phrenic neurogram under baseline conditions, confirmed that perfusion of the brainstem and carotid

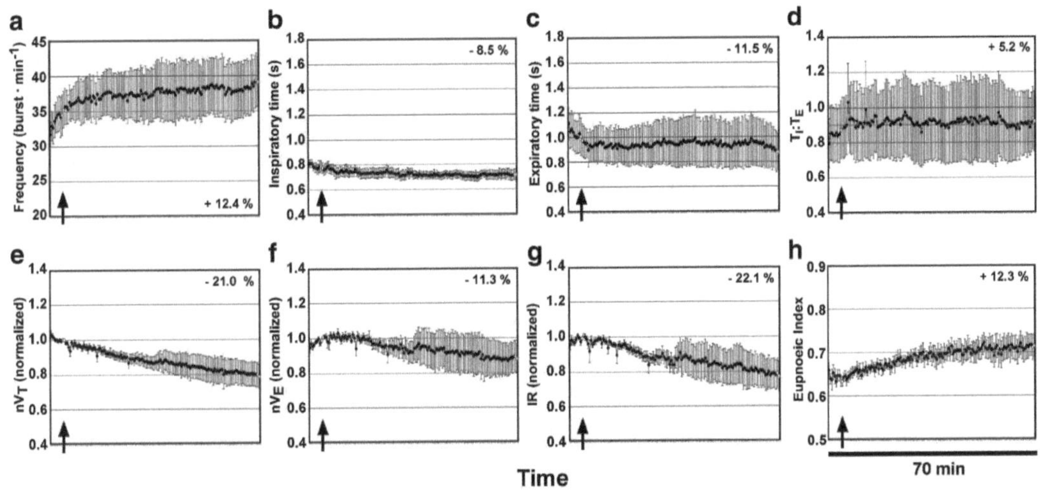

Fig. 3. *Time control protocol illustrating the stability of phrenic variables over a 70-min time period. Arrows* indicate the initial baseline point (min 4–5) used for comparison throughout the study (n = 5). Percent (%) changes on each graph represent the difference between this point and the end point (min 69–70 at the end of the protocol). Gases equilibrated in perfusate held constant throughout: brainstem, 30 mmHg PCO_2, balance O_2; carotid body, 40 mmHg PCO_2, 100 mmHg PO_2, balance N_2. nV_T, nV_E, and IR data are normalized to initial baseline (Reproduced with kind permission from (1)).

bodies are independent, and characterized the systemic PCO_2 chemoresponse in order to compare it to other decerebrate preparations and intact rats. What follows is a brief description of this characterization (see Day and Wilson (1) for more details).

4.1. Stability of the DPP

Firstly, we characterized any changes that may occur in phrenic variables under baseline conditions over time, which is a prerequisite for using the DPP to study the effects of chemostimuli or pharmacological perturbations. For this, we allowed preparations 70–75 min after the initiation of artificial descending aorta perfusion to recovery from dissection and then recorded respiratory variables over the subsequent 70 min, the "protocol period." The following conditions were used throughout: brainstem perfusate equilibrated with 30 mmHg PCO_2, balance O_2; carotid body perfusate equilibrated with 40 mmHg PCO_2, 100 mmHg PO_2, balance N_2. Review of the data revealed that over the first 5 min of the protocol period, timing variables were still recovering from dissection (Fig. 3). However, after this period, T_I:T_E and burst frequency stabilized considerably. Comparing the 5th and 70th minute of the protocol period, we found phrenic frequency increased by 12.4%, and nV_T and nV_E decreased by 21% and 11.3%, respectively. The phrenic bursts retained their eupneic-like shape throughout, having a stable eupneic index and a T_I of 0.7–0.8 s (Fig. 3).

4.2. Independence of Perfusion Circuits

The level of decerebration should have transected the circle of Willis at its caudal end, thereby eliminating the anastomosis between the internal carotid and basilar perfusion circuits (Fig. 1).

That the circle of Willis was removed was confirmed by visual inspection of the ventral surface of the decerebrated brain. However, we were concerned that other anastomoses may be present that might allow peripheral (carotid) perfusate to "leak" into the brainstem and stimulate brainstem chemoreceptors. This would negate the utility of the DPP for studies of chemoreceptor interaction. Therefore, we functionally tested the possibility that peripheral perfusion may directly influence the central chemoreceptors.

4.2.1. Protocol and Analysis

In six preparations, under baseline conditions, the brainstem and carotid body perfusates were equilibrated with 28 mmHg PCO_2 (average), balance O_2 and 28 mmHg PCO_2 (average), 100 mmHg PO_2, balance N_2, respectively. We chose a PCO_2 of ~28 mmHg for both compartments to maximize sensitivity of chemoreceptors to increases in CO_2 without causing apnea. Firstly, we confirmed the presence of phrenic responses to specific carotid body chemostimuli using the following protocol: 5-min baseline recording, 5-min specific carotid body hypoxia (60 mmHg PO_2, PCO_2 unchanged), 5-min washout, and 5-min specific carotid body hypercapnia (60 mmHg CO_2, PO_2 unchanged). Secondly, after sectioning the carotid sinus nerves, (CSN) the above protocol was repeated: the second hypoxic carotid body perturbation tested whether the carotid sinus nerve transection was successful, and the second hypercapnic carotid body perturbation tested for "leak" of peripheral perfusate onto brainstem chemoreceptors. Finally, we tested whether the brainstem was still sensitive to PCO_2 perturbation using 5 min of specific brainstem hypercapnia (60 mmHg PCO_2, balance PO_2). Two of the six preparations developed apneusis following CSN transection and were not included in subsequent analysis.

4.2.2. Main Findings

At the beginning of the experiment, peripheral hypoxia and peripheral hypercapnia induced increases in mean nV_E. Transecting the CSN eliminated both responses but did not eliminate the effects of brainstem hypercapnia (Fig. 4). As transecting the CSN abolished respiratory responses to changes in peripheral perfusate PCO_2, we can conclude that the peripheral perfusate has no effect on the perfusion of the brainstem. Thus, these results confirm the independence of central and peripheral perfusion circuits in the DPP.

4.3. Characterization of the Systemic CO_2 Chemoresponse of the DPP

Next, we tested the dynamic range of the DPP in response to respiratory chemochallenges and compared it with other decerebrate, vagotomized preparations, a necessary prerequisite for using the DDP for studies of chemosensitivity.

4.3.1. Protocol and Analysis

To assess the systemic PCO_2 sensitivity of the DPP, PCO_2 perturbations were applied to nine preparations randomly in ascending (n = 4) or descending (n = 5) order to both central and peripheral

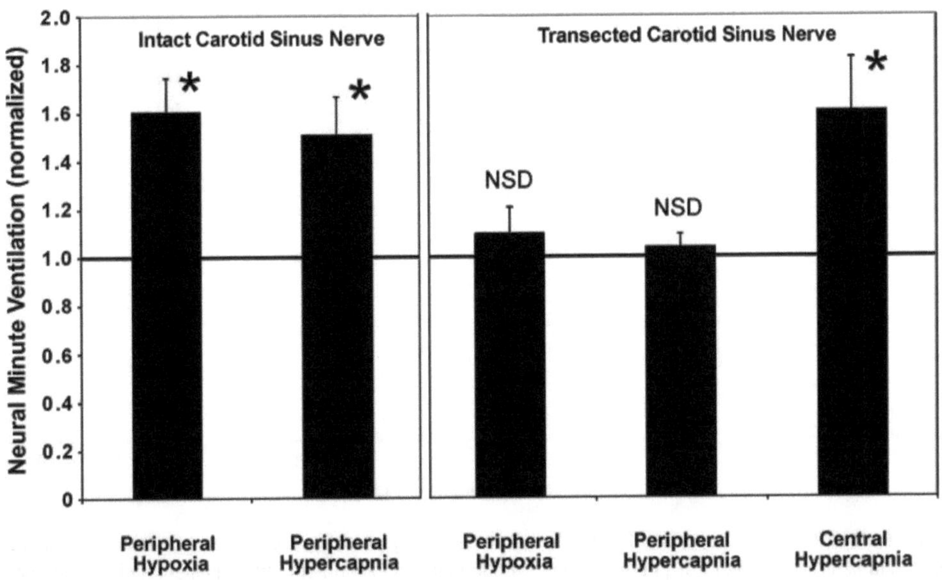

Fig. 4. *Independence of central and peripheral perfusion circuits. Left panel*: before carotid sinus nerve (CSN) transection, peripheral perfusion with hypoxia (60 mmHg PO_2, 28 mmHg PCO_2, balance N_2) and hypercapnia (60 mmHg PCO_2, 100 mmHg PO_2, balance N_2) produced robust nV_E responses. *Right panel*: after CSN transection, nV_E responses to peripheral perfusion of hypoxic and hypercapnic solutions are absent. Absence of hypoxia response demonstrates that the transection was successful (NSD), whereas absence of response to the peripheral hypercapnia demonstrates that peripheral perfusion does not influence brainstem chemoreceptors (nonsignificant difference, NSD). Note that the response to the central hypercapnia (60 mmHg PCO_2, balance O_2) confirms that the preparation retained central respiratory chemosensitivity. Baseline central and peripheral PCO_2: 28 mmHg ($n = 4$). * Indicates a response that is significantly different ($P < 0.05$) from the baseline immediately preceding challenge. *Error bars* represent SEM (Reproduced with kind permission from (1))

perfusion circuits in unison. The protocol consisted of a 10-min baseline period at 35 mmHg PCO_2 followed by six 10-min PCO_2 perturbations of 15, 25, 35, 45, 55, and 65 mmHg followed by a washout at 35 mmHg PCO_2. Throughout this protocol, the PO_2 at central chemoreceptors was maintained above 550 mmHg, whereas the carotid bodies were kept at a PO_2 of 100 mmHg. The apneic threshold was determined at the end of each protocol as the level of PCO_2 at or below which apnea could be maintained (42, 43).

4.3.2. Results

The mean apneic threshold of the DPP was 22.44 ± 1.78 mmHg with all nine preparations having phrenic activity above 35 PCO_2 (Fig. 5). An increase in PCO_2 above the apneic threshold increased f_R, nV_T, nV_E, and IR and decreased T_E and T_I:T_E. These effects topped out above 50 mmHg. The middle of the dynamic range was approximately 30 mmHg PCO_2. PCO_2 had no effect on the eupneic index, demonstrating the shape of phrenic bursts was not affected (Fig. 5).

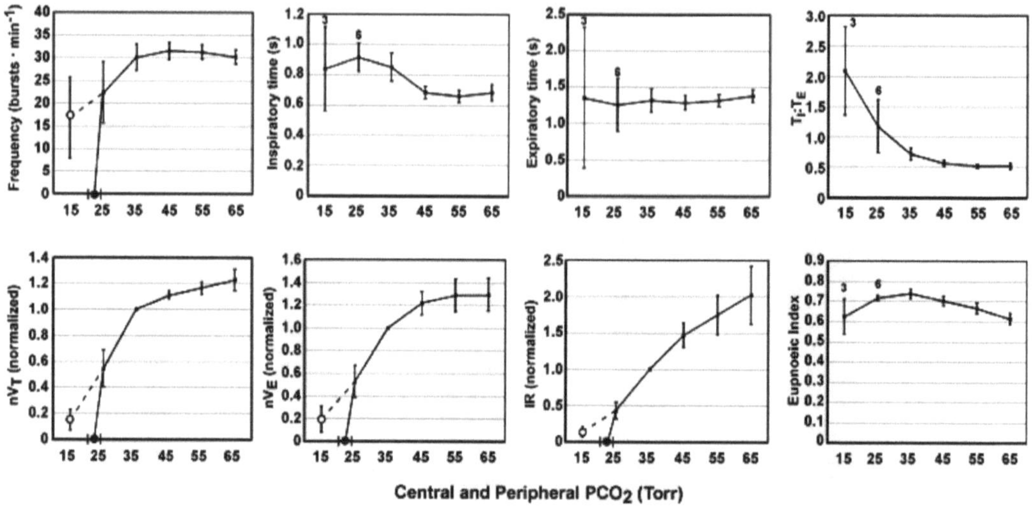

Fig. 5. *Systemic PCO_2 chemoresponse of the DPP*. Graphs showing phrenic variables as a function of systemic (i.e., peripheral and central matched) PCO_2 perturbations. Throughout this experiment, the brainstem and carotid bodies had a PO_2 of >550 mmHg and 100 mmHg, balance N_2, respectively. Unless indicated by numbers above data points, n = 9. For phrenic shape variables, T_I, T_E, T_I:T_E, and EI, we excluded animals from the means once they developed apnea to prevent values of zero from distorting the mean. *Solid circle* (•) on the abscissa indicates the apneic threshold (22.44 ± 1.78 SEM, n = 9). *Open circle* (o) represents mean for f_R, nV_T, nV_E, and IR at 15 mmHg PCO_2, including zero values from apneic animals. Note: nV_T, nV_E, and IR data are normalized within animal to the value at 35 mmHg PCO_2 (Reproduced with kind permission from (1)).

5. Key Advantages of the DPP

By modifying the "Paton-type" WHBP, we developed a novel rat preparation in which central and peripheral chemoreceptor compartments are intact and independently perfused. This model allows the study of respiratory chemosensitivity independent of cardiovascular dynamics, vagal feedback, or descending cortical input while avoiding the nonspecific, depressive effects of anesthetics and paralytic agents on respiratory chemoreceptors (40, 44, 45). The DPP provides an important intermediate between in vivo chemoreceptor interaction studies in goats, dogs, and cats and recent studies in anesthetized rodents at the cellular level (46, 47). Unlike in vivo preparations used to study the interaction of central and peripheral chemoreceptors that breathe room air or inspired gases, we are able to maintain the PCO_2 within the brainstem below eupneic levels. Thus, the chemoreflex resulting from activation of one chemoreceptor can be explored across the entire physiological range, while the other chemoreceptor is exposed to a constant PO_2 and PCO_2.

One might argue that only by understanding the pure interaction between brainstem and carotid body chemoreceptors will we be able to understand the influences of vagal feedback, suprapontine structures, and other influences that modulate chemosensitivity. From here, it might be possible to start adding some of these factors back into the system in a controlled manner. For instance, it would be possible, if not technically challenging, to leave the vagi intact and phasically inflate the lungs in phase with the phrenic neurogram.

6. Critique of the DPP

No preparation is without limitation. One of the biggest limitations of the DPP is also its strength, that being the degree of reductionism. For example, one should note that in vivo, blood is likely to play an important role in determining systemic chemoresponses, being involved in the transport of gases and a host of other biologically active factors, like the buffering of pH and the regulation of the microvasculature. Furthermore, the brainstem of the DPP is hyperoxic and has a tissue pH that matches the perfusate. Although these features are an improvement over other reduced preparations allowing precise control of chemical perturbations to central chemoreceptors, hyperoxia represents a departure from the in vivo conditions. In vivo, mammalian brainstem tissue typically has a PO_2 of ~40 mmHg (and likely approaches values < 10 mmHg close to active cells), and the tissue PCO_2 is approximately 8–10 mmHg higher than blood (see also Chaps. 1-Trapp and Ballanyi (38), 2-Ruangkittisakul et al. (15), and 12-De Curtis et al. (39)). Although not supported by pilot experiments, we remain open minded as to whether hyperoxia has an impact on phrenic output.

When considering the DPP as an alternative to other preparations used for the study of chemoreceptor interaction, a number of other caveats must also be considered: the DPP is decerebrate, vagotomized, and slightly hypothermic (~33°C). In addition, sympathetic activation produced by carotid body stimulation could cause changes in brainstem perfusion, which in turn might change brainstem tissue PO_2 and PCO_2. These issues are briefly discussed below.

6.1. Possible Effects of Decerebration

Our preparation is decerebrated at the midcollicular level, which removes all inputs from the thalamus, hypothalamus, and cortex. Thus, the animal is rendered "neurologically anesthetized," without the potential chemosensitivity depressive effects of anesthetic agents (44, 45). However, important changes in respiratory rhythm and pattern occur following decerebration. Decerebration has been shown in several preparations to reduce the apneic threshold, shifting the CO_2 chemoresponse curve in the hypocapnic direction (26, 48, 49). For example, Hayashi and Sinclair (26) used anemically decerebrate rats to investigate chemoresponses and showed sustained increases in f_R, V_T, and V_E despite a corresponding drop in arterial PCO_2 from 41 to 25 mmHg breathing room air. This study suggests that breathing in these animals is being "driven" at substantially lower levels of CO_2 than experienced when intact. When exposed to hypercapnia (5% inspired CO_2), intact animals increased V_E by 87% compared to when breathing room air, whereas decerebrate animals increased V_E by only 8% over baseline. During this perturbation, intact animals showed an increase in arterial PCO_2 from 46 to 60 mmHg, whereas decerebrate animals rose from 29 to 55 mmHg. This may indicate an attenuated chemoresponse or a leftward-shifted chemoresponse curve. In a study using decerebrate dogs, Nielsen et al. (49) found a significant reduction in the PCO_2 apneic threshold following decerebration, from 38 mmHg in anesthetized animals to 25 mmHg. Further, this leftward shift was accompanied by reductions in the slopes (sensitivity) for f_R, V_T, and V_E obtained from linear regression analysis of the data. Both intact and decerebrate dogs responded to hypercapnia with phrenic amplitude augmentation, but changes in frequency were attenuated and have greater variability in decerebrate dogs. Thus, taken together, these studies suggest that suprapontine structures may modulate stimulus-induced changes in both the timing and amplitude components of the phrenic neurogram. Further, these data suggest that the maximum chemoresponse after decerebration occurs at a substantially lower level of PCO_2. In the only study published applying CO_2 perturbations to the WHBP, St.-John and Paton (9) illustrate the hyperoxic CO_2 response for both vagally intact and vagotomized preparations. They demonstrate that CO_2 perturbations increases minute phrenic activity linearly from an apneic threshold, for which they do not report a mean value. However, their graphical representation of normalized neural minute ventilation as a function of perfusate PCO_2 shows a mean activity of approximately 50% that of baseline (35 mmHg) at 20 mmHg, indicating a shift in the hypocapnic direction.

We characterized the systemic (both brainstem and carotid bodies in unison) PCO_2 chemoresponse of the DPP and found it was also shifted in the hypocapnic direction. The apneic threshold was ~10 mmHg lower than in intact anesthetized rats of the same species (42). The asymptotes of the chemoresponse curves for f_R, nV_T, and nV_E were ~45 mmHg. Consequently, in future studies designed to test the effects of specific carotid body chemostimulation, holding the brainstem at 30 mmHg PCO_2 ensured the preparation is within its chemoresponsive range.

6.2. Possible Effects of Vagotomy

The vagus nerve carries afferents from various visceral organs, the aortic sinus, aortic body and tonic and phasic information from stretch receptors in the lung. Inhibition of respiration from increased activity in the vagus nerve from lung inflation is referred to as the Hering-Breuer inflation reflex (30, 35, 50). Vagal input has powerful effects on respiratory timing. The termination of the inspiratory ramp of the phrenic neurogram is modulated by off-switch mechanisms reaching a critical threshold of afferent input from pulmonary stretch reflexes (51). Indeed, bilateral vagotomy in decerebrate newborn rats and conscious or anesthetized rats, rabbits, and kittens causes significant increases in T_E and slowing of respiratory frequency (52). Wang and colleagues (53) recorded the phrenic neurograms of neonatal, 0–13-day-old rats and found that in both anesthetized and decerebrate animals, T_E was significantly longer and respiratory frequency was significantly slower following vagotomy. Similarly, being vagotomized, the DPP has a baseline respiratory frequency of approximately 30–40/min.

Vagal feedback from pulmonary stretch receptors may also modulate chemosensitivity. Kinkead and colleagues (54) studied the effect of phasic electrical stimulation of the pulmonary branch of the vagus nerve on respiratory chemosensitivity in an in vitro superfused bullfrog brainstem-spinal cord preparation. They found that phasic electric stimulation increased respiratory responses to pH perturbations threefold (54). More recently, Moreira and colleagues (50) demonstrated that there are anatomical connections between the vagus nerve carrying information from slowly adapting lung receptors, neurons of the nucleus of the solitary tract, and chemosensitive cells within the retrotrapezoid nucleus in the rat. Further, Moreira and colleagues demonstrated that a background of positive end expiratory pressure of 2–6 cmH_2O (tonic inflation) *inhibited* the sensitivity of putative central respiratory chemoreceptors and phrenic output in response to inspired CO_2 in intact anesthetized rats (50). This study did not address phasic lung inflation, but it did demonstrate the existence of anatomical and functional connections between vagal feedback and central chemosensitivity.

In the DPP, the lungs are deflated and the vagi are cut. In the WHBP, from which the DPP is derived, the lungs are also deflated, but the vagi are left intact. St.-John and Paton (9) investigated the possibility that tonic vagal inputs from the deflated lung of the WHBP influence respiratory burst activity. Vagotomy had no effect on the systemic CO_2 chemoresponse of the WHBP. However, Harris and St.-John (35) demonstrated that phrenic-triggered phasic lung inflation in the WHBP increased baseline frequency 2–3-fold over that without lung inflation, bringing the frequency closer to in vivo values (35).

6.3. Possible Effects of Temperature

In addition to decerebration and vagotomy, temperature also has an impact on respiration. Using anesthetized rats, Maskrey (55) lowered body temperature to ~35°C using abdominally implanted heat exchangers. When breathing room air, hypothermic rats had increased respiration compared to normothermic rats (55). Given that CO_2 is more soluble in water (and blood) at lower temperatures, for a given PCO_2 and bicarbonate concentration, the pH will be slightly more acidic at cooler temperatures than 37°C (56). This may have implications for cellular functions that involve proteins with imidazole-histidine residues, where the ionization status of these residues, affected by temperature in addition to $[H^+]$, may be more important for protein function rather than the absolute pH (i.e., alphastat hypothesis) (57).

6.4. Possible Effects of Sympathetic Activation on Brainstem Perfusion

Sympathetic activity produced by carotid body stimulation might conceivably vasoconstrict the cerebral vasculature, causing tissue hypoxia and hypercapnia. While systemic hypoxic and hypercapnic blood has direct effects on the cerebral vasculature (58, 59), the central effects of sympathetic activation on the cerebral vasculature caused by carotid body stimulation are controversial (60–63). In all likelihood, these effects do not affect brainstem regions (60). However, carotid body stimulation does cause considerable vasoconstriction in the peripheral vasculature through sympathetic activation (64). In the current study, we used a feedback controller to clamp the perfusion pressure of the central circuit supplying the brainstem, forelimbs, and trunk of the preparation. Using this system, changes in peripheral vascular resistance in the forelimbs and trunk are compensated at the level of the descending aorta by changes in flow rate delivered by the peristaltic pump. Thus, brainstem perfusion is expected to remain constant. During hypoxic challenges delivered through the carotid artery, the eupneic index decreased only slightly (from 0.73 ± 0.026 to 0.69 ± 0.034). Furthermore, T_I also decreased, an effect inconsistent with brainstem hypoxia. These considerations, along with the data from the independent perfusion experiment, suggest that hypoxic stimulation delivered through the peripheral circuit was specific to the carotid body and did not cause any secondary brainstem hypoxia due to cross contamination of perfusion circuits or sympathetic effects on brainstem vasculature.

7. Conclusion

Here we have described a unique perfused preparation for studying chemosensory integration within the respiratory control system. Perfused preparations eliminate many of the feedback loops that confound the study of breathing in vivo while retaining many of the salient features of the respiratory motor pattern that are lost when brainstems are isolated. Moreover, unlike isolated rat and mouse brainstem preparations that are limited to the study of neonatal animals (e.g., owing to the presence of substantial anoxic cores), perfused brainstem preparations are oxygenated throughout and can be used to study brainstem circuits from neonates to juveniles. However, the preparation we describe requires technical and surgical proficiency and suffers from limitations resulting from decerebration, vagotomy, and reduced temperature such as a more hypocapnic apneic threshold, a leftward-shifted chemoresponse curve, and blunted chemoreflexes compared with intact animals. These factors need to be carefully considered in interpreting data. Thus, as with all preparations, the DPP likely represents but one of many important stepping stones towards understanding respiratory control.

Acknowledgments

This work was supported by grants from CIHR and AIHS (RJAW) and CIHR and MRU Internal Grants Fund (TAD). Development of the DPP was made possible by the generosity of Dr. Julian Paton who taught us the WHBP on which the DPP was based.

References

1. Day TA, Wilson RJA (2005) Specific carotid body chemostimulation is sufficient to elicit phrenic post-stimulus frequency decline in a novel in situ dual perfused rat preparation. Am J Physiol Reg Int Comp Physiol 289: R532–R544
2. Day TA, Wilson RJA (2007) Brainstem PCO_2 modulates phrenic responses to specific carotid body hypoxia in an in situ dual perfused rat preparation. J Physiol 578:843–857
3. Day TA, Wilson RJA (2009) A negative interaction between central and peripheral respiratory chemoreceptors modulates peripheral chemoreflex magnitude. J Physiol 587:883–896
4. Suzue T (1984) Respiratory rhythm generation in the in vitro brain stem-spinal cord preparation of the neonatal rat. J Physiol 354:173–183
5. Onimaru H, Arata A, Homma I (1987) Localization of respiratory rhythm-generating neurons in the medulla of brainstem-spinal cord preparations from newborn rats. Neurosci Lett 78:151–155
6. Smith JC, Ellenberger HH, Ballanyi K, Richter DW, Feldman JL (1991) Pre-Bötzinger complex: a brainstem region that may generate respiratory rhythm in mammals. Science 254:726–729
7. Brockhaus J, Ballanyi K, Smith JC, Richter DW (1993) Microenvironment of respiratory neurons in the in vitro brainstem-spinal cord of neonatal rats. J Physiol 462:421–445
8. Okada Y, Mückenhoff K, Holtermann G, Acker H, Scheid P (1993) Depth profiles of pH and PO_2 in the isolated brain stem-spinal cord of the neonatal rat. Respir Physiol 93:315–326
9. St.-John WM, Paton JFR (2000) Characterization of eupnea, apneusis and gasping in a perfused rat preparation. Respir Physiol 123:201–213
10. Ballanyi K, Onimaru H, Homma I (1999) Respiratory network function in the isolated brainstem-spinal cord of newborn rats. Progr Neurobiol 59:583–634
11. Ballanyi K (2004) Neuromodulation of the perinatal respiratory network. Curr Neuropharmacol 2:221–243
12. Gray PA, Janczewski WA, Mellen N, McCrimmon DR, Feldman JL (2001) Normal breathing requires preBötzinger complex neurokinin-1 receptor-expressing neurons. Nat Neurosci 4:927–930
13. Mellen NM, Janczewski WA, Bocchiaro CM, Feldman JL (2003) Opioid-induced quantal slowing reveals dual networks for respiratory rhythm generation. Neuron 37:821–826
14. Onimaru H, Homma I (2003) A novel functional neuron group for respiratory rhythm generation in the ventral medulla. J Neurosci 23:1478–1486
15. Ruangkittisakul A, Panaitescu B, Secchia L, Bobocea N, Kantor C, Kuribayashi J, Iizuka M, Ballanyi K (2012) Isolated brainstem respiratory centers from perinatal rodents. In

Isolated Central Nervous System Circuits (Ed K Ballanyi), Neuromethods Series Vol. 73 (Ed W Walz). Springer Science+Business Media, LLC, New York, NY, pp 61–124

16. Vasilakos K, Wilson RJ, Kimura N, Remmers JE (2005) Ancient gill and lung oscillators may generate the respiratory rhythm of frogs and rats. J Neurobiol 62:369–385
17. Leusen IR (1954) Chemosensitivity of the respiratory center; influence of CO_2 in the cerebral ventricles on respiration. Am J Physiol 176:39–44
18. Fencl V, Miller TB, Pappenheimer JR (1966) Studies on the respiratory responses to disturbances of acid-base balance, with deductions concerning the ionic composition of cerebral interstitial fluid. Am J Physiol 210:459–472
19. Smith CA, Jameson LC, Mitchell GS, Musch TI, Dempsey JA (1984) Central-peripheral chemoreceptor interaction in awake cerebrospinal fluid-perfused goats. J Appl Physiol 56:1541–1549
20. Busch MA, Bisgard GE, Forster HV (1985) Ventilatory acclimatization to hypoxia is not dependent on arterial hypoxemia. J Appl Physiol 58:1874–1880
21. Smith CA, Saupe KW, Henderson KS, Dempsey JA (1995) Ventilatory effects of specific carotid body hypocapnia in dogs during wakefulness and sleep. J Appl Physiol 79:689–699
22. Berkenbosch A, Heeringa J, Olievier CN, Kruyt EW (1979) Artificial perfusion of the ponto-medullary region of cats. A method for separation of central and peripheral effects of chemical stimulation of ventilation. Respir Physiol 37:347–364
23. Richerson GB, Getting PA (1987) Maintenance of complex neural function during perfusion of the mammalian brain. Brain Res 409:128–132
24. Richerson GB, Getting PA (1990) Preservation of integrative function in a perfused guinea pig brain. Brain Res 517:7–18
25. Morin-Surun MP, Denavit-Saubie M (1989) Rhythmic discharges in the perfused isolated brainstem preparation of adult guinea pig. Neurosci Lett 101:57–61
26. Hayashi F, Sinclair JD (1991) Respiratory patterns in anesthetized rats before and after anemic decerebration. Respir Physiol 84:61–76
27. Paton JF (1996) A working heart-brainstem preparation of the mouse. J Neurosci Methods 65:63–68
28. Dutschmann M, Wilson RJ, Paton JF (2000) Respiratory activity in neonatal rats. Auton Neurosci 84:19–29
29. St. Jacques R, St-John WM (2000) Sensitivities of eupnea and gasping to alterations in temperature of in vivo and perfused rat preparations. Respir Physiol 123:215–224
30. Harris MB, St.-John WM (2003) Tonic pulmonary stretch receptor feedback modulates both eupnea and gasping in an in situ rat preparation. Am J Physiol Regul Integr Comp Physiol 285:R215–R221
31. Wilson RJ, Remmers JE, Paton JF (2001) Brain stem PO_2 and pH of the working heart-brain stem preparation during vascular perfusion with aqueous medium. Am J Physiol Regul Integr Comp Physiol 281:R528–R538
32. Smith CA, Rodman JR, Chenuel BJ, Henderson KS, Dempsey JA (2006) Response time and sensitivity of the ventilatory response to CO_2 in unanesthetized intact dogs: central vs. peripheral chemoreceptors. J Appl Physiol 100:13–19
33. Rodman JR, Curran AK, Henderson KS, Dempsey JA, Smith CA (2001) Carotid body denervation in dogs: eupnea and the ventilatory response to hyperoxic hypercapnia. J Appl Physiol 91:328–335
34. Lahiri S, DeLaney RG (1975) Stimulus interaction in the responses of carotid body chemoreceptor single afferent fibers. Respir Physiol 24:249–266
35. Harris MB, St.-John WM (2005) Phasic pulmonary stretch receptor feedback modulates both eupnea and gasping in an in situ rat preparation. Am J Physiol Regul Integr Comp Physiol 289:R450–R455
36. Bouckaert JJ, Heymans C (1933) Carotid sinus reflexes. Influence of central blood-pressure and blood supply on respiratory and vaso-motor centres. J Physiol 79:49–66
37. Comroe JH (1939) The location and function of the chemoreceptors of the aorta. Am J Physiol 127:176–191
38. Trapp S, Ballanyi K (2012) Autonomic nervous system in vitro: studying tonically active neurons controlling vagal outflow in rodent brainstem slices. In Isolated Central Nervous System Circuits (Ed K Ballanyi), Neuromethods Series Vol. 73 (Ed W Walz). Springer Science+Business Media, LLC, New York, NY, pp 1–59
39. De Curtis M, Lilbrizzi L, Uva L, Gnatkovsky V (2012) Neuronal networks in the in vitro isolated guinea pig brain. In Isolated Central Nervous System Circuits (Ed K Ballanyi), Neuromethods Series Vol. 73 (Ed W Walz). Springer Science+Business Media, LLC, New York, NY, pp 357–383

40. Wyon N, Joensen H, Yamamoto Y, Lindahl SG, Eriksson LI (1998) Carotid body chemoreceptor function is impaired by vecuronium during hypoxia. Anesthesiology 89:1471–1479
41. Eldridge FL (1971) Relationship between phrenic nerve activity and ventilation. Am J Physiol 221:535–543
42. Boden AG, Harris MC, Parkes MJ (1998) Apnoeic threshold for CO_2 in the anaesthetized rat: fundamental properties under steady-state conditions. J Appl Physiol 85:898–907
43. Dempsey JA, Skatrud JB (1986) A Sleep induced apnoeic threshold and its consequences. Am Rev Respir Dis 133:1163–1170
44. Czapla MA, Zadina JE (2005) Reduced suppression of CO_2-induced ventilatory stimulation by endomorphines relative to morphine. Brain Res 1059:159–166
45. Faber JE, Harris PD, Wiegman DL (1982) Anesthetic depression of microcirculation, central hemodynamics, and respiration in decerebrate rats. Am J Physiol 243: H837–H843
46. Mulkey DK, Stornetta RL, Weston MC, Simmons JR, Parker A, Bayliss DA, Guyenet PG (2004) Respiratory control by ventral surface chemoreceptor neurons in rats. Nat Neurosci 7:1360–1369
47. Takakura AC, Moreira TS, Colombari E, West GH, Stornetta RL, Guyenet PG (2006) Peripheral chemoreceptor inputs to retrotrapezoid nucleus (RTN) CO_2-sensitive neurons in rats. J Physiol 572:503–523
48. Mitchell GS (1990) Phrenic nerve responses to lung inflation and hypercapnia in decerebrate dogs. Pflüger's Arch 416:580–585
49. Nielsen AM, Bisgard GE, Mitchell GS (1986) Phrenic nerve responses to hypoxia and CO_2 in decerebrate dogs. Respir Physiol 65:267–283
50. Moreira TS, Takakura AC, Colombari E, West GH, Guyenet PG (2007) Inhibitory input from slowly adapting lung stretch receptors to retrotrapezoid nucleus chemoreceptors. J Physiol 580:285–300
51. Von Euler C (1983) On the central pattern generator for the basic breathing rhythmicity. J Appl Physiol 55:1647–1659
52. Zhou D, Huang Q, Fung M-L, Li A, Darnall RA, Nattie EE, St.-John WM (1996) Phrenic response to hypercapnia in the unanesthetised, decerebrate, newborn rat. Respir Physiol 104:1–22
53. Wang W, Fung ML, Darnall RA, St. John WM (1996) Characterizations and comparisons of eupnoea and gasping in neonatal rats. J Physiol 490:277–292
54. Kinkead R, Filmyer WG, Mitchell GS, Milsom WK (1994) Vagal input enhances responsiveness of respiratory discharge to central changes in pH/CO_2 in bullfrogs. J Appl Physiol 77:2048–2051
55. Maskrey M (1990) Body temperature effects on hypoxic and hypercapnic responses in awake rats. Am J Physiol 259:R492–R498
56. Kaila K, Ransom BR (1998) Concept of pH and its importance in neurobiology. Chapter 1. In: Kaila K, Ransom BR (eds) pH and brain function. Wiley-Liss, New York, pp 3–10
57. Nattie E (1999) CO_2 brainstem chemoreceptors and breathing. Prog Neurobiol 59: 299–331
58. Steinback CD, Poulin MJ (2008) Cardiovascular and cerebrovascular responses to acute isocapnic and poikilocapnic hypoxia in humans. J Appl Physiol 104:482–489
59. Ainslie PN, Duffin J (2009) Integration of cerebrovascular CO_2 reactivity and chemoreflex control of breathing: mechanisms of regulation, measurement, and interpretation. Am J Physiol Regul Integr Comp Physiol 296: R1473–R1495
60. Heistad DD, Marcus ML, Ehrhardt JC, Abboud FM (1976) Effect of stimulation of carotid chemoreceptors on total and regional cerebral blood flow. Circ Res 38:20–25
61. Miyabe M, Jones MD, Koehler RC, Traystman RJ (1989) Chemodenervation does not alter cerebrovascular response to hypoxic hypoxia. Am J Physiol 257:H1413–H1418
62. Vatner SF, Priano LL, Rutherford JD, Manders WT (1980) Sympathetic regulation of the cerebral circulation by the carotid chemoreceptor reflex. Am J Physiol 238: H594–H598
63. Anwar M, Kissen I, Weiss HR (1990) Effect of chemodenervation on the cerebral vascular and microvascular response to hypoxia. Circ Res 67:1365–1373
64. Prabhakar NR, Dick TE, Nanduri J, Kumar GK (2007) Systemic, cellular and molecular analysis of chemoreflex-mediated sympathoexcitation by chronic intermittent hypoxia. Exp Physiol 92:39–44

Chapter 14

Defined Medium Organotypic Cultures of Spinal Cord Put 'Pain in a Dish'

James E. Biggs, Van B. Lu, Helena J. Kim, Aaron Lai, Kathryn G. Todd, Klaus Ballanyi, William F. Colmers, and Peter A. Smith

Abstract

Nerve injury and/or disease are known to provoke the release of various neurotrophic and inflammatory agents from spinal microglia and astrocytes. Prolonged exposure of spinal neurons to these 'nociceptive mediators' invokes pathophysiological changes that lead to the 'central sensitization' which characterizes chronic pain. In order to investigate their actions, it is necessary to expose spinal neural networks to individual putative mediators for prolonged periods as would occur following injury or in the presence of disease. Because this is not feasible using acutely isolated spinal cord slices, we developed a defined medium organotypic spinal cord culture model which retains the structural and synaptic organization of the spinal cord in vivo. This 'pain in a dish' system represents an exciting new methodology for the investigation of the pathophysiological processes that underlie chronic pain.

Key words: Calcium imaging, Chronic pain, Dorsal horn, Electrophysiology, Immunohistochemistry, Neurons, Nociception, Organotypic, Spinal cord

List of Abbreviations

ACSF	Artificial cerebrospinal fluid
AdGFP	Adenovirus containing green fluorescent protein
AM	AcetoxyMethyl
aMCM	Activated microglia-conditioned medium
ANOVA	Analysis of variance assay
AraC	Cytosine-b-D-arabinofuranoside
AUC	Area under curve
BDNF	Brain-derived neurotrophic factor
CCI	Chronic constriction injury
CMV	Cytomegalovirus
DMEM	Dulbecco's modified Eagle's medium
DMOTC	Defined medium organotypic cultures
DRG	Dorsal root ganglion

Klaus Ballanyi (ed.), *Isolated Central Nervous System Circuits*, Neuromethods, vol. 73,
DOI 10.1007/978-1-62703-020-5_14,

ELISA	Enzyme-linked immunosorbent assay
FBS	Fetal bovine serum
GAD	Glutamic acid decarboxylase
GcaMP	GFP-calmodulin-protein
GFP	Green fluorescent protein
HBSS	Hanks buffered saline solution
IR-DIC	Infrared differential interference contrast
I_m	Membrane current
LPS	Lipopolysaccharide
NGF	Nerve growth factor
OTC	Organotypic
PBS	Phosphate-buffered saline
ROI	Region of interest
siRNA	Small interfering ribonucleic acid
vGlut	Glutamate vesicular transporter
V_m	Membrane potential

1. Introduction

Pain serves the fundamental biological purpose of protecting an individual from tissue damage or injury. Once the harmful stimulus has been removed, or the tissue has healed, the pain should abate. When this does not happen, pain becomes chronic and can therefore present a major health problem. Many types of chronic pain involve an enduring increase in the excitability of the dorsal horn of the spinal cord that is known as 'central sensitization' (1, 2). Although the underlying pathophysiology of central sensitization in the dorsal horn is poorly understood, it is recognized that a host of pain mediators including cytokines, growth factors, adenosine triphosphate (ATP) and neuropeptides are released (3). The continued presence of these mediators for periods of several days to weeks brings about the long-term changes that underlie central sensitization (4, 5).

The ability to study the long-term effects of potential mediators in isolation is therefore mandatory to gain insight into the pathophysiology of chronic pain. This is not readily accomplished using conventional acutely isolated spinal cord slices as these do not retain viability beyond 1 day, at least not when obtained from mature mammals. This may be too brief a period for analysing functional changes brought about by processes such as altered protein expression, axonal sprouting, dendritic retraction and synapse formation or elimination. For this reason, we have developed 'defined medium organotypic cultures' (DMOTC) of spinal cord slices (6, 7). Since the cultures remain viable for >6 weeks, this allows us to expose intact neurons to mediators or drugs for periods of days or weeks.

DMOTC retain the basic structural and connective organization of their tissue of origin (7–10), hence the term 'organotypic'. Neurons in DMOTC display a remarkable physiological, morphological and developmental similarity to neurons in intact animals. In the case of the spinal cord (6, 7), the synaptic pharmacology and connectivity of DMOTC spinal cord slices remain intact and the various morphological and electrophysiological phenotypes of dorsal horn neurons are preserved (11, 12). Because individual neurons in the thin DMOTC can be visualized, this in vitro model is especially amenable to electrophysiological recordings using infrared differential interference contrast (IR-DIC) optics and to dynamic imaging of the free cytosolic Ca^{2+} concentration which provides information on both the activity and morphology of neuronal (and glial) populations (5, 13) (see also Chaps. 1-Trapp and Ballanyi (14), 2-Ruangkittisakul et al. (15), 11-Kantor et al. (16), 12-De Curtis et al. (17) and 15-Mandadi et al. (18)). Organotypic slices are also suited to immunohistochemical procedures and can be modified genetically by adenoviral infection (8). For all these reasons, spinal cord DMOTC provide a unique model to study regulation of neural circuits in the dorsal horn by defined concentrations of putative mediators of chronic pain for extended time period (5, 10–13, 19). For example, we exposed such slices to brain-derived neurotrophic factor (BDNF) for periods of 1 week or more. This putative pain mediator (4) produced a specific pattern of changes in excitatory synaptic transmission to five different identified neuronal phenotypes found in lamina II of spinal DMOTC (13). This 'electrophysiological signature' is very similar to that observed in acutely isolated rat spinal cord slices after chronic constriction injury (CCI) of the sciatic nerve (13, 20). Changes induced by both BDNF and CCI included increased excitatory synaptic drive to putative excitatory neurons and decreased excitatory drive to putative inhibitory neurons (5, 21). This and attenuation of the inhibitory actions of GABA and glycine (4) contribute to the overall increase in dorsal horn excitability that underlies central sensitization.

It is now accepted that both microglia of non-neural origin and astrocytic glial cells play an indispensible role in the genesis and maintenance of chronic pain (4, 22, 23). Consequently, we can effectively 'put chronic pain in a dish' by exposing spinal DMOTC to medium that has been collected from a culture containing 'activated microglia-conditioned medium' (aMCM). In this system, effects can be studied of drugs used to treat chronic pain, such as gabapentin and pregabalin, which need several days to exert their actions (24, 25). Spinal cord slices in DMOTC therefore afford an excellent opportunity to study mechanisms of currently used pain-relieving agents and to develop novel more effective neuron- and/or glia-specific therapeutic strategies (26).

The present chapter presents the detailed methodology used in the preparation of DMOTC of rat spinal cord slices. We will review the similarity between neurons in acute slices and those in DMOTC and show how the latter can be used as a pain in a dish system to study chronic pain mechanisms. We will also illustrate that microglial cells appear to be activated in acutely isolated slices and that this may imply that they are undergoing an injury reaction. By contrast, microglia appear to be quiescent in DMOTC (19). Thus, it might be argued that the physiological status of control neurons in the culture is, at least in some regards, more like that of uninjured neurons in vivo than those in acutely isolated slices.

2. Materials

2.1. Generation of Defined Medium Organotypic Cultures

DMOTC spinal cord slices are maintained in culture for up to 6 weeks after isolation form rat fetuses. To ensure that these slices remain viable throughout this time course, only fetuses from dams that are designated specific pathogen free are used. Consequently, all slices generated should also be specific pathogen-free. We have excellent experience in that regard with Charles River (Wilmington, MA, USA; www.criver.com), although there are other companies. To ensure that the DMOTC spinal cord slices are aseptic and are not contaminated during their generation or maintenance, all involved procedures are to be performed using aseptic techniques. The caesarean section is carried out in an appropriately sterilized surgical suite with appropriately sterilized surgical tools, sterilized either by baking the tools or through autoclaving. Slices are sectioned, isolated and mounted in a Class II Biohazard Cabinet NSF St No. 49 (EnvirCo, Sanford, NC, USA; www.envirco.com). All slice maintenance such as media exchanges and drug treatments is carried out in a Laminar Flow Biohazard Cabinet (Pure Aire Corporation, Simmy Valley, CA, USA; www.pureair.com). Any certified and appropriately sterilized Biohazard cabinet would be suitable for the generation and maintenance of DMOTC spinal cord slices.

For obtaining DMOTC, acute spinal cord slices are generated using a 'McIlwain' tissue chopper (Warner Instruments, Hamden, CT, USA; www.warneronline.com), as this allows for faster sectioning of the spinal cord which aids in keeping the slices aseptic. Since experiments are performed at least 2 weeks after slice generation, it is assumed that the slices will have recovered from any trauma imposed by the use of tissue chopper. It is important to note that vibrating microtomes can also be used to generate slices although this process is slower and thus increases the chances of contaminating slices.

Slices are visualized using a Wild M3 light microscope (Leica Biosystems, Nussloch, Germany; www.leica-microsystems.com), fitted with × 10 magnification eyepiece objectives, but any light microscope could be used that is easily sterilized. After mounting slices on to glass coverslips, the slice and coverslip are placed into a flat-bottomed tissue culture tube (Nunc. Cat. no. 156–758) containing 1 ml of medium. The tissue culture tubes are then placed in a roller drum, e.g. Model TC8 (New Brunswick Scientific Company, Edison, NJ, USA; www.nbsc.com). All media ingredients are obtained from established suppliers. We have found that Alomone Labs (Jerusalem, Israel; Israel; www.alomone.com) are the best supplier for neurotrophins, e.g. nerve growth factor (NGF).

2.2. Electrophysiological Recordings

DMOTC spinal cord slices are placed in a submerged-type acrylic recording chamber with a glass bottom (see also Chaps. 1-Trapp and Ballanyi (14) and 2-Ruangkittisakul et al. (15)) and continuously perfused in artificial cerebrospinal fluid (ACSF, for composition, see Sect. 3.3). The small volume (<1 ml) of this chamber allows for rapid solution exchange. Solutions are administered via a gravity-driven system, adjusted to provide constant flow rate and removed via a bent needle attached to a vacuum system. ACSF and all other superfusates are 'carbogenated' (i.e. gassed with carbogen comprised of 95% O_2, 5% CO_2) and exchanged at a rate of 1 ml/min. The tubing (e.g. Tygon®) used to transport the superfusates is impermeable to gases, i.e. O_2 and CO_2, to ensure that superfusate PO_2 and pH do not change during transport to the recording chamber.

For 'whole-cell' recording, patch electrodes with outer tip diameters of 1.5–3 μm are produced from (thin-walled) 1.5/1.12-mm outer/inner diameter borosilicate glass capillaries, e.g. TW140F-4 (World Precision Instruments, WPI, Sarasota, FL, USA; www.wpiinc.com). For fabricating these electrodes, we use the PP-830 microelectrode puller (Narishige, Tokyo, Japan; www.usa.narishige-group.com). Other pullers include the P-1000 (Sutter Instrument Company, Novato, CA, USA; www.sutter.com).

Electrical stimulation of fibre tracts in spinal cord slices is reliably achieved using bipolar tungsten electrodes; these are 3" stereotaxic electrodes with a 0.345-mm shaft diameter, 1-mm tip with 100–140 mm between the two tips and ~0.5 MΩ nominal electrical impedance (item # TST33A05KT, WPI) (see also Chaps. 1-Trapp and Ballanyi (14), 2-Ruangkittisakul et al. (15), 10-Luhmann and Kilb (27), 11-Kantor et al. (16) and 12-De Curtis et al. (17)). The electrodes are placed inside a glass patch electrode, which is cut (with a diamond-tipped cutter) to allow the tip of the tungsten electrode to fit through the tip aperture of the glass electrode. The electrode is then sealed into place with epoxy glue (note this should be fast setting (5 min)); we typically use LePage Speed Set Epoxy syringe glue available from any good crafts store. Both ends of the glass electrode are glued to prevent fluid entering the shaft and to insulate the electrical stimulus. Glue should be allowed to set for at least 24 h before use. We have also used large (broken) patch pipettes (inner tip diameter 5–10 μm) filled with standard superfusate (Sect. 3.3) (see also Chap. 1-Trapp and Ballanyi (14)) as stimulating electrodes. Stimulating and recording microelectrodes are positioned using micromanipulators. The patch electrode is positioned with a MX7600R motorized manipulator (Siskiyou Corporation, Grants Pass, OR, USA; www.siskiyou.com). The tungsten bipolar stimulating electrode is placed with a coarser mechanical manipulator, WR-88 Three Dimensional Water Robot Micromanipulator (Narishige).

Membrane current or potential (I_m and V_m respectively) are recorded with a SEC-05LX single electrode voltage-clamp amplifier (NPI electronic, Tamm, Germany; www.npielectronic.com). Electrical signals are sampled into a computer via the interface and acquisition software ('Clampex') and are analysed by its sister software Clampfit (both Axon Instruments, Molecular Devices Inc, Sunnyvale, CA, USA; www.moleculardevices.com). Recordings of spontaneous and miniature excitatory (and inhibitory) postsynaptic currents are analysed using Mini Analysis software (v6, Synaptosoft, Decatour, GA, USA; www.synaptosoft.com).

For whole-cell recording, neurons in DMOTC spinal cord slices are viewed with a Axioskop FS upright microscope (Zeiss, Oberkochen, Germany; www.zeiss.com/smt) equipped with IR-DIC optics. This microscope requires a water immersion lens (usually 40× or 60× magnification) with a working distance >1 mm to facilitate pipette positioning and should also be equipped with IR-DIC optics (see also Chap. 1-Trapp and Ballanyi (14)). Precision is further improved by attaching an IR-sensitive charge-coupled device (CCD) video camera (NC-70, Dage-MTI, Michigan City, IN, USA; www.dagemti.com) to the microscope and sending the captured images to a video monitor.

All equipment for the recordings is mounted on a vibration isolation table to avoid that floor vibrations disrupt the measurements. For more information about these tables see also Chap. 1-Trapp and Ballanyi (14).

2.3. Ca^{2+} Imaging

The basic setup for the recording changes in the free cytosolic Ca^{2+} concentration in populations of DMOTC neurons (or adjacent glia) is very similar to that for recording V_m or I_m from individual DMOTC neurons, i.e. slices are placed in a recording chamber and neurons visualized with an upright microscope (Sect. 2.2). In this section, we will only list the slight technical and equipment differences between the two methods.

The DMOTC spinal cord slice is placed into a recording chamber and submerged in ACSF, unlike with electrophysiological recordings, the superfusate is administered via a peristaltic roller pump, the Watson-Marlow Alitea (Sin-Can, Calgary, AB, Canada; www.sincaninc.com). The recording chamber is also larger, with a volume of 1.5 ml, as such the flow rate of the superfusate is also greater, at a rate of 4 ml/min.

Dynamic imaging of fluorescence signals is achieved using an FV300 laser scanning confocal microscope equipped with an argon (488 nm) laser and 510–530-nm bandpass filters and a 20× XLUMPlanFl-NA-0.95 water immersion objective (Olympus, Center Valley, PA, USA; www.olympusamerica.com). Fluorescent images are acquired and analysed using Fluoview software (for details, see Chap. 2 Ruangkittisakul et al. (15)).

2.4. Morphology of Dorsal Horn Neurons

Neuronal morphology was determined by adding 0.2% biocytin (Tocris Bioscience, Ellisville Mo, Catalog #3349) to the whole-cell patch electrode solution. At the end of the recording, the electrode is slowly removed and the DMOTC slice fixed in cold 4% paraformaldehyde. Streptavidin conjugated to Texas Red (Invitrogen Canada, Burlington ON Catalog # S6370) is then used to label the recorded neurons. Neurons are visualized with a Zeiss LSM 510 confocal laser scanning microscope, equipped with the appropriate laser (HeNe1, wavelength 543 nm) and Texas Red filter. Confocal images and 3D reconstructions were acquired using Zeiss LSM image browser software.

2.5. Other Materials

Sources for materials such as drugs or fluorescent dyes that cannot be obtained from established suppliers like Sigma-Aldrich (Canada; www.sigmaaldrich.com), VWR (Radnor, PA, USA; www.vwr.com), Fisher Scientific (Canada; www.fishersci.ca) or Tocris Biosciences (Ellisville, Missouri, USA; www.tocris.com), are referred to specifically in the text.

3. Methods

3.1. Preparation of Slice Cultures

All experimental procedures have been approved by the Animal Care and Utilization Committee for Health Sciences at the University of Alberta and are in accordance with the guidelines of the Canadian Council for Animal Care.

3.1.1. Embryonic Dissection

It is imperative that dissection and preparation of cultures proceed quickly because delays will result in deterioration of tissue and will therefore impact the quality of the cultures obtained. In order to minimize the deterioration, all dissection solutions are kept cold

(~5°C). Some laboratories carry out the entire procedure on ice. Since the procedures are carried out at low temperatures to depress metabolism, we believe it is not necessary to carbogenate the solution.

Embryonic day (E) 13–14 rat fetuses are delivered by caesarean section from timed-pregnant Sprague-Dawley rats (Sect. 2.1) under isoflurane anaesthesia. Attainment of the appropriate level of surgical anaesthesia in the pregnant dam is evaluated by the loss of paw withdrawal and ocular reflexes. Once this is attained, the 8–14 fetuses present in each dam are exposed by cutting into the abdomen and creating a 2–3-cm incision, which can be enlarged if necessary. To avoid damaging the fetuses, the skin is grasped with tweezers and gently lifted up. The lifted skin is cut, taking care to cut away from the fetuses. Once access to the fetuses is achieved, tweezers are used to grasp the embryonic sac in the space between two individual fetuses (so as to not damage them) and to gently pull them up through the incision in the abdomen. The lower section of the uterus is cut to release the embryonic sac from the abdomen. The anaesthetized dam is then euthanized by decapitation using a small animal guillotine. Under aseptic conditions, the entire embryonic sac is then placed immediately in chilled Hanks' balanced salt solution (HBSS) containing (in mM) 138 NaCl, 5.33 KCl, 0.44 KH_2PO_4, 0.5 $MgCl_2$–6.H_2O, 0.41 $MgSO_4$–7.H_2O, 4 $NaHCO_3$, 0.3 Na_2HPO_4, 5.6 D-glucose and 1.26 $CaCl_2$. This and all other solutions are prepared with 'HPLC' grade water and filtered under a sterile tissue culture hood.

Prior to further fine dissection, the embryos are transferred to fresh HBSS and cleaned of excess tissue and blood. Each fetus is removed from the embryonic sac by cutting a longitudinal slit and exposing the amniotic sac. The amniotic sac is then punctured and the fetus removed into a new dish with cold HBSS. The fetus is released by teasing it gently out through the opening in the embryonic sac. The backs of each fetus are isolated with three cuts: one to decapitate the fetus, one to cut the hind limbs and tail and one to trim the internal organs ventral to the spinal cord (Fig. 1a).

The next step is to slice the tissue. Because this must be done within ~5 min, we use a tissue chopper (Sect. 2.1). All surfaces of the tissue chopper are sterilized with 70% ethanol. To produce transverse slices, the spinal cord is placed dorsal side up on a sterilized plastic cover (vinyl cut to the size of the tissue chopper stage and sterilized in 70% ethanol) on the chopping stage of the tissue chopper, perpendicular to the chopper blade. Any excess solution from around the tissue is removed to prevent it from sliding around the stage. However, tissues must remain slightly moist to prevent them from sticking to the chopper's blade. Also, the spinal cord must lie flat on the cutting stage. Several spinal cords can be chopped on the same plastic vinyl cover. Each spinal cord is

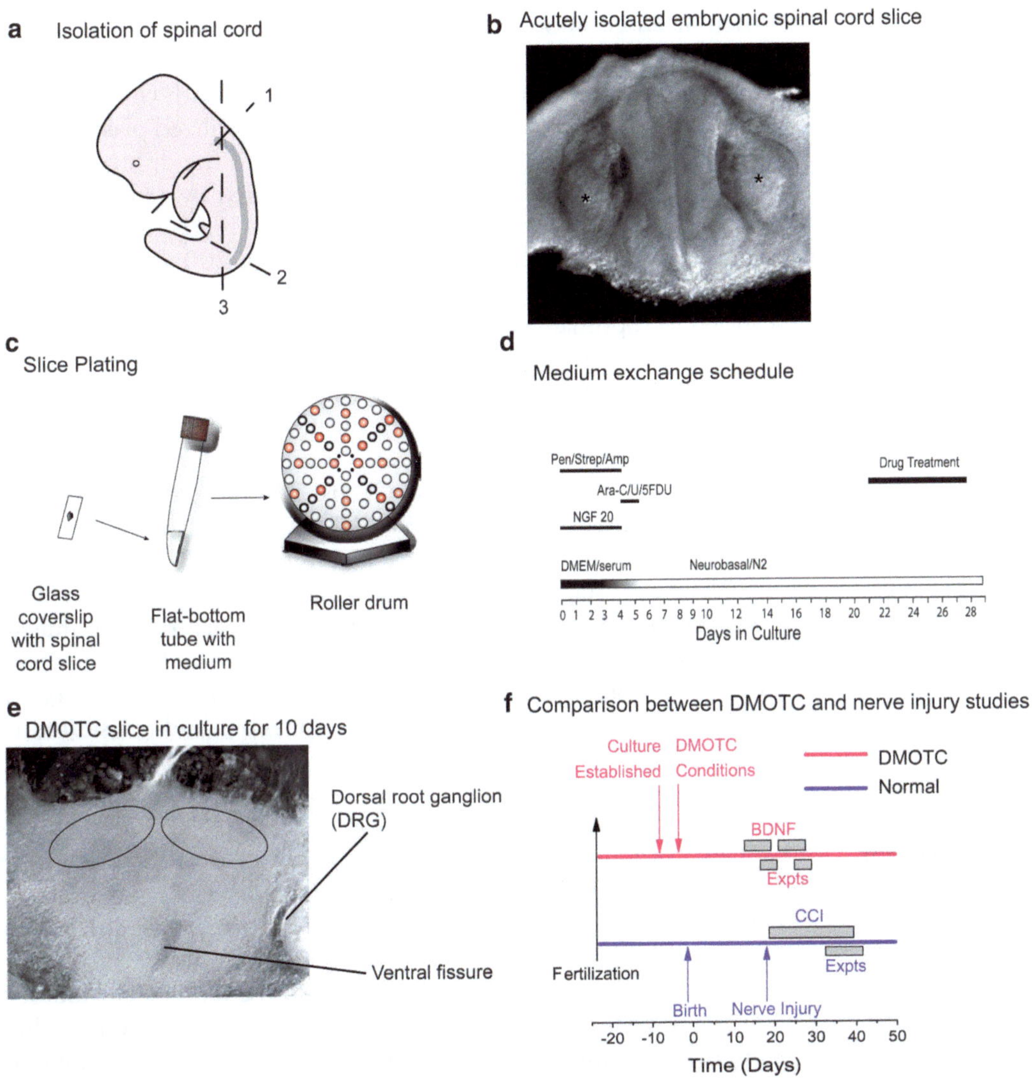

Fig. 1. *Preparation and maintenance of defined medium organotypic cultures (DMOTC) of rat spinal cord.* (**a**) Illustration of the cuts required to isolate the fetal spinal cords. The first cut is to decapitate the fetus, the second to remove the tail and hind limbs and the third to trim the internal organs ventral to the spinal cord, which is drawn as a *thick grey line*. (**b**) An acutely isolated 300 μm thin transverse slice of the embryonic spinal cord; dorsal side is along the *top* of the picture. Asterisks indicate the dorsal root ganglia at either side of the cord. (**c**) Slice plating procedure. The glass coverslip with a spinal cord attached is placed into a flat-sided tube filled with serum medium and tightly sealed before placing on a drum at 120 rotations per hour in a dry heat (36°C) incubator. (**d**) Medium exchange schedule for DMOTC slices. Day 0 marks the start of cultures. Antibiotics (Pen/Strep/Amp) and nerve growth factor (NGF, 20 ng/ml) are present in the DMEM (Dulbecco's Modified Eagle's Medium)/serum medium for the first 4 days. Antimitotics (AraC/U/5FDU) are applied for 24 h to prevent overgrowth of glial cells. After 4 days in culture, the serum medium is progressively switched to the serum-free medium. First, the serum medium is diluted to 50%, and then it is completely switched to a serum- and neurotrophin-free medium. DMOTC slices are treated with drugs (e.g. gabapentin or brain-derived neurotrophic factor [BDNF]) after 2 weeks in culture. (**e**) A DMOTC slice after 10 days in vitro. Dorsal side is along the *top* of the picture, and the *black circles* on the dorsal half indicate where the recordings were obtained from. The dorsal root ganglion and ventral fissure are labelled. (**f**) Timeline comparison of nerve injury studies and DMOTC studies in rats (Panels **b–d** are reproduced with kind permission from (11)).

then sliced into nominally 300 μm thick (effectively 275–325 μm) transverse slices. Each spinal cord yields ~30 slices. Once the tissue is cut, the slices are transferred to a fresh dish with chilled HBSS. The easiest way to do this is to pick up the sterilized plastic covering with the tissue attached and invert it into the dish containing the HBSS. The slices can usually be dislodged by gently bending the plastic covering; if necessary, they can be gently rinsed off with HBSS. Principally, slices can be cut between 200 and 700 μm to suit individual needs. We prefer a thickness of 300 μm as this allows for optimal visualization of individual neurons during electrophysiological and Ca^{2+}-imaging experiments. Slices thinner than 300 μm are more fragile and may eventually contain fewer viable neurons, whereas thicker slices are easier to handle, but adequate diffusion of oxygen and nutrients to the centre of the slice may be restricted (see also Chaps. 1-Trapp and Ballanyi (14), 2-Ruangkittisakul et al. (15), 10-Luhmann and Kilb (27), 11-Kantor et al. (16) and 12-De Curtis et al. (17)).

The next stage is to select slices for culturing. This step is very important as it allows the quality of the sections to be examined. For our purposes, only lumbar spinal cord slices with two attached dorsal root ganglia (DRGs) are chosen (Fig. 1b). Excess ventral tissue is trimmed, using as few cuts as possible, and transferred to fresh HBSS and chilled for 1 h at 4°C to allow recovery from the slicing procedure.

3.1.2. Slice Mounting

The slices are mounted on pre-cleaned glass coverslips which were soaked for 24 h in 0.5 M HCl and then thoroughly rinsed in distilled water prior to placing them in a beaker of 100% ethanol for 1 h. Coverslips are thoroughly rinsed again with distilled water and dried in an oven prior to sterilization at 200°C for 2 h. The method of attachment must allow the slice to adhere to the coverslip during the culturing procedure without impediment of adequate oxygen and nutrient diffusion to the slice. We use a plasma clot to attach slices to the coverslips as this seems more effective than using collagen-coated coverslips. We have obtained consistently good results from chicken plasma obtained from Sigma (catalogue # P-3266). The plasma is obtained lyophilized and reconstituted in HBSS according to the manufacturer's instructions and then centrifuged at 3,500 rpm at 4°C for 30 min to separate fats. The thrombin (Sigma, catalogue #T-4648) is also reconstituted in HBSS and used at a concentration of 200 units/ml. Coverslips are plated in sterile disposable Petri dishes (we use the lids of the dishes as this makes it easier to remove the coverslips when required), and two drops, one of plasma (20 μl) and one of thrombin (30 μl), are added onto opposite ends of each coverslip. One slice of spinal cord tissue is added to each drop of plasma, and the thrombin is then mixed well with the plasma. The slices should be positioned close to the centre of each coverslip. The excess plasma clot is then spread over the entire coverslip and removed with the remainder left to coagulate for an hour at room temperature.

Curved forceps are used to place the coverslip in a flat-bottomed culture tube (Nunc. 156–758). The tube contains 1 ml of culture medium, composed of 82% Dulbecco's Modified Eagle Medium (DMEM; high glucose, Na pyruvate; 'Glutamax'), 10% fetal bovine serum and 8% sterile water to adjust osmolarity (all from Gibco (Invitrogen, Carlsbad, CA, USA; invitrogen.com)). The medium is supplemented with 20 ng/ml NGF (Alomone Labs) for the first 4 days. Antibacterial and antimycotic drugs (5 units/ml penicillin G, 5 units/ml streptomycin and 12.5 ng/ml amphotericin B: all from Gibco/Invitrogen (Grand Island, NY, USA; www.invitrogen.com)) are also included in the medium during the first 4 days of culture. The coverslip should be completely immersed in the medium. It is sometimes necessary for the coverslip to be gently pushed to the bottom of the tube, by gently tapping the tube against the surface or pushing the coverslip to the bottom with a spatula. Culture tubes are then placed in a roller drum, set to 120 rotations per hour, in an incubator at 36°C (Fig. 1c). The volume of the medium means that the cultures will be immersed in medium during half a rotation in the roller drum and be out of the medium during the other half rotation, simulating oxygenation and washing away cell debris on the surface of the slice. Carbogenation is not used at any stage during the preparation of cultures, and slices in the culture tube/roller drum are incubated in normal air. Also, since the tubes are sealed shut, there is no need to incubate slices in a humidified chamber.

By comparison with organotypic cultures from other brain regions which sometimes form a monolayer of cells (8, 10), spinal cord slices thin only slightly in DMOTC. The thickness of six slices at their centres was ~300 μm which corresponds to the thickness of freshly cut slices. Although there was obvious thinning of the cultures towards their rims, landmarks, such as the DRG and ventral fissure, are preserved (Fig. 1e). These cytoarchitectural landmarks are still visible and recognizable up to 5 weeks in culture (11) and are used to orientate the slices, e.g. for identifying regions of interest (ROIs) in Ca^{2+}-imaging experiments (Sect. 3.6).

3.1.3. Defined Medium Protocol

Fetal bovine serum (FBS) is useful for buffering the pH of culture medium and also provides non-specific trophic support for the survival of the cells. However, the undefined factors and mediators present in FBS can confound the interpretation of effects induced by long-term application of neurotrophins, pain mediators or other drug treatment. For this same reason, we gradually remove NGF and FBS from the medium, such that the cultures are grown and maintained in a defined, neurotrophin- and serum-free medium (see below and Fig. 1d for details). After 4 days in culture, the serum medium is progressively switched to the defined, neurotrophin- and serum-free medium consisting of Neurobasal medium with N-2 supplement and 5 mM Glutamax-1 (all from Gibco). The serum medium is first diluted 50:50 with the NGF concentration reduced from 20 ng/ml to 2.5 ng/ml. During this time period, the DMOTC slices are also treated with a 10 μM antimitotic drug cocktail consisting of uridine, cytosine-b-D-arabinofuranoside (AraC) and 5-fluorodeoxyurindine (all from Sigma), to prevent the overgrowth of glial cells. After 24 h, the medium is fully exchanged to the defined medium, i.e. with no serum or NGF added, DMOTC slices are maintained in this medium with fresh medium exchanges taking place every 3–4 days (Fig. 1d). Medium is freshly prepared prior to use and warmed up to 37°C before the exchange. The old medium from flat-bottom tubes is completely aspirated and then replaced with 1 ml of new medium. Only ten tubes are exchanged at a time to prevent excess cooling of the slices while removed from the incubator. All media preparations and exchanges take place in a tissue culture hood, and proper sterile and aseptic techniques are used to minimize contamination of cultures. To check for possible growth of contaminating organisms, we place 1 ml of medium in an empty, control culture tube in the incubator along with the DMOTC.

3.2. Treatment of DMOTC Slices with Putative Pain Mediators

3.2.1. BDNF

While DMOTC slices can be maintained successfully in culture for >6 weeks, we typically maintain them for 3–4 weeks prior to experiments, allowing them to stabilize and mature to an age compatible with previous ex vivo experiments in our laboratory (Fig. 1f) (5, 20, 28). Because of the sustained viability of the spinal DMOTC slices, we were able to investigate the long-term ('chronic') effects of neurotrophins, transport inhibitors and drugs on dorsal horn neurons in this model (5, 13, 19, 26). Our initial studies were performed to investigate the effect of long-term BDNF treatment on dorsal horn neurons (Sect. 1). Consequently, the treatment protocol was chosen to parallel previous whole animal studies (20) so that exposure to BDNF would match the time course of BDNF elevation following nerve injury. For this, the DMOTC slices were treated after 15–21 days in vitro for a time period of 5–6 days with 50–200 ng/ml BDNF (Alomone) in the serum-free medium described previously (Fig. 1f). The BDNF medium was then exchanged on the third treatment day. Age-matched naïve DMOTC slices served as controls.

We have also used DMOTC slices to examine the long-term actions of the anti-allodynic drugs, gabapentin and pregabalin (26). The treatment protocol is the same as that for BDNF, with gabapentin (Tocris) being applied to DMOTC slices for 5–6 days and medium being exchanged on the third day.

3.2.2. Activated Microglia-Conditioned Medium and Pain in a Dish

Nerve injury alters the state of activation of spinal microglia. This prompts them to release a plethora of inflammatory mediators that activate and sensitize spinal dorsal horn neurons (4, 29–31).

Microglia can be isolated in primary culture and be activated with lipopolysaccharide (LPS), an antigenic component of bacterial cell walls. Subsequently, the aMCM (Sect. 1) is harvested and applied to the DMOTC slices. Since aMCM likely contains the inflammatory mediators that are responsible for dorsal horn sensitization, the procedure effectively puts pain in a dish (Sect. 1). Microglial cells are isolated using a method adapted from Siao and Tsirka (32). In brief, brains of postnatal day 1 Sprague-Dawley rats are dissected in HBSS without divalent cations to prepare mixed glial cultures. Brains are dissociated by 15 min enzymatic digestion with 0.25% trypsin–ethylene-diamine-tetraacetic acid (EDTA) followed by mechanical titrations in DMEM/F-12 containing 10% FBS. Poly-L-lysine-coated 100-mm dishes are used to plate out the mixed glia in DMEM/F-12 with 10% FBS at a density of 8×10^6 cells per dish. After 8–14 days of culture, 15 mM lidocaine is added for 5 min to inhibit adhesion molecules. Cultures are shaken on a rotator at 100 r.p.m. for 30 min. The medium is collected and centrifuged at 2000*g* for 5 min. The supernatant is discarded and the pellet resuspended in DMEM and plated into 24-well tissue culture plates. After 45-min incubation, microglial cells are washed with DMEM twice to remove non-adhering cells. After 24-h incubation, half of the wells containing microglia cells are activated with 100 ng/ml LPS for 6 h, the other half of the wells receive phosphate-buffered saline (PBS). LPS is removed after 6 h with 3 PBS washes and then replaced with the defined, neurotrophin- and serum-free culture medium. After 24 h, medium is harvested and used to culture 14–21 days in vitro DMOTCs, for an additional 5–6 days with the medium being exchanged for fresh activated microglia-conditioned medium on the third day. Medium from non-activated microglia serves as control.

3.3. Electrophysiology of Dorsal Horn Neurons

A glass coverslip with an attached DMOTC slice is cut using a diamond-tipped pen, in order to fit it into the recording chamber. To prevent detachment from the coverslip, the slice is held down with a 'harp' made of strands of nylon attached to platinum wire (see Chap. 1-Trapp and Ballanyi (14)). The recording chamber is continually perfused at a flow rate of 1 ml/min with carbogenated ACSF containing (in mM) 127 NaCl, 2.5 KCl, 1.2 NaH_2PO_4, 1.3 $MgSO_4$, 26 $NaHCO_3$, 25 D-glucose and 2.5 $CaCl_2$. Patch pipettes are pulled from thin-walled borosilicate glass (Sect. 2.2) to a tip diameter resulting in 5–10 MΩ DC resistances when filled with an internal recording solution containing (in mM) 130 K-gluconate, 1 $MgCl_2$, 2 $CaCl_2$, 10 4-(2-hydroxyethyl)-1-piperazineethanesulfonic acid (HEPES), 10 ethylene glycol tetraacetic acid (EGTA), 4 Mg-ATP and 0.3 Na-GTP (pH 7.2 and 280–310 mOsm). We use an SEC-05LX amplifier in bridge balance or in discontinuous, single electrode, current or voltage-clamp mode for V_m or I_m recording (Sect. 2.2).

DMOTC spinal cord slices are positioned under visible light using the 10× objective of the upright microscope before viewing dorsal horn neurons under IR-DIC optics using a 40× water immersion objective, and a CCD camera is used to capture the image of neurons which is viewed on a video monitor (Sect. 2.2). To orient the slices, in order to identify the dorsal half of the spinal cord, markers such as the ventral fissure and positions of the DRG are used as they are usually easily identifiable (Fig. 1e). In some experiments, synaptic responses are evoked by stimulation of the dorsal root entry zone or the dorsal root ganglia with bipolar tungsten electrodes (Sect. 2.2). Following the perturbation of the slices associated with electrode positioning, they are allowed to recover for ~5 min prior to initiating stimulation protocols.

Neurons for recording are selected from the dorsal half of the spinal cord. We assume that neurons encountered between 0.5 and 2 mm from the edge of the slice lie in a region that is analogous to the *substantia gelatinosa* area in which central sensitization for chronic pain occurs (Sect. 1) (Fig. 1e). Healthy neurons possess a pear-shaped or fusiform cell body with a flat smooth appearance, while unhealthy neurons look swollen and rounded with a 'wrinkled' appearance. In acute slices from juvenile rats, neurons can be visualized, using IR-DIC optics, to a depth of up to 100 μm, whereas this distance typically decreases to 40–50 μm in >28-day-old animals. In DMOTC spinal cord slices, neuronal visibility is generally better with excellent visualization of mature neurons to depths of 100 μm or more. Upon finding a viable neuron, the patch pipette electrode is lowered until it is just above that cell. Weak positive pressure is applied via a syringe to clear surrounding tissue and prevent the electrode from becoming blocked while it is lowered into the tissue. We find that injecting ~0.2 ml of air from the syringe into the sealed electrode holder - electrode system produces an appropriate amount of positive pressure. When pressure is released, a small portion of the cell membrane is 'sucked' into the electrode. Optimal 'gigaohm-seal' formation is achieved with the amplifier in voltage-clamp mode and by applying a negative voltage (−70 mV) command to the electrode. Further gentle negative pressure is applied via suction to achieve a seal resistance of >1GΩ between the cell and electrode. We find that suction via mouth gives far better control than can be achieved via a syringe. The cell membrane patch is then ruptured by applying quick bursts of negative pressure, thus achieving intracellular access.

3.4. Morphology of Dorsal Horn Neurons

We routinely fill neurons in either acute or DMOTC spinal cord slices with biocytin (0.2%) during electrophysiological recordings for post hoc identification of cellular morphology.

Biocytin is added to the patch pipette internal solution allowing easy access to the inside of the neuron. A low concentration is used as biocytin can interfere with postsynaptic drug responses at

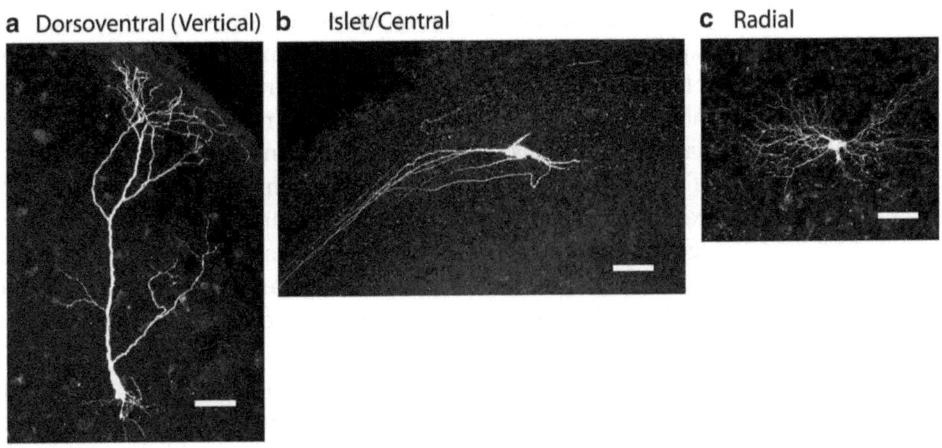

Fig. 2. *Confocal images of DMOTC neurons*. Images include (**a**) dorsoventral, (**b**) islet/central and (**c**) radial cells from cultured organotypic slices that were filled with biocytin via the whole-cell-recording patch electrode. Dorsal side is the *top edge* of pictures. Scale bar is 50 μm (V.B. Lu and P.A. Smith unpublished observations).

higher concentrations (33). At the completion of recording, the patch pipette is slowly withdrawn from the cell and the slice transferred to cold (4°C) 4% paraformaldehyde in PBS, pH 7.4, and stored overnight at 4°C. Slices are rinsed three times with PBS and transferred to a 24-well tissue culture plate for staining. Slices are incubated in 0.3% Triton X-100 and streptavidin–Texas Red conjugate (1:50 dilution, Molecular Probes) for 50 min on a 3D-rotator Model 4630 (Labline Instruments Inc., Melrose Park, IL, USA or Mumbai, India; www.labline.in). Slices are thoroughly rinsed with distilled water, transferred to slides, allowed to dry for 5 min and coverslipped with Cytoseal-60 (Richard-Allan Scientific, Kalamazoo, MI, USA; www.thermoscientific.com). A confocal laser scanning microscope is used to examine the tissue and generate 3D reconstructions (Sect. 2.4).

Although the rostrocaudal spread of dendrites is reduced by the plane of spinal cord transection, a variety of neuronal morphologies are observed in DMOTC (5), and these correspond to those previously described in intact dorsal horn (34, 35). Thus, islet/central neurons, which have a rostrocaudal orientation; radial neurons that send processes in all directions; and vertical neurons, which have a dorsoventral orientation can be identified in spinal cord DMOTC (Fig. 2a–c). This supports the idea that the processes that determine morphological phenotype seem to develop normally in these cultures (7). Moreover, the relationship between electrophysiological and morphological features of dorsal horn neurons is preserved. For example, islet/central cells in both acute and DMOTC slices often display a 'tonic' discharge pattern, whereas neurons with a radial morphology often display a 'delay' discharge pattern (5).

We are currently using immunohistochemical procedures to explore the relationship between morphology and neurotransmitter phenotype in the cultures. Preliminary data identify the glutamate vesicular transporter (VGlut) in radial neurons and glutamic acid decarboxylase (GAD) in islet/central neurons. The presence of VGlut in radial neurons is consistent with their putative role as excitatory interneurons, and the presence of GAD in islet/central neurons is consistent with their putative inhibitory phenotype. These relationships between neurotransmitter content and morphological phenotype are similar to those seen in vivo (36).

3.5. Microglia Activation in Acute Versus Organotypic Slices

Microglial cells, as part of the immune system, are the resident macrophages of the brain and spinal cord. They can be activated by exposure to LPS (Sect. 3.2.2). Given their established role as detectors of injury or infection (37, 38) and our procedure to apply aMCM in order to mimic the onset of neuropathic pain (Sect. 3.2.2), we compared untreated cultures with acute slices investigated to look for evidence of microglial activation. We used the microglia-specific Ca^{2+}-binding protein, Iba1 as a marker (39). Although this protein is present in resting microglia, it is upregulated when they are activated (38).

Unless otherwise mentioned, all steps for microglial cell marking are carried out at room temperature, and there are three 'washes' with PBS between each step. Acutely dissected spinal cords from 20- to 40-day-old Sprague-Dawley rats (for detailed protocol see (20)) and DMOTC slices are fixed in 10% formalin solution for 10 min. Acute slices are fixed ~80 min after removal from the animal and subject to the normal isolation and stabilization processes used in our whole-cell recording protocol. DMOTC slices are fixed after 4 weeks in culture. Fixed slices are then blocked in 10% horse serum diluted in 0.25% PBS-Triton for 1 h, then incubated with rabbit anti-Iba1 (1:1,000 in 1% horse serum and 0.25% PBS-Triton) overnight in a cold room. They are then incubated with biotinylated rabbit anti-IgG (1:200, in 1% horse serum-PBS) for 1 h, followed by incubation with the ABC-peroxidase solution (1:200 in 1% horse serum-PBS) for 1 h. Finally, the labelling is visualized by incubation with diaminobenzidine for 1 h.

Figure 3 illustrates photomicrographs from an acute (Fig. 3a) and a DMOTC (Fig. 3b) slice labelled with the microglial marker Iba-1. The acutely isolated spinal cord slice displays numerous rod-shaped, amoeboid and round cells positive for Iba-1 indicating microglia/macrophages at various stages of activation. By contrast, the organotypic section displayed shows very few Iba-1-positive cells suggesting little microglial/macrophage activation. Thus, in DMOTC slices, there will be little interference from inflammatory mediators released from activated microglia,

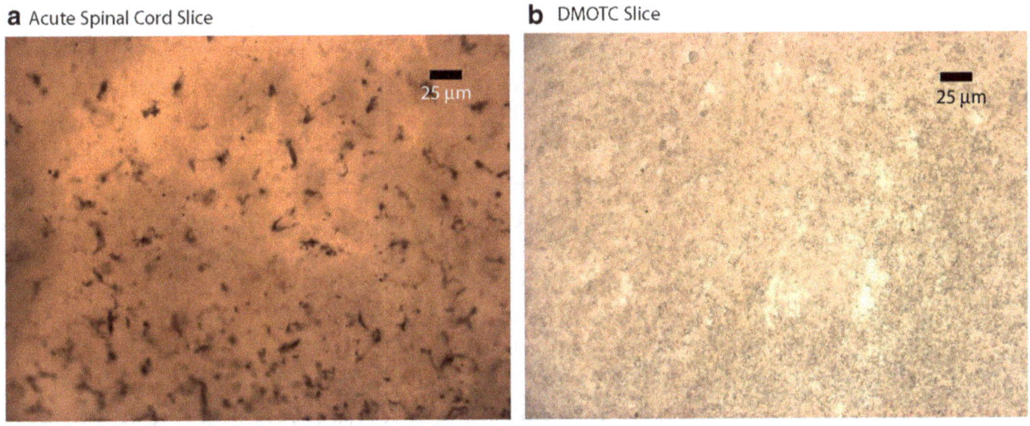

Fig. 3. *Confocal images of Iba-1-labelled DMOTC and acute spinal cord slices.* (**a**) Immunoreactivity to the microglial marker Iba-1 (Ionized calcium-binding adapter molecule 1) in an acutely dissected spinal cord slice from a 20-day-old rat. (**b**) Iba-1 immunoreactivity in a DMOTC spinal cord after 4 weeks in culture (Reproduced with kind permission from (19)).

when compared to acutely dissected spinal cords. This is another clear advantage, in addition to long-term exposure to mediators, of DMOTC slices over acutely isolated slices, where responses obtained from neurons could be affected by microglial activation as a consequence of cell damage associated with the slicing procedure.

3.6. Ca^{2+} Imaging in Dorsal Horn of DMOTC Slices

Dynamic imaging of cytosolic Ca^{2+} takes advantage of Ca^{2+}-sensitive fluorescent indicators which can be loaded into neurons in vitro. Fluorescent dyes can be delivered directly into cells through an intracellular pipette. Alternatively, lipophilic membrane-permeant acetoxymethyl (AM) versions of indicators can be applied extracellularly. Once inside the cell, intracellular esterases free the carboxyl group, unmasking the AM-type Ca^{2+} indicator and enabling it to bind Ca^{2+}.

Since fluorescence intensity of the Ca^{2+} indicators is altered by the binding to Ca^{2+} ions, they can be used to provide an index of the free cytosolic Ca^{2+} concentration. Monitoring changes in fluorescence intensity therefore allows for analysis of Ca^{2+} dynamics, although the kinetics may be distorted based on the binding and dissociation properties of the specific dye used (40). Many fluorescent Ca^{2+} indicators are based on the EGTA homologue, 1,2-bis (o-aminophenoxy)ethane-N,N,N′,N′-tetraacetic acid (BAPTA), which displays high selectivity for Ca^{2+} ions versus Mg^{2+} ions. Other indicators include Fura-2, calcium-crimson and Fluo-4. For more details, the reader is directed to recent reviews (41) which discusses common Ca^{2+} indicators and their applications in intracellular Ca^{2+} imaging (see also Chaps. 1-Trapp and Ballanyi (14), 2-Ruangkittisakul et al. (15), 11-Kantor et al. (16) and 12-De Curtis et al. (17)).

We utilize the Ca^{2+} indicator dye Fluo-4-AM (TEF Labs Inc., Austin, Texas, USA; www.teflabs.com) to investigate the effects of various pain mediators on Ca^{2+} in individual presumptive neurons in spinal cord DMOTC. It is important to note that although the exact intracellular Ca^{2+} concentration cannot be determined, the relative increase in Fluo-4 fluorescence intensity still provides a good measure of relative changes of Ca^{2+} concentration in response to various perturbations (42). Fluo-4-AM is dissolved in a mixture of dimethyl sulfoxide and 20% pluronic acid to a 5 mM stock solution, further diluted to 0.5 mM with ACSF, and kept frozen until used. Note that the dye is light-sensitive and must thus be kept in the dark at all times. Before each use, the dye is thoroughly mixed by vortexing and sonicating; this ensures that the dye is dissolved.

Up until now, our protocol has involved incubation of a single DMOTC slice for 1 h with 10 μl of Fluo-4-AM stock solution resulting in a final concentration of 5 μM prior to imaging. The incubation conditions for the dye are kept constant across different slices to avoid uneven dye loading. In the rare cases where DMOTC slices are weakly fluorescent or highly fluorescent after the dye-loading procedure, they are discarded. We have found recently however, that localized injection of Fluo-4-AM produces better dye loading than can be obtained with incubation. Because the dye enters the slice only by about 40–50 μm upon bath-application, this limits recording plane within a narrow depth from slice surface. However, the injected dye enters deeper into the tissue and will thus likely produce a larger area of loaded cells in the focal plane. Our most recent approach has involved backfilling a patch pipette (tip diameter 5–10 μm) with 0.5 mM Fluo-4-AM and pressure injecting (0.7 psi, 10 min) at 50 μM below the surface of the slice. This approach is identical with that used by the Ballanyi laboratory to load respiratory neurons in brainstem slices (40) (see also Chap. 2-Ruangkittisakul et al. (15)).

Resting fluorescence depends on both the extent of dye loading and the amount of laser power applied. Our data were sampled from cells exhibiting 200–400 arbitrary fluorescent units at rest using 1–2% of maximum power of the argon laser (Sect. 2.3). After dye loading, the slice is transferred to a recording chamber and perfused with ACSF (20°C, flow rate 4 ml/min). Note that the slice is prevented from detaching from the coverslip by using the 'harp' used for electrophysiological studies (Sect. 3.3), and the dorsal horn is identified using the structural markers as described below. Changes in the Ca^{2+} fluorescence intensity in dorsal horn neurons to stimuli are measured with a confocal microscope (Sect. 2.3) that acquires full frame images (512 × 512 pixels) in a fixed xy plane acquired at a scanning time of 1.08 s/frame (40). In some experiments, images are 'cropped' in the y-axis to accommodate faster scan rates.

We use three different methods to stimulate the DMOTC spinal cord slices: (*i*) bath-application of high K^+ solution (20, 35 or 50 mM) – we typically use a 90-s application as this gives reliable results; (*ii*) repetitive electrical stimulation of the dorsal root entry zone or dorsal root ganglia; (*iii*) bath-application of pharmacological agents, such as glutamate, or agonists of ionotropic glutamate receptors such as N-methyl-D-aspartate (NMDA) or α-amino-3-hydroxy-5-methyl-4-isoxazolepropionic acid (AMPA). Selected ROIs are then drawn on the computer screen around cell bodies of presumptive neurons and fluorescence intensity traces are generated with FluoView software (Sects. 2.4 and 3.1.2).

We have assumed that measurement of changes in Ca^{2+}-induced fluorescence in response to stimulation serves as an index of overall dorsal horn excitability (5, 13, 19). Thus, a net increase in excitatory synaptic transmission and/or a net decrease in inhibitory transmission and/or an increase in intrinsic neuronal firing properties would be expected to increase the stimulation-induced increase in Ca^{2+}-dependent fluorescence intensity.

4. Results

4.1. Dorsal Horn Neuron Types

Organotypic cultures are sometimes criticized on the grounds that the developmental processes operating therein may differ from those occurring in vivo. This could lead to changes in neuronal phenotype in the long-term cultures (43). Therefore, the electrophysiological properties of dorsal horn neurons in DMOTC slices were compared with those found in acutely isolated spinal cord slices (5, 11). Neurons were characterized by their action potential firing pattern in response to a series of 2 pA incremental depolarizing DC current steps from a preset membrane potential of −60 mV. In our most recent study of acute slices (5), we have characterized six types of neurons on the basis of their firing pattern as tonic, delay irregular, delay tonic, irregular, phasic and transient (Fig. 4a). Similar populations of neurons are found in spinal cord DMOTC slices except that no delay irregular firing patterns are observed (Fig. 4b). One possible explanation for this difference is that action potentials observed in delay irregular neurons in acute slices reflect passive invasion of spikes into the somata. This is possible because the cell bodies of some dorsal horn neurons have been reported to lack Na^+ channels (44). By contrast in culture, slight alterations in neuronal morphology may allow for improved invasion of the soma. Thus, for comparison of the proportion of cell types in acute and DMOTC slices (Fig. 4c), all delay neurons in acute slices were placed in one category. When this was done, no significant differences were observed in the

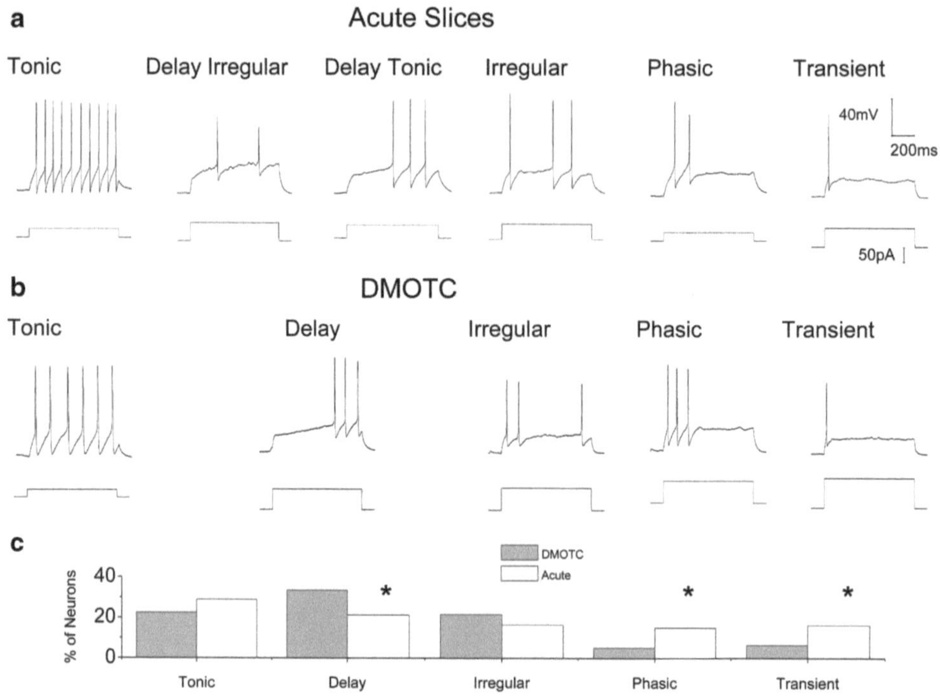

Fig. 4. *Comparison of neuron types in acutely isolated spinal cord slices with those in DMOTC.* (**a**) Illustrations of firing patterns of tonic, delay irregular, delay tonic, irregular, phasic and transient neurons in acute slices. *Lower traces*=current, *Upper traces*=voltage. Calibrations refer to all panels in **a** and **b**. (**b**) Illustrations of firing patterns of tonic, delay, irregular, phasic and transient neurons in DMOTC slices. (**c**) Bar graphs comparing percentage of neurons exhibiting each firing pattern in acute slices and DMOTC. For acute slices, numbers of delay irregular and delay tonic neurons were pooled in the tonic category. Data from 620 neurons in acute slices and 134 in DMOTC. * $P<0.01$ (χ^2 test) (S. Balasubramanyan, V.B. Lu, Y. Chen and P.A. Smith unpublished observations).

proportions of tonic and irregular neurons, compared to acute slices. By contrast, there are fewer phasic and transient neurons in the cultures and fewer delay neurons in the acute slices. Despite these qualitative differences, the processes that determine the electrophysiological phenotype seem to function normally in the DMOTC spinal cord slices.

A full description of the neural phenotypes found in the superficial dorsal horn, their connections and physiological properties is outside the scope of this review. For additional information, the reader is referred to a recent review (45).

4.2. Ca^{2+} Imaging of Dorsal Horn Excitability

4.2.1. Responses to High K^+

As already mentioned (Sect. 1), activation of spinal microglia is one of the early consequences of peripheral nerve injury and likely leads to the central sensitization underlying neuropathic pain (4, 22, 23, 37, 46–48). Consequently, we exposed DMOTC spinal cord slices to aMCM (Sects. 1 and 3.2.2.) and investigated its effects on dorsal horn excitability. In these experiments, we measured the increase in Ca^{2+}-dependent fluorescence intensity in response to exposure to solutions containing 20, 35 or 50 mM K^+. Typical images are shown in Fig. 14.5a–c. It is important to note that in our experiments, dye saturation is unlikely to occur as there is a progressive increase in the Ca^{2+} signal amplitude at higher K^+ concentrations (Fig. 5d). Generally, there are between 4 and 20 responding cells and an average of 10 responding cells in each field of view on the slice. Responses reflect likely primarily depolarization-induced opening of voltage-gated Ca^{2+} channels as well as the release of glutamate as they are attenuated but not completely eliminated by a mixture of the blockers of NMDA-type and kainate/AMPA-type ionotropic glutamate receptors ((2R)-amino-5-phosphonovaleric acid (AP-5, 10 μM) and 6-cyano-7-nitroquinoxaline-2,3-dione (CNQX, 10 μM), respectively). Conversely, these responses are augmented upon blockade of the glutamate transporter EAAT2 with 100 μM dihydrokainate.

The amplitude and area under the curve (AUC) of K^+-evoked Ca^{2+} rises in DMOTC spinal cord slices treated with aMCM are significantly larger than compared to controls (Fig. 5d–i) (see also (5)). This pain in a dish system provides an excellent opportunity for examining which molecules, such as BDNF (Sect. 1), may be responsible for central sensitization. We have therefore examined BDNF actions (5, 12, 13) and found a larger K^+-evoked Ca^{2+} rise in BDNF-treated slices compared to untreated controls. To test the possibility that BDNF, which is released from activated microglia and thus present in aMCM, is responsible for the observed increase in Ca^{2+} signal amplitude following high K^+ stimulation, a BDNF-binding protein, TrkB-d5, was applied.

TrkB-d5, which binds and inactivates all neurotrophins that interact with TrkB receptors (49), prevented the increase in excitability produced by aMCM (Fig. 5g–i). This suggests that BDNF is a likely mediator of the excitatory actions of aMCM and strongly implicate it in the overall increase in dorsal horn activity associated with central sensitization (5). The actions of BDNF reflect increased excitatory drive to excitatory neurons and decreased excitatory synaptic drive to inhibitory neurons (5, 13) as well as attenuation of inhibition, likely mediated by both glycine and γ-aminobutyric acid (4)

4.2.2. Responses to Electrical Stimulation

One criticism of the latter approach for testing dorsal horn excitability indirectly via Ca^{2+} imaging is that application of a high K^+ solution is not a physiological stimulus. To address this problem, we used electrical stimulation of the dorsal root entry zone and/or DRGs (Sect. 2.2, 3.1.1 and 3.3). Although electrical stimulation itself is also non-physiological, this technique makes use of the fact that afferent fibres are activated which are located remote to the Ca^{2+}-imaged cells and which causes physiology-like release of neurotransmitter within the network of these cells (5) (see also Chaps. 1-Trapp and Ballanyi (14), 2-Ruangkittisakul et al. (15), 10-Luhmann and Kilb (27), 11-Kantor et al. (16) and 12-De Curtis et al. (17)).

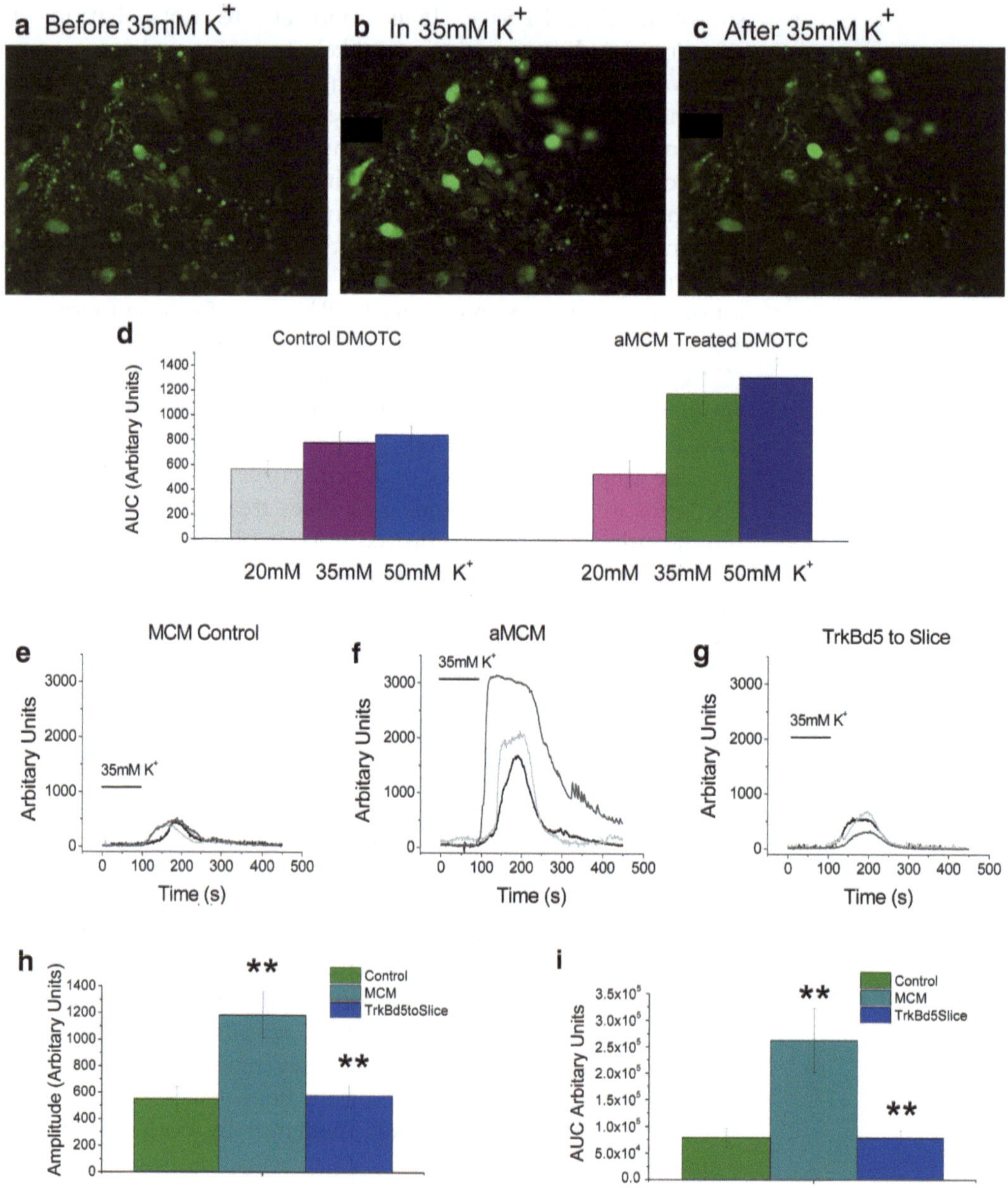

Fig. 5. *Effect of high K^+-induced depolarization on intracellular Ca^{2+} in substantia gelatinosa neurons.* (**a–c**) Images of the same area of culture before, during and after depolarization with a solution containing 35 mM K^+. Note that fluorescence intensity of the Ca^{2+}-sensitive dye Fluo-4-AM increases when the intracellular Ca^{2+} concentration in cells increases. (**d**) Bar graphs to demonstrate concentration dependence of the effect of K^+ on Fluo-4-AM fluorescence in cells from control DMOTC and from those treated with activated microglia-conditioned medium (aMCM). Fluorescence expressed as area under the curve (AUC) in arbitrary fluorescent units. (**e**) Quantification and time course of Ca^{2+} rises in response to 35 mM K^+ recorded from 3 typical neurons in a control DMOTC spinal cord slice. (**f**) Time course of Ca^{2+} increases in response to 35 mM K^+ recorded from 3 typical neurons in a culture treated with aMCM. Note enhancement of responses in presence of this medium. (**g**) As in **f** except that the BDNF-binding protein TrkBd5 was added to the culture. (**h** and **i**) Pooled data from 22 neurons in control slices, 28 neurons from aMCM-treated slices and 28 neurons from aMCM + TrkBd5-treated slices. (**h**) Shows mean ± SEM of peak amplitude responses such as those shown in **e–g**. (**i**) Shows AUC. ** = $P<0.05$ (One-way Anova with Tukey–Kramer multiple comparisons test) (Panels **e–i** were reproduced with kind permission from (5)).

DMOTC are stimulated for 5 s at 50 Hz using a 100 μs single pulse width and at twice the threshold intensity required to produce a robust increase in Ca^{2+} fluorescence intensity. Because the responses to stimulation change with time (5), these trains of stimuli are applied at a regular interval of several minutes to investigate the effects of various substances on overall dorsal horn excitability. Specifically, the first train of stimulus pulses is applied 30 s into the Ca^{2+} imaging with the second applied 5 min later. The protocol is performed four times, once every 10 min, and an average from the third recording, representing the fifth and sixth responses, is taken. This is because the response amplitude changes little during these successive stimuli (Fig. 6a). As for stimulation with high K^+ (Sect. 4.2.1), there are between 4 and 20 responding cells and on average 10 responders in each field of view on the slice. The peak amplitude of stimulation-evoked Ca^{2+} responses were attenuated by AP-5 (50 μM), but the AUC was little changed (Fig. 6d, e). Although further and significant attenuation of both response amplitude and AUC were seen upon addition of CNQX (10 μM), responses were not completely eliminated by that concentration (Fig. 6d–e). This raised the possibility that group-I-type metabotropic glutamate receptors, which signal through G_q, may be involved in the generation of the stimulation-evoked Ca^{2+} response. In support of this hypothesis, we observed almost complete blockade of stimulation-evoked Ca^{2+} responses by combining the mGluR5 receptor antagonist, 2-methyl-6-(phenylethynyl)-pyridine MPEP (10 μM), with AP-5 (50 μM) and CNQX (10 μM). By contrast, the mGluR1 receptor antagonist, LY367385, 50 μM was without effect (H.J. Kim, K. Ballanyi and P.A. Smith, unpublished observations). Thus stimulation-evoked Ca^{2+} responses appear to require the activation of kainate/AMPA, NMDA and mGluR5 receptors.

Figure 6d, e illustrates the effect of 5–6 days treatment of spinal cord DMOTC slices with BDNF. At 200 ng/ml, BDNF significantly increases both the amplitude of the electrically evoked Ca^{2+} rise and the AUC similar to the results with K^+ stimulation (13). These findings indicate that the overall excitability of dorsal horn neurons is increased after such 'chronic' BDNF treatment of the DMOTC slices. This long-term exposure of slices to BDNF mimics the time course of the nerve injury-induced BDNF elevation (50, 51) and provides additional evidence for its role as a mediator of central sensitization.

Since 200 ng/ml of BDNF is about twice the maximally effective concentration used in other studies (52), we also examined effects of 50 ng/ml BDNF. This lower concentration did not significantly increase the area under the curve or the amplitude of the electrically evoked Ca^{2+} signal ($P > 0.05$ two-tailed *T*-test, Fig. 6b, c). We therefore used a standard ELISA (enzyme-linked immunosorbent assay) to measure actual concentrations of BDNF

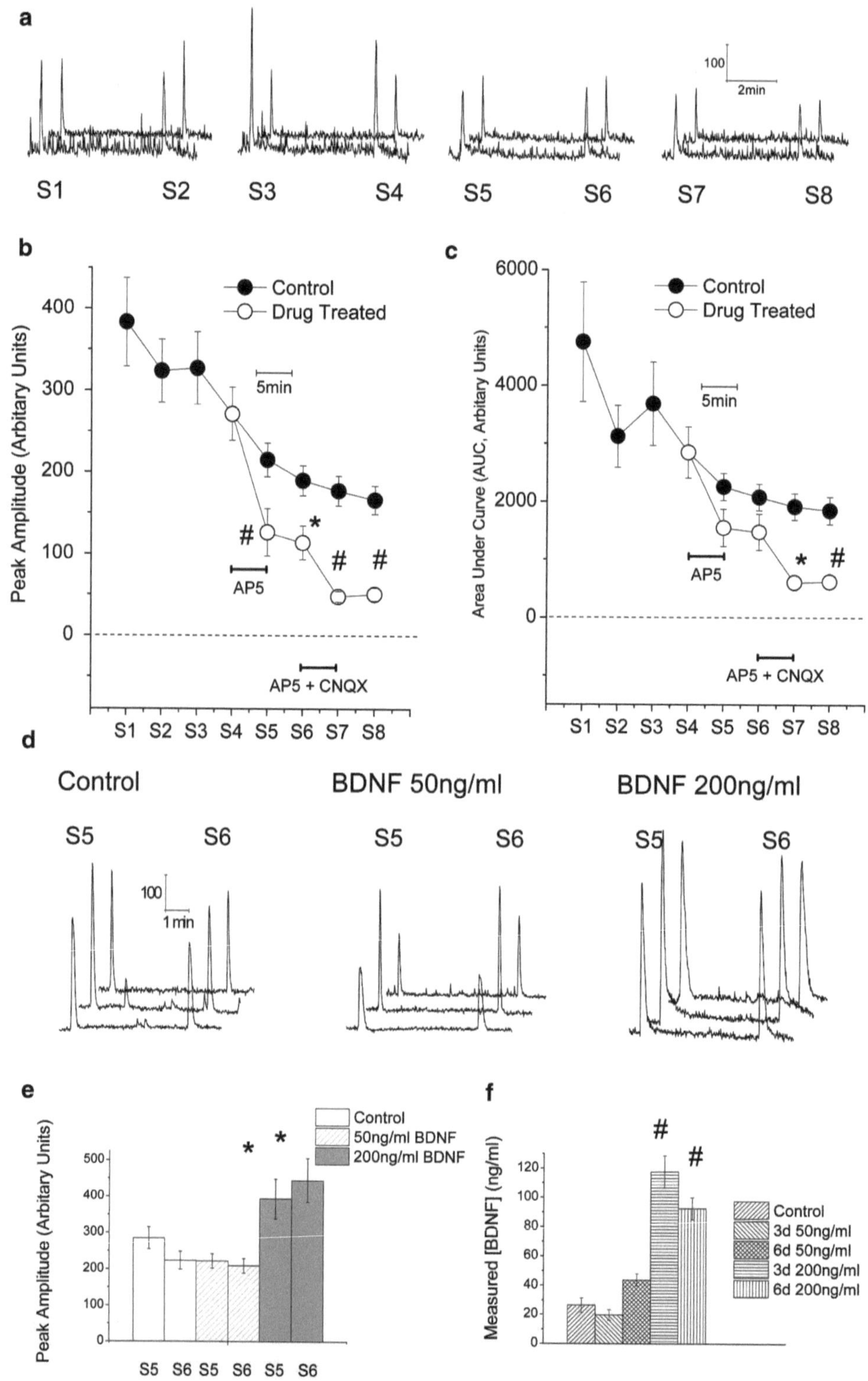

Fig. 6. *Effect of tetanic electrical stimulation on intracellular Ca^{2+} in substantia gelatinosa neurons.* (**a**) Response to a series of tetanic (5 s train at 50 Hz using a 100 μs pulse width) stimuli (S1-S8) delivered at 5 min intervals. Superimposed traces from two neurons. Note that despite the variability, responses S5 and S6 displayed fairly consistent amplitudes. These two responses were used for further comparisons. (**b** and **c**) *Grey circles* show time course of reduction of stimulus evoked Ca^{2+} responses with time (**b** shows peak amplitude and **c** shows area under curve). *Black circles* show additional

in control cultures and how this is altered by the addition of exogenous BDNF. The results are presented in Fig. 6d. We found that control cultures contained 26 ng/ml BDNF and that no significant increase in BDNF levels was seen following inclusion of 50 ng/ml BDNF in the culture medium for 3 or 6 days. This may have reflected break down of the exogenous neurotropin in the cultures and could account for the lack of effect of 50 ng/ml BDNF on the Ca^{2+} signal (Fig. 6d, e). By contrast, 3 or 6 days inclusion of 200 ng/ml in the cultures caused an increase in neurotrophin levels that was significant compared to both control levels and to levels achieved by adding 50 ng/ml exogenous BDNF (ANOVA with Tukey's post hoc test $P < 0.01$, Fig. 6f). The exogenous addition of 200 ng/ml yielded a measured BDNF concentration of 117.5 ng/ml after 3 days and a slightly lower concentration 92.4 ng/ml after 6 days (Fig. 6f). This suggests there was some breakdown or catabolism of the neurotrophin in the cultures, presumably the resting level of 26 ng/ml BDNF in control cultures is maintained by ongoing synthesis (21).

4.2.3. Adenovirus Infection and Expression of Genetically Encoded Proteins

Protein expression in organotypic slice cultures can be manipulated using an adenovirus vector to introduce modified genetic material (53–55). To demonstrate the potential of utilizing viral tools for studying central sensitization, DMOTC spinal cord slices were infected with a viral vector encoding GFP (Green Fluorescent Protein) driven by a CMV (cytomegalovirus) promoter, AdGFP (Fig. 7). Slices were exposed to 30 pfu/cell AdGFP for 4 h, with the AdGFP added directly to the slice in the culture tube and placed back in the incubator for the duration of treatment. The slices were then gently washed in sterile PBS twice before fresh defined medium is added. (Note that 30pfu/cell is an approximation based on the estimate that there are 2.5×10^6 cells in each slice.) The expression of GFP was then monitored over a period of 72 h, by viewing under a fluorescent microscope. The GFP was clearly expressed 24 h after infection with the AdGFP (Fig. 7).

Fig. 6. (continued) reduction in responses seen during and after superfusion of 50 μM AP-5 followed by 10 μM CNQX + 50 μM AP-5. AP-5 produced a significant decrease in the response amplitude, and subsequent application of AP-5 + CNQX produced significant decreases in both response amplitude and area under curve. (* = $P < 0.05$ compared to untreated controls; *grey circles*; # = $P < 0.01$. Student's unpaired *T*-test, n = 40 for controls, n = 30 for neurons form antagonist treated slices.) It should be noted that the combined bath-application of ionotropic glutamate receptor blockers (AP-5 + CNQX) never eliminated the stimulation-evoked Ca^{2+} response. (**d**) Typical S5 and S6 responses sampled from neurons in control slices and in slices treated with 50 or 200 ng/ml BDNF. (**e**) Summary of data from 32 neurons in control slices, 41 neurons from slices treated with 50 ng/ml BDNF and 22 neurons from slices treated with 200 ng/ml BDNF. Note enhanced responses in 200 ng/ml BDNF, * = $P < 0.05$ (One-way Anova with Tukey–Kramer multiple comparisons test). (**f**) Bar graph to show measured concentrations of BDNF in control cultures as well as those maintained in medium containing 50 or 200 ng/ml BDNF for 3 or 6 d. # = $P < 0.01$ compared to control and to 50 ng/ml medium (One-way Anova with Tukey–Kramer multiple comparisons test). Panels **a, d** and **e** modified with permission from (21) (Panels **b** and **c** are J. E. Biggs' unpublished observations; Panel **f**, A. Lai, K.G. Todd and J.E. Biggs' unpublished observations).

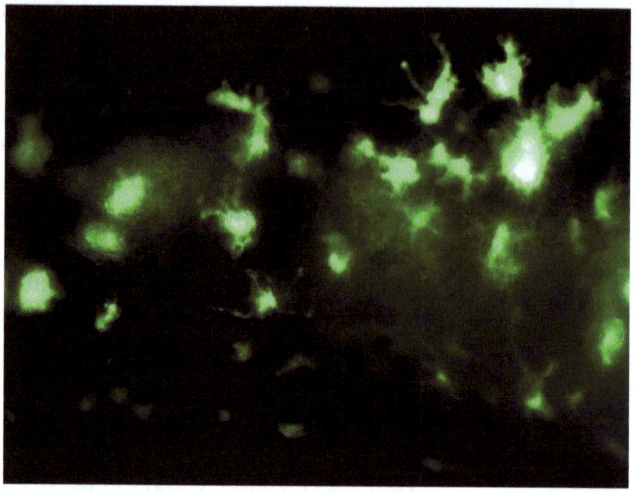

Fig. 7. *DMOTC slice infected with AdGFP virus.* Photomicrograph showing a 3-week-old spinal DMOTC slice 24 h after infection with an AdGFP virus. This adenovirus provides genetic material for the expression of GFP (green fluorescent protein) driven by the cytomegalovirus (CMV) promoter. Cells that have been infected by the adenovirus express GFP and therefore display fluorescence (J. E. Biggs, P. E. Light and P. A. Smith, unpublished observations).

This preliminary finding is very exciting as the ability to infect DMOTC slices with adenovirus provides an opportunity to knock down or overexpress genes that are upregulated by nerve injury.

5. Potential Pitfalls and Future Directions

5.1. Pitfalls

The potential methodological pitfalls of preparing DMOTC are self-evident and are similar to those encountered with any preparation of isolated brain tissues. The main practical problem with long-term maintenance of the cultures is the growth of bacteria in the medium if contamination occurs during tissue isolation and dissection. This is minimized by the rigorous application of aseptic techniques, during dissection, culture preparation and medium exchanges.

The neurons in our DMOTC spinal cord slices are of the same age as those used in our acute slice studies (Fig. 1d) but are not subject to the trauma resulting from acute tissue isolation. Indeed, Fig. 3 shows that microglial activation in the cultures is far less pronounced than in acutely isolated spinal cord slices. Although the neurons appear to exhibit similar morphological and electrophysiological phenotypes in both DMOTC and acute slices (Figs. 2 and 4), it is difficult to ascertain that they exhibit the same degree of synaptic connectivity in both situations. For example, BDNF-treated DMOTC spinal cord slices almost always exhibit

Table 1
Comparison of properties of acutely isolated and DMOTC slices of spinal cord

	DMOTC slice	Acute slice
Developmental status	Close to normal	Normal
Neuronal morphology	May recover from initial injury during sectioning process, but neuronal orientation may be altered during lateral spread of slices	May be disturbed by sectioning process
Viability	6 weeks	8 h
Microglial status	Microglia quiescent	Microglia activated
Visibility of neurons in IR-DIC	Superior to acute slices	Acceptable
Ca^{2+}-imaging studies	Feasible	Difficult
Use for term studies of drug/neuromodulator/neurotrophin action	Feasible for days or weeks	Only feasible for periods <6 h
Use of adenoviral infection for genetic manipulation	Feasible	Difficult

synchronized oscillations in intracellular Ca^{2+} (13) and oscillations are seen in 20% of control cultures. In both cases, oscillations are synchronized to extracellularly recorded field potential activity. By contrast, there is no evidence for oscillatory activity in acutely isolated slices, including those isolated from animals subject to CCI. Does this mean that aberrant synaptic connections develop in the cultures or that the acutely isolated slices fail to oscillate due to the acute neuronal trauma provoked by tissue isolation? Whereas spontaneous action potential activity is frequently observed in cultures, this is not seen in neurons in acutely isolated spinal cord slices. Does this mean that the cultures, in which the neurons have had weeks to recover from trauma, actually a better model of the intact spinal cord? Although Ca^{2+} oscillations do not occur spontaneously in acutely isolated spinal cord slices, they can be induced by treatment with 4-aminopyridine (56) or high K^{+} (57), suggesting that appropriate circuitry for the generation of oscillations is present ex vivo. From this, it might be argued that oscillations seen in DMOTC simply reflect healthy, physiological network activity rather than the appearance of non-physiological phenomena as a result of modifications to neural circuitry during growth of cultures. Table 1 compares the characteristics of DMOTC slices with those of acutely isolated slices. Another issue with DMOTC slices and transverse spinal cord slices in general is that the neurons therein do not receive input from adjacent segments of the spinal cord or from descending pathways from the periaqueductal grey and rostroventral medulla. Some laboratories use acute parasagittal spinal cord slices, and these may better preserve the function of islet cells which project in the rostrocaudal direction (4, 58). We have not as yet attempted to prepare DMOTC of parasagittal spinal cord slices, but this may be useful for further work. It would be more difficult to prepare brainstem spinal cord DMOTC's that would preserve the integrity of descending pathways. A third issue which we are currently addressing is the origin of the Ca^{2+} signals such as those illustrated in Fig. 5a–c as it is possible that these may originate from astrocytes (59) and/or microglia as well as neurons. These experiments involve examination of the actions of specific astrocyte activators such as P2X-type purinergic receptor ligands (59–61) (see also Chaps. 2-Ruangkittisakul et al. (15) and 11-Kantor et al. (16)).

5.2. Future Directions

The ease of infection of DMOTC with viral vectors and the ability to express virally encoded proteins such as GFP (Fig. 7) pave the way to many exciting further studies using the pain in a dish system. For example, adenovirus or adeno-associated viral vectors could be used to deliver genetically encoded Ca^{2+} indicators such as GcaMPs (62–64). Since GCaMP expression can be driven by various cell type-specific promoters, we have the option of specifically monitoring Ca^{2+} signals in astrocytes, inhibitory neurons, excitatory neurons and primary afferents, thereby helping to resolve the above mentioned issue of the origin of Ca^{2+} signals. Other experiments might involve examination of neuron-glial signalling that may be especially important to understanding neuropathic pain mechanisms. Other genetic approaches involve expression of dominant negative isoforms of cellular proteins of interest as well as siRNA and optogenetic technologies. Introduction of light-activated ionophores such as photorhodopsin will allow selective stimulation of individual neurons and assessment of their contribution to network activity in pain-related situations.

6. Summary

This chapter provides an overview of the techniques involved in the preparation of DMOTC spinal cord slices and also demonstrates some of the practical uses of these slices in the study of central sensitization in chronic pain. We suggest that concerns relating to the physiological significance of the use of cultures are very much outweighed by the experimental advantages they offer. In fact, neurons in organotypic culture may exist in a more physiological state than those in freshly isolated brain slices (10).

Acknowledgements

Supported by grants from the Canadian Institutes of Health Research, Alberta Heritage Foundation for Medical Research, Canadian Institutes for Innovation, Advanced Education Alberta and the Pfizer Canada Neuropathic Pain Research Awards Program.

References

1. Woolf CJ (1983) Evidence for a central component of post-injury pain hypersensitivity. Nature 306:686–688
2. Sandkühler J (2009) Models and mechanisms of hyperalgesia and allodynia. Physiol Rev 89:707–758
3. Millan MJ (1999) The induction of pain: an integrative review. Prog Neurobiol 57:1–164
4. Coull JA, Beggs S, Boudreau D, Boivin D, Tsuda M, Inoue K, Gravel C, Salter MW, De Koninck Y (2005) BDNF from microglia causes the shift in neuronal anion gradient underlying neuropathic pain. Nature 438: 1017–1021
5. Lu VB, Biggs JE, Stebbing MJ, Balasubramanyan S, Todd KG, Lai AY, Colmers WF, Dawbarn D, Ballanyi K, Smith PA (2009) BDNF Drives the changes in excitatory synaptic transmission in the rat superficial dorsal horn that follow sciatic nerve injury. J Physiol 587:1013–1032
6. Ballerini L, Galante M (1998) Network bursting by organotypic spinal slice cultures in the presence of bicuculline and/or strychnine is developmentally regulated. Eur J Neurosci 10:2871–2879
7. Avossa D, Rosato-Siri MD, Mazzarol F, Ballerini L (2003) Spinal circuits formation: a study of developmentally regulated markers in organotypic cultures of embryonic mouse spinal cord. Neuroscience 122:391–405
8. Gähwiler BH (1981) Organotypic monolayer cultures of nervous tissue. J Neurosci Methods 4:329–342
9. De Simoni A, Griesinger CB, Edwards FA (2003) Development of rat CA_1 neurones in acute versus organotypic slices: role of experience in synaptic morphology and activity. J Physiol 550:135–147
10. Gähwiler BH, Capogna M, Debanne D, McKinney RA, Thompson SM (1997) Organotypic slice cultures: a technique has come of age. Trends Neurosci 20:471–477
11. Lu VB, Moran TD, Balasubramanyan S, Alier KA, Dryden WF, Colmers WF, Smith PA (2006) *Substantia Gelatinosa* neurons in defined-medium organotypic slice culture are similar to those in acute slices from young adult rats. Pain 121:261–275
12. Lu VB, Colmers WF, Smith PA (2009) Long-term effects of brain-derived neurotrophic factor on the frequency of inhibitory synaptic events in the rat superficial dorsal horn. Neuroscience 161:1135–1143
13. Lu VB, Ballanyi K, Colmers WF, Smith PA (2007) Neuron type-specific effects of brain-derived neurotrophic factor in rat superficial dorsal horn and their relevance to 'central sensitization'. J Physiol 584:543–563
14. Trapp S, Ballanyi K (2012) Autonomic nervous system in vitro: studying tonically active neurons controlling vagal outflow in rodent brainstem slices. In Isolated Central Nervous System Circuits (Ed K Ballanyi), Neuromethods Series Vol. 73 (Ed W Walz). Springer Science+Business Media, LLC, New York, NY, pp 1–59
15. Ruangkittisakul A, Panaitescu B, Secchia L, Bobocea N, Kantor C, Kuribayashi J, Iizuka M, Ballanyi K (2012) Isolated brainstem respiratory centers from perinatal rodents. In Isolated Central Nervous System Circuits (Ed K Ballanyi), Neuromethods Series Vol. 73 (Ed W Walz). Springer Science+Business Media, LLC, New York, NY, pp 61–124
16. Kantor C, Panaitescu B, Kuribayashi J, Ruangkittisakul A, Jovanovic I, Leung V, Lee TF, MacTavish D, Jhamandas JH, Cheung PY, Ballanyi K (2012) Spontaneous neural network oscillations in hippocampus, cortex and *locus coeruleus* of newborn rat and piglet brain slices. In Isolated Central Nervous System Circuits (Ed K Ballanyi), Neuromethods Series Vol. 73 (Ed W Walz). Springer Science+Business Media, LLC, New York, NY, pp 315–356
17. De Curtis M, Lilbrizzi L, Uva L, Gnatkovsky V (2012) Neuronal networks in the in vitro isolated guinea pig brain. In Isolated Central Nervous System Circuits (Ed K Ballanyi), Neuromethods Series Vol. 73 (Ed W Walz).

Springer Science+Business Media, LLC, New York, NY, pp 357–383
18. Mandadi S, Nakanishi ST, Han P, Humphreys J, Whelan PJ (2012) The use of rodent isolated spinal cord preparations to examine motor output. In Isolated Central Nervous System Circuits (Ed K Ballanyi), Neuromethods Series Vol. 73 (Ed W Walz). Springer Science+Business Media, LLC, New York, NY, pp 437–464
19. Gustafson-Vickers SL, Lu VB, Lai AY, Todd KG, Ballanyi K, Smith PA (2008) Long-term actions of interleukin-1beta on delay and tonic firing neurons in rat superficial dorsal horn and their relevance to central sensitization. Mol Pain 4:63
20. Balasubramanyan S, Stemkowski PL, Stebbing MJ, Smith PA (2006) Sciatic chronic constriction injury produces cell-type specific changes in the electrophysiological properties of rat *Substantia Gelatinosa* neurons. J Neurophysiol 96:579–590
21. Biggs JE, Lu VB, Stebbing MJ, Balasubramanyan S, Smith PA (2010) Is BDNF sufficient for information transfer between microglia and dorsal horn neurons during the onset of central sensitization? Mol Pain 6:44
22. Tsuda M, Inoue K, Salter MW (2005) Neuropathic pain and spinal microglia: a big problem from molecules in 'small' glia. Trends Neurosci 28:101–107
23. Echeverry S, Shi XQ, Zhang J (2007) Characterization of cell proliferation in rat spinal cord following peripheral nerve injury and the relationship with neuropathic pain. Pain 135:37–47
24. Hendrich J, Van Minh AT, Heblich F, Nieto-Rostro M, Watschinger K, Striessnig J, Wratten J, Davies A, Dolphin AC (2008) Pharmacological disruption of calcium channel trafficking by the alpha/2delta ligand gabapentin. Proc Natl Acad Sci USA 105: 3628–3633
25. Bauer CS, Nieto-Rostro M, Rahman W, Tran-Van-Minh A, Ferron L, Douglas L, Kadurin I, Sri Ranjan Y, Fernandez-Alacid L, Millar NS, Dickenson AH, Lujan R, Dolphin AC (2009) The increased trafficking of the calcium channel subunit alpha2/delta-1 to presynaptic terminals in neuropathic pain is inhibited by the alpha2/delta ligand pregabalin. J Neurosci 29:4076–4088
26. Biggs JE, Balasubramanyan S, Ballanyi K, Smith PA (2008) Chronic Gabapentin acts intracellularly to reduce dorsal horn excitability. Soc Neurosci Abs Abstract # 569.6/LL6
27. Luhmann HJ, Kilb W (2012) Intact in vitro preparation of the neonatal rodent cerebral cortex-analysis of cellular properties and network activity. In Isolated Central Nervous System Circuits (Ed K Ballanyi), Neuromethods Series Vol. 73 (Ed W Walz). Springer Science+Business Media, LLC, New York, NY, pp 301–314
28. Chen Y, Balasubramanyan S, Lai AY, Todd KG, Smith PA (2009) Effects of sciatic nerve axotomy on excitatory synaptic transmission in rat substantia gelatinosa. J Neurophysiol 102:3203–3215
29. Watkins LR, Maier SF (2002) Beyond neurons: evidence that immune and glial cells contribute to pathological pain states. Physiol Rev 82:981–1011
30. Milligan ED, Twining C, Chacur M, Biedenkapp J, O'Connor K, Poole S, Tracey K, Martin D, Maier SF, Watkins LR (2003) Spinal glia and proinflammatory cytokines mediate mirror-image neuropathic pain in rats. J Neurosci 23:1026–1040
31. Milligan ED, Zapata V, Chacur M, Schoeniger D, Biedenkapp J, O'Connor KA, Verge GM, Chapman G, Green P, Foster AC, Naeve GS, Maier SF, Watkins LR (2004) Evidence that exogenous and endogenous fractalkine can induce spinal nociceptive facilitation in rats. Eur J Neurosci 20:2294–2302
32. Siao CJ, Tsirka SE (2002) Tissue plasminogen activator mediates microglial activation via its finger domain through annexin II. J Neurosci 22:3352–3358
33. Eckert WA III, Willcockson HH, Light AR (2001) Interference of biocytin with opioid-evoked hyperpolarization and membrane properties of rat spinal substantia gelatinosa neurons. Neurosci Lett 297:117–120
34. Todd AJ, Lewis SG (1986) The morphology of Golgi-stained neurons in lamina II of the rat spinal cord. J Anat 149:113–119
35. Grudt TJ, Perl ER (2002) Correlations between neuronal morphology and electrophysiological features in the rodent superficial dorsal horn. J Physiol 540:189–207
36. Todd AJ, McKenzie J (1989) GABA-immunoreactive neurons in the dorsal horn of the rat spinal cord. Neuroscience 31:799–806
37. Scholz J, Woolf CJ (2007) The neuropathic pain triad: neurons, immune cells and glia. Nat Neurosci 10:1361–1368
38. Lai AY, Todd KG (2008) Differential regulation of trophic and proinflammatory microglial effectors is dependent on severity of neuronal injury. Glia 56:259–270

39. Ito D, Imai Y, Ohsawa K, Nakajima K, Fukuuchi Y, Kohsaka S (1998) Microglia-specific localisation of a novel calcium binding protein, Iba1. Brain Res Mol Brain Res 57:1–9
40. Ruangkittisakul A, Schwarzacher SW, Secchia L, Poon BY, Ma Y, Funk GD, Ballanyi K (2006) High sensitivity to neuromodulator-activated signaling pathways at physiological [K^+] of confocally imaged respiratory center neurons in on-line-calibrated newborn rat brainstem slices. J Neurosci 26: 11870–11880
41. Paredes RM, Etzler JC, Watts LT, Zheng W, Lechleiter JD (2008) Chemical calcium indicators. Methods 46:143–151
42. Gee KR, Brown KA, Chen WN, Bishop-Stewart J, Gray D, Johnson I (2000) Chemical and physiological characterization of fluo-4 Ca^{2+}-indicator dyes. Cell Calcium 27:97–106
43. Perrier JF, Noraberg J, Simon M, Hounsgaard J (2000) Dedifferentiation of intrinsic response properties of motoneurons in organotypic cultures of the spinal cord of the adult turtle. Eur J Neurosci 12:2397–2404
44. Safronov BV, Wolff M, Vogel W (1997) Functional distribution of three types of Na^+ channel on soma and processes of dorsal horn neurones of rat spinal cord. J Physiol 503: 371–385
45. Todd AJ (2010) Neuronal circuitry for pain processing in the dorsal horn. Nat Rev Neurosci 11:823–836
46. Hashizume H, De Leo JA, Colburn RW, Weinstein JN (2000) Spinal glial activation and cytokine expression after lumbar root injury in the rat. Spine 25:1206–1217
47. Lee J, Kim M, Back S, Uhm C, Hong S, Na H (2004) Activation of spinal microglia following peripheral nerve injury is related to the generation of mechanical allodynia in a rat model of neuropathic pain. Abstract Viewer/ Itinerary Planner Washington, DC: Society for Neuroscience, Online 2004; Program No. 518.8
48. Clark AK, Gentry C, Bradbury EJ, McMahon SB, Malcangio M (2006) Role of spinal microglia in rat models of peripheral nerve injury and inflammation. Eur J Pain 11:223–230
49. Banfield MJ, Naylor RL, Robertson AG, Allen SJ, Dawbarn D, Brady RL (2001) Specificity in Trk receptor:neurotrophin interactions: the crystal structure of TrkB-d5 in complex with neurotrophin-4/5. Structure 9:1191–1199
50. Cho HJ, Kim JK, Zhou XF, Rush RA (1997) Increased brain-derived neurotrophic factor immunoreactivity in rat dorsal root ganglia and spinal cord following peripheral inflammation. Brain Res 764:269–272
51. Fukuoka T, Kondo E, Dai Y, Hashimoto N, Noguchi K (2001) Brain-derived neurotrophic factor increases in the uninjured dorsal root ganglion neurons in selective spinal nerve ligation model. J Neurosci 21: 4891–4900
52. Pezet S, Cunningham J, Patel J, Grist J, Gavazzi I, Lever IJ, Malcangio M (2002) BDNF modulates sensory neuron synaptic activity by a facilitation of GABA transmission in the dorsal horn. Mol Cell Neurosci 21:51–62
53. Wilkemeyer MF, Smith KL, Zarei MM, Benke TA, Swann JW, Angelides KJ, Eisensmith RC (1996) Adenovirus-mediated gene transfer into dissociated and explant cultures of rat hippocampal neurons. J Neurosci Res 43: 161–174
54. Cho ES, Lee SY, Park JY, Hong SG, Ryu PD (2007) Organotypic slice culture of the hypothalamic paraventricular nucleus of rat. J Vet Sci 8:15–20
55. Wilkemeyer MF, Angelides KJ (1995) Adenovirus-mediated expression of a reporter gene in thalamocortical cocultures. Brain Res 703:129–138
56. Ruscheweyh R, Sandkühler J (2005) Long-range oscillatory Ca^{2+} waves in rat spinal dorsal horn. Eur J Neurosci 22:1967–1976
57. Asghar AU, Cilia La Corte PF, LeBeau FE, Al Dawoud M, Reilly SC, Bühl EH (2005) Whittington MA, King AE Oscillatory activity within rat substantia gelatinosa in vitro: a role for chemical and electrical neurotransmission. J Physiol 562:183–198
58. Labrakakis C, Lorenzo LE, Bories C, Ribeiro-da-Silva A, De Koninck Y (2009) Inhibitory coupling between inhibitory interneurons in the spinal cord dorsal horn. Mol Pain 5:24
59. Bardoni R, Ghirri A, Zonta M, Betelli C, Vitale G, Ruggieri V, Sandrini M, Carmignoto G (2010) Glutamate-mediated astrocyte-to-neuron signalling in the rat dorsal horn. J Physiol 588:831–846
60. Ballanyi K, Panaitescu B, Ruangkittisakul A (2010) Control of breathing by "nerve glue". Sci Signal 3:e41
61. Huxtable AG, Zwicker JD, Alvares TS, Ruangkittisakul A, Fang X, Hahn LB, Posse de Chaves E, Baker GB, Ballanyi K, Funk GD (2010) Glia contribute to the purinergic modulation of inspiratory rhythm-generating networks. J Neurosci 30:3947–3958
62. Tian L, Hires SA, Mao T, Huber D, Chiappe ME, Chalasani SH, Petreanu L, Akerboom J, McKinney SA, Schreiter ER, Bargmann CI,

Jayaraman V, Svoboda K, Looger LL (2009) Imaging neural activity in worms, flies and mice with improved GCaMP calcium indicators. Nat Methods 6:875–881

63. Akerboom J, Rivera JD, Guilbe MM, Malavé EC, Hernandez HH, Tian L, Hires SA, Marvin JS, Looger LL, Schreiter ER (2009) Crystal structures of the GCaMP calcium sensor reveal the mechanism of fluorescence signal change and aid rational design. J Biol Chem 284: 6455–6464

64. Mao T, O'Connor DH, Scheuss V, Nakai J, Svoboda K (2008) Characterization and subcellular targeting of GCaMP-type genetically-encoded calcium indicators. PLoS One 3:e1796

Chapter 15

Rodent Isolated Spinal Cord Preparations to Examine Motor Output

Sravan Mandadi, Stan T. Nakanishi, Pengcheng Han, Jennifer M. Humphreys, and Patrick J. Whelan

Abstract

Understanding mammalian neural networks such as the central pattern generator (CPG) for locomotion requires the use of multiple experimental models. In vitro approaches provide easy access to the spinal cord, thus allowing more experimental tools to be used. In this chapter, some established and new tools are outlined that are being used to advance our understanding of CPG function at both cellular and network levels. A variety of in vitro models exist including the isolated spinal cord, brainstem-spinal cord, and spinal cord leg-attached preparations. There are also a variety of ways to activate spinal CPGs using these preparations. Choosing which preparation and mode of CPG activation will vary depending on the question being asked. This chapter will review some of the common approaches used and also examine optogenetic, genetic, and Ca^{2+} imaging approaches that are being used to dissect the structure and function of CPGs.

Key words: CPG, Locomotion, Mouse, Spinal cord

1. Introduction

In vitro preparations enable the use of a rich set of tools to examine spinal functions with the major advantages of easy modification of the extracellular microenvironment and ease of access to neural networks at the cellular level. This ease of access allows the use of high-resolution analytical techniques such as visually guided whole-cell patch-clamp techniques, or optical imaging of neurons expressing genetically engineered fluorescent tags. While these advantages are principally common to all in vitro preparations, there are several advantages of using in vitro techniques when considering spinal motor output. Firstly, one can leave a limb attached to the lower

Klaus Ballanyi (ed.), *Isolated Central Nervous System Circuits*, Neuromethods, vol. 73,
DOI 10.1007/978-1-62703-020-5_15, © Springer Science+Business Media New York 2012

body segment containing exposed spinal cord tissue, thus allowing recording of muscle activity using electromyograms (EMGs) and selective (electrical) activation of primary afferents (1–5). Secondly, isolated brainstem-spinal cord preparations allow the activation and inactivation of supraspinal centers, providing a relatively simple method for assessing the role of endogenous modulators on motor function (6–9).

This chapter is divided into four sections dealing with (1) properties and limitations of most commonly used in vitro spinal cord preparations, (2) traditional (electrophysiological) approaches to examine motor function using in vitro preparations, (3) techniques to analyze data with focus on spinal motor networks, and (4) novel (imaging and molecular) approaches. Here, we emphasize the use of mouse and rat models, though many techniques described here can equally apply to spinal cord preparations from other species. Because this chapter reviews diverse rodent models for analyzing spinal motor output from various laboratories, no separate sections are provided for specific materials and methods (for this, see, e.g., Chaps. 1-Trapp and Ballanyi (10), 2-Ruangkittisakul et al. (11), and 14-Biggs et al. (12)).

2. Historical Background

The use of in vitro spinal cord preparations to study electrophysiological responses is not new. The first study that we are aware of used an isolated hemisegment of a spinal cord from a frog to examine ventral root responses recorded following dorsal root stimulation (13). This work was extended to the mammalian rat spinal cord preparation in the early 1980s (14). Around the same time, Cohen and colleagues demonstrated the use of an isolated spinal cord preparation from the lamprey to examine locomotor function (15). Jack Feldman and Jeff Smith adapted the methods of Toshihiko Suzue (14) to demonstrate the use of the isolated brainstem-spinal cord newborn rat preparation to examine locomotor output (16). Their subsequent paper led to the adoption of this model by several labs around the world (1). Later on, it was demonstrated that similar pharmacological and electrical stimulation approaches could be used in the in vitro mouse as well as the rat preparation to evoke bouts of locomotion (5, 17, 18). Today, many laboratories make use of isolated spinal cord preparations, which has naturally led to new advances, good-natured competition, and the formation of international collaborations. This has led to a wide number of methodologies used to activate locomotor networks, for example, chemically, electrically, or optically. In this chapter, we will introduce these preparations and techniques and discuss the pros and cons associated with their use.

By taking advantage of genetic tools coupled with the use of in vitro techniques, the architecture of the network that underlies locomotion is starting to crystallize. Most progress to date has been made in understanding the role of commissural inhibitory interneurons (CINs) that control the left-right alternating activity that classically defines locomotion (19–21). A subset of interneurons that project onto these CINs has been identified, and work published to date suggests that these subsets are responsible for regulating gait during high speeds (21–23). While much work needs to be done, the use of genetically engineered mice shows excellent promise as a tool to decipher neural connectivity within the spinal cord locomotor network. The reader is referred to a number of recent reviews on this topic (24–26). Models of vertebrate locomotor networks have provided a skeleton on which to test hypotheses based on data from a variety of preparations (27–29). Rybak and McCrea have proposed a model that elaborates on the classic half-center model (30) and can reproduce many aspects of the stepping pattern. It is a multilayer model composed of a distributed rhythm-generating layer projecting to a patterning layer that ultimately sculpts unique discharges of motoneuronal pools during stepping (Fig. 1). While it must be emphasized that other solutions to this problem are possible, and indeed likely (31, 32), models are useful tools for both the computational and experimental neuroscientist. Throughout this chapter, we will use the term central pattern generator (CPG) to refer to the spinal locomotor network. The CPG is defined here as a network of neurons that can produce rhythmic output in the absence of phasic excitatory input.

3. In Vitro Models for Analysis of Spinal Motor Output

There are various in vitro spinal cord preparations currently in use, each of which has advantages and disadvantages associated with them. Choosing which one to use depends on the question being asked, and complex projects may require the use of more than one type of preparation.

3.1. Isolated Spinal Cord Preparations

3.1.1. Basic Features

The isolated spinal cord preparation is widely used. There are several reasons for this including a relatively simple dissection and recording setup. The dissection procedure has been described before in numerous publications and will only be touched on briefly here (5). The animal is first anesthetized. After decapitation and evisceration, a ventral laminectomy is performed in solution mimicking the ionic composition of artificial cerebrospinal fluid ("ACSF") (see also Chaps. 1-Trapp and Ballanyi (10), 2-Ruangkittisakul et al. (11), and 8-Sanchez-Vives (33)), the dorsal and ventral roots are cut, and the

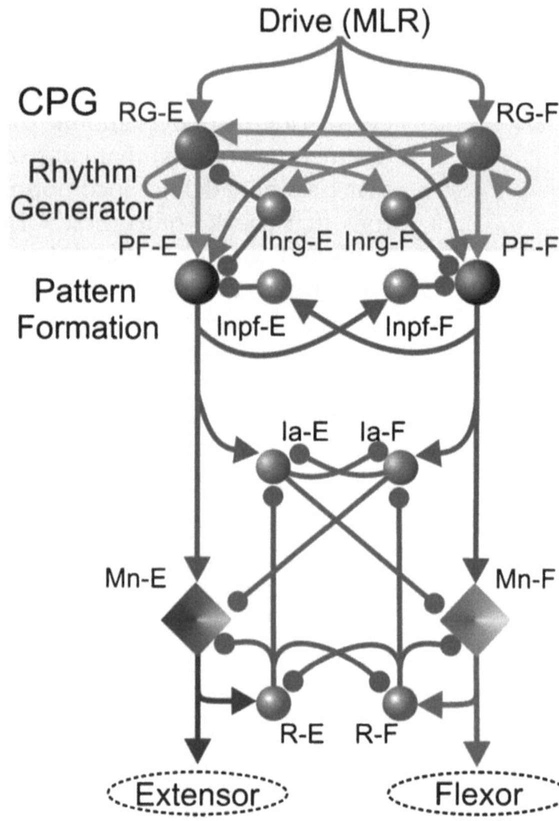

Fig. 1. Locomotor network model by Rybak and McCrea. Populations of interneurons are indicated by spheres, while motoneurons are represented by diamonds. This three-layer model consists of a rhythm-generating layer comprised of Extensor (*RG-F*) and Flexor (*RG-F*) interneurons which have both recurrent excitatory connections. These interneurons, in turn, receive mutually inhibitory input from Inrg cells. The drive projects to a pattern formation layer (*PF*), which acts through mutually inhibitory connections (*Inpf* cells) to sculpt the pattern which is then output to the extensor and flexor motoneurons. The final output of the motoneurons is modulated by a final layer of Ia inhibitory interneurons (*Ia-E, Ia-F*) and Renshaw cells (*R-E, R-F*). *Arrows* indicate excitatory drive, while filled circles indicate inhibitory drive (Reproduced with permission from (123)).

spinal cord is gently removed from the vertebral column. Spinal shock associated with the isolation generally lasts for <1 h, and so the preparation is allowed to recover for at least that period of time before recordings are attempted. By correlating the electroneurograms (ENGs) of the ventral root motor output with flexor and extensor muscle EMGs and/or peripheral nerve ENGs, it was found that a signature of locomotor-like activity can be established by recording ENGs from lumbar (L) ventral roots, L1 or 2 and L5 or 6 (3, 5). The L1–2 ENGs reveal increased discharge during the flexor phase of the step cycle, while the L5–6 ENGs are most active during extension. It should be recognized that both L1–2 and L5 regions contain a heterogeneous population of extensor and flexor motoneurons (34).

A limitation of superfused in vitro preparations is that the viability of the tissue often declines with the age of the animal, although there are exceptions such as brain slices from suprachiasmatic nucleus that can survive for several days (see also Chap. 3-Moore et al. (35)). Preparations can be obtained from perinatal rodents up to around postnatal days (P) 4–5 in both the mouse and rat. Any older than that, it progressively gets more difficult to reliably evoke fictive locomotor activity. In our laboratory, we like to restrict the use of postnatal animals to ages between P1 and P3. We found that in the mouse preparations at P3, an anoxic core develops whose diameter increases when metabolic demands increase (36). This is an important consideration when rhythmogenic drugs are applied since they activate most neurons in the spinal cord. This anoxic core is a concern since neurons located within it will cease to fire. As the diameter of the spinal cord increases with age, the anoxic core will increase, eventually encompassing interneurons that form the CPG. There are ways in which to minimize the formation of an anoxic core, such as maximizing the flow rate where possible and minimizing the volume of ACSF within the recording chamber (36). There is also the possibility of using only sacral tissue that is thin enough to remain viable. However, these preparations have been used so far to examine cellular motoneuronal properties and not (locomotor) network functions (37, 38).

3.1.2. CPG Activation

Since isolated spinal cord preparations lack descending projections which normally provide the background excitation for the CPG to operate, excitability has to be increased by other means. The primary routine for eliciting activity has been the combined use of bath-applied monoamines, specifically serotonin (5-HT), dopamine, and N-methyl-D-aspartate (NMDA) (5, 39, 40). There are several advantages to this neurochemical method. First of all it is easy – one simply has to add the appropriate concentrations of drugs and more often than not a long-lasting stable locomotor-like rhythm can be recorded. Secondly, it allows selective activation of different spinal cord segments using compartmented recording chambers (Fig. 2) (41). However, there are also disadvantages associated with the use of bath-applied rhythmogenic drugs. Firstly, different fictive locomotor patterns are produced with different combinations of drugs (3, 42–44). Therefore, the rhythms produced and the networks activated may not be comparable between laboratories using different "cocktails" of the activators. There is a clear need for researchers in this field to decide for one specific combination of CPG-activating drugs to allow future sharing of databases. This is especially important for the development of computer models to understand locomotor network function. Modeling is an essential tool to conceptually understand how networks function, and data obtained from numerous labs are often pooled to build appropriate models (29, 31, 45).

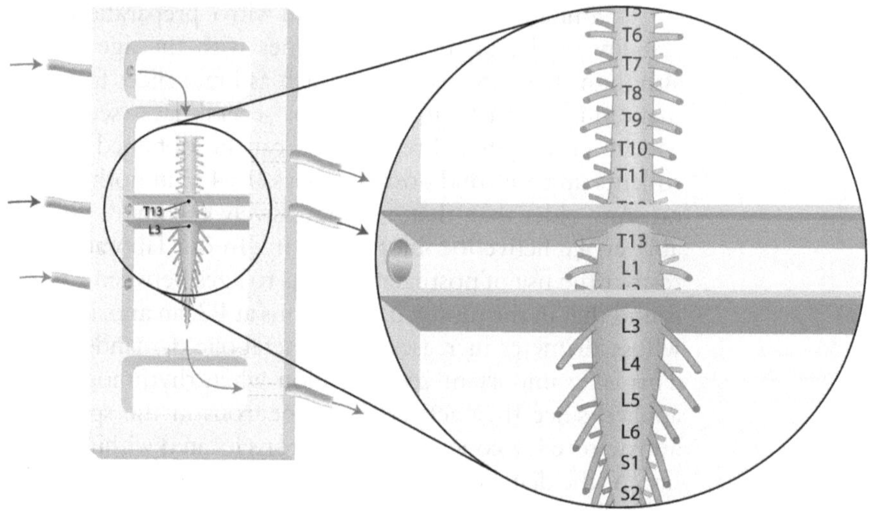

Fig. 2. *Compartmentalized recording chambers for separately superfusing sections of isolated perinatal rodent spinal cord preparations used as models for studying fictive locomotion.* This schematic shows a three-way split allowing three compartments of the spinal cord to be superfused separately. One peristaltic ("roller") pump with three heads and thus three lines of tubing can be used for this. The temperature within each compartment is controlled separately and is usually maintained at 27°C.

Another approach that has been adopted in the mouse and rat is the use of afferent stimulation to elicit bouts of locomotion (2, 5, 46, 47). This approach has the advantage of using a more physiologically relevant method of activating locomotor networks. Electrically stimulating sacral or lumbar dorsal roots with a 4-Hz, 10-s train with stimulation intensities between 1 and 2 × threshold for eliciting a just observable response in the ventral root ENGs produce excellent bouts of rhythmic alternating fictive locomotor activity (5). Another advantage of electrical stimulation is that short sequences of locomotion can be evoked reliably which is perfect for imaging studies, where the shutter can be triggered to open only during the stimulated sequences (48). While the CPG can be activated by stimulating afferents from different spinal segments, stimulation of sacral dorsal roots appears to be particularly effective (5). Such stimulation activates sacrocaudal afferents (SCAs) which project to interneurons in the sacral segments that send projections in several funiculi (49). In addition, there is a direct pathway from high-threshold afferents that ascend primarily in the dorsolateral funiculus. Presumably, these direct and indirect ascending collaterals project onto interneurons comprising the CPG at multiple lumbar segments. The SCA system is probably analogous to the escape reflex observed in many animal species. A noxious pinch of the tail induces a rapid activation of locomotor bouts which can be reproduced in animals in vivo with a full thoracic transection of the spinal cord (50). Evidence from both the mouse and rat indicates that stimulation of the SCA system recruits C and Aδ afferents that signal noxious stimuli (51, 52). These are not likely to be the only classes of afferents recruited, but it is clear that they contribute substantially to the excitation of the CPG (52). Like the situation with the use of drug cocktails, techniques to elicit afferent-evoked locomotion differ between laboratories. Some groups prefer to stimulate the *cauda equina*, which consists of a "clump" of afferents beyond the end of the cord (5). While this produces excellent bouts of locomotion, it is difficult to precisely target afferents that innervate a given dorsal root. A robust rhythm can be elicited by electrical stimulation of a single lumbosacral dorsal root. In particular, stimulation of the S4 or Co1 dorsal root is attractive since the SCA pathway to the lumbar segments has been partly worked out (49). The main disadvantage of electrical afferent fiber stimulation is that this cannot be used to elicit bouts of locomotor rhythm beyond time periods of 10–20 s. This is likely due to short-term depression of presynaptic afferent transmission that is particularly evident in neonates (53), though, if stimulus trains are separated by ~3 min, full recovery takes place and each train can produce a reliable bout of activity over time periods of between 10 and 20 s (5, 47). Using this approach, the data from each bout can be pooled, allowing relatively long experiments to be performed. For example, in our lab, we found that it was possible to stimulate every 3 min over the course of 90 min, allowing for the collection of 30 trials (7, 54). There are other methods to electrically evoke locomotion, such as stimulation of the ventrolateral funiculus (55). Presumably, this method recruits a mixture of propriospinal and descending fibers which project to the CPG. Short bursts of locomotor-like activity can be produced by using a standard patch-clamp electrode (2–5-μm outer diameter) inserted into the cervical ventrolateral funiculus (stimulus parameters; 5 × threshold; 50-Hz, 2-s trains). Finally, a recent technique involves stimulation of L5 or L6 ventral roots with stimulus trains (4-Hz train for 10 s) (56, 57). The choice of method used rests on the experimental question being addressed. However, in our hands, the most reliable electrical method to activate CPGs is by stimulation of the sacral dorsal roots.

The isolated spinal cord preparation is extremely useful, but there are occasions when greater amount of control over the afferent subgroups stimulated and more information regarding the pattern produced by the central pattern generator is needed. This is an important issue since changes in individual muscle activity patterns cannot always be detected by recording from ventral roots in isolated cords (43). The leg-attached preparation offers many of the advantages conferred by the isolated spinal cord model, yet with the ability to record EMGs or ENGs from nerves (2, 5, 58). The dissection for the latter model differs slightly from that for the isolated spinal cord preparation. In particular, care must be taken when performing the ventral laminectomy not to damage the ventral roots and the spinal nerves that run close to the lateral edge of the

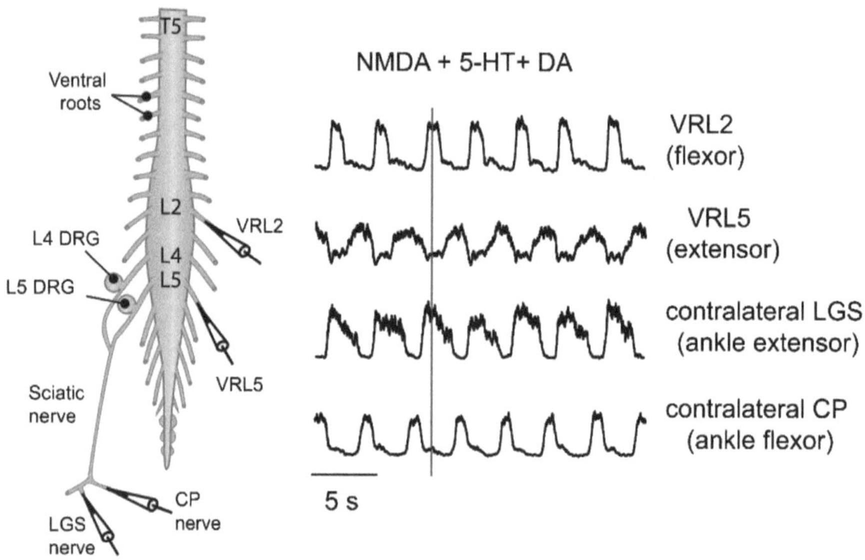

Fig. 3. *Chemical activation of locomotor rhythm in isolated spinal cords.* The schematic on the *left* shows a preparation where the nerves are dissected free from the hind limb. Recording of fictive locomotor activity can be made from various combinations of nerves. In this figure, recordings are highlighted from the lateral gastrocnemius (*LGS*) and soleus nerve, which are ankle extensors, and from the common peroneal (*CP*) nerve which supplies the tibialis anterior and extensor digitorum longus nerves. The common peroneal nerve provides an index of flexor activity. Plastic suction electrodes are used to suck up the nerves. The peripheral nerves are much smaller than the ventral roots, and extra care needs to be exercised so as not to damage them. The *traces on the right* illustrate recordings obtained with this arrangement. Fictive locomotion was evoked by bath application of dopamine (*DA*), N-methyl-D-aspartate (*NMDA*), and serotonin (5-HT). Notice the long duration of the LGS locomotor bursts compared to the CP ENG. This pattern is similar to what is observed in vivo.

vertebral column. Also a dorsal laminectomy has to be performed to ensure that the spinal cord is adequately superfused. A drawback of the leg-attached spinal cord preparation is that the extra tissue hampers fast exposure of the cord to bath-applied agents. Therefore, it is necessary to funnel the superfusate flow directly onto the spinal cord portion of the preparation. Even with these precautions, we have found that a rhythm can at times be difficult to evoke or may just activate flexors without extensors being activated. One way to ensure that viability is not compromised is to isolate the nerve of interest from the tissue, leaving an isolated spinal cord with a nerve attached (Fig. 3). This type of experiment can allow for whole-cell recordings since the isolated spinal cord is stable.

3.2. Brainstem-Spinal Cord Preparations

While the spinal cord contains sufficient circuitry for producing a well-coordinated locomotor pattern, brainstem circuits are critical for modulating these patterns and providing tonic excitatory drive to activate the CPG (59). A brainstem-spinal cord preparation can be dissected from perinatal rodents, in which the brainstem can be stimulated to evoke locomotor activity. Typically, the brainstem is transected around the exit of cranial nerve VII (9). In our laboratory, this has proven to be the best approach for eliciting brainstem-evoked locomotor activity in the newborn mouse. In this animal species, it is difficult to record coordinated cervical and lumbar rhythms when the CPG is activated pharmacologically (60), contrary to stable rhythms upon electrical brainstem stimulation (61).

Electrical stimulation of the brainstem readily evokes coordinated lumbar rhythms that have the signature of locomotor-like activity (8). The stimulating electrode is manufactured from polyethylene tubing and cut at an angle of 45° to allow a maximum amount of the tissue to be in contact with the electrode and to minimize damage to the brainstem. We manufacture our stimulating and recording electrodes using plastic tubes (Intramedic PE50; Becton Dickinson Diagnostic Systems, Franklin Lakes, NJ, USA; www.bd.com) pulled-out over an alcohol flame (61); however, others have used glass electrodes with equal effect (8). Stimulus currents between 0.1 and 0.5 mA evoke stable bouts of locomotion. The optimal placement for the electrode in our hands is the midline of the medulla around cranial nerve XII. A train of stimuli applied over the course of 100 s is usually effective in producing a stable sequence of rhythmicity. The rhythm usually takes about 10 s to develop after onset of stimulation. There are though important caveats to consider. The most important is that an anoxic core exists in the medulla of both rat and mouse brainstem-spinal cord preparations (62–64) (see also Chap. 2-Ruangkittisakul et al. (11)). However, in rats, this anoxic area lies about 800 μm below the surface; so many areas will remain well oxygenated.

3.3. Sensory Control of In Vitro Locomotor Activity

One of the issues with isolated spinal cord preparations is that they are not well suited for examining the role of sensory input in the control of locomotion. This is because electrical stimulation of dorsal roots activates multiple classes of afferents originating from disparate parts of the hind limb. There are several solutions to this problem. For example, in our lab, we designed a single-legged preparation with spinal cord attached. A ventral and dorsal laminectomy is first performed to expose the cord and preserve the spinal roots. Then, the ventral roots are cut, leaving the dorsal roots intact. Subsequently, excess tissue is trimmed away and the opposite leg removed. We designed a special chamber to accommodate the preparation. We made a split bath constructed from thin plastic obtained from standard lab weigh boats (Fig. 4). To ensure that the temperature containing the spinal cord remains constant, we insulate the sides of the plastic with a layer of Vaseline. We separately superfuse the hind limb to selectively activate thermoreceptors

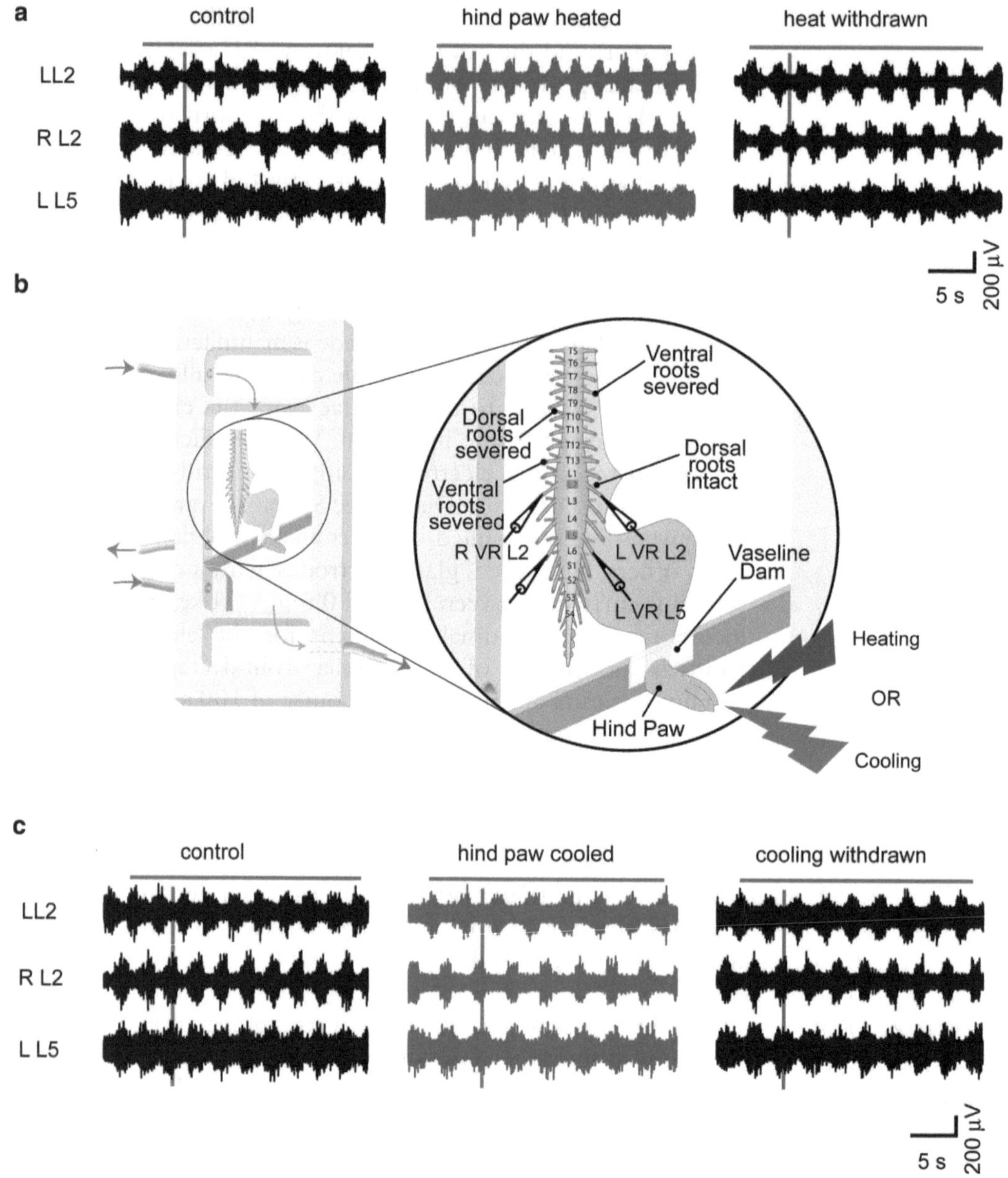

Fig. 4. *Effect of cooling and heating on the frequency of the in vitro locomotor rhythm*. (**a**) Ventral root electroneurograms (ENGs) show that increasing the temperature of the hind limb increases the frequency of locomotor-like rhythm. (**b**) Schematic of the chamber used to separate the hind limb from the spinal cord. Each chamber is superfused separately with artificial cerebrospinal fluid (ACSF). A layer of Vaseline is used to seal and provide an insulating blanket between the two chambers. (**c**) Cooling the hind limb decreases the frequency of the pharmacologically evoked rhythm. (**a**, **c**) Rhythm evoked using 50 µM DA, 10 µM 5-HT, and 5 µM NMDA.

that detect noxious as well as cool temperatures (65). This is done by heating or cooling the ASCF that superfuses the hind limb compartment or by directly injecting substances into the paw which activate thermoreceptors such as capsaicin.

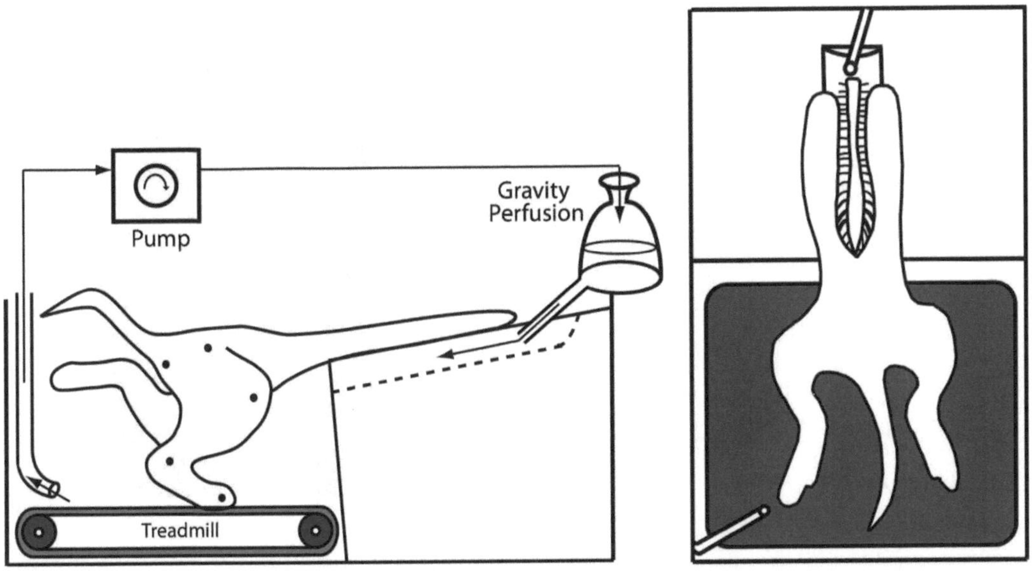

Fig. 5. Recording chamber and perfusion system used to record kinematic and electrophysiological data from in vitro spinal cord with leg-attached preparations (Figure adapted with permission (67)).

It is well recognized that sensory feedback from mechanoreceptors in the skin and muscle spindles and Golgi tendon organs in muscle contributes to the timing and the pattern of fictive stepping in adult preparations, mainly from cat (66). However, there has been a paucity of papers that have examined this issue in neonates. Recently, Shawn Hochman and colleagues have developed an apparatus for recording hind limb activity in the neonatal rat which overcomes some of the limitations of the isolated spinal cord preparation (67). Essentially, this is a leg-attached spinal cord model with the legs suspended over a miniature treadmill belt (Fig. 5). This allows kinematic analyses of the moving limb to be recorded along with EMGs. The authors have shown that changing the speed of the treadmill belt alters the cycle period of the fictive locomotor rhythm. This simple, yet powerful, experiment demonstrates that sensory input from the limb can modulate locomotor circuits already during neonatal development. Furthermore, swing assistance stabilized an irregular rhythm in that novel model (67). At the same time, the preparation affords the advantage common to all in vitro preparations of being able to modify the bathing solution surrounding the spinal cord. Also, if the spinal cord is suitably stabilized, intracellular recording from spinal neurons is feasible even while the legs are moving (68). This is a unique advantage since movements of the limbs usually make it impossible to obtain stable intracellular recording in vivo. Overall, this preparation appears quite promising, although care must be taken to adequately superfuse it.

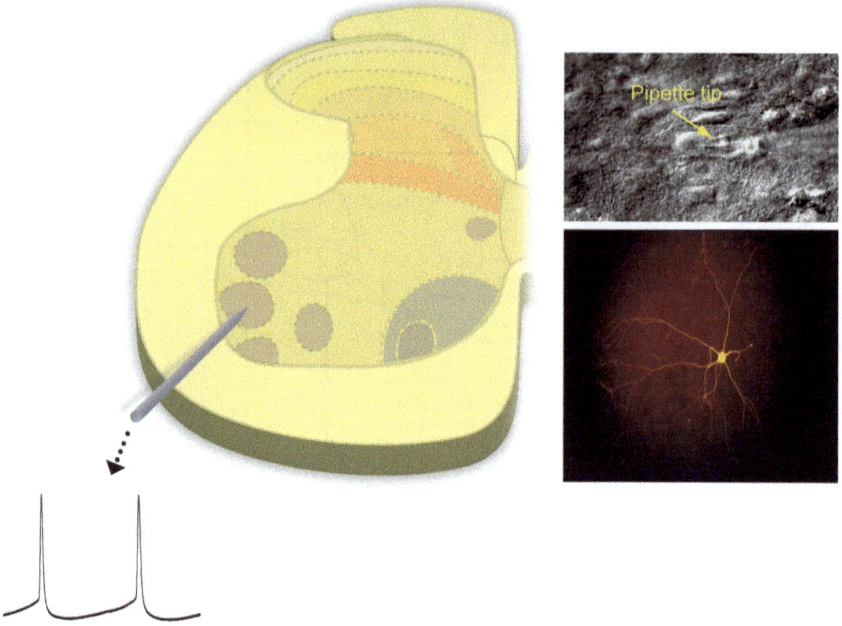

Fig. 6. *Patch recording from locomotor CPG neurons in spinal cord slices.* Schematic on the *left* showing visually guided patch-clamp techniques. Perinatal rodent spinal cord slices are cut transversely. The cells, which are potentially either backfilled with dye or have XFP (FP fluorescent protein; X indicates generic form (e.g., green fluorescent protein [GFP], yellow fluorescent protein [YFP]) constructs that are expressed), are visualized using a fluorescent microscope and marked. The user then switches to infrared differential interference contrast optics for optimal visualization of marked cells for patching. The *upper right* schematic shows a pipette tip attached to a motoneuron. The patch microelectrode typically contains biocytin that allows for post hoc identification of the cell (*bottom right panel*).

3.4. Arterially Perfused Preparations

In contrast to superfused preparations, arterially perfused in vitro preparations have been developed which use the existing vasculature to deliver ACSF to the brainstem and spinal cord. These preparations are perfused either transcardially or by cannulating the descending aorta (69–71). These preparations provide many of the advantages of isolated spinal cord preparations with the additional advantage that juvenile or adult preparations can be used. Stable recording from spinal cord neurons has been obtained using whole-cell patch-clamp approaches, although reliable locomotor output has yet to be reported (69). While several labs have demonstrated the utility of these preparations in understanding functions of neural networks in the brainstem such as chemosensitive or respiratory CPGs (70, 72, 73), they have not been widely adopted for examining spinal cord networks (71). A modification of this approach by transcardial perfusion of the left ventricle shows promise (69).

3.5. Spinal Cord Slices

One major advantage of using brain slices is increased access, making it relatively easy to identify neurons using visually guided patch-clamp techniques (Fig. 6) (see also, e.g., Chap. 1-Trapp and Ballanyi (10)). Slices have been used for many years in a variety of brain tissues (74, 75). Typically, they are useful when examining connectivity between pairs of neurons (76) or for recording synaptic or intrinsic properties of neurons (77–79). As one drawback of spinal cord slices, they contain a limited part of the locomotor network and the slicing procedure can injure neurons. This is an especially important consideration when recording from motoneurons that are especially susceptible to injury due to their size, extensive dendritic arborization, and susceptibility to ischemia and other insults (38, 80–83). To address this issue, it is important to cool the tissue and ensure that the ACSF used for the generation of spinal cord slices contains sucrose which blocks voltage-gated Na^+ channels (ice-cold sample solution (in mM): 25 NaCl, 188 sucrose, 1.9 KCl, 10 MgSO4, 1.2 Na_2HPO_4, 26 $NaHCO_3$, and 25 D-glucose) (77) (see also, e.g., Chaps. 1-Trapp and Ballanyi (10) and 11-Kantor et al. (10)). We have recently found that transcardial perfusion of mice with cooled ACSF further improves the viability of the slices. Brain slices are made in most cases using a vibrating blade microtome in which the block of spinal cord tissue is embedded in agar and cut so that, for example, 250–350-μm-thick slices are created. We and others have found that it is worth investing in a good microtome. The VT1200 (Leica Biosystems, Nussloch, Germany; www.leica-microsystems.com) series has produced satisfactory results for us since it minimizes vertical displacement of the blade which leads to tissue damage. For our purposes, the maximal thickness of the slice is <500 μm since we use infrared differential interference contrast (IR-DIC) microscopy to get an impression on the three-dimensional structure of the cells which is important for correct alignment of the pipette tip onto the cell of interest. Slices are then immersed in a "recovery" solution (in mM): 119 NaCl, 1.9 KCl, 1 $CaCl_2$, 10 $MgSO_4$, 1.2 Na_2HPO_4, 26 $NaHCO_3$, 10 D-glucose, 1.5 kyurenic acid, 2 lactic acid, and 3% dextran) for 1 h during which time the solution is maintained at 36°C. Finally, they are stored in a holding chamber before being transferred to the recording chamber (77).

Where slice preparations come into their own is when they are used in combination with fluorescent markers to identify cells (84, 85). Lumbar motoneurons can be back labeled with dye directly injected into the hind limb of the intact neonatal mouse (85). Two to three days later, spinal cord slices can be harvested and motoneurons expressing the dye can be identified. Using either genetically designed mice or by transfecting viruses into the tissue of interest, fluorescent XFP markers can be expressed in distinct classes of interneurons (26, 86, 87). For either tracer or fluorescent proteins, the cells are first identified using epifluorescent approaches. In our case, one can simply mark the cell using a plastic protector placed on the video/computer screen. Then, the user switches to IR-DIC optical mode, identifies the cell that is fluorescent, and "patches"

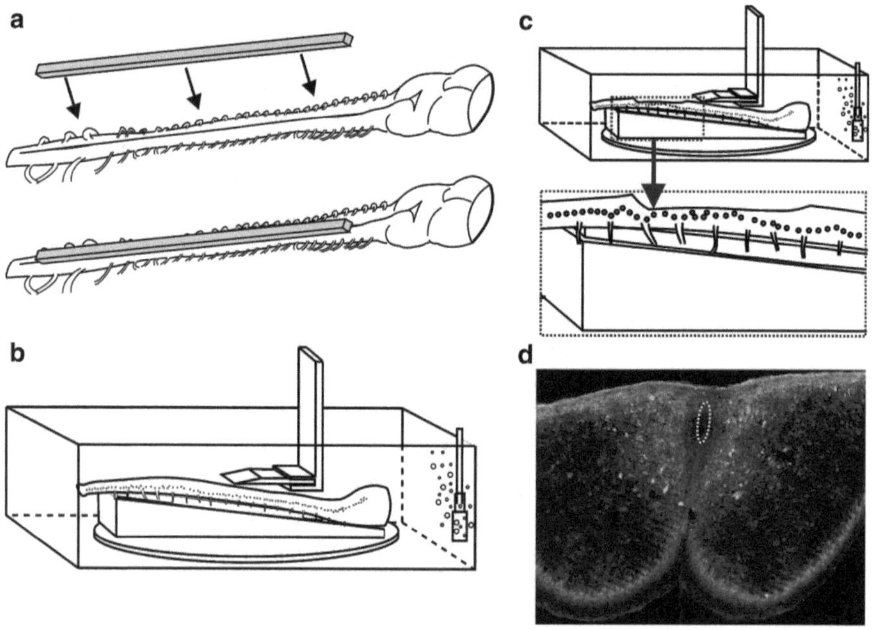

Fig. 7. *Technique for visualizing and accessing ventral horn interneurons in newborn rodents.* (**a**) Thin strip of agarose is attached to ventral midline of spinal cord, taking care to avoid ventral roots. (**b**) Agarose strip is attached to angled agarose block with caudal end higher than rostral end and placed in the sectioning tray of a vibratome. Dorsal surface of the cord is exposed to vibratome blade. (**c**) Notch is cut with vibratome and spans from L1 to L6 segments. *Bottom panel* is an enlargement of region within dashed box. Circles represent labeled neurons located near the central canal, close to the cut surface. (**d**) Coronal cryostat section cut from notched region of spinal cord. Central canal encircled by dashed oval. GFP expressing cells (white) are located in lamina VIII (Text and figure reproduced with permission (89)).

the cell under visual guidance. These techniques for targeting specific interneurons allow for the recording of intrinsic properties from classes of cells that may form part of the CPG.

To examine locomotor activity, one needs to have more of the network intact than is contained within a slice. Some time ago, Kiehn and colleagues found that the dorsal portion of the spinal cord was not necessary to chemically evoke locomotor CPG activity (88). Taking advantage of this fact, Simon Gosgnach and colleague (Fig. 7) removed the dorsal spinal cord tissue using a microtome (89). The exposed region allows ready access to ventral horn interneurons. Using similar approaches as described for the slice, they can identify neurons labeled with tracers or with green fluorescent protein (GFP) and whole-cell record from them. A caveat with this procedure is that while it is possible to elicit a *rhythmic* alternating pattern recorded from ventral roots, it has not been established by recording from muscle nerves whether the *pattern* is changed as a consequence of removal of the dorsal tissue. Another approach is to make a transverse cut of the spinal cord at a particular segment. This method has the advantage of allowing Ca^{2+} imaging of exposed transverse populations to be done (90–92). For best results, a microtome should be used to slice the cord. Finally, a midsagittal section can be made and neurons recorded using either electrophysiological or imaging techniques (87). This technique has the advantage of providing access to neurons in lamina VIII which may be important for generating the rhythmic pattern.

3.6. Recording and Stimulating Considerations

Many investigators choose Molecular Devices/Axon Instruments A/D boards and software (www.moleculardevices.com) to stimulate and record ventral root ENGs since they frequently use Molecular Devices intracellular amplifiers as well. Some possible combinations are Molecular Devices Digidata 1440 which includes Axoscope 10 which is adequate for simple stimulus protocols. For more complex requirements, including intracellular recordings, it is necessary to add a software package such as pClamp10. For those who use MatLab (www.mathworks.com), more flexibility and automation can be built in by using a National Instruments DAQ board (www.ni.com/). Recording a stream of data from several analog channels is surprisingly easy since MatLab is a high-level programming language. We suggest a National Instruments USB DAQ (e.g., NI USB-6221 BNC), which will work with the data acquisition toolbox in MatLab. Another option is the use of Spike2 software along with a Micro1401 or Power1401 data acquisition board from Cambridge Electronic Devices (Cambridge, UK; www.ced.co.uk). Data can be collected right out of the box with no programming. The advantage of Spike2 is that it provides a built in scripting language and a greater variety of analysis routines, allowing more flexibility for the user.

4. Analyzing Locomotor Output

Over the years, numerous techniques have been employed to analyze rhythmic patterns from ENGs obtained from ventral root and nerve recordings and EMG recordings from muscles. All the methods described here require proper conditioning of the analog voltage signal. Since digital filtration of the digitized signal is easy, it is ideal to use a DC-coupled amplifier combined with just a low-pass filter (1 kHz) of the analog signal, for example, via a Dagan EX4-400 (Dagan Corporation, Minneapolis, MI, USA: www.dagan.com). Recording of subthreshold activity as well as spike activity is possible since the distance from the recorded root and the soma is short. DC recordings allow to record from a premotor interneuron and use spike-triggered averaging techniques to examine connectivity onto motoneurons, eliminating the need to dual intracellular recordings (20, 93). If ENGs are obtained from a strip of white matter, the sub- and suprathreshold activity of a population of interneurons can be monitored (85, 94).

DC recordings can be prone to drift, but this can be minimized by chloriding the electrode wire that contacts the ACSF. If drift is a problem, then it is worth considering using the lowest high-pass filter frequency possible.

Perhaps the simplest approach to data analysis is to select a stretch of data and determine the onsets and offsets of the (locomotor) bursts. This method has the advantage of being intuitive and also of allowing a comparison of duty cycles and cycle periods. The cycle period is the time from the start of one burst to the start of the next burst and defines a complete fictive step cycle. The duty cycle is a ratio and provides a measure of the proportion of the cycle period that a burst occupies. The regularity of the rhythm can be determined by calculating a "coefficient of variance." The coefficient of variance is the ratio of the standard deviation to the mean. The closer the value is to 0, the more regular the rhythm is. A key disadvantage of detecting individual bursts is that it is arbitrary, especially if performed manually, and can be time consuming. This type of analysis works well if the rhythm is regular, but it can be very difficult to analyze irregular sequences.

Other methods include using time series analysis techniques such as auto and cross-correlation (49, 54, 95–97). These analytical tools have the advantage of being able to automatically detect the cycle period, the phase, and the regularity of the rhythmic pattern. For the approaches listed below, it is usually necessary to condition the signal. Firstly, the analog locomotor signal is digitally band pass-filtered at 0.03–1 kHz and rectified. The ENGs are conditioned so that the rectified trace is heavily smoothed, for example, with Savitzky Golay filter (maintains features of the burst such as width better than a moving average). Following smoothing, any trend for the amplitude values to drift in either the positive or negative direction is removed. Then any remaining offset is removed, and the data can be then analyzed using correlogram functions. A property of correlograms is that one signal is shifted one point at a time and their similarity examined. The higher the correlation coefficient value, the more similar the traces are. If the signals are 180° out of phase, the correlation coefficient will be negative. As the reference signal is shifted at some point, the signals will realign. An index of stability can be obtained by examining the peak to trough cross-correlation (Fig. 8) (98). An autocorrelation is simply a specific type of cross-correlation where the second signal is a copy of the first signal. The phase lag between selected traces can be extracted quite easily using these methods. The cycle period for each ENG can be extracted using the autocorrelogram. The distance from lag 0 to the next peak will represent the average cycle period (Fig. 8). The disadvantages of this approach are that it is not suitable for analysis of rhythms that show large variations in cycle periods. This can be overcome by dividing the trace into segments which are analyzed separately (98).

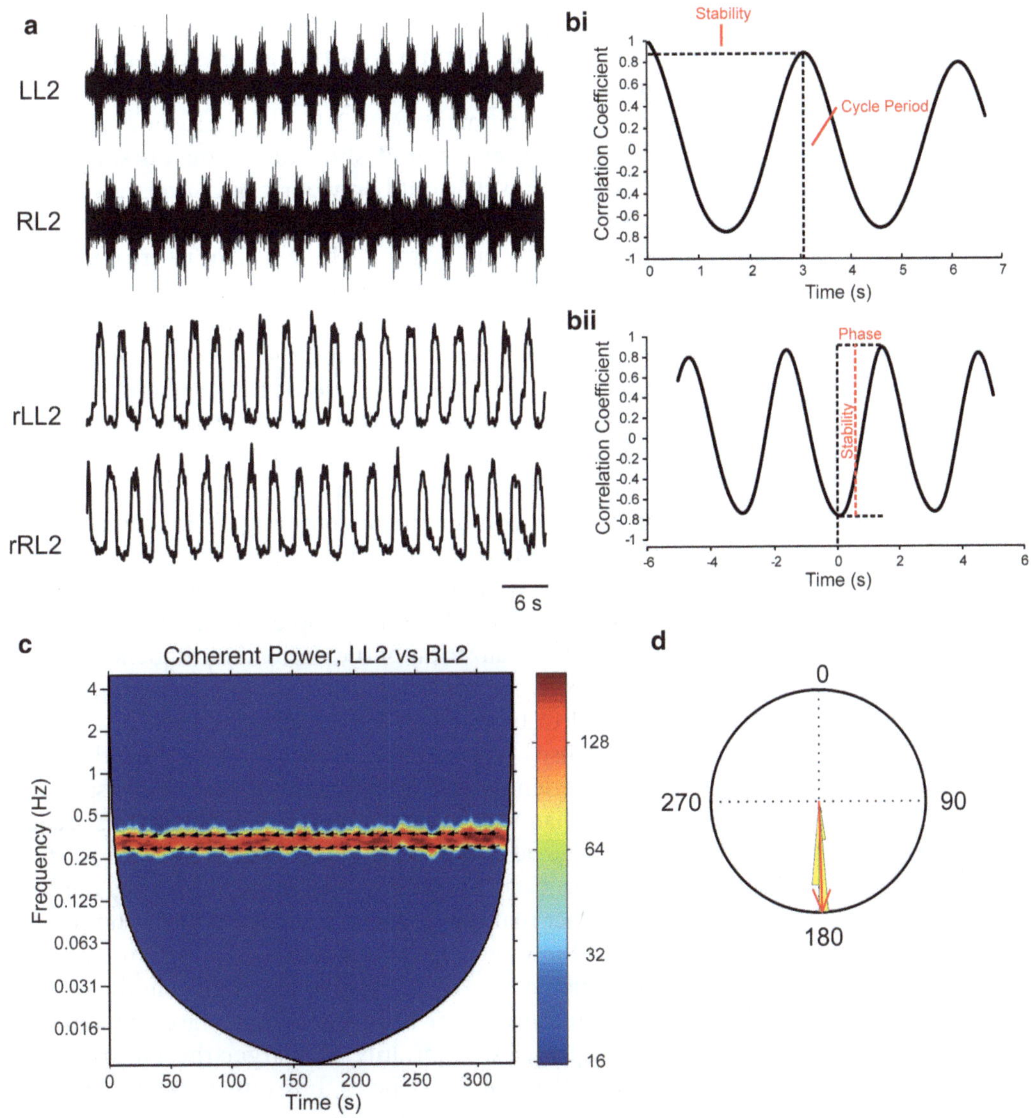

Fig. 8. *Automated methods to quantify fictive locomotor activity.* Two methods are described to analyze data from the original suction electrode recordings in **a**, **b**i, ii. Rectified and filtered (5-Hz digital low-pass filter, resampled at 30 Hz) data are used to generate autocorrelograms and cross-correlograms. The cycle period for each ENG can be obtained from the autocorrelogram and corresponds to the number of lags of the first peak following the zero point (which is always 1). The regularity of rhythm can be estimated by measuring the correlation coefficient from the first peak. The more regular the rhythm, the closer this value will be to 1. The cross-correlogram measures the degree to which the two channels are coordinated. The phase of the second channel compared to the first is obtained by measuring the lag of the first positive peak following zero. If this value is close to 0.5 (as is the case for the data in **a**), it suggests that the lagged channel is alternating compared to the reference channel. The regularity of the rhythm can be estimated by measuring the peak-to-trough correlation coefficient. The closer this value is to 2, the more coordinated and regular the rhythm is. (**c**) Methods to handle nonstationary signals using wavelets coupled with coherence. The graph indicates the coherent power spectra of the rhythm. The color-coded calibration on the right provides a logarithmic scale of the power of the signal. As can be seen, there is a strong dominant band around 0.35 Hz. These data indicate that the oscillations observed between the two signals are strongly related at that frequency band which agrees with an inspection of the data. The value of this approach is that the power can be examined at any point in time rather than relying on an average value using other approaches (e.g., **b**). (**d**) Phase analyses generated from the spectra in (**c**). A region of interest in (**c**) bounds the data analyzed. Within the dominant frequency band, the data are clearly alternating. Data analyzed using "SpinalCore" that was a gift from Dr. Aharony Lev-Tov.

However, dividing a trace into smaller segments reduces the ability to detect slower rhythms. Another issue is that this method is not ideal when examining traces that contain rhythms that show greater than 1:1 entrainment (i.e., 2:1, 3:1, etc.). For example, a 2:1 entrainment pattern would show one side bursting every second, but the other side would only burst once every 2 s. Finally, it can be difficult to examine rhythmic traces that contain stimulus artifacts not related to the rhythm. These situations are commonly found when stimulating afferents and are a special issue with brainstem stimulation where large currents are used to activate brainstem centers. We have found that a digital cancellation routine can significantly reduce the amplitude of the artifacts, making it possible to use auto and cross-correlation techniques (7).

Fictive locomotor rhythms can also be analyzed with spectral analysis techniques that operate in the frequency domain. This is useful if more than one dominant frequency is contained in the ENG. This can occur if the signal is contaminated by trains of stimulus artifacts or both a slow and a fast rhythm are being produced by spinal circuits (97). For example, power spectral density analysis allows the visualization of the dominant frequencies within an ENG. Power spectral density is used on a single trace, and when two signals need to be compared to examine interrelationships, a cross-spectral density is used. "Coherence" is a method that takes the cross-spectral density and normalizes it to the power spectrum density of both channels. If the coherence is 1, the signals are perfectly related, whereas a value of 0 means that there is no relationship between the two. As good as these methods are, they still suffer from the fact that they assume that the rhythm is not changing over time. One way to get around this issue is to use a short-time Fourier analysis by dividing the trace into multiple bins. However, a large number of bins will compromise the ability to detect slow frequencies. Contrarily, the larger the bin, the more reliable is the analysis of frequencies, but unfortunately one loses the time component. Recently, Aharon Lev-Tov and colleagues introduced a hybrid method that uses analyses in both the time and frequency domain for nonstationary data from locomotor CPGs (97). The data can be automatically quantified, and robust statistics can be used to infer whether data sequences differ from each other. The approach uses a Morlet wavelet approach to analyze locomotor rhythms. Wavelets have been used extensively in signal processing. Essentially, they are minioscillatory waveforms that can be used to resolve discrete frequencies within a multifrequency signal. Wavelet analyses allow power to be assigned to different frequencies and combined with coherence analysis provide data concerning the interrelationship between two rhythmic signals. The advantage of this approach is that no assumptions of stationarity are made. A disadvantage is that interpretation of the spectrograms can be difficult.

5. Novel Experimental Approaches

5.1. Genetic Approaches

Identifying the locomotor CPG in mammals is challenging because the classes of spinal interneurons constituting the CPG apparently comprise a heterogeneously distributed set of cells (24, 25). Identifying locomotor interneurons has been largely restricted to cells receiving afferent or descending input, which excludes many locally projecting interneurons or propriospinal interneurons. Over the last decade, data from many laboratories have converged to provide evidence that molecularly identified classes of interneurons contribute to CPG function (23, 78, 99–102). Spinal interneurons involved in locomotor functions are derived from postmitotic interneurons expressing transcription factors that can be used to identify them later in development. There are five main cell classes within the ventral spinal cord that have been identified. The V0 class contains a subpopulation which consists of commissural inhibitory interneurons that participate in left-right locomotor coordination (100) and a population of cholinergic interneurons that projects ipsilaterally onto motoneurons (103). V1 interneurons are a large class of inhibitory interneurons that project ipsilaterally (101). Renshaw cells and Ia inhibitory interneurons are two well-known cell types that form part of this population. V2 interneurons can be split into two main groups. The V2a interneurons are excitatory, project ipsilaterally, and may contribute to projections onto the commissural inhibitory interneurons (24, 26, 104). Inactivation of these cells disrupts left-right coordination as well as changing the switching of locomotor patterns during changes in gait (21, 23). The role for the V2b class has not been determined. But, since the neurons project caudally and are inhibitors, they could contribute to control of reciprocal inhibition between flexors and extensors (26). The V3 class contains excitatory glutamatergic commissural interneurons, with a subpopulation of cells that also project ipsilaterally. Disruption of this interneuronal class interferes with the regularity and robustness of rhythm (105). Finally, there are interneurons located around the midline of the spinal cord in lamina VIII that express the Hb9 transcription factor (78, 85, 106). These interneurons express characteristics which suggest that they may form part of the CPG, but their precise role has not been defined (107).

Genetic tools allow the cells to be visualized using GFP or to be inactivated, for example, using allatostatin. Allatostatin is a G-protein-coupled receptor that activates an ion channel for hyperpolarizing the specific cell type that has been engineered to express this protein (101, 105, 108). If more precise temporal and spatial control is required, such as perturbation of the rhythm for a single cycle for example, then the use of recently established optogenetic approaches is best suited (see below). While genetic approaches comprise an excellent strategy per se, several caveats need to be borne in mind. Firstly, when genetically engineered interneurons comprise multiple cell classes, the data are difficult to interpret. For example, only 25% of the V1 population have been identified as Renshaw cells or Ia inhibitory interneurons. Secondly, during development, some cells may express GFP but not the transcription factor identifying the population. However, this problem can be solved in most cases with the use of appropriate controls. As a final caveat, the rhythmogenic recurrently connected excitatory CPG interneurons seem to comprise a rostrocaudally distributed network that may include multiple cell classes. Reversibly silencing classes of interneurons would be expected to result in "graceful degradation" of the rhythm (109). Graceful degradation is a term used in computer science where loss of one element does not eliminate output but instead may slow it down. For these reasons, reversible silencing of neurons to demonstrate necessity may be challenging. Use of these approaches to examine changes in pattern will likely lead to more robust conclusions (21, 23, 100).

5.2. Optogenetics

Optogenetic strategies use light to activate channel opsins which are expressed in interneurons (86, 110). There are two major types of opsins, firstly, Channelrhodopsin 2 (ChR2) which is derived from a bacterium and can be engineered using viral vectors to be inserted into the plasmalemmal membrane to express ion channel kinetics similar to "AMPA"-type ionotropic glutamate receptors. ChR2 can be excited by single photon blue light delivered by a laser tuned to 470 nm. Recently, it was demonstrated that ChR2 can be coupled with a promotor for the vesicular glutamate transporter "vglut2," which is expressed in glutamatergic spinal and hind brain interneurons (111). In the latter study, exposure of the brainstem or spinal cord to blue light produced bouts of locomotion in the isolated brainstem-spinal cord preparation. However, the potential of these approaches extends beyond this especially since ChR2 can be expressed in interneurons that express Cre recombinase. This allows V-class interneurons, for example, to be selectively targeted with ChR2 and tests of sufficiency can be performed.

Halorhodopsin is the second major type of opsin. It is activated by yellow light and alters Cl^- pumps leading to a hyperpolarization of cells expressing this protein (86). Using these approaches, interneurons can be transiently hyperpolarized, thus providing an excellent tool to test the necessity of a class of interneurons in producing locomotor rhythm. Of interest to locomotor network analysis is the recent development of light-activated G-protein-coupled receptors which enable, for example, transient activation of monoaminergic neural systems (112). A potential issue with the use of light-activated approaches is ensuring that all the desired tissue is activated to control against false negative results. In these cases, the use of pharmacological genetic approaches such as allatostatin (see above) may be preferred.

5.3. Imaging Techniques

Examining the cellular activity of a single spinal interneuron, while useful, does not provide much information regarding its possible function within the locomotor network. In this regard, imaging techniques provide a complementary approach since they are ideally suited for examining the activity of populations of neurons. Work by Michael O'Donovan pioneered the approach of imaging spinal cord networks using the embryonic chick (113, 114). Essentially, there are two approaches for dye-loading of cells. The first one is to bath-apply or microinject the membrane-permeant acetoxymethyl (AM) ester form of a fluorescent dye, such as the Ca^{2+}-sensitive fluorescent dye Fura2-AM, which enters the cell where it is de-esterified and thus trapped. Extracellular dye is then washed out after ~1 h of exposure (in the dark), and cells can be imaged. The fact that the entire preparation is exposed to the dye may affect the network activity because, for example, Ca^{2+} dyes act as buffers for cytosolic Ca^{2+} and may thus affect the activity of Ca^{2+}-dependent processes or ion channels, such as Ca^{2+}-activated K^+ channels. However, bath application of the dye only stains superficial tissue layers of up to 60 μm into the spinal cord, thus unaffecting (rhythmogenic) interneurons located in deeper layers. To image from targeted neurons in deeper layers, multiphoton imaging can be combined with focal injection of dye using a (broken) patch electrode (87, 115, 116). This results in labeling of interneurons (and also glial cells) in a radius of ~150 μm and >100 μm deep into the tissue which can principally be imaged up to that depth using multiphoton microscopy techniques. Details of this approach for imaging from supraspinal neural networks are given in Chaps. 2-Ruangkittisakul et al. (11) and 11-Kantor et al. (117).

Another approach is to apply a dextran conjugated to a Ca^{2+}-sensitive dye to cut axons (113). High concentrations of the dye (25–50 mM) can be applied to cut ventral roots using suction electrodes containing the dye to label motoneurons. Within 3–6 h, the dye is transported to motoneuronal somata (Fig. 9) (118, 119). Alternatively, subpopulations of ventral horn interneurons can be labeled by cutting a slip of ventrolateral funiculus, although the time taken to label these cells is longer than for motoneurons (8–24 h). This back-labeling approach has the advantage of only labeling discrete populations of cells such as motoneurons. The main disadvantage is the time taken to label the cells of interest. To overcome these issues, O'Donovan and colleagues used electroporation to rapidly deliver dye into cells of interest (120). Electroporation occurs by using two stimulus electrodes located on either side of the spinal cord. The dye itself is placed extracellularly in the region of interest. A transient stimulus (10 s) makes the membrane porous to the dye. A disadvantage of this approach is that the delivery of the dye is accomplished by disrupting the integrity of the plasma membrane. While this does result in transient depression of neuronal activity, it is still possible to elicit (alternating) rhythms after the loading

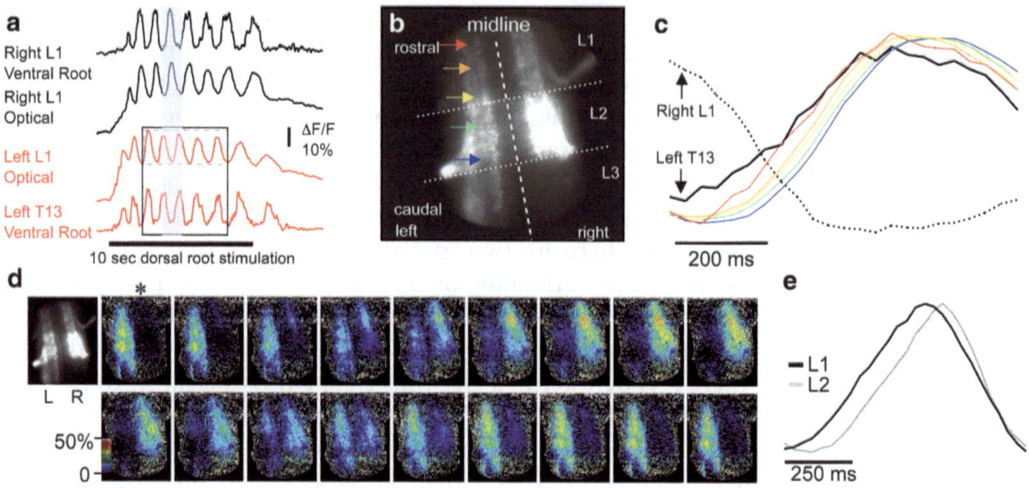

Fig. 9. *Ca^{2+} imaging of locomotor activity in the isolated mouse spinal cord.* (**a**) Rostrocaudal propagation of motoneuron optical signals recorded during fictive locomotion evoked by electrical stimulation of a dorsal root (30 μA, 4 Hz). The time course of the optical signals measured from the labeled motoneurons on both sides of the L1 segment (shown in **b**) was compared with that of electrical activity recorded from the right L1 and left T13 ventral roots. The electrical and optical activity alternates between the left and right sides (highlighted). (**b**) Videomicrograph of labeled motoneurons (L1–L3) viewed through the ventral white matter. The *colored arrows* indicate the center of the regions of interest. The midline (*dashed line*) and the limits between L1, L2, and L3 segments (*dotted lines*) are indicated. (**c**) Average of five successive fluorescence transients (indicated by the dotted box in **a**) obtained from the regions marked in **b** (same color code) compared with the averaged discharge recorded from the right L1 and the left T13 ventral roots (cycles indicated by the boxed area in **a**). (**d**) Rostrocaudal spread of activity during a single cycle of rhythmic activity (indicated by gray bar in **a**). The first panel is the same as **b**, and subsequent panels are different images normalized to the quiescent baseline. The trough in the normalized activity of each side was subtracted from the resulting images to demonstrate the propagation of activity during the rhythmic component of the signal. In this example, the rostrocaudal wave starts on the right side (R) marked by the asterisk and progresses rostrocaudally. The first row shows the propagation on the right side, and the second row shows the propagation on the left side (see accompanying video). (**e**) Average of eight successive bursts of electrical activity in L1 and L2 ventral roots obtained during a train of dorsal root stimuli (25 μA, 4 Hz) (Reproduced with permission (119)).

procedure (120). All of these approaches can be combined with the use of XFPs (family of genetically engineered fluorescent markers which include GFP) to examine activity in targeted classes of interneurons that express these markers (87).

A major issue regarding Ca^{2+}-sensitive dyes is that they are not useful for detecting subthreshold events, and while the signal is relatively high, the temporal resolution is low. Voltage-sensitive dyes have the advantage of being able to detect fast events over the millisecond range and are sensitive within the subthreshold range for spike generation (121, 122), though these dyes have not been widely adopted for the examination of spinal cord networks [125], largely because the signal-to-noise ratio for the currently used agents is very small (see also Chap. 2-Ruangkittisakul et al. (11)). This issue can be resolved by using a photomultiplier diode array; however, this is at the expense of spatial resolution. New voltage-sensitive dyes and also Ca^{2+} dyes are being developed all the time, and the hope is that better dyes will be able to overcome the current limitations of the method.

6. Conclusions

The goal of this chapter is to orientate the reader to experimental models and tools that are available to examine the structure and function of the mammalian locomotor CPGs. Over the last decade, there has been renewed interest into understanding the function of these complex networks. In part, this interest has been fueled by the growth of genetic strategies that use the mouse as a model. However, at the same time, there has been a parallel maturation of electrophysiological techniques, dissection approaches, and methods to quantify spinal network output. The combination of these approaches allows the experimentalist unprecedented flexibility in designing experiments that unlock the black box of the locomotor CPG.

Acknowledgments

We acknowledge the excellent technical assistance of Michelle Tran. Support from the Canadian Institutes of Health Research, Christopher Reeve Paralysis Foundation, the Heart and Stroke Foundation of Canada, and the Natural Sciences and Engineering Research Council is gratefully acknowledged. Dr. Patrick Whelan is a Senior Scholar of the Alberta Heritage Foundation for Medical Research (AHFMR).

References

1. Smith JC, Feldman JL (1987) In vitro brain-stem-spinal cord preparations for study of motor systems for mammalian respiration and locomotion. J Neurosci Methods 21:321–333
2. Smith JC, Feldman JL, Schmidt BJ (1988) Neural mechanisms generating locomotion studied in mammalian brain stem-spinal cord in vitro. FASEB J 2:2283–2288
3. Kiehn O, Kjaerulff O (1996) Spatiotemporal characteristics of 5-HT and dopamine-induced rhythmic hindlimb activity in the in vitro neonatal rat. J Neurophysiol 75: 1472–1482
4. Iizuka M, Kiehn O, Kudo N (1997) Development in neonatal rats of the sensory resetting of the locomotor rhythm induced by NMDA and 5-HT. Exp Brain Res 114: 193–204
5. Whelan P, Bonnot A, O'donovan MJ (2000) Properties of rhythmic activity generated by the isolated spinal cord of the neonatal mouse. J Neurophysiol 84:2821–2833
6. Atsuta Y, Garcia-Rill E, Skinner RD (1990) Characteristics of electrically induced locomotion in rat in vitro brain stem-spinal cord preparation. J Neurophysiol 64:727–735
7. Gordon IT, Whelan PJ (2008) Brainstem modulation of locomotion in the neonatal mouse spinal cord. J Physiol 586:2487–2497
8. Zaporozhets E, Cowley KC, Schmidt BJ (2004) A reliable technique for the induction of locomotor-like activity in the in vitro neonatal rat spinal cord using brainstem electrical stimulation. J Neurosci Methods 139:33–41
9. Liu J, Jordan LM (2005) Stimulation of the parapyramidal region of the neonatal rat brain stem produces locomotor-like activity involving spinal 5-HT7 and 5-HT2A receptors. J Neurophysiol 94:1392–1404
10. Trapp S, Ballanyi K (2012) Autonomic nervous system *in vitro*: studying tonically active

neurons controlling vagal outflow in rodent brainstem slices. In Isolated Central Nervous System Circuits (Ed K Ballanyi), Neuromethods Series Vol. 73 (Ed W Walz). Springer Science+Business Media, LLC, New York, NY, pp 1–59

11. Ruangkittisakul A, Panaitescu B, Secchia L, Bobocea N, Kantor C, Kuribayashi J, Iizuka M, Ballanyi K (2012) Isolated Brainstem Respiratory Centers from Perinatal Rodents. In Isolated Central Nervous System Circuits (Ed K Ballanyi), Neuromethods Series Vol. 73 (Ed W Walz). Springer Science+Business Media, LLC, New York, NY, pp 61–124
12. Biggs JE, Lu VB, Kim H, Lai A, Todd KG, Ballanyi K, Colmers WF, Smith PA (2012) Defined medium organotypic cultures of spinal cord put 'pain in a dish'. In Isolated Central Nervous System Circuits (Ed K Ballanyi), Neuromethods Series Vol. 73 (Ed W Walz). Springer Science+Business Media, LLC, New York, NY, pp 405–435
13. Curtis DR, Phillis JW, Watkins JC (1961) Actions of amino acids on the isolated hemisected spinal cord of the toad. Br J Pharmacol Chemother 16:262–283
14. Suzue T, Jessell T (1980) Opiate analgesics and endorphins inhibit rat dorsal root potential in vitro. Neurosci Lett 16:161–166
15. Cohen AH, Wallén P (1980) The neuronal correlate of locomotion in fish. "Fictive swimming" induced in an in vitro preparation of the lamprey spinal cord. Exp Brain Res 41:11–18
16. Smith JC, Feldman JL (1985) Motor patterns for respiration and locomotion generated by an in vitro brainstem-spinal cord from neonatal rat. Soc Neurosci Abstracts, p 11:24
17. Bonnot A, Morin D, Viala D (1998) Genesis of spontaneous rhythmic motor patterns in the lumbosacral spinal cord of neonate mouse. Brain Res Dev Brain Res 108:89–99
18. Bonnot A, Morin D (1998) Hemisegmental localisation of rhythmic networks in the lumbosacral spinal cord of neonate mouse. Brain Res 793:136–148
19. Butt SJB, Harris-Warrick RM, Kiehn O (2002) Firing properties of identified interneuron populations in the mammalian hindlimb central pattern generator. J Neurosci 22:9961–9971
20. Butt SJB, Kiehn O (2003) Functional identification of interneurons responsible for left-right coordination of hindlimbs in mammals. Neuron 38:953–963
21. Crone SA, Quinlan KA, Zagoraiou L, Droho S, Restrepo CE, Lundfald L, Endo T, Setlak J, Jessell TM, Kiehn O, Sharma K (2008) Genetic ablation of V2a ipsilateral interneurons disrupts left-right locomotor coordination in mammalian spinal cord. Neuron 60:70–83
22. Zhong G, Droho S, Crone SA, Dietz S, Kwan AC, Webb WW, Sharma K, Harris-Warrick RM (2010) Electrophysiological characterization of V2a interneurons and their locomotor-related activity in the neonatal mouse spinal cord. J Neurosci 30:170–182
23. Crone SA, Zhong G, Harris-Warrick R, Sharma K (2009) In mice lacking V2a interneurons, gait depends on speed of locomotion. J Neurosci 29:7098–7109
24. Kiehn O (2011) Development and functional organization of spinal locomotor circuits. Curr Opin Neurobiol 21:100–109
25. Whelan PJ (2010) Shining light into the black box of spinal locomotor networks. Philos Trans R Soc Lond B Biol Sci 365: 2383–2395
26. Goulding M (2009) Circuits controlling vertebrate locomotion: moving in a new direction. Nat Rev Neurosci 10:507–518
27. Cangiano L, Grillner S (2003) Fast and slow locomotor burst generation in the hemispinal cord of the lamprey. J Neurophysiol 89: 2931–2942
28. Bem T, Cabelguen JM, Ekeberg O, Grillner S (2003) From swimming to walking: a single basic network for two different behaviors. Biol Cybern 88:79–90
29. McCrea DA, Rybak IA (2008) Organization of mammalian locomotor rhythm and pattern generation. Brain Res Rev 57:134–146
30. Graham-Brown T (1911) The intrinsic factors in the act of progression in the mammal. Proc R Soc B Biol Sci 84:308–319
31. Prinz AA, Bucher D, Marder E (2004) Similar network activity from disparate circuit parameters. Nat Neurosci 7:1345–1352
32. Norris BJ, Wenning A, Wright TM, Calabrese RL (2011) Constancy and variability in the output of a central pattern generator. J Neurosci 31:4663–4674
33. Sanchez-Vives MV (2012) Spontaneous rhythmic activity in the adult cerebral cortex *in vitro*. In Isolated Central Nervous System Circuits (Ed K Ballanyi), Neuromethods Series Vol. 73 (Ed W Walz). Springer Science+Business Media, LLC, New York, NY, pp 263–284
34. McHanwell S, Biscoe TJ (1981) The localization of motoneurons supplying the hindlimb muscles of the mouse. Philos Trans R Soc Lond B Biol Sci 293:477–508
35. Nakamura TJ, Michel S, Block GD, Colwell CS (2012) Neural circuits underlying circadian oscillations in mammals: clocks in a dish.

In Isolated Central Nervous System Circuits (Ed K Ballanyi), Neuromethods Series Vol. 73 (Ed W Walz). Springer Science+Business Media, LLC, New York, NY, pp 183–210
36. Wilson R, Chersa T, Whelan P (2003) Tissue PO2 and the effects of hypoxia on the generation of locomotor-like activity in the in vitro spinal cord of the neonatal mouse. Neuroscience 117:183–196
37. Jiang MC, Heckman CJ (2006) In vitro sacral cord preparation and motoneuron recording from adult mice. J Neurosci Methods 156: 31–36
38. Carp JS, Tennissen AM, Mongeluzi DL, Dudek CJ, Chen XY, Wolpaw JR (2008) An in vitro protocol for recording from spinal motoneurons of adult rats. J Neurophysiol 100:474–481
39. Sqalli-Houssaini Y, Cazalets JR, Clarac F (1993) Oscillatory properties of the central pattern generator for locomotion in neonatal rats. J Neurophysiol 70:803–813
40. Jiang Z, Carlin KP, Brownstone RM (1999) An in vitro functionally mature mouse spinal cord preparation for the study of spinal motor networks. Brain Res 816:493–499
41. Cazalets JR, Borde M, Clarac F (1995) Localization and organization of the central pattern generator for hindlimb locomotion in newborn rat. J Neurosci 15:4943–4951
42. Cazalets JR, Grillner P, Menard I, Cremieux J, Clarac F (1990) Two types of motor rhythm induced by NMDA and amines in an in vitro spinal cord preparation of neonatal rat. Neurosci Lett 111:116–121
43. Cowley KC, Schmidt BJ (1994) A comparison of motor patterns induced by N-methyl-D-aspartate, acetylcholine and serotonin in the in vitro neonatal rat spinal cord. Neurosci Lett 171:147–150
44. Klein DA, Tresch MC (2010) Specificity of intramuscular activation during rhythms produced by spinal patterning systems in the in vitro neonatal rat with hindlimb attached preparation. J Neurophysiol 104:2158–2168
45. Prinz AA (2006) Insights from models of rhythmic motor systems. Curr Opin Neurobiol 16:615–620
46. Marchetti C, Beato M, Nistri A (2001) Alternating rhythmic activity induced by dorsal root stimulation in the neonatal rat spinal cord in vitro. J Physiol 530:105–112
47. Lev-Tov A, Delvolvé I, Kremer E (2000) Sacrocaudal afferents induce rhythmic efferent bursting in isolated spinal cords of neonatal rats. J Neurophysiol 83:888–894
48. Bonnot A, Whelan PJ, Mentis GZ, O'Donovan MJ (2002) Locomotor-like activity generated by the neonatal mouse spinal cord. Brain Res Brain Res Rev 40:141–151
49. Strauss I, Lev-Tov A (2003) Neural pathways between sacrocaudal afferents and lumbar pattern generators in neonatal rats. J Neurophysiol 89:773–784
50. Norreel J, Pflieger J, Pearlstein E, Simeoni-Alias J, Clarac F, Vinay L (2003) Reversible disorganization of the locomotor pattern after neonatal spinal cord transection in the rat. J Neurosci 23:1924–1932
51. Blivis D, Mentis GZ, O'donovan MJ, Lev-Tov A (2007) Differential effects of opioids on sacrocaudal afferent pathways and central pattern generators in the neonatal rat spinal cord. J Neurophysiol 97:2875–2886
52. Mandadi S, Nakanishi ST, Takashima Y, Dhaka A, Patapoutian A, McKemy DD, Whelan PJ (2009) Locomotor networks are targets of modulation by sensory transient receptor potential vanilloid 1 and transient receptor potential melastatin 8 channels. Neuroscience 162:1377–1397
53. Li Y, Burke RE (2002) Developmental changes in short-term synaptic depression in the neonatal mouse spinal cord. J Neurophysiol 88:3218–3231
54. Gordon IT, Whelan PJ (2006) Monoaminergic control of cauda-equina-evoked locomotion in the neonatal mouse spinal cord. J Neurophysiol 96:3122–3129
55. Magnuson DS, Trinder TC (1997) Locomotor rhythm evoked by ventrolateral funiculus stimulation in the neonatal rat spinal cord in vitro. J Neurophysiol 77:200–206
56. Bonnot A, Chub N, Pujala A, O'Donovan MJ (2009) Excitatory actions of ventral root stimulation during network activity generated by the disinhibited neonatal mouse spinal cord. J Neurophysiol 101:2995–3011
57. Mentis GZ, Alvarez FJ, Bonnot A, Richards DS, Gonzalez-Forero D, Zerda R, O'Donovan MJ (2005) Noncholinergic excitatory actions of motoneurons in the neonatal mammalian spinal cord. Proc Natl Acad Sci USA 102:7344–7349
58. Cazalets JR, Sqalli-Houssaini Y, Clarac F (1992) Activation of the central pattern generators for locomotion by serotonin and excitatory amino acids in neonatal rat. J Physiol 455:187–204
59. Jordan LM, Liu J, Hedlund PB, Akay T, Pearson KG (2008) Descending command systems for the initiation of locomotion in mammals. Brain Res Brain Res Rev 57: 183–191
60. Gordon IT, Dunbar MJ, Vanneste KJ, Whelan PJ (2008) Interaction between developing

spinal locomotor networks in the neonatal mouse. J Neurophysiol 100:117–128
61. Dunbar MJ, Tran MA, Whelan PJ (2010) Endogenous extracellular serotonin modulates the spinal locomotor network of the neonatal mouse. J Physiol 588:139–156
62. Whelan PJ, Wilson RJA (2001) Depth profiles of PO_2 within the isolated in vitro mouse spinal cord during locomotor-like activity. In: Society for neuroscience. p Program No. 297.3. San Diego
63. Okada Y, Mückenhoff K, Holtermann G, Acker H, Scheid P (1993) Depth profiles of pH and PO_2 in the isolated brain stem-spinal cord of the neonatal rat. Respir Physiol 93: 315–326
64. Brockhaus J, Ballanyi K, Smith JC, Richter DW (1993) Microenvironment of respiratory neurons in the in vitro brainstem-spinal cord of neonatal rats. J Physiol 462:421–445
65. Mandadi S, Whelan PJ (2009) A new method to study sensory modulation of locomotor networks by activation of thermosensitive cutaneous afferents using a hindlimb attached spinal cord preparation. J Neurosci Methods 182:255–259
66. Whelan PJ (1996) Control of locomotion in the decerebrate cat. Prog Neurobiol 49:481–515
67. Hayes HB, Chang Y, Hochman S (2009) An in vitro spinal cord-hindlimb preparation for studying behaviorally relevant rat locomotor function. J Neurophysiol 101:1114–1122
68. Hayes HB, Chang Y, Hochman S (2009) Afferent signaling and interneuronal activity during non-fictive locomotion in the in vitro spinal cord – hindlimb rat preparation. In: Society for neuroscience. p Program no. 564.10. Chicago
69. Pickering AE, Paton JFR (2006) A decerebrate, artificially-perfused in situ preparation of rat: utility for the study of autonomic and nociceptive processing. J Neurosci Methods 155:260–271
70. Paton JF (1996) A working heart-brainstem preparation of the mouse. J Neurosci Methods 65:63–68
71. Chizh BA, Headley PM, Paton JF (1997) An arterially-perfused trunk-hindquarters preparation of adult mouse in vitro. J Neurosci Methods 76:177–182
72. Paton JF (1996) The ventral medullary respiratory network of the mature mouse studied in a working heart-brainstem preparation. J Physiol 493:819–831
73. Day TA, Wilson RJA (2007) Brainstem PCO2 modulates phrenic responses to specific carotid body hypoxia in an in situ dual perfused rat preparation. J Physiol 578:843–857
74. Li CL, McILWAIN H (1957) Maintenance of resting membrane potentials in slices of mammalian cerebral cortex and other tissues in vitro. J Physiol 139:178–190
75. Collingridge GL (1995) The brain slice preparation: a tribute to the pioneer Henry McIlwain. J Neurosci Methods 59:5–9
76. Jonas P, Bischofberger J, Sandkühler J (1998) Corelease of two fast neurotransmitters at a central synapse. Science 281:419–424
77. Miles GB, Dai Y, Brownstone RM (2005) Mechanisms underlying the early phase of spike frequency adaptation in mouse spinal motoneurons. J Physiol 566:519–532
78. Wilson JM, Hartley R, Maxwell DJ, Todd AJ, Lieberam I, Kaltschmidt JA, Yoshida Y, Jessell TM, Brownstone RM (2005) Conditional rhythmicity of ventral spinal interneurons defined by expression of the Hb9 homeodomain protein. J Neurosci 25:5710–5719
79. Zhong G, Díaz-Ríos M, Harris-Warrick RM (2006) Intrinsic and functional differences among commissural interneurons during fictive locomotion and serotonergic modulation in the neonatal mouse. J Neurosci 26: 6509–6517
80. Nohda K, Nakatsuka T, Takeda D, Miyazaki N, Nishi H, Sonobe H, Yoshida M (2007) Selective vulnerability to ischemia in the rat spinal cord: a comparison between ventral and dorsal horn neurons. Spine 32:1060–1066
81. von Lewinski F, Keller BU (2005) Ca^{2+}, mitochondria and selective motoneuron vulnerability: implications for ALS. Trends Neurosci 28:494–500
82. Duggal N, Lach B (2002) Selective vulnerability of the lumbosacral spinal cord after cardiac arrest and hypotension. Stroke 33:116–121
83. Carriedo SG, Yin HZ, Weiss JH (1996) Motor neurons are selectively vulnerable to AMPA/kainate receptor-mediated injury in vitro. J Neurosci 16:4069–4079
84. Zhong G, Díaz-Ríos M, Harris-Warrick RM (2006) Serotonin modulates the properties of ascending commissural interneurons in the neonatal mouse spinal cord. J Neurophysiol 95:1545–1555
85. Han P, Nakanishi ST, Tran MA, Whelan PJ (2007) Dopaminergic modulation of spinal neuronal excitability. J Neurosci 27:13192–13204
86. Zhang F, Gradinaru V, Adamantidis AR, Durand R, Airan RD, de Lecea L, Deisseroth K (2010) Optogenetic interrogation of neural circuits: technology for probing mammalian brain structures. Nat Protoc 5:439–456
87. Wilson JM, Dombeck DA, Díaz-Ríos M, Harris-Warrick RM, Brownstone RM (2007)

Two-photon calcium imaging of network activity in XFP-expressing neurons in the mouse. J Neurophysiol 97:3118–3125
88. Kjaerulff O, Kiehn O (1996) Distribution of networks generating and coordinating locomotor activity in the neonatal rat spinal cord in vitro: a lesion study. J Neurosci 16: 5777–5794
89. Dyck J, Gosgnach S (2009) Whole cell recordings from visualized neurons in the inner laminae of the functionally intact spinal cord. J Neurophysiol 102:590–597
90. Christie KJ, Whelan PJ (2005) Monoaminergic establishment of rostrocaudal gradients of rhythmicity in the neonatal mouse spinal cord. J Neurophysiol 94:1554–1564
91. O'Donovan MJ, Bonnot A, Mentis GZ, Arai Y, Chub N, Shneider NA, Wenner P (2008) Imaging the spatiotemporal organization of neural activity in the developing spinal cord. Dev Neurobiol 68:788–803
92. Ritter A, Wenner P, Ho S, Whelan PJ, O'donovan MJ (1999) Activity patterns and synaptic organization of ventrally located interneurons in the embryonic chick spinal cord. J Neurosci 19:3457–3471
93. Wenner P, O'donovan MJ (1999) Identification of an interneuronal population that mediates recurrent inhibition of motoneurons in the developing chick spinal cord. J Neurosci 19:7557–7567
94. Fedirchuk B, Wenner P, Whelan PJ, Ho S, Tabak J, O'Donovan MJ (1999) Spontaneous network activity transiently depresses synaptic transmission in the embryonic chick spinal cord. J Neurosci 19:2102–2112
95. Kremer E, Lev-Tov A (1997) Localization of the spinal network associated with generation of hindlimb locomotion in the neonatal rat and organization of its transverse coupling system. J Neurophysiol 77:1155–1170
96. Madriaga MA, McPhee LC, Chersa T, Christie KJ, Whelan PJ (2004) Modulation of locomotor activity by multiple 5-HT and dopaminergic receptor subtypes in the neonatal mouse spinal cord. J Neurophysiol 92:1566–1576
97. Mor Y, Lev-Tov A (2007) Analysis of rhythmic patterns produced by spinal neural networks. J Neurophysiol 98:2807–2817
98. Pearson SA, Mouihate A, Pittman QJ, Whelan PJ (2003) Peptidergic activation of locomotor pattern generators in the neonatal spinal cord. J Neurosci 23:10154–10163
99. Kullander K, Butt SJB, Lebret JM, Lundfald L, Restrepo CE, Rydström A, Klein R, Kiehn O (2003) Role of EphA4 and EphrinB3 in local neuronal circuits that control walking. Science 299:1889–1892
100. Lanuza GM, Gosgnach S, Pierani A, Jessell TM, Goulding M (2004) Genetic identification of spinal interneurons that coordinate left-right locomotor activity necessary for walking movements. Neuron 42: 375–386
101. Gosgnach S, Lanuza GM, Butt SJB, Saueressig H, Zhang Y, Velasquez T, Riethmacher D, Callaway EM, Kiehn O, Goulding M (2006) V1 spinal neurons regulate the speed of vertebrate locomotor outputs. Nature 440: 215–219
102. Dougherty KJ, Kiehn O (2010) Firing and cellular properties of V2a interneurons in the rodent spinal cord. J Neurosci 30:24–37
103. Zagoraiou L, Akay T, Martin JF, Brownstone RM, Jessell TM, Miles GB (2009) A cluster of cholinergic premotor interneurons modulates mouse locomotor activity. Neuron 64: 645–662
104. Lundfald L, Restrepo CE, Butt SJB, Peng C, Droho S, Endo T, Zeilhofer HU, Sharma K, Kiehn O (2007) Phenotype of V2-derived interneurons and their relationship to the axon guidance molecule EphA4 in the developing mouse spinal cord. Eur J Neurosci 26:2989–3002
105. Zhang Y, Narayan S, Geiman E, Lanuza GM, Velasquez T, Shanks B, Akay T, Dyck J, Pearson K, Gosgnach S, Fan C, Goulding M (2008) V3 spinal neurons establish a robust and balanced locomotor rhythm during walking. Neuron 60:84–96
106. Hinckley CA, Hartley R, Wu L, Todd AJ, Ziskind-Conhaim LH (2005) Locomotor-like rhythms in a genetically distinct cluster of interneurons in the mammalian spinal cord. J Neurophysiol 93:1439–1449
107. Kwan AC, Dietz SB, Webb WW, Harris-Warrick RM (2009) Activity of Hb9 interneurons during fictive locomotion in mouse spinal cord. J Neurosci 29:11601–11613
108. Lechner HAE, Lein ES, Callaway EM (2002) A genetic method for selective and quickly reversible silencing of Mammalian neurons. J Neurosci 22:5287–5290
109. Duchcherer M, Kottick A, Wilson RJ (2010) Evidence for a distributed respiratory rhythm generating network in the goldfish (carssius auratus). In: Homma I (ed) Advances. Springer
110. Deisseroth K (2011) Optogenetics. Nat Methods 8:26–29
111. Hägglund M, Borgius L, Dougherty KJ, Kiehn O (2010) Activation of groups of excitatory neurons in the mammalian spinal cord or hindbrain evokes locomotion. Nat Neurosci 13:246–252

112. Airan RD, Thompson KR, Fenno LE, Bernstein H, Deisseroth K (2009) Temporally precise in vivo control of intracellular signalling. Nature 458:1025–1029
113. O'Donovan MJ, Ho S, Sholomenko G, Yee W (1993) Real-time imaging of neurons retrogradely and anterogradely labelled with calcium-sensitive dyes. J Neurosci Methods 46:91–106
114. O'Donovan M, Ho S, Yee W (1994) Calcium imaging of rhythmic network activity in the developing spinal cord of the chick embryo. J Neurosci 14:6354–6369
115. Stosiek C, Garaschuk O, Holthoff K, Konnerth A (2003) In vivo two-photon calcium imaging of neuronal networks. Proc Natl Acad Sci USA 100:7319–7324
116. Ruangkittisakul A, Schwarzacher SW, Secchia L, Poon BY, Ma Y, Funk GD, Ballanyi K (2006) High sensitivity to neuromodulator-activated signaling pathways at physiological $[K^+]$ of confocally imaged respiratory center neurons in on-line-calibrated newborn rat brainstem slices. J Neurosci 26:11870–11880
117. Kantor C, Panaitescu B, Kuribayashi J, Ruangkittisakul A, Jovanovic I, Leung V, Lee TF, MacTavish D, Jhamandas JH, Cheung PY, Ballanyi K (2012) Spontaneous neural network oscillations in hippocampus, cortex and *locus coeruleus* of newborn rat and piglet brain slices. In Isolated Central Nervous System Circuits (Ed K Ballanyi), Neuromethods Series Vol. 73 (Ed W Walz). Springer Science+Business Media, LLC, New York, NY, pp 315–356
118. Fields RD, O'Donovan MJ (2001) Imaging nervous system activity. Curr Protoc Neurosci. Chapter 2. Unit 2 3
119. Bonnot A, Whelan PJ, Mentis GZ, O'Donovan MJ (2002) Spatiotemporal pattern of motoneuron activation in the rostral lumbar and the sacral segments during locomotor-like activity in the neonatal mouse spinal cord. J Neurosci 22:RC203
120. Bonnot A, Mentis GZ, Skoch J, O'donovan MJ (2005) Electroporation loading of calcium-sensitive dyes into the CNS. J Neurophysiol 93:1793–1808
121. Salzberg BM, Davila HV, Cohen LB (1973) Optical recording of impulses in individual neurones of an invertebrate central nervous system. Nature 246:508–509
122. Arai Y, Mentis GZ, Wu J, O'Donovan MJ (2007) Ventrolateral origin of each cycle of rhythmic activity generated by the spinal cord of the chick embryo. PLoS One 2:e417
123. Rybak IA, Shevtsova NA, Lafreniere-Roula M, McCrea DA (2006) Modelling spinal circuitry involved in locomotor pattern generation: insights from deletions during fictive locomotion. J Physiol 577: 617–639
124. Akay T, Acharya HJ, Fouad K, Pearson KG (2006) Behavioral and electromyographic characterization of mice lacking EphA4 receptors. J Neurophysiol 96:642–651
125. Wenner P, Tsau Y, Cohen LB, O'donovan MJ, Dan Y (1996) Voltage-sensitive dye recording using retrogradely transported dye in the chicken spinal cord: staining and signal characteristics. J Neurosci Methods 70:111–120

Index

A

B

C

D

E

Klaus Ballanyi (ed.), *Isolated Central Nervous System Circuits*, Neuromethods, vol. 73,
DOI 10.1007/978-1-62703-020-5, © Springer Science+Business Media New York 2012

T

V

W

MIX
Papier aus verantwortungsvollen Quellen
Paper from responsible sources
FSC® C105338

If you have any concerns about our products,
you can contact us on
ProductSafety@springernature.com

In case Publisher is established outside the EU,
the EU authorized representative is:
Springer Nature Customer Service Center GmbH
Europaplatz 3, 69115 Heidelberg, Germany

Printed by Libri Plureos GmbH
in Hamburg, Germany